Geophysical Monograph Series

Including

IUGG Volumes
Maurice Ewing Volumes
Mineral Physics Volumes

Geophysical Monograph Series

Geophysical Monograph 141

Solar Variability and Its Effects on Climate

Judit M. Pap and Peter Fox
Editors

**Claus Frohlich, Hugh S. Hudson,
Jeffrey Kuhn, John McCormack,
Gerald North, William Sprigg,
and S. T. Wu**
Contributing Editors

American Geophysical Union

Library of Congress Cataloging-in-Publication Data

Solar variability and its effects on climate / Judit M. Pap, Peter Fox, editors.
      p. cm.-- (Geophysical monograph ; 141)
   Includes bibliographical references.
   ISBN 0-87590-406-8
      1. Solar oscillations. 2. Climatic changes--Effect of solar activity on. I. Pap, Judit M.
II. Fox, Peter A. III. Series.

QC883.2.S6S638 2003
523.7--dc22   2003066285

ISSN 0065-8448
ISBN 0-87590-406-8

CONTENTS

CONTENTS

PREFACE

This monograph presents a state-of-the-art description of the most recent results on solar variability and its possible influence on the Earth's climate and atmosphere. Our primary goal in doing so is to review solar energy flux variations (both electromagnetic and particle) and understand their relations to solar magnetic field changes and global effects, their impact on different atmospheric layers, and—as a collaboration of scientists working on solar-terrestrial physics—to note unresolved questions on an important interdisciplinary area.

One of the highest-level questions facing science today is whether the Earth's atmosphere and climate system changes in a way that we can understand and predict. The Earth's climate is the result of a complex and incompletely understood system of external inputs and interacting parts. Climate change can occur on various time scales as a consequence of natural variability—including solar variability—or anthropogenic causes, or both. The Sun's variability in the form of sunspots and related magnetic activity has been the subject of careful study ever since the earliest telescopic observations. High precision photometric observations of solar-type stars clearly show that year-to-year brightness variations connected with magnetic activity are a widespread phenomenon among such stars. As our nearest star, the Sun is the only star where we can observe and identify a variety of structures and processes which lead to variations in the solar energy output, in both radiative and particle fluxes. Studying even tiny changes in solar energy flux variations may teach us about internal processes taking place in the Sun's convective zone and below.

Because this monograph covers a range of topics, including the Sun's interior and atmosphere, energy transport and radiation processes, measurements of solar energy flux variations and their interpretation, and terrestrial effects related to solar changes, it will be an important resource for both scientists and students with various interests. Toward that end, we have divided the book into four sections. Section One features papers that describe the physical processes in the Sun's interior and atmosphere, and basic physical processes characterizing the Earth's atmosphere. Section Two describes the possible mechanisms that prompt or lead to irradiance variations, measurements of solar and particle fluxes, modeling efforts, and composite reference solar spectra. Section Three focuses on long-term climate changes and descriptions of solar signals found in atmospheric and climate data. Part of this section also describes atmospheric changes caused by irradiance and particle variations. In section Four, we summarize results presented in various papers of the monograph, identify problem areas for future research, outline measurement requirements, and recommend strategies to achieve our goal: to understand and predict solar changes and related atmospheric and climate variations.

We realized the need for the present monograph during our preparation for a symposium entitled "Solar Variability, Climate, and Space Weather," which took place in June 2001 in Longmont, Colorado, under the auspices of the International Solar Cycle Study (ISCS) of the Scientific Commission of Solar-Terrestrial Physics (SCOSTEP). It must be emphasized, however, that the book is not a conference proceedings; it is the result of a collective effort to describe the current situation in our respective fields.

We gratefully acknowledge the effort of each author who contributed to this monograph. Since each paper went through a formal review process, we also thank the reviewers. We are especially thankful to Drs. N. Andronova, S. Antiochos, E. Avrett, S. Bailey, J. Beer, P. Brekke, P. Damon, L. Floyd, P. Fox, C. Fröhlich, M. Giampapa, J. Haigh, T. Hoeksema, M. DeLand, H. Hudson, S. Jordan, S. Kahler, J. Kuhn, K. Labitzke, E. Marsch, J. McCormack, R. Muscheler, G. North, J. Pap, D. Reames, M. Schlesinger, D. Siskind, S. Sofia, S. Walton, T. Woods, and S.T. Wu for their useful comments and suggestions on various papers. We also extend our gratitude for financial support from

Solar Variability and its Effects on Climate
Geophysical Monograph 141
Copyright 2004 by the American Geophysical Union
10.1029/141GM00

SCOSTEP, NASA, NSF, and NCAR/HAO which made it possible to carry out the ISCS 2001 meeting where the topics presented in this Monograph were widely discussed. We are also very thankful for technical support from the NCAR/High Altitude Observatory and NASA/Goddard Space Flight Center which made it possible to edit this book. The editors invested considerable effort into making the work uniform throughout. We are responsible for any errors that may have been introduced in this proceses.

Judit M. Pap
Greenbelt, MD, USA

Peter Fox
Boulder, CO, USA

Claus Fröhlich
Davos, Switzerland

Hugh S. Hudson
Berkeley, CA, USA

Jeffrey Kuhn
Honolulu, HI, USA

John McCormack
Washington, DC, USA

Gerald North
College Station, TX, USA

William Sprigg
Tucson, AZ, USA

S.T. Wu
Huntsville, AL, USA

Section 1

Fundamentals

Section 1. Fundamentals of The Solar Interior and Atmosphere

Jeffrey R. Kuhn

Institute for Astronomy, University of Hawaii, Honolulu, Hawaii

John McCormack

E. O. Hulburt Center for Space Research, Naval Research Laboratory, Washington D.C.

The practical, and immediately useful, questions related to predicting how terrestrial climate may change are completely dependent on the theoretical (and perhaps more fundamental) questions about how the solar cycle varies – what causes it, and how it changes the solar outputs which affect the Earth. For example, we obviously don't have a chance of accurately predicting whether the next solar cycle will be dangerously high or low if we don't have a useful physical model for it.

As we have learned from decades of solar-terrestrial research, it is not easy to find a "useful physical" model of the solar cycle. Here "useful" is synonymous with "predictive." Until recently we've only been able to view the cycle using magnetic observables with retrospective models. Interestingly, over about the last cycle, we have developed new accurate photometric and helioseismic tools. These photometric and acoustic observables hold the promise of finally elucidating the solar cycle physics.

Interpreting the new observables requires a better understanding of the convection zone and the connection between convectve energy transport and magnetic fields. The papers by *Sofia and Li*, and *Kuhn and Armstrong* address these questions.

While acoustic and photometric solar cycle changes hold important clues to the cycle mechanisms and are critical to understanding the terrestrial climate connections, we are steady in our belief that the solar cycle is fundamentally a magnetic oscillation involving global solar toroidal and poloidal magnetic fields. Unfortunately, no magnetic

dynamo model accounts for all the new observations. These new data seem to teach us that the simplest and most elegant dynamo models are fundamentally incomplete. The paper by *Schussler and Schmitt* illustrates the current state of such models and takes a stab at describing how their physical principles can be combined to build a long-term predictive solar cycle model.

The solar cycle is visible at the photosphere and in the Earth's atmosphere. We have learned to study it in the solar interior using acoustic tools. Above the photosphere, in the corona and the interplanetary medium, we depend on magnetohydrodynamical models to understand how magnetic fields in the photosphere and chromosphere could affect the Earth. The paper by *Low and Zhang* explores the state of our understanding of fields above the photosphere.

Confidence in predictive models ultimately depends on observational tests. Over timescales as long as the solar cycle or longer we may not have the opportunity for empirical tests. Since the Sun is an "average" star we can use astronomical observations of distant stars to educate and constrain our solar cycle modeling efforts. For the longest timescales this effort is critical – studies of an ensemble of Sun-like stars are our only chance to sample a landscape that tells us the range in behavior a solar cycle could ultimately take. The paper by *Radick* nicely illustrates what we have learned from stellar-cycle observations.

The ability to predict long-term solar variability, however, is only part of the problem. Understanding the effect of solar variability on the Earth's climate also requires fundamental knowledge of the processes that drive the atmospheric circulation. In particular, it is necessary to understand how these processes change in response to both external (i.e., solar) and internal forcings over a wide range of time scales. The review by *Haigh* describes the Sun's importance

Solar Variability and its Effects on Climate
Geophysical Monograph 141
Copyright 2004 by the American Geophysical Union
10.1029/141GM01

as the primary energy source for the Earth-atmosphere system, and shows how relatively small changes in solar output can have an observable impact on atmospheric composition and structure. The exact mechanism linking solar variability and climate has yet to be determined. One current theory is that changes in solar irradiance may interact with one or more internal modes of variability in atmospheric circulation. These modes of variability are outlined in the review by *Haigh*, and discussed further in Chapter 3.

Jeffrey Kuhn, Institute for Astronomy, University of Hawaii, Honolulu, HI 96822, USA (kuhn@pelea.ifa.hawaii.edu)

John McCormack, E. O. Hulburt Center for Space Research Naval Research Laboratory, Code 7641.5, Washington, D.C. 20375, USA (mccormack@uap2.nrl.navy.mil)

Long-Term Solar Variability: Evolutionary Time Scales

Richard R. Radick

Air Force Research Laboratory, Space Vehicles Directorate
National Solar Observatory, Sunspot, New Mexico

1. INTRODUCTION

Galileo, who played a central role in the modern discovery of sunspots, may have wondered whether the Sun varies. Certainly, his 17th century contemporaries did. The Sun itself all but answered this question a few decades later when it nearly stopped forming sunspots as it entered what is now known as the Maunder Minimum. Herschel's speculation that the price of wheat might be related to the number of sunspots indicates that the possibility of solar variability was firmly established in the scientific thought of the late 18th century [*Eddy*, 1983]. In the mid-19th century, the ~11-yr variation in sunspot number was recognized, apparently first by Schwabe. Indeed, it sometimes seems even today that solar variability is all but defined as this variation in sunspot number. One of the first efforts to get beyond counting sunspots was Abbot's determined program of ground-based solar radiometry in the early years of the 20th century [e.g., *Abbot*, 1934], which foundered mainly on the difficulty of correcting accurately for atmospheric extinction. Successful detection of the solar cycle in the Sun's radiative outputs did come, however, when measurements of 10.7 cm radio flux variations in the mid-20th century opened what might be called the modern era in the study of solar variability.

The final decades of the 20th century brought remarkable advances in our knowledge of solar variability. In 1974, observations to follow the cyclic variation of chromospheric Ca II K-line emission were begun at the National Solar Observatory [*White et al.*, 1998]. Eddy's studies of historical records of solar activity stimulated renewed interest in solar variability and its possible effect on terrestrial climate

Solar Variability and its Effects on Climate
Geophysical Monograph 141
This paper not subject to U.S. copyright
Published in 2004 by the American Geophysical Union
10.1029/141GM02

[*Eddy*, 1976]. Measurements from space, however, took center stage. Starting in 1978, variations in solar total irradiance were detected and monitored by a series of spacecraft, among which the ACRIM experiment aboard the Solar Maximum Mission (SMM) satellite produced a pivotal series of measurements between 1980 and 1989 [*Willson*, 1997]. Theory closely followed measurement, and by the close of the century a picture of solar radiative variability on time scales ranging from days to years had emerged.

2. SOLAR MAGNETIC ACTIVITY AND VARIABILITY

The solar atmosphere harbors a broad range of non-thermal phenomena collectively called activity. On the global scale, this activity is responsible for the sharp temperature increase above the solar photosphere. On the local scale it produces a variety of discrete features, including the dark sunspots and bright faculae of the photosphere, the emission plages and network of the chromosphere, and the intricate structures of the solar corona.

The solar atmosphere is permeated by magnetic fields. It is generally accepted that much (though perhaps not all) of what we call solar activity arises from the generation, evolution, and annihilation of these magnetic fields. The most prominent magnetic structures on the solar surface are the active regions, with their spots, faculae, and overlying chromospheric plages. Active regions erupt and decay with lifetimes of a few months. As an active region ages, its spots disappear. Its faculae disperse, merge with the surrounding network, and gradually also disappear. Large sunspots can live for several weeks, a time scale which accidentally corresponds to the ~25 day solar rotation period. The faculae of an active region may persist, at least collectively, for a few months. The Sun characteristically shows longitudinal asymmetries in the distribution of its active regions, which it maintains by its tendency to create successive active regions in relatively fixed locations. These activity complexes can last for many months or even years—much longer than the lifetime of individual active regions.

Many aspects of solar activity are accompanied by temporal variability; indeed, the very word "activity" strongly suggests variation. Solar irradiance variability is widely, but not universally, attributed to flux deficits produced by dark sunspots, and excess flux produced by bright faculae in both active regions and the network [e.g., *Foukal and Lean*, 1988; *Lean et al.*, 1998]. Dissenters argue that part of the variability, at least on the cycle time scale, arises from a non-facular and possibly global component, perhaps small changes in the surface temperature [e. g., *Kuhn and Libbrecht*, 1991; *Li and Sofia*, 2001]. All substantially agree, however, that sunspots and faculae cause solar irradiance fluctuations amounting to 0.1% or so on time scales of days to weeks, with sunspots producing the most obvious effects. These variations arise both from the emergence and decay of individual solar active regions, and from solar rotation that carries the active regions into and out of view on the solar disk. The relative prominence of the sunspot signature in the initial SMM/ACRIM radiometry helped create a widespread (but mistaken) expectation that the Sun would become brighter as its 11-yr activity cycle waned and sunspots grew scarce [see, e.g., *Eddy*, 1983]. In fact, sustained measurements showed that the Sun's brightness varies directly with the rise and fall of the activity cycle (Figure 1). In terms of the spot/facular model, this implies that solar irradiance variability on the 11-year activity cycle time scale (unlike shorter time scales) is dominated by excess radiation from faculae and bright magnetic network features, rather than the sunspot deficit. The dissenters enter here.

Our understanding of solar variability on time scales longer than the 11-yr activity cycle remains considerably more sketchy. Studies of radioisotopes such as ^{14}C and ^{10}Be provide valuable, albeit indirect, insight about solar variability on time scales of centuries to millennia [e.g., *Beer*, 2000]. On even longer (i.e., evolutionary) time scales, contemporary solar observations, paleoclimatic studies, and analyses of lunar and meteoritic samples offer only teasing hints. Fortunately, the behavior of stars similar to the Sun offers a powerful approach for exploring the likely past history, present state, and future course of solar variability.

This chapter considers solar and stellar brightness variations, primarily on the time scale of years, and the evolution of these variations during the main-sequence lifetime of a star like the Sun. Only variability arising from magnetic activity is considered. Neither transient effects caused by flare-like events nor the long-term secular luminosity

increase associated with interior evolution will be further noted. Before proceeding, however, we should perhaps remind ourselves how pervasively the Sun influences our thinking and expectations. The Sun provides our vocabulary: spots, faculae, plages, active regions, chromospheres, coronae—such terms have all been borrowed from solar astronomy, and underlie the conceptual framework used to interpret a wide range of stellar observations. The Sun also suggests possible mechanisms for variability, such as rotational modulation, active region evolution, and activity cycles. We must beware, lest familiarity with the Sun lure us into traps of circular reasoning.

3. STELLAR PROXIES FOR THE SUN

3.1. Physical Determinants of Stellar Activity and Variability

Contemporary astrophysics has accumulated considerable evidence indicating that magnetic activity following a predictable evolution is a normal feature of lower main-sequence stars, including the Sun. Furthermore, the magnetic

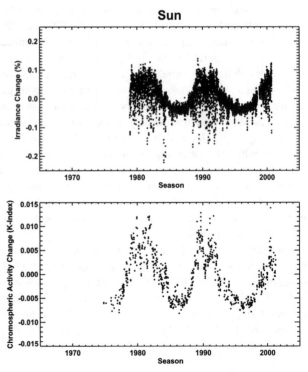

Figure 1. Solar variability. Upper panel: solar radiometry (1978-present) from spaceborne radiometers (courtesy C. Fröhlich); Lower panel: solar Ca II K-line spectroscopy (1974-present) from NSO/KP (courtesy W. Livingston).

activity of such stars is apparently governed primarily by rotation and mass. In one important sense, mass is absolutely decisive: only stars with convective envelopes appear to harbor magnetic activity. Beyond this, rotation seems to be the more dominant determinant. Both observations and theory argue that the surface rotation rate of a lower main-sequence star quickly loses its memory of initial conditions, after which it gradually slows in a manner that depends primarily on stellar age. Accordingly, for a specific star such as the Sun, the evolution of its magnetic activity and variability is very much the story of its rotation.

3.2. The Influence of Rotation on Stellar Magnetic Activity

Although it was first proposed over eighty years ago that the surface magnetic fields of the Sun are produced by dynamo action [Larmor, 1919], our understanding of the solar dynamo remains annoyingly incomplete. Theory suggests that the dynamo is seated near the base of the Sun's convective envelope. The key ingredients of the dynamo are rotation and convection, which interact to produce both differential rotation and helicity. Differential rotation readily converts poloidal magnetic flux into toroidal flux and amplifies it, while helicity twists toroidal flux, thereby regenerating poloidal field. Something resembling this oscillatory, self-sustaining process is generally believed to underlie the 11-yr solar activity cycle [e.g., Weiss, 1994].

Dynamo theory predicts that magnetic-field amplification should depend strongly and directly on stellar rotation rate, and observations agree that rapidly rotating stars show relatively more vigorous magnetic activity. Thus, the existence of a causal relation between rotation and the magnetic activity of lower main-sequence stars is widely accepted. In practice, the rotation-activity relation is indirectly expressed in terms of chromospheric or coronal emission rather than magnetic flux or field strength, because direct measurements of stellar magnetic fields are difficult and therefore scarce. Strictly speaking, commonly observed activity indicators, such as chromospheric Ca II H + K line-core emission or coronal soft X-ray emission, prove only that a star's outer atmosphere is being heated by non-thermal processes. Solar studies, however, have demonstrated that regions of enhanced chromospheric and coronal emission coincide with underlying concentrations of photospheric magnetic flux, and that emission strength scales directly with mean field flux in these regions [Skumanich, Smyth and Frazier, 1975; Schrijver et al., 1989]. Although temperature, surface gravity, and chemical composition probably all affect the complex and incompletely understood connection between surface magnetic flux and its radiative indicators, these indicators are, nevertheless, widely accepted as reliable diagnostics of stellar magnetic activity.

On both theoretical and experimental grounds, it seems likely that the basic convective pattern in the outer envelope of a lower main-sequence star changes, as its rotation rate decreases, from something unlike anything observed on the solar surface, namely, a regular distribution of elongated rolls aligned parallel to the stellar rotation axis, to a more irregular, cellular pattern at least vaguely reminiscent of solar granulation and other modes of solar convection. Recent work suggests that this transition is both more complicated and less abrupt than once thought [Miesch, 2000], but opinion still appears to agree that the transition does occur. Such a dramatic change must surely affect the behavior of the highly nonlinear stellar dynamo. Thus, there is reason to suspect that the magnetic activity and the attendant brightness variability of young, rapidly rotating stars may not be simply more vigorous versions of what is observed on the Sun. Indeed, young stars tend to show strong, non-cyclic activity with relatively complicated temporal behavior, quite unlike what we see on the present-day Sun. These differences will be described in more detail subsequently.

Early investigations of stellar activity tended to focus on its relation to stellar age, rather than rotation. In fact, both mean activity and surface rotation rate decline with age in nearly identical ways among lower main-sequence stars. The classic Skumanich power-law dependence (Ca II H+K emission and rotation rate are both proportional to the inverse square root of stellar age for stars of similar mass: [Skumanich, 1972]) remains a simple and useful rule for lower main-sequence stars throughout much of the age range considered here, despite the fact that it fails for very young stars as well as for certain active binaries. Therefore, it seems natural to ask whether stellar activity is more fundamentally linked to rotation rate or age. The clearest answer comes from a class of interacting binary stars, the RS CVn systems. These systems contain aging, evolved stars that have maintained rapid rotation through tidal coupling with their nearby companions. They also exhibit very high levels of activity, proving thereby that the causal determinant for stellar activity is rotation, rather than age.

Whereas the magnetic activity and variability of ordinary lower main-sequence stars depend strongly on rotation, rotation itself is governed by age and mass. For a star with a convective envelope, an inescapable fact of life is that its rotation will slow as its magnetically coupled wind carries away angular momentum [Schatzman, 1962]. This magnetic braking is strongly dependent on rotation rate, so the deceleration is, initially, very rapid. Theory suggests that the rotational velocity of a solar mass star becomes independent (to a few percent) of initial angular momentum within a few hundred Myr [*Kawaler*, 1988]. Measured rotation periods for G-type stars some 700 Myr old in the Hyades and Coma Berenices clusters, which show little scatter at any given color, support this conclusion.

The past 20 years have witnessed remarkable progress in our empirical and theoretical understanding of the angular momentum evolution of young solar-type stars [e.g., *Krishnamurthi et al.*, 1997]. A number of factors involving both the structure of the star as well as its interaction with its environment enter the story. As a star settles onto the main sequence, the formation of a dense, radiative core decreases its moment of inertia, which tends to spin it up, while interactions with remnants of the stellar accretion disk oppose this tendency. To complicate matters further, the mechanism for angular momentum loss apparently saturates above some critical rotation rate, which hampers the efficacy of magnetic braking. The growing radiative core also decouples to some degree from its surrounding convective envelope, which further alters the accounting. Depending on its specific initial conditions, an infant sunlike star may settle down more-or-less directly onto the main sequence, with a rotation period of a few days, or it may undergo a more dramatic spin-up, spin-down episode which, nevertheless, leads the star to almost the same rotational state after 100 Myr or so. Which path the young Sun took is probably unknowable but, in any case, left no strong mark on its subsequent evolution.

The surface rotation rate of F- to K-type main-sequence stars several hundred Myr old shows some dependence on stellar mass, declining from ~7 km/sec to ~ 4 km/sec as mass decreases by 30% or so [*Radick et al.*, 1987]. This trend in mean rotation rate is probably a relic of early evolution, because the temporal evolution of rotation for such stars does not appear to be strongly mass dependent after an age of a few hundred Myr.

3.3. The Influence of Mass on Stellar Magnetic Activity

Stellar mass should also affect magnetic activity. As mentioned above, mass is absolutely decisive in one important sense: stars more than about 50% more massive than the Sun (i.e., earlier than mid F-type), which lack deep convective envelopes, do not experience activity cycles (or, for that matter, pronounced rotational slowdown). It should be noted, however, that early F-type stars do show evidence of magnetic activity [e.g., *Giampapa and Rosner*, 1984]—what is absent, apparently, are the large active regions so characteristic of solar magnetic activity. On theoretical

grounds, mass should also strongly affect the behavior of the stellar dynamo. Specifically, the dynamo number, which parameterizes the efficiency of the hydromagnetic dynamo in mean field theory, depends as strongly on the time scale for convective turnover at the base of the stellar envelope as it does on rotation rate [*Durney and Latour*, 1978]. This time scale in turn depends strongly on stellar mass, increasing monotonically with decreasing mass among lower main-sequence stars, and rising especially steeply between mid F- and mid G-type main-sequence stars [*Gilliland*, 1985]. Thus, lower-mass stars should be relatively more active, among otherwise comparable stars. Empirically, this seems to be the case: K-type stars in the Hyades, for example, tend to show more evident activity than do F-type stars, and G-type stars tend to be in the middle, but more closely resemble the K-type stars [*Radick et al.*, 1987].

3.4. Other Factors Affecting Stellar Magnetic Activity: Composition and Companions

Like stellar mass, chemical composition strongly affects convective zone structure, and should therefore influence magnetic activity. The identification and measurement of the real and apparent effects of composition on stellar magnetic activity, however, remains largely an unfinished task, in part because the nearby stars seem to be fairly homogeneous in composition, in part because it is difficult to separate composition effects from other factors.

In brief, a metal-rich star will have a deeper convective zone and a longer convective turnover time scale than a metal-poor star of the same mass. In this respect, metal enhancement mimics lower mass, and a metal-rich star should be relatively more active than its metal-poor counterpart of equal mass. In practice, it is difficult to test this hypothesis, because it is not easy to identify stars of strictly equal mass. Furthermore, stellar masses are often estimated from intermediaries, such as photometric color indices, that are sensitive to metallicity, so it is difficult to avoid circularity.

Even if chemical composition had no intrinsic effect on stellar magnetic activity, it would enhance the apparent activity of metal-rich stars through its influence on radiative activity diagnostics such as chromospheric Ca II H+K emission. Calculations suggest that Ca II H+K emission depends strongly and directly on metallicity. The Mount Wilson S-index, which measures the ratio of the combined fluxes in two 1-A continuum windows centered on the H and K lines to the sum of the fluxes in two 20-A continuum windows on either side of the doublet, could be even more sensitive because of enhanced line-blanketing effects outside the cores of the lines.

Several decades ago, starspots were invoked to explain the photometric variability of the RS CVn binaries. Their extreme activity (relative to the Sun) and the ability of tidal coupling to enforce rapid rotation indicates that the presence of a close, interacting companion can dramatically affect the rotational evolution and therefore the magnetic activity of a star.

4. WHAT MAKES A SUNLIKE STAR?

The Sun appears to be a rather ordinary star, a conclusion that is reinforced by the fact that a single theory of stellar structure and evolution suffices for both the Sun and other stars. This alone, however, is probably insufficient to guarantee that the Sun is completely normal in terms of its rotation, its magnetic activity, or the variability that accompanies that activity. Both theory and observation also agree, however, that the history of rotation for solar-type stars is a story of rapidly convergent evolution. We may therefore be fairly confident that the young Sun rotated more rapidly than does the present-day Sun, and that the Sun will continue to slow as it ages. More provocatively, we may assert that a single (or, more generally, a tidally non-interacting) star of the same mass, chemical composition, and age as the Sun will be found to rotate at about the same rate as the Sun, and that its activity and variability will also closely resemble those of the Sun. It is this expectation that valid stellar proxies for the Sun (i.e., sunlike stars) exist that underlies the branch of contemporary astrophysics known as the solar-stellar connection. Furthermore, good proxies for the early Sun should be available among young solar-type stars, and solar-type stars from kinematically old populations and subdwarfs should be reasonable proxies for what the Sun is likely to become as it ages further.

5. VARIABILITY PATTERNS OF SUNLIKE STARS

In 1966, Olin Wilson began to monitor the Ca II H+K emission for about 100 stars using the 100 inch (2.54 m) Hooker Telescope at Mount Wilson Observatory [*Wilson*, 1978]. Wilson's observations have been continued as part of the Mount Wilson HK program, at the 60 inch (1.5 m) telescope between 1977 and 1995 [*Baliunas et al.*, 1995], and at the refurbished Hooker telescope after 1995.

Wilson's goal was "to answer the general question, Does the chromospheric activity of main-sequence stars vary with time, and if so, how?" The observational quantity that Wilson used to measure chromospheric emission is known as the S-index, from which *Noyes et al.* [1984] developed their chromospheric emission ratio, R'_{HK}, defined as the

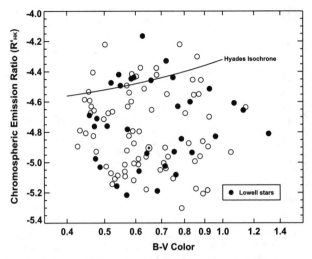

Figure 2. The activity-color plot for the augmented Wilson sample. Activity increases upward, and mass (B-V color is related to mass) decreases rightward. The position of the Sun in the lower left center of the diagram is indicated by its symbol. The Hyades isochrone marks the locus of young stars about 15% the age of the Sun. Filled symbols designate the subset observed photometrically from Lowell Observatory.

chromospheric HK emission of a star normalized by its bolometric luminosity. The conversion to R'_{HK} also corrects for a color effect present in the S-index and removes the photospheric contribution to the line core emission. In Wilson's sample, R'_{HK} ranges (in logarithmic units) from about –4.1 to –5.3; the solar value is –4.94. Figure 2 shows the distribution of Wilson's sample (including a few stars added later to the Mount Wilson HK program) on an activity-color plot.

Wilson did not set out to study only sunlike stars. It is now clear that his sample includes several subgiants: Wilson himself recognized the probability that his sample contained some evolved stars. The release of the Hipparcos parallaxes has now enabled calculation of absolute magnitudes for the stars, independent of their MK spectral classifications. Figure 3 shows the HR diagram for the Wilson sample.

We may quantitatively assess each star of Wilson's sample as a solar analog by measuring its distance from the Sun in the three-dimensional M, B-V, R'_{HK} manifold. For example, if we weight equally distances of 1.0 mag in M_V, 0.1 mag in B-V, and 0.2 in log R'_{HK} (about 0.05 in the S-index), we find that the best solar analog among Wilson's stars is HD 126053 (0.28 units) whose variability, it turns out, does not much resemble that of the Sun [*Baliunas et al.*, 1995; G. W. Henry, 2001, private communication]. By the same weighting, the current favorite candidate for solar twin, HD

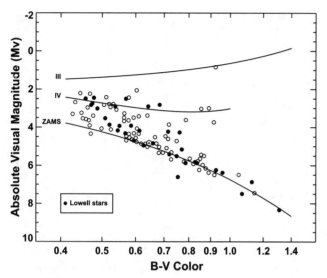

Figure 3. The Hertzsprung-Russell (color-luminosity) diagram for the augmented Wilson sample. The position of the Sun in this diagram is indicated by its symbol. Filled symbols designate the subset observed photometrically from Lowell Observatory.

146233 (18 Sco), is 0.02 units distant. In anticipation, it may also be noted that the closest solar analogs in the Lowell subset of the Wilson stars are HD 143761 (0.96 units) and HD 114710 (1.29 units). It is somewhat disappointing to learn in hindsight that there are no very good solar twins in either Wilson's sample or, especially, the Lowell subset. Despite these limitations, the Wilson stars are the core of our heritage of time-series measurements, and will remain so until results from more narrowly selected samples of stars, now being observed at the Mount Wilson and Fairborn Observatories, become available.

The temporal variation of stellar activity may be classified into three categories [*Baliunas et al.*, 1997], as shown in Figure 4. Young, rapidly rotating stars tend to vary erratically, rather than in a smooth cycle like the Sun. About 80% of older, more slowly rotating stars, including the Sun, tend to show regular activity cycles, and the remaining 20% show little or no variation at all. The inactivity of these latter stars is not necessarily a consequence of great age; the one demonstrably old star in the Wilson sample, the subdwarf shown well below the ZAMS in Figure 3, currently has a relatively vigorous cycle. Rather, these inactive stars may be temporarily in low activity states analogous to the Sun's Maunder Minimum.

Stellar activity cycles, when present, often seem to be about a decade in length. Finding a simple explanation for this fact has become somewhat like the quest for the Holy Grail, and an equally elusive goal [e.g., *Saar and*

Brandenburg, 1999]. Among older stars, those more massive than the Sun tend to show low amplitude cycles, whereas those less massive than the Sun often have strong cycles. In fact, stars with strong, "sunlike" cycles (selected subjectively from *Baliunas et al.,* [1995]) tend to occupy a fairly narrow range in an activity-color plot, as shown in Figure 5, with the Sun near the high-mass end of the distribution. Apparently, the Sun has a fairly prominent activity cycle, as traced by its HK emission, for a star of its mass.

The photometric variability of ordinary, lower main-sequence stars was first detected among members of young clusters such as the Pleiades and Hyades. These studies helped inspire a program, begun at Lowell Observatory in 1984, to study the long-term photometric variability of lower main-sequence field stars [*Lockwood et al.,* 1997; *Radick et al.,* 1998]. The Lowell program sampled stars bracketing the Sun in temperature and average activity level. It included 41 program stars, 34 of which were selected from the stars of the Mount Wilson HK program. From these measurements, we now know that the amplitude of the year-to-year photometric variation for young, active stars (including those in young clusters) is typically several percent. It decreases dramatically, typically by a factor of thirty or so, to a level approaching the detection limit of the measurements (about 0.001 mag ≈ 0.1%) among stars similar to the Sun in age and average activity. In contrast, the corresponding decrease in chromospheric Ca HK variation is only about a factor of three. Figures 6 and 7 show representative time series for a young star, probably several hundred Myr old (HD 1835), and an older star, probably comparable to the Sun in age (HD 10476). A catalog of time series plots like the two illustrated may be found in *Radick et al.* [1998]. In addition to the differences in scale, a careful look at the

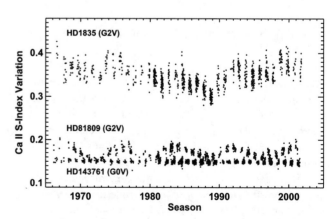

Figure 4. HK time series for three stars (1) HD 1835, a young, active star, (2) HD 81809, an older star with a sunlike activity cycle, and (3) HD 143761, an inactive older star.

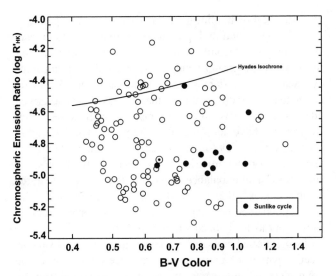

Figure 5. The incidence of pronounced activity cycles among the stars of the Wilson sample. With the possible exception of HD 81809 (a subgiant), the stars with prominent, Sun-like cycles are less massive than the Sun.

time series shown in Figures 6 and 7 reveals another interesting fact: whereas the year-to-year brightness changes of HD 10476 tend to correlate directly with the activity variations shown by Ca HK emission, just as for the Sun, the time series for HD 1835 show anticorrelated variations. In fact, this pattern appears to distinguish young, active stars from older, more sunlike stars in general, as shown in Figure 8.

In 1993, measurements similar to the Lowell program were begun at the Fairborn Observatory. Although detailed comparison of the observations has only begun, it is already clear that these newer observations validate the broad outline of variability among sunlike stars sketched in the preceding paragraph. In general, photometric variation is fairly common across the entire lower main sequence, with about half of the combined sample of several hundred stars showing detectable variability on the year-to-year time scale [*Lockwood et al., 1997; Henry, 1999*].

6. EVIDENCE FOR MAUNDER MINIMA STATES AMONG SUNLIKE STARS

What insight into states of solar activity such as the Maunder Minimum, which have not been observed on the Sun during the modern era, do stellar observations provide? As mentioned above, among older, less active stars like the Sun, about 20% show little or no variability at all. It has been suggested that these stars may currently be in activity states resembling the Maunder Minimum [*Baliunas et al., 1997*]. The numbers support this suggestion, if the Sun itself

spends about 20% of its time in such states. More interesting, however, would be to watch a star making the transition between a cycling state and no variation, or vice-versa. Sufficient stars have been observed long enough that one might expect to find a few cases of this behavior in the presently-available database. In fact, the search turns up no compelling examples: what is found are several stars whose cycle amplitude is clearly changing (generally, decreasing) in a secular fashion. HD 10476, shown in Fig. 7, is a good example of this. It also appears in this case that the photometric brightness amplitude is decreasing along with the HK emission. Whether HD 10476 is actually entering a Maunder Minimum state, however, remains to be determined. On several occasions, the Sun itself has presented us with sequences of several activity cycles that decrease (or increase) regularly in amplitude, without entering a Maunder Minimum state. We may simply have to watch stars like HD 10476 for several more decades before it becomes clear what is happening.

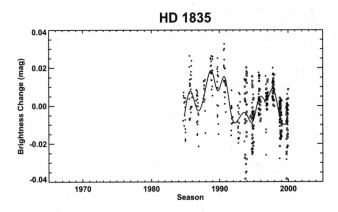

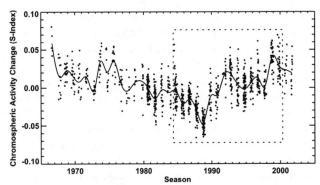

Figure 6. Photometric (top) and chromospheric HK (bottom) time series for HD 1835, a young, active star. The dotted-line box in the lower panel encloses the portion of the Mount Wilson Ca HK emission observations during which photometric measurements were also made at Lowell Observatory.

HD 10476

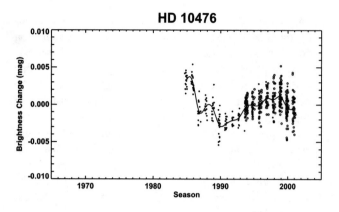

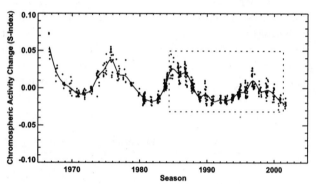

Figure 7. Photometric (top) and chromospheric HK (bottom) time series for HD 10476, an older star with a prominent activity cycle. Note that the scale of the upper panel is 4x that of the corresponding panel of Figure 6, whereas the lower panels have the same scale.

7. THE STELLAR MID-LIFE CRISIS

Twenty years ago, *Vaughan and Preston* [1980] noted a relative lack of F- and G-type stars with intermediate activity (i.e., $\log R'_{HK} \sim -4.7$). The suggestion that this "Vaughan-Preston gap" might represent an abrupt change in the nature of stellar chromospheric emission at an age of perhaps 2 Gyr prompted a rebuttal [*Hartmann et al.,* 1984], which argued that the gap is a statistical artifact created by saturation at high levels of emission and a floor at low levels. Since then the Vaughan-Preston gap has lingered, mainly, it seems, as a curiosity. And since then, evidence has continued to grow that something rather dramatic happens to the activity and variability of sunlike stars at an age of about 2 Gyr. Stars younger than this seem rarely to show simple, regular, well-defined activity cycles like that of the Sun. Rather, they vary irregularly, and some appear to vary with at least two comparably-long periodicities operating simultaneously [*Baliunas et al.,* 1995]. There seems to be a real, qualitative difference in the character of variation, recalling the old idea

that the highly nonlinear stellar dynamo may operate in more complex and multiple modes as rotation rate increases. There is a second distinction between young and old stars: at an age greater than ~1 Gyr, a sunlike star apparently switches its variability, from a strongly spot dominated mode to a rather different form that is weakly faculae-driven, like the present-day Sun. The transition seems to be abrupt, because we have not found an example of an active sunlike star (as indicated by Ca HK emission, for example) that does not also vary photometrically. In other words, there seems always to be an imbalance between dark and bright features, which suggests that the tendency to create faculae rather than spots does not gradually evolve as a star's rotation slows, but instead appears rather suddenly.

As mentioned earlier, there is opinion that the basic convective pattern of a lower main-sequence star changes as its rotation rate decreases, from elongated rolls aligned parallel to the stellar rotation axis to a cellular pattern more reminiscent of solar granulation. It is tempting to speculate that the Vaughan-Preston gap marks this transition in convective modes, and that this transition affects the operation of the dynamo strongly enough to change the phenomenology of stellar variability.

If there is, indeed, a stellar midlife crisis at an age of about 2 Gyr (and $\log R'_{HK} \sim -4.7$), then HD 114710 (see Figure 8) must be near or just through it. The photometric and Ca HK time series for this star, shown in Figure 9, sug-

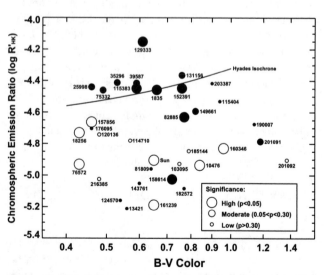

Figure 8. Correlation between year-to-year photometric brightness and Ca HK emission variations, displayed on activity-color axes. The size of each symbol indicates the significance of the correlation. Open symbols are used to represent stars (like the Sun) that become brighter as their HK emission increases. Filled symbols represent anti-correlated behavior.

HD 114710

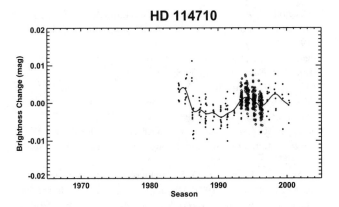

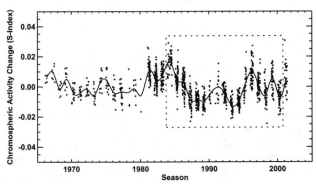

Figure 9. Photometric and Ca HK time series for HD 114710, a star of intermediate age.

gest that it indeed has not yet quite made up its mind about how to behave. According to *Baliunas et al.* [1995], it is a dual-period star, with a 16.6 yr cycle superposed on a 9.6 yr cycle. Its photometric variation, while strong relative to the Sun's, is nevertheless less robust than that of a young star like HD 1835 (Figure 6). Figure 9 also clearly suggests that the photometric brightness and Ca HK emission of HD 114710 vary in step, like the Sun and unlike a young star. (Figure 8, which implies no strong correlation for this particular star, is based on data between 1988 and 1995. Figure 9 indicates that the sense of the correlation is unclear when only that restricted time span is considered.)

8. SUMMARY: THE SUN IN TIME

The story of the activity and variability of a sunlike star on evolutionary time scales is a story of decay. The inexorable decrease in the star's rotation rate as it loses angular momentum through its magnetic wind decrees this fate.

As a young star, the Sun probably resembled HD 1835. Its rotation period slowed, from a few days at an age of 100 Myr to perhaps a week at an age of 700 Myr, and its level of magnetic activity also declined, probably by a compara-

ble factor. Its activity level, nevertheless, remained several times greater than it is today. Its activity varied strongly, but irregularly, on time scales of days to years. The corresponding brightness variations could have been 30–50 times stronger than the Sun's are today. This variability was driven by dark features, probably spots, rather than by bright features as today. Whether these spots were larger or simply more numerous than the spots on the present-day Sun is hard to say. Spot models and Doppler images of extremely active stars (often interacting binaries) tend to imply small numbers of enormous spots, often with one at a rotational pole, reminiscent of the continents on Earth. In a provocative paper, however, *Eaton et al.* [1996] demonstrated that a large number of moderately sized spots, placed randomly in space and time on a differentially rotating star, explain equally well the observed variability of active stars, including the appearance of pseudo-cycles in the time series.

At an age of perhaps 2 Gyr, by the time its rotation had slowed by another factor of two, the Sun apparently realized, perhaps gradually but maybe suddenly, that it was no longer a young star. It began dressing differently—spots were out, faculae were in, at least in preference. Gone also was the intemperance of youth—rather than varying erratically, it now found a dignified, regular cycle more to its taste. Apparently, it also began taking naps, shutting off its cyclic behavior more-or-less completely every few centuries to relax in episodes like the Maunder Minimum. It had become a middle-aged star.

The stars suggest that the Sun's future is likely to resemble its present, at least during the remaining 4 Gyr or so of its main-sequence lifetime. Its rotation will continue to slow, perhaps by another 30%, and its average activity will also decline. It may very well continue its present pattern of activity cycles, interrupted by quiescent episodes. Among sunlike stars, middle age lasts a very long time.

Acknowledgments. As always, it is a pleasure to thank my long-term colleagues, Sallie Baliunas and Wes Lockwood, for generously giving me free use of material prior to publication elsewhere. W. Livingston provided me with the NSO/KP solar K-line data. The solar radiometry data has been made available by C. Fröhlich on the Web. This work was supported at AFRL by the Air Force Office of Scientific Research.

REFERENCES

Abbot, C. G. 1934. The sun and the welfare of man, *Smithsonian Sci. Ser.* Vol. 2, (New York: Smithsonian Scientific Series).

Baliunas, S. L., Donahue, R. A., Soon, W. H., and Henry, G. W., 1997. Activity cycles in lower main sequence and post main sequence stars: The HK project, in *Cool Stars, Stellar Systems, and the Sun*, eds. Donahue, R. A., and Bookbinder, J. A., *ASP Conf. Ser.*, 154, pp. 153-172.

Baliunas, S. L., Donahue, R. A., Soon, W. H., Horne, J. H., Frazer, J., Woodard-Eklund, L., Bradford, M., Rao, L. M., Wilson, O. C., Zhang, Q., Bennett, W., Briggs, J., Carroll, S. M., Duncan, D. K., Figueroa, D., Lanning, H. H., Misch, A., Mueller, J., Noyes, R. W., Poppe, D., Porter, A. C., Robinson, C. R., Russell, J., Shelton, J. C., Soyumer, T., Vaughan, A. H., and Whitney, J. H., 1995. Chromospheric variations in main-sequence stars. II, *Ap. J.*, 438, pp. 269-287.

Beer, J. 2000. Long-term indirect indices of solar variability, *Space Sci. Rev.*, 94, pp. 53-66.

Durney, B. R., and Latour, J. 1978. On the angular momentum loss of late-type stars, *Geophys. Astrophys. Fluid Dyn.*, 9, pp. 241-255.

Eaton, J. A., Henry, G. W., and Fekel, F. C. 1996. Random spots on chromospherically active stars, *Ap. J.*, 462, pp. 888-893.

Eddy, J. A. 1976. The Maunder Minimum, *Science*, 192, pp. 1189-1202.

Eddy, J. A. 1983. Keynote address: An historical review of solar variability, weather, and climate, in *Weather and Climate Responses to Solar Variations*, ed. B. M. McCormac (Boulder, Colorado Assoc. Univ. Press), pp. 1-15.

Foukal, P., and Lean, J., 1988. Magnetic modulation of solar luminosity by photospheric activity, *Ap. J.*, 328, pp. 347-357.

Giampapa, M. S., & Rosner, R. 1984. The appearance of magnetic flux on the surfaces of the early main-sequence *F* stars, *Ap. J.*, 286, pp. L19-L22.

Gilliland, R. L. 1985. The relation of chromospheric activity to convection, rotation, and evolution off the main sequence, *Ap. J.*, 299, pp. 286-294.

Hartmann, L., Soderblom, D. R., Noyes, R. W., Burnham, N., and Vaughan, A. H. 1984. An analysis of the Vaughan-Preston survey of chromospheric emission, *Ap. J.*, 276, pp. 254-265.

Henry, G. W., 1999. Techniques for automated high-precision photometry of sun-like stars, *Pub. Astr. Soc. Pac.*, 111, pp. 845-860.

Kawaler, S. D. 1988. Angular momentum loss in low-mass stars, *Ap. J.*, 333, pp. 236-247.

Krishnamurthi, A., Pinsonneault, M. H., Barnes, S., and Sofia, S. 1997. Theoretical models of the angular momentum evolution of solar-type stars, *Ap. J.*, 480, pp. 303-323.

Kuhn, J. R., & Libbrecht, K. G. 1991. Nonfacular solar luminosity variations, *Ap. J.*, 381, pp. L35-L37.

Larmor, J. 1919. How could a rotating body such as the Sun become a magnet? *Engl. Mech.*, 110, pp. 113-114.

Lean, J. L., Cook, J., Marquette, W., & Johannesson, A. 1998. Magnetic sources of the solar irradiance cycle, *Ap. J.*, 492, pp. 390-401.

Li, L. H., & Sofia, S. 2001. Measurements of solar irradiance and effective temperature as a probe of solar interior magnetic fields, *Ap. J.*, 549, pp. 1204-1211.

Lockwood, G. W., Skiff, B. A., and Radick, R. R., 1997. The photometric variability of sun-like stars: Observations and results, 1984-1995, *Ap. J.*, 485, pp. 789-811.

Miesch, M. S., 2000. The coupling of solar convection and rotation, *Sol. Phys.*, 192, pp.59-89.

Noyes, R. W., Hartmann, L. W., Baliunas, S. L., Duncan, D. K., & Vaughan, A. H. 1984. Rotation, convection, and magnetic activity in lower main-sequence stars, *Ap. J.*, 279, pp. 763-777.

Radick, R. R., Lockwood, G. W., Skiff, B. A., and Baliunas, S. L., 1998. Patterns of variation among sun-like stars, *Ap. J. Suppl. Ser.*, 118, pp. 239-258.

Radick, R. R., Thompson, D. T., Lockwood, G. W., Duncan, D. K., and Baggett, W. E., 1987. The activity, variability, and rotation on lower main-sequence Hyades stars, *Ap. J.*, 321, pp. 459-472.

Saar, S. H., and Brandenburg, A., 1999. Time evolution of the magnetic activity cycle period. II. Results for an expanded stellar sample, *Ap. J.*, 524, pp. 295-310.

Schatzman, E. 1962. A theory of the role of magnetic activity during star formation, *Ann. Astrophys.*, 25, pp. 18-29.

Schrijver, C. J., Coté, J., Zwaan, C., and Saar, S. H. 1989. Relations between the photospheric magnetic field and the emission from the outer atmosphere of cool stars. I. The solar Ca II K line core emission, *Ap. J.*, 337, pp. 964-976.

Skumanich, A. 1972. Time scales for Ca II emission decay, rotational braking, and lithium depletion, *Ap. J.*, 171, pp. 565-567.

Skumanich, A., Smythe, C., and Frazier, E. N. 1975. On the statistical description of inhomogeneities in the quiet solar atmosphere. I. Linear regression analysis and absolute calibration of multichannel observations of the Ca+ emission network, *Ap. J.*, 200, pp. 47-764.

Vaughan, A. H., and Preston, G. W. 1980. A survey of chromospheric Ca II H and K emission in field stars of the solar neighborhood, *Pub. Astr. Soc. Pac.*, 92, pp. 385-391.

Weiss, N. O. 1994. Solar and stellar dynamos, in *Lectures on Solar and Planetary Dynamos*, eds. M. R. E. Proctor and A. D. Gilbert (Cambridge: Cambridge Univ. Press), pp. 59-95.

White, O. R., Livingston, W. C., Keil, S. L., and Henry, T. W., 1998. Variability of the solar Ca II K line over the 22-year Hale cycle, in *Synoptic Solar Physics*, eds. Balasubramaniam, K. S., Harvey, J. W., and Rabin, D. M., *ASP Conf. Ser.*, 140, pp. 293-300.

Willson, R. C., 1997. Total solar irradiance trend during solar cycles 21 and 22, *Science*, 277, pp. 1963-1965.

Wilson, O. C., 1978. Chromospheric variations in main-sequence stars, *Ap. J.*, 226, pp. 379-396.

Richard R. Radick, Air Force Research Laboratory, Phillips Lab, National Solar Observatory, Sacramento Peak, Sunspot, NM 88349.

Solar Variability Caused by Structural Changes of the Convection Zone

S. Sofia and L. H. Li

Yale University, New Haven, Connecticut

A varying magnetic field in the solar interior will result in variations of the solar internal structure, and consequently it will affect all the global solar parameters, including the luminosity. Modulation mechanisms of the total iriadiance based on surface phenomena will be in addition to the luminosity changes produced by the structural variations. Here we expand the standard theory of stellar structure and evolution to include the effects of varying magnetic fields to model the structural variations of the Sun over the activity cycle. This new category of solar models, especially after being generalized to two dimensions, has the potential to explain all the observed solar cycle variations, including the luminosity, effective temperature, radius, p-mode oscillation frequencies, sound speed, and their latitudinal variations, in addition to the irradiance. Conversely, these studies should provide a much better understanding of the dynamics (including dynamo) in the convection zone of the Sun.

1. INTRODUCTION

Active regions clearly cause most of the short-term variations of the total irradiance. It was natural to assess if these same features could account for the 11 year magnetic activity cycle timescale variability of vradiance. There are two opposite viewpoints on this regard. *Gray and Livingston* [1997] observed variations of the photospheric temperature that could account by themselves for the entire variation of the total solar irradiance over the activity cycle. On the other hand, *Lean et al.* [1998], *Solanki and Unruh* [1998], *Spruit* [2000] argue that the model with only the magnetic surface effects such as sunspot darkening and faculai brightening can account for the irradiance cycle without photospheric temperature variation. The first viewpoint allows (or even requires) various variations of the solar interior structure which cause changes of the global solar parameters such as the luminosity, radius, and effective temperature. The second

viewpoint, on the other hand, assumes that none of these changes occur. We will show here that this second viewpoint is inconsistent with some observations, at least for the activity cycle and longer timescales, and that a structure variation mechanism is both needed and naturally expected from a varying magnetic field in the interior of the Sun.

The difficulties encountered by the exclusive magnetic activity picture include (i) the inability to observationally confirm the contrast of the different features needed to explain the irradiance variations, (ii) the existence of observations in support of variations of the photospheric temperature, of the solar diameter, and of the spectrum of solar oscillations. Those observations indicate that the Sun undergoes structural variations over an activity cycle, which should affect the solar luminosity. Such variations, presumably caused by a change of the internal (dynamo) magnetic field, will inevitably produce a variation of luminosity above which the undeniable activity-related variations are superposed.

Helioseismic data from SOHO/MDI are so precise that accurate solar models at different phases of the solar activity cycle are needed to confront theory with observation. The

Solar Variability and its Effects on Climate
Geophysical Monograph 141
Copyright 2004 by the American Geophysical Union
10.1029/141GM03

comparison of the frequency cycle variations is likely to be crucial. If there were no changes, the irradiance cycle variations would be totally produced by the magnetic surface effects. If the amplitude of the irradiance cycle variations required to explain the oscillation frequency changes were equal to the observed amplitude of the irradiance variations, then the entire variations of the total irradiance would be due to changes of the luminosity. Intermediate cases would, of course, require both types of change, and the variations of the total irradiance would be due in part to luminosity changes, and in part to the magnetic surface effects. Therefore, a systematic theory of solar structure and evolution including varying magnetic fields and turbulence is demanded. In this chapter we will formulate the theoretical foundation to compute the effects of a varying magnetic field in the solar interior on the global parameters of the Sun as a function of the physical properties (magnitude, depth, shape, etc.) of the field.

2. EVIDENCE IN SUPPORT OF SOLAR STRUCTURE VARIATIONS

2.1. Variations of Solar Effective Temperature

The solar effective temperature was measured by *Gray and Livingston* [1997] from ratios of spectral line depths of

$$\text{CI}(\lambda 5380)/\text{FeI}(\lambda 5379)$$

and

$$(\text{CI}(\lambda 5380)/\text{TiII}(\lambda 5381)$$

The excitation potentials of these lines are different from each other:

$$\text{CI}=7.68 \text{ eV},$$

$$\text{FeI}=3.69 \text{ eV},$$

$$\text{TiII}=1.57 \text{ eV}.$$

The consistency of results indicates that the effective temperature (T_{eff}) they measure is the photospheric temperature. Although the calibration of the temperature variations determined by *Gray and Livingston* [1997] over the period from 1978 to 1992 is being questioned [*Caccin and Penza* 2002], it is clear that an important component to the variations of the solar luminosity (hence irradiance) is omitted in the activity models.

2.2. Variations of Solar Oscillations

Solar-cycle related variations on solar oscillation frequencies were first determined by *Woodard* [1987]. The first evidence of frequency dependent changes in accoustic splittings were found by *Kuhn* [1988]. Recently, *Bhatnagar et al.* [1999] presented a correlation analysis of GONG p-mode frequencies with nine solar activity indices for the period from August 1995 to August 1997. A decrease of $0.06 \, \mu\text{Hz}$ in frequency during the descending phase of solar cycle 22 and an increase of $0.04 \, \mu\text{Hz}$ in the ascending phase of solar cycle 23 are observed. This analysis further confirms that the temporal behavior of the solar frequency shifts closely follow the phase of the solar activity cycle. Besides, the analysis given by *Howe et al.* [1999] suggests that the solar cycle related variation of the oscillation frequencies is not due to contamination of observed Doppler shifts by the surface magnetic fields.

2.3. Radius Variations

Ground-based measurements of the solar radius exist over three centuries, but the results are controversial and inconsistent. At the present time, there are two methods that can provide high precision data. One is space-based measurements (e.g. the Solar Disk Sextant [SDS, *Sofia et al.*, 1994] and SOHO/MDI [*Emilio et al.*, 2000]). The other is based on solar oscillations data, in particular using the f-modes [*Antia et al.*, 2000; *Dziembowski*, 2001]. Although in no instance a complete cycle has been measured, they all show variations indicating structural changes which occur within the solar interior.

3. LIMITATION OF THE STANDARD SOLAR MODEL

Standard solar models, (SSM) use pressure, temperature, radius, and luminosity as the stellar structure variables and predict no observable variations on the timescales of years, decades, and centuries. For example, SSMs predict the relative temporal variation rates of the last three variables at the solar surface at the present age of the Sun to be,

$$\frac{d\ln T}{dt} = 4.0 \times 10^{-12} \text{ yr}^{-1},$$

$$\frac{d\ln R}{dt} = 3.3 \times 10^{-11} \text{ yr}^{-1},$$

$$\frac{d\ln L}{dt} = 8.3 \times 10^{-11} \text{ yr}^{-1}.$$

However, the observed cyclic variations of solar effective temperature, solar radius, and the total solar irradiance are equal to about 1.5 K, 21 milli arc second (mas), and 0.1% respectively. The corresponding yearly variations are

$$\frac{d\ln T}{dt} = 4.5 \times 10^{-5}\,\text{yr}^{-1},$$

$$\frac{d\ln R}{dt} = 4.0 \times 10^{-6}\,\text{yr}^{-1},$$

$$\frac{d\ln L}{dt} = 2.0 \times 10^{-4}\,\text{yr}^{-1}.$$

These variations cannot be completely explained by the surface features of magnetic activity sunspots, faculae, and magnetic network. In particular, the surface features will not affect the photospheric temperature or solar radius.

In any event, regardless of the final steps in which the rate of energy flow (and direction) is modulated, the ultimate source of the variation is the solar interior. Thus understanding the origin of the solar radiation striking the Earth requires understanding the variations of the internal structure of the Sun.

We want to trace back all these variations to a common origin. The obvious candidate link is magnetic fields. The reason is that solar dynamo generates magnetic fields which vary on timescales relevant to climate, and which can, in principle, produce significant structural variations.

4. INCLUSION OF MAGNETIC FIELDS IN SOLAR MODELING

4.1. Definition of Magnetic Variables

To follow the behavior of the solar model in response to a variable magnetic field in its interior, $B = (B_t, B_p)$, it is necessary to formulate the equations of stellar structure and evolution including magnetic fields. SSMs use pressure, temperature, radius and luminosity as the structure variables, and mass M_r interior to a radius r, as the independent variable. When magnetic fields are present, we introduce two magnetic variables since a magnetic field is a vector: the magnetic energy per unit mass, χ, and the effective ratio of specific heats due to the magnetic perturbations, γ:

$$\chi = (B^2/8\pi)/\rho, \tag{1}$$

$$\gamma = 1 + 2B_t^2/B^2, \tag{2}$$

where $B^2 = B_t^2 + B_p^2$, ρ is the density. The former describes the magnetic perturbation strength, and the latter describes

the tensor feature of the magnetic pressure. In general, the determination of χ and γ requires a comprehensive understanding of turbulent dynamics in the solar convection zone, an undertaking that is impractical at present. Therefore, we specify χ and γ as functions of time t and the mass depth M_D:

$$M_D = \log(1 - M/M_\odot). \tag{3}$$

4.2. Radius Variation

Magnetic fields produce a magnetic pressure,

$$P_B = (\gamma - 1)\chi\rho \tag{4}$$

in addition to the gas pressure P_0. If the unperturbed density is ρ_0 with only the gas pressure P_0, where $\rho_0 = \rho_0(P_0, T)$ is the equation of state for the gas, the magnetic pressure will redistribute the gas so that the gas density becomes

$$\rho = \rho(P, T, \chi, \gamma), \tag{5}$$

where $P = P_0 + P_B$ is the total pressure.

By assuming that the gas is ideal and that the magnetic field does not affect the temperature T, we have

$$\rho \approx \rho_0/(1 + P_B/P). \tag{6}$$

Using this approximate relation, we can estimate the radius variations due to the magnetic pressure P_B by integrating the mass conservation equation,

$$\Delta R \approx \int_0^{M_\odot} \frac{P_B\,dM}{4\pi r^2 \rho_0 P}. \tag{7}$$

Since the gas in the Sun is not ideal, especially near the surface, and the magnetic fields will affect the temperature gradient, as we will show below, Eq. (7) may somewhat overestimate the radius variation of the Sun due to the changing magnetic field in the solar interior.

4.3. Temperature Variation

Magnetic fields affect the temperature gradient in the convection zone. This effect can be implemented by using an effective adiabatic temperature gardient (see Appendix B):

$$\nabla'_{ad} = \nabla_{ad}\left[1 - 3\left(\nu\nabla_\chi + \nu'\nabla_\gamma\right)/\mu\right], \tag{8}$$

where

$$\nu = -\left(\partial \ln \rho / \partial \ln \chi\right)_{T,P,\gamma} \qquad \nabla_\chi = \partial \ln \chi / \partial \ln P$$

$$\nu' = -\left(\partial \ln \rho / \partial \ln \gamma\right)_{T,P,\chi} \qquad \nabla_\gamma = \partial \ln \gamma / \partial \ln P$$

are the derivatives relevant to the magnetic field. μ is the compressibility coefficent at a constant temperature and a constant magnetic field.

The change of the temperature gradient can be estimated by the difference between the effective and actual adiabatic gradients:

$$\Delta \nabla \approx -3\left(\nu \nabla_\chi + \nu' \nabla_\gamma\right)\nabla_{\text{ad}} / \mu, \tag{9}$$

where $\nabla = \partial \ln T / \partial \ln P$ is the actual temperature gradient. The relative variation of the surface (or effective) temperature can be estimated by

$$\Delta \ln T_{\text{eff}} \approx -\int_0^{M_\odot} \frac{GM}{4\pi r^4 P}\left(\Delta \nabla\right)\mathrm{d}M. \tag{10}$$

4.4. Luminosity Variation

The luminosity at the surface is determined by both the radius R and the effective temperature T_{eff}:

$$L = 4\pi R^2 \sigma T_{\text{eff}}^4, \tag{11}$$

where σ is the Stefan constant. Therefore, the observed luminosity variation can be estimated by

$$\Delta \ln L = 2\Delta \ln R + 4\Delta \ln T_{\text{eff}}. \tag{12}$$

4.5. Summary

The above magnetic influences can be modeled by modifying the equation of state,

$$P = P_0 + (\gamma - 1)\chi \rho, \tag{13}$$

where χ and γ are taken as thermodynamic variables such as P and T. In this case $\rho = \rho(P, T, \chi, \gamma)$. This modifies the temperature gradient in the convection zone

$$\nabla_c = \nabla'_{\text{ad}} + \left(y / V\gamma_0^2 C\right)(1 + y/V), \tag{14}$$

see Appendix B for the definitions of the quantities appearing on the right hand side.

The stellar structure equations approximately take on the same form as for the standard stellar structure equations:

$$\partial \log P / \partial s = -GM^2 / 4\pi P r^4, \tag{15}$$

$$\partial \log T / \partial s = \nabla \cdot \left(\partial \log P / \partial s\right), \tag{16}$$

$$\partial \log r / \partial s = M / 4\pi \rho r^3, \tag{17}$$

$$\partial L / \partial s = \ln 10 \cdot \left(M / L_\odot\right)\left(\varepsilon - T\mathrm{d}S / \mathrm{d}t\right), \tag{18}$$

where $s = \log M$ is chosen to be the independent variable. $\nabla = \nabla_c$ in the convection zone, while $\nabla = \nabla_{\text{rad}}$ in the radiation zone. The neglected terms are one order of magnitude smaller than the modification to the equation of state. All units are in cgs, except for the luminosity (L) which is in solar units.

5. NUMERICAL IMPLEMENTATION

There are three requirements for the numerical implementation in order to resolve the cycle variation of the Sun:

1. The time step must be as short as 1 year,
2. The numerical accuracy must be much smaller than 4×10^{-6},
3. The envelope mass must be smaller than $10^{-9}\, M_\odot$.

These requirements are not trivial. In appendix A we describe a method that can satisfy these requirements and provide the necessary details (and references) so that the readers can use them to reproduce our results.

6. NUMERICAL EXPERIMENTS

Using the technique described above, and assuming the observed cyclic variations given by *Fröhlich and Lean* [1998] for irradiance, by *Gray and Livingston* [1997] for the effective temperature, by *Emilio et al.* [2000] for the radius, we want to know: what magnetic field variations, at what solar depth, can produce the observed variations?

In order to accommodate the various magnetic effects described above, we must first specify χ and γ. We assume $\chi = \chi_m(t)F(M_D)$. γ is fixed to be 2 to maximize the magnetic effects. The maximum magnetic energy density $\chi_m(t)$, can be determined from solar activity indices. The yearly-averaged sunspot number, R_Z, is the most widely used solar activity index. From numerical experiments, we find that the results are sensitive to the function form of B_m on R_Z. If the maximum magnetic field in the solar interior, B_m, is related to R_Z via

$$B_m = B_0\left\{190 + \left[1 + \log_{10}\left(1 + R_Z\right)\right]^5\right\}, \tag{19}$$

then by adjusting B_0, we can nearly match the measured cyclic variations of irradiance and effective temperature. The reason why such a functional form of B_m is chosen is that $B \sim 20$ kG, at a depth of $M_D = -4.25$ ($r = 0.96 R_\odot$) in 1996. This result is inferred from helioseismology [*Antia* et al., 2000] when R_Z was at a minimum. Using this prescription for B_m, the value of B_m is about twice as large at the maximum of solar cycle, as it is at the minimum.

$F(M_D)$ specifies the distribution of χ, and must be determined by fitting the measured irradiance and effective temperature variations. $F(M_D)$ has infinite degrees of freedom and thus cannot be determined uniquely by observational results, which have finite degrees of freedom. However, we can remove this degeneracy by assuming a field shape of the form,

$$F\left(M_D; M_{\mathrm{Dc}}, \sigma\right) = \exp\left[-\frac{1}{2}\left(M_D - M_{\mathrm{Dc}}\right)^2 / \sigma^2\right], \qquad (20)$$

where M_{Dc} specifies the location and s specifies its width. This gaussian profile allows us to pinpoint the location of the required magnetic field, by using observations of cyclic variations of irradiance and effective temperature.

Figure 1 shows three possible configurations of solar internal magnetic fields at the minimum of the solar activity. Figure 2 compares the measured (*dot-dashed*) with the calculated irradiance variations, and Figure 3 compares the measured (*dot-dashed*) with the calculated effective temperature variations. Figure 4 shows how the variations of these global parameters are caused by solar internal structure changes. The predicted solar radius variations are small, as depicted in Figure 5.

Now we have the answer for the question proposed at the beginning of this subsection: various combinations of strengths of depths of the magnetic fields can reproduce the total irradiance variation, and the photospheric temperature variation observed over the 11-year activity cycle. If the cyclic changes of the radius and the effective temperature are assumed to be small (20 mas and 1.5°K), the internal solar magnetic field of 20–50 KGauss would peak at the depth $r = 0.96R_\odot$.

7. DISCUSSIONS OF THE RESULTS

The example presented earlier indicates the type of data that we need, and the implications that our method can

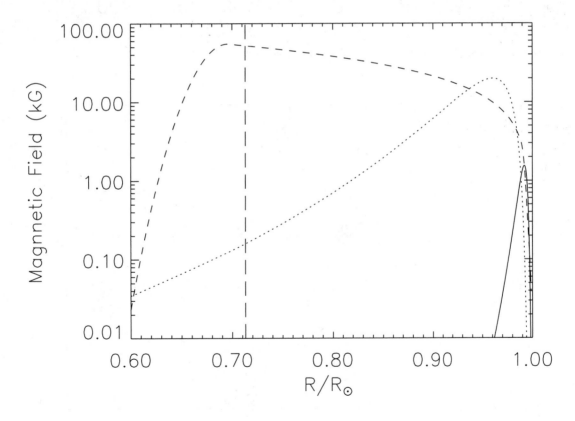

Figure 1. Three possible distributions of inferred magnetic fields in the solar interior at the minimum (e.g., 1996) according to the measured irradiance and photospheric temperature cyclic variations. The vertical line indicates the base of the convection zone.

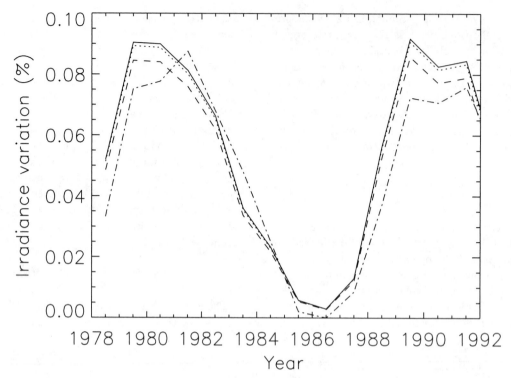

Figure 2. Comparison between measured (dot-dashed) and calculated irradiance variations.

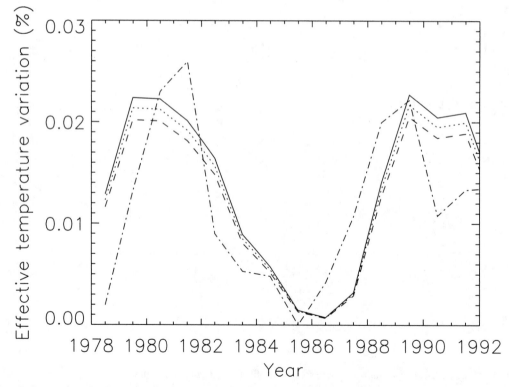

Figure 3. Comparison between measured (dot-dashed) and calculated photospheric temperature variations.

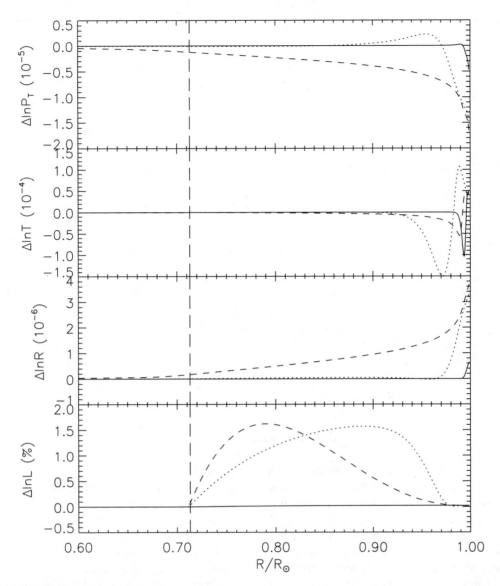

Figure 4. The internal structure changes.

derive. We can uniquely identify the properties (size, magnitude, shape, depth) of the variation of a magnetic field which can produce a given variation of all the solar global parameters. Moreover, helioseismology can test (through the direct method) the accuracy of the derived model. Of course, in order to obtain such a model, it is necessary to carry out near simultaneous observations of all the global parameters and oscillations for at leat 11 years.

7.1. Observational Issues

Perhaps surprisingly, the best ongoing set of measurements in this sense is that of the solar oscillations.

Continuing operation of a number of ground-based networks, notably GONG, but also IRIS and BISON, together with various experiments on SOHO and on other upcoming satellites from the US and Europe, will likely insure continuity of these data for the foreseable future.

The next measurement that may have the longest continuous record is the total solar irradiance. Starting in late of 1978 with the ERB experiment on Nimbus 7, it was soon followed by radiometers on SMM, ERBE, UARS, SOHO, and ACRIMSAT, and it is currently planned to continue in a variety of forthcoming satellites both in the US and in Europe. Continuity of total irradiance measurements has, fortunately, become a goal of various space agencies.

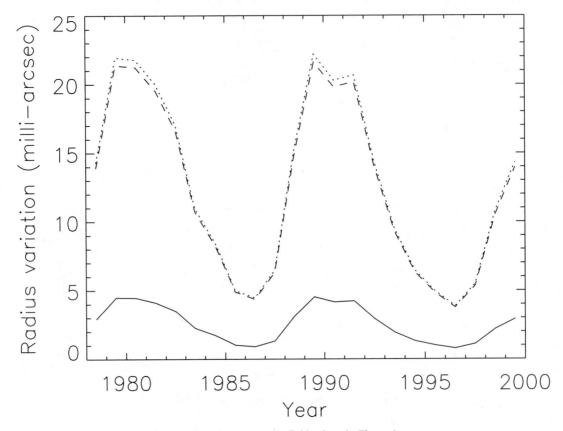

Figure 5. Calculated solar radius variations using magnetic fields given in Figure 1.

Because from time to time the continuity is threatened, it is imperative to recognize that, in view of the current absolute radiometric accuracy, maintaining continuity of measurement is an absolute science requirement. Even under optimal continuity circumstances, and additional problem with this measurement is the great difficulty of calibrating, and maintaining calibration while in space. Despite carefult and cleverly devised measurement strategies (i.e. flying three independent radiometers for intercalibration), poorly understood instrumental degradation often occurs that even this strategy cannot uniquely resolve. To solve this difficulty, it is often necessary to use proxy models whose validity for long timescales can be questioned. The most unfortunate consequence of this is the development of circular arguments, in which the measurements are corrected with models which are subsequently tested on the basis of these same measurements. It is this precarious situation which demands alternate means of monitoring total solar irradiance, critically needed to understand the solar input in climate change, and other solar-terrestrial effects.

Measuring the photospheric (also called effective) temperature variations is another difficult problem. The

method devised by *Gray and Livingston* [1997] is extremely clever and sensitive. However, it depends on a calibration coefficient relating the variation of T_{eff} with the variation of the equivalent depth of the line. They obtain this correlation coefficient empirically from observations of six stars with colors identical to the Sun. However, *Caccin and Penza* [2002] note that the g for all these stars is not the same, and through theoretical calculations they find a g dependence of the coefficient, where g is the gravitational acceleration on the surface of stars. This leaves an uncertainty which affects the amplitude of the 11 year variation of the temperature. The problem is that their calculation produces a substantially higher value of the coefficient indicating a larger amplitude of the variation of effective temperature. This poses a problem, since in the absence of a significant counteractive radius change, the Gray-Livingston results could account for nearly the totality of the 11 year variation of the total irradiance, whereas some significant contribution from the network is expected. The new results would substantially increase the T_{eff} variation, and make the problem worse. Obviously, important work remains to be done on the subject, and it is imperative that the *Gray and Livingston* type observations continue for some years into the future.

There is direct photometric evidence of photospheric (or effective) temperature variations [*Kuhn et al.*, 1988; *Wilson and Hudson*, 1991]. Such a changing also provides a plausible explanation for solar cycle variations in helioseismic splitting coefficients and mode centroid frequencies [*Kuhn*, 1989]. The evidence is supported by lacking of the significant limb brigtening evident in active region visible light faculae [*Kuhn and Libbrecht*, 1991], by the fact that facular irradiance contributions can be positive and negative [*Kuhn et al.*, 1999], and by the SOHO/MDI observations that confirm the latitudinal surface brightness variation seen in early limb observations, while simultaneous magnetic field observations established that faculae and sunspot magnetic fields are not responsible for the latitudinal effective temperature changes. However, *Spruit* [2000] argued that the observations are not free from magnetic surface effects and whether there is anything left to be explained after the surface effects are taken out is not clear at the moment. Of course, our current 1D model cannot address latitudinal variations or any other 2D effects. We are currently generalizing our code to the 2D case [*Li et al.*, 2002b], and this will allow us to sort out this kind of problems.

The parameter with most precarious determination is the radius. The complete lack of agreement between near simultaneous ground-based measurements at different locations (often with similar instruments) suggests that atmospheric contamination is so severe that it prevents any meaningful measurement of solar diameter variations at the expected level from the ground. The measurements from space are very few, and some are still being interpreted. The SOHO/MDI determinations have been few and require substantial corrections since the design of the instrument was not optimized for astrometry. The SDS, on the other hand, was specifically designed for this purpose. Like SOHO/MDI, it soon showed the sensitivity to milli arc s changes during a single flight [*Sofia et al.*, 1994]. However, since the SDS is balloon-borne, between different flights it is subjected to instrumental variations produced during the landings following each flight. The principle of design of the instrument allowed for the calibration of these changes. However, the complexity of this calibration is very high, and only recently we have implemented two different analysis schemes to accomplish this [*Egidi et al.*, 2002; *Sofia et al.*, 2002]. The SDS has had flights in fall 1992, 1994, 1995, 1996 and 2001. Of course, like all balloon-borne experiments, we have many measurements (upwards of 150,000) at best one day per year, with occasional (much) larger intervals. The forthcoming measurements of the French

PICARD experiment, which will measure the solar diameter from space in the time frame of 2006–2008, will help to solve this problem [*Thuillier*, 2002].

An alternative way of determining variations of the solar radius is by helioseismology, particularly of the f-modes of oscillation. Scarce measurements indicate small changes in opposite phase with the solar activity. By contrast, the SOHO/MDI results indicate equally small changes, but in phase with activity cycle, and the early results of the SDS indicate changes in opposite phase with the activity cycle, but substantially larger than the helioseismological ones. It remains to complete all the analyses, carry them out for a longer period of time, and understand any remaining discrepancy on physical terms. Obviously, a lot remains to be done.

Finally, we must realize that whereas our models produce luminosity variations, the direct observations provide total irradiance variations, and that they are linked by means much more complicated than simple geometrical considerations. In particular, they are affected by active regions and network (as described by *Fox* [this volume]), and these effects must be properly taken into account.

In summary, we need to carry out for a decade or longer a number of observations/measurements whose interpretation requires additional (often ongoing) refinements. Only with these results we can interpret and understand what portion of the variation of the total solar irradiance is due to structural internal adjustments produced by variable magnetic fields. The advantages of this process include the likelihood that it is the most important for climate change, and because we will understand the physics of the process, we may be able to address its predictability.

7.2. Theoretical Issues

The calculations that we have carried out to date, and we have presented above, have two shortcomings:

1. They are one-dimensional.
2. They are based on the mixing length theory of convection.

The above shortcomings may affect the accuracy of the results in potentially significant ways. The 1-D limitation requires that the dynamo magnetic field be a shell rather than a toroidal structure. This means that all convective flows must cross the field without having the possibility of going around the field as it may happen in in a toroidal field configuration.

The mixing length theory of convection cannot by itself deal with magnetic fields, turbulence, and other processes that occur in real stars. Of course, all those effects are small, but so are the effects we are seeking to understand.

Our current effort involves developing a static [Li et al., 2002b] and a hydrodynamic [Deupree et al., 2002] 2D stellar structure and evolution code. Although the mixing length formalism will still be used, we are developing modifications based on large eddy numerical simulations of convection which allow us to include the effects that the pure MLT cannot handle.

1. APPENDIX A: A HIGH PRECISION SOLVER FOR THE EQUATIONS OF STELLAR STRUCTURE AND EVOLUTION

This method follows up Prather [1976] with some modifications to accommodate magnetic fields [Endal et al., 1985; Lydon and Sofia, 1995; Li and Sofia, 2001; Sofia and Li, 2001] or turbulence [Li et al., 2002a].

1.1. Linearization of the Stellar Structure Equations

The construction of a stellar model begins by dividing the star into N mass shells which are assigned a value $s_i = \log M_i$, where M_i is the interior mass at the midpoint of shell i. A starting model is supplied with a run of ($\log P_i$, $\log T_i$, $\log r_i$, L_i) for i=1 to N. The differential equations of stellar structure are then linearized with respect to first-order changes in the dependent variables. A set of corrections, ($\delta \log P_i$, $\delta \log T_i$, $\delta \log r_i$, $\delta \log L_i$) for i=1 to N, is then calculated and applied, the mass and composition remaining fixed at each point. The procedure is iterated until a numerically or physically suitable convergence is reached.

For this purpose, one needs to set up and solve the difference equations for corrections to the dependent variables of the starting models. The first step is to define a set of functions for every pair of adjacent mass points,

$$F_P^i \equiv \left(P_i' - P_{i-1}'\right) - \frac{1}{2}\Delta s_i \cdot \left(\mathcal{P}_i + \mathcal{P}_{i-1}\right) \quad \text{(A.1)}$$

$$F_T^i \equiv \left(T_i' - T_{i-1}'\right) - \frac{1}{2}\Delta s_i \cdot \left(\mathcal{T}_i + \mathcal{T}_{i-1}\right) \quad \text{(A.2)}$$

$$F_R^i \equiv \left(R_i' - R_{i-1}'\right) - \frac{1}{2}\Delta s_i \cdot \left(\mathcal{R}_i + \mathcal{R}_{i-1}\right) \quad \text{(A.3)}$$

$$F_L^i \equiv \left(L_i - L_{i-1}\right) - \frac{1}{2}\Delta s_i \cdot \left(\mathcal{L}_i + \mathcal{L}_{i-1}\right) \quad \text{(A.4)}$$

where $\Delta s_i \equiv (s_i - s_{i-1})$ and i=2 to N. $P' = \log P$, $T' = \log T$, $R' = \log r$. ($\mathcal{P}$, $\mathcal{T}$, $\mathcal{R}$, $\mathcal{L}$) represent the right hand sides of Eqs. (15)–(18), respectively. One wants then to solve for the set of (P_i, T_i, r_i, L_i) such that $F_P^i = F_T^i = F_R^i = F_L^i = 0$. The linearization of Eqs. (A.1)–(A.4) with respect to ($\delta P_i'$, ($\delta T_i'$,

($\delta R_i'$, ($\delta L_i'$) yields 4N–4 equations for the 4N unknowns. The 4 additional equations are supplied by the boundary conditions at the center,

$$F_R^1 \equiv R_1' - \frac{1}{3}\left[s_1 - \log\left(4\pi\rho_1/3\right)\right] \quad \text{(A.5)}$$

$$F_L^1 \equiv L_1 - \mathcal{L}_1 / \ln 10, \quad \text{(A.6)}$$

and those at the surface (see §3),

$$F_R^{N+1} \equiv R_N' - a_1 P_N' - a_2 T_N' - a_3 \quad \text{(A.7)}$$

$$F_L^{N+1} \equiv L_N \ln 10 \cdot \left(\log L_N - a_4 P_N' - a_5 T_N' - a_6\right). \quad \text{(A.8)}$$

The F equations are linearized,

$$\sum_{j=1}^{N}\left(\frac{\partial F^i}{\partial R_j'}\delta R_j' + \frac{\partial F^i}{\partial L_j}\delta L_j + \frac{\partial F^i}{\partial P_j'}\delta P_j' + \frac{\partial F^i}{\partial T_j'}\delta T_j'\right) = -F^i \quad \text{(A.9)}$$

where i=2 to N and the summation over j has non-zero terms only for j=i-1, i. Including the boundary equations, one now calculates the corrections to the previous model by solving a system of 4N equations in 4N unknowns.

The partial derivatives of the differential equations can be expressed by the partial derivatives of ($\mathcal{P}$, $\mathcal{T}$, $\mathcal{R}$, $\mathcal{L}$). In order to ensure a high numerical accuracy, we use the analytical derivatives rather than the numerical derivatives. By defining the shorthand notation $\partial_X Y \equiv \partial Y/\partial X$, one can calculate the derivatives as follows.

$$\partial_{R'}\mathcal{P} = -4\ln 10 \cdot \mathcal{P}$$
$$\partial_{P'}\mathcal{P} = -\ln 10 \cdot \mathcal{P} \quad \text{(A.10)}$$
$$\partial_{T'}\mathcal{P} = \partial_L \mathcal{P} = 0$$

$$\partial_{R'}\mathcal{T}_c = \ln 10 \cdot \left(\partial \ln \nabla_c / \partial \ln r - 4\right) \cdot \mathcal{T}_c$$
$$\partial_L \mathcal{T}_c = 0 \qquad \text{(convective)}$$
$$\partial_{P'}\mathcal{T}_c = -\ln 10 \cdot \left(1 - \partial \ln \nabla_c / \partial \ln P\right) \cdot \mathcal{T}_c \quad \text{(A.11)}$$
$$\partial_{T'}\mathcal{T}_c = \ln 10 \left(\partial \nabla_c / \partial \ln T\right) \cdot \mathcal{T}_c$$

$$\partial_{R'}\mathcal{T}_r = -4 \cdot \ln 10 \cdot \mathcal{T}_r$$
$$\partial_L \mathcal{T}_r = \mathcal{T}_r / L \qquad \text{(radiative)}$$
$$\partial_{P'}\mathcal{T}_r = \ln 10 \cdot \left(\partial \ln \kappa / \partial \ln P\right)_T \cdot \mathcal{T}_r \quad \text{(A.12)}$$
$$\partial_{T'}\mathcal{T}_r = \ln 10 \left[\left(\partial \ln \kappa / \partial \ln T\right)_P - 4\right] \cdot \mathcal{T}_r$$

$$\partial_{R'}\mathcal{R} = -3 \cdot \ln 10 \cdot \mathcal{R}$$
$$\partial_L \mathcal{R} = 0$$
$$\partial_{P'}\mathcal{R} = -\ln 10 \cdot \mu \cdot \mathcal{R} \quad \text{(A.13)}$$
$$\partial_{T'}\mathcal{R} = -\ln 10 \cdot q \cdot \mathcal{R}$$

$$\partial_{P'}\mathcal{L} = \ln^2 10 \frac{M_r}{L_\odot}\left[\left(\frac{\partial\varepsilon}{\partial\ln P}\right)_T + \left(\frac{\partial\tilde{S}}{\partial\ln P}\right)_T / \Delta t\right]$$

$$\partial_R\mathcal{L} = \partial_L\mathcal{L} = 0$$

$$\partial_{T'}\mathcal{L} = \ln^2 10 \frac{M_r}{L_\odot}\left[\left(\frac{\partial\varepsilon}{\partial\ln T}\right)_P + \left(\frac{\partial\tilde{S}}{\partial\ln T}\right)_P / \Delta t\right] \quad (A.14)$$

The formulas for the various partial derivatives of the physical quantities will be presented in the following subsections. The equation of state calculates ρ, $\mu = (\partial\ln\rho/\partial\ln P)_{T,\chi,\gamma}$, q, c_P, ∇_{ad} and the pressure and temperature derivatives of these quantities (see Appendix D). μ and Q are used in Eq. (A.13). The opacity tables provide $\log\kappa$ vs. ($\log\rho$, $\log T$), in order to calculate

$$\left(\partial\ln\kappa/\partial\ln T\right)_P = \left(\partial\ln\kappa/\partial\ln T\right)_\rho$$
$$+\left(\partial\ln\kappa/\partial\ln\rho\right)_T\left(\partial\ln\rho/\partial\ln T\right)_P$$
$$\left(\partial\ln\kappa/\partial\ln P\right)_T = \left(\partial\ln\kappa/\partial\ln\rho\right)_T\left(\partial\ln\rho/\partial\ln T\right)_P$$

which are used in Eq. (A.12), one needs $(\partial\ln k/\partial\ln T)\rho$ and $(\partial\ln\kappa/\partial\ln\rho)_T$ (see [*Iglesias and Rogers*, 1996; *Alexander and Ferguson*, 1994]). Energy generation rate ϵ is a function of ρ and T, too. So $(\partial\epsilon/\partial T)_P$ and $(\partial\epsilon/\partial P)_T$ used in Eq. (A.14) can be expressed by $(\partial\epsilon/\partial\ln T)_\rho$ and $\partial\epsilon/\partial\ln\rho)_T$ (see Appendix D):

$$\left(\partial\varepsilon/\partial\ln T\right)_P = \left(\partial\varepsilon/\partial\ln T\right)_\rho$$
$$+\left(\partial\varepsilon/\partial\ln\rho\right)_T\left(\partial\ln\rho/\partial\ln T\right)_P$$
$$\left(\partial\varepsilon/\partial\ln P\right)_T = \left(\partial\varepsilon/\partial\ln\rho\right)_T\left(\partial\ln\rho/\partial\ln T\right)_P$$

The derivatives of the convective gradient ∇_c which are used in Eq. (A.11) are presented in Appendix B.

The entropy term in Eq. (A.14) contains the only explicit reference to any time-dependence in the stellar structure equations. It can be reformulated as follows:

$$\tilde{S} = \left(Pq/\rho\right)\ln 10 \cdot \left(\Delta T'/\nabla_{ad} - \Delta P'\right)$$

$$\left(\partial\tilde{S}/\partial\ln T\right)_P = \tilde{S}\left[-q+\left(\partial\ln q/\partial\ln T\right)_P\right]$$
$$+\left(Pq/\rho\nabla_{ad}\right)\left[1-\left(\partial\ln\nabla_{ad}/\partial\ln T\right)_P\ln 10\cdot\Delta T'\right]$$

$$\partial\tilde{S}/\partial\ln P_T = \tilde{S}\left[1-\mu+\left(\partial\ln q/\partial\ln P\right)_T\right]$$
$$-\left(Pq/\rho\right)\left[1+\left(\partial\ln\nabla_{ad}/\partial\ln P\right)_T\ln 10\cdot\Delta T'/\nabla_{ad}\right]$$

where $(\Delta P', \Delta T')$ are the changes between successive models.

1.2. Solution of the Linearized Equations

The linearized system consists of 4N algebraic equations. So the coefficient is a 4N×4N matrix. However, only 8×4N elements are non-zero at most. Figure 6 shows Eq. (A.9) for a star in 3 mass shells, in which the coefficient matrix is defined in Figure 7 and Figure 8 by using the partial derivatives given above.

The matrix is reduced in a forward direction ($i = 2 \to$ N) as the coefficients are defined and is then solved in the backward direction ($i =$ N→1) for the corrections ($\delta P'_i$, $\delta T'_i$, $\delta R'_i$, δL_i). Te reduction procedure begins: (i) using the boundary conditions, eliminate the first two columns of Figure 7; (ii) continue diagonalizing the four bottom rows; (iii) store the right-hand side and the elements in the rightmost columns, see Figure 9b. After this reduction is completed, the bottom two rows of the first part of the coefficient matrix become the "boundary equations" for the F equations of the next pair of mass points. The method is repeatedly applied until the surface is reached, whereupon the surface boundary conditions complete the set of 4N equations, see Figures 9b–c. For the back solution (i) the values of ($\delta P'_N$, $\delta T'_N$) are first calculated, (ii) then the values of ($\delta R'_i$, δL_i, $\delta P'_{i-1}$, $\delta T'_{i-1}$) for i=N to 2 are calculated using the stored elements of the array and ($\delta P'_i$, $\delta T'_i$), (iii) and finally the values of ($\delta R'_1$, δL_1 are computed from the central boundary conditions and the values of ($\delta P'_1$, $\delta T'_1$), see Figures 9d–f.

$$\begin{pmatrix} 1\ 0\ X X \\ 0\ 1\ X X \\ X\ 0\ X\ 0\ X\ 0\ X\ 0 \\ X X X X X X X X \\ X\ 0\ X X X\ 0\ X X \\ 0\ {-1}\ X X\ 0\ 1\ X X \\ \qquad X\ 0\ X X X\ 0\ X X \\ \qquad 0\ {-1}\ X X\ 0\ 1\ X X \\ \qquad X\ 0\ X\ 0\ X\ 0\ X\ 0 \\ \qquad X X X X X X X X \\ \qquad\qquad 1\ 0\ 1\ X \\ \qquad\qquad 0\ 1\ X X \end{pmatrix} \bullet \begin{pmatrix} \delta R_1 \\ \delta L_1 \\ \delta P_1 \\ \delta T_1 \\ \delta R_2 \\ \delta L_2 \\ \delta P_2 \\ \delta T_2 \\ \delta R_3 \\ \delta L_3 \\ \delta P_3 \\ \delta T_3 \end{pmatrix} = - \begin{pmatrix} F_R^1 \\ F_L^1 \\ F_P^2 \\ F_T^2 \\ F_R^2 \\ F_L^2 \\ F_R^3 \\ F_L^3 \\ F_P^3 \\ F_T^3 \\ F_R^4 \\ F_L^4 \end{pmatrix}$$

Figure 6. Linearized equations consisting of 3 mass shells. The matrix block is denoted by 0's, 1's, and X's are non-zero.

$$
\begin{array}{cccccccc}
1 & 0 & \tfrac{1}{3}\left(\tfrac{\partial \ln \rho}{\partial \ln P}\right)_{T,\chi,\gamma,1} & \tfrac{1}{3}\left(\tfrac{\partial \ln \rho}{\partial \ln T}\right)_{P,\chi,\gamma,1} & 0 & 0 & 0 & 0 \\[4pt]
0 & 1 & -\ln 10 \cdot \partial_{P'}\mathcal{L}_1 & -\ln 10 \cdot \partial_{T'}\mathcal{L}_1 & 0 & 0 & 0 & 0 \\[4pt]
\sigma \cdot \partial_{R'}\mathcal{P}_1 & 0 & \sigma \cdot \partial_{P'}\mathcal{P}_1 - 1 & 0 & \sigma \cdot \partial_{R'}\mathcal{P}_2 & 0 & \sigma \cdot \partial_{P'}\mathcal{P}_2 + 1 & 0 \\[4pt]
\sigma \cdot \partial_{R'}\mathcal{T}_1 & \sigma \cdot \partial_{L}\mathcal{T}_1 & \sigma \cdot \partial_{P'}\mathcal{T}_1 & \sigma \cdot \partial_{T'}\mathcal{T}_1 - 1 & \sigma \cdot \partial_{R'}\mathcal{T}_2 & \sigma \cdot \partial_{L}\mathcal{T}_2 & \sigma \cdot \partial_{P'}\mathcal{T}_2 & \sigma \cdot \partial_{T'}\mathcal{T}_2 + 1 \\[4pt]
\sigma \cdot \partial_{R'}\mathcal{R}_1 - 1 & 0 & \sigma \cdot \partial_{P'}\mathcal{R}_1 & \sigma \cdot \partial_{T'}\mathcal{R}_1 & \sigma \cdot \partial_{R'}\mathcal{R}_2 + 1 & 0 & \sigma \cdot \partial_{P'}\mathcal{R}_2 & \sigma \cdot \partial_{T'}\mathcal{R}_2 \\[4pt]
0 & -1 & \sigma \cdot \partial_{P'}\mathcal{L}_1 & \sigma \cdot \partial_{T'}\mathcal{L}_1 & 0 & 1 & \sigma \cdot \partial_{P'}\mathcal{L}_2 & \sigma \cdot \partial_{T'}\mathcal{L}_2
\end{array}
$$

$$\sigma = -\tfrac{1}{2}(s_2 - s_1)$$

Figure 7. The first part of the coefficient matrix.

$$
\begin{array}{cccccccc}
\sigma \cdot \partial_{R'}\mathcal{P}_2 & 0 & \sigma \cdot \partial_{P'}\mathcal{P}_2 - 1 & 0 & \sigma \cdot \partial_{R'}\mathcal{P}_3 & 0 & \sigma \cdot \partial_{P'}\mathcal{P}_3 + 1 & 0 \\[4pt]
\sigma \cdot \partial_{R'}\mathcal{T}_2 & \sigma \cdot \partial_{L}\mathcal{T}_2 & \sigma \cdot \partial_{P'}\mathcal{T}_2 & \sigma \cdot \partial_{T'}\mathcal{T}_2 - 1 & \sigma \cdot \partial_{R'}\mathcal{T}_3 & \sigma \cdot \partial_{L}\mathcal{T}_3 & \sigma \cdot \partial_{P'}\mathcal{T}_3 & \sigma \cdot \partial_{T'}\mathcal{T}_3 + 1 \\[4pt]
\sigma \cdot \partial_{R'}\mathcal{R}_2 - 1 & 0 & \sigma \cdot \partial_{P'}\mathcal{R}_2 & \sigma \cdot \partial_{T'}\mathcal{R}_2 & \sigma \cdot \partial_{R'}\mathcal{R}_3 + 1 & 0 & \sigma \cdot \partial_{P'}\mathcal{R}_3 & \sigma \cdot \partial_{T'}\mathcal{R}_3 \\[4pt]
0 & -1 & \sigma \cdot \partial_{P'}\mathcal{L}_2 & \sigma \cdot \partial_{T'}\mathcal{L}_2 & 0 & 1 & \sigma \cdot \partial_{P'}\mathcal{L}_3 & \sigma \cdot \partial_{T'}\mathcal{L}_3 \\[4pt]
0 & 0 & 0 & 0 & 1 & 0 & -a_1 & -a_2 \\[4pt]
0 & 0 & 0 & 0 & 0 & 1 & -2a_4 L_N \ln 10 & -a_5 L_N \ln 10
\end{array}
$$

$$\sigma = -\tfrac{1}{2}(s_3 - s_2)$$

Figure 8. The second part of the coefficient matrix.

```
1 0 X X                    A          1 0 X X                    A          1 0 X X                    A
0 1 X X                    A          0 1 X X                    A          0 1 X X                    A
X 0 X 0 X 0 X 0            A          0 0 1 0 0 0 Y Y            B          0 0 1 0 0 0 X X            A
X X X X X X X X            A          0 0 0 1 0 0 Y Y            B          0 0 0 1 0 0 X X            A
X 0 X X X 0 X X            A          0 0 0 0 1 0 Y Y            B          0 0 0 0 1 0 X X            A
0 -1 X X 0 1 X X           A          0 0 0 0 0 1 Y Y            B          0 0 0 0 0 1 X X            A
        X 0 X X X 0 X X    A                  X X X X X X X X    A                  0 0 1 0 0 0 Y Y    B
        0 -1 X X 0 1 X X   A                  X X X X X X X X    A                  0 0 0 1 0 0 Y Y    B
        X 0 X 0 X 0 X 0    A                  X X X X X X X X    A                  0 0 0 0 1 0 Y Y    B
        X X X X X X X X    A                  X X X X X X X X    A                  0 0 0 0 0 1 Y Y    B
              1 0 1 X      A                        1 0 1 X      A                        1 0 1 X      A
              0 1 X X      A                        0 1 X X      A                        0 1 X X      A

           (a)                                   (b)                                   (c)


1 0 X X                    A          1 0 X X                    A          1 0 X X                    A
0 1 X X                    A          0 1 X X                    A          0 1 X X                    A
0 0 1 0 0 0 X X            A          0 0 1 0 0 0 X X            A          0 0 1 0 0 0 0 0            B
0 0 0 1 0 0 X X            A          0 0 0 1 0 0 X X            A          0 0 0 1 0 0 0 0            B
0 0 0 0 1 0 X X            A          0 0 0 0 1 0 X X            A          0 0 0 0 1 0 0 0            B
0 0 0 0 0 1 X X            A          0 0 0 0 0 1 X X            A          0 0 0 0 0 1 0 0            B
        0 0 1 0 0 0 X X    A                  0 0 1 0 0 0 0 0    B                  0 0 1 0 0 0 0 0    A
        0 0 0 1 0 0 X X    A                  0 0 0 1 0 0 0 0    B                  0 0 0 1 0 0 0 0    A
        0 0 0 0 1 0 X X    A                  0 0 0 0 1 0 0 0    B                  0 0 0 0 1 0 0 0    A
        0 0 0 0 0 1 X X    A                  0 0 0 0 0 1 0 0    B                  0 0 0 0 0 1 0 0    A
              0 0 1 0      B                        0 0 1 0      A                        0 0 1 0      A
              0 0 0 1      B                        0 0 0 1      A                        0 0 0 1      A

           (d)                                   (e)                                   (f)
```

Figure 9. Schematic Henyey solution for a 3-point star. The notation is the same as in Figure 6. The right hand side is denoted by A, the elements changed through pivoting, by Y and B. The final reduction to the identity matrix is not shown.

2. APPENDIX B: CONVECTIVE TEMPERATURE GRADIENTS

2.1. Flux Conservation

The convective temperature gradient ∇_c is determined by the requirement that the total energy flux F_{total} equals the sum of the radiative flux F_{rad} and the convective flux F_{conv},

$$F_{total} = F_{rad} + F_{conv}, \tag{B.1}$$

The total flux at any given layer in the star is determined by the photon luminosity L_r,

$$F_{total} = \frac{L_r}{4\pi r^2} = \frac{4acG}{3} \frac{T^4 M_r}{\kappa P_T r^2} \nabla_{rad}. \tag{B.2}$$

The radiative flux is determined by the convective temperature gradient:

$$F_{rad} = \frac{4acG}{3} \frac{T^4 M_r}{\kappa P_T r^2} \nabla_{conv}. \tag{B.3}$$

The convective flux is determined by the convective velocity v_{conv} and the heat excess DQ_T:

$$F_{conv} = \rho v_{conv} DQ_T. \tag{B.4}$$

When the convective velocity is much smaller than the sound speed of the medium, the process can be considered to be of constant pressure. In this case, the heat excess can be obtained from the first law given by:

$$DQ_T = c_p DT + \left[1 - \frac{P_T q v}{\rho \mu \chi}\right] D\chi - \frac{P_T q v'}{\rho \mu \gamma} D\gamma. \tag{B.5}$$

2.2. Mixing Length Approximation

Using the mixing length approximation, DT, $D\chi$ and $D\gamma$ can be expressed by the mixing length l_m as follows:

$$
\begin{aligned}
DT/T &= (1/T)\partial(DT)/\partial r (l_m/2) \\
&= (\nabla_{conv} - \nabla_e)(l_m/2)(1/H_p), \\
D\chi/\chi &= (1/\chi)\partial(D\chi)\partial r (l_m/2) = 0, \\
D\gamma/\gamma &= (1/\gamma)\partial(D\gamma)\partial r (l_m/2) = 0,
\end{aligned}
\tag{B.6}
$$

where $H_p = -P_T(dr/dP_T)$ is the pressure scale height. In order to determine v_{conv}, the MLT assumes that half of the work done by half of the radial buoyancy force acted over half the mixing length goes into the kinetic energy of the element ($v_{conv}^2/2$). Since the radial buoyancy force per unit mass is related to the density difference by:

$$k_r = -g(D\rho/\rho), \tag{B.7}$$

and since the process is in pressure equilibrium, we obtain

$$v_{conv}^2 = -q(\nabla_{conv} - \nabla_e)\frac{l_m^2 g}{8H_p}, \tag{B.8}$$

where $g = GM_r/r^2$ is the gravitational acceleration.

2.3. Radiative and Turbulent Losses

An additional relation is required to close the MLT:

$$
\begin{aligned}
(dQ_T/dr)_e &= (\text{radiative and turbulent losses}) \\
&\quad + \{\text{change of } \chi\},
\end{aligned}
\tag{B.9}
$$

which can be expressed as:

$$
\begin{aligned}
&(2acT^3)/(\rho c_p v_{conv})\left(\omega/(1+\tfrac{1}{3}\omega^2)(\nabla_{conv} - \nabla_e)\right) \\
&= (\nabla_e - \nabla_{ad}) + 3(\nabla_{ad}/\mu)(\nu\nabla_\chi + \nu'\nabla\gamma),
\end{aligned}
\tag{B.10}
$$

where $\omega = \kappa \rho l_m$.

2.4. Solutions

Solving Eqs. (B.1) and (B.10), we obtain

$$\nabla_{conv} = \nabla'_{ad} + (y/V\gamma_0^2 C)(1 + y/V), \tag{B.11}$$

where y is the solution of the following equation

$$2Ay^3 + Vy^2 + V^2 y - V = 0. \tag{B.12}$$

A, γ_0, C, and V are defined by

$$A = (9/8)\left[\omega^2/(3 + \omega^2)\right],$$

$$\gamma_0 = \left[(c_p\rho)/(2acT^3)\right]\left[(1 + \tfrac{1}{3}\omega^2)/\omega\right],$$

$$C = (g/l_m^2 \mu')/8H_p,$$

$$V = 1/\left[\gamma_0 C^{1/2}(\nabla_{rad} - \nabla'_{ad})^{1/2}\right].$$

The radiative gradient is

$$\nabla_{rad} = \frac{3L_\odot}{16\pi acG}\frac{\kappa LP}{MT^4}. \tag{B.13}$$

Eq. (B.12) is equivalent to

$$F(y) \equiv a_3 y^3 + y^2 + a_1 y - 1 = 0, \qquad (B.14)$$

where $a_1 = V$ and $a_3 = 2A/V$. An initial estimate of the root y is made and a second-order Newton-Raphson correction is applied,

$$\Delta y = -F(y)/F'(y) \left\{ 1 + \frac{1}{2}[F(y)/F'(y)]F''(y)/F'(y) \right\}$$

The initial estimate of y is $y = 1/a_1$, unless $a_3 > 10^3$ in which case $y = (1/a_3)^{1/3}$ which follows the asymptotic behavior of the solution in either limit. Given the solution y, the convective gradient is computed as follows

$$\nabla_{conv} = \nabla'_{ad} + (\nabla_{rad} - \nabla'_{ad}) y(y + a_1) \qquad (B.15)$$

2.5. Derivatives of the Convective Gradient

In the linearization of the equation of energy transport, we need the derivatives of the convective (and radiative) gradient with respect to log P, log T, log R and L. This can be done by using Eqs. (B.14) and (B.15), see *Prather* [1976] for the details.

3. APPENDIX C: THE BOUNDARY CONDITIONS

The central boundary conditions is based upon a first-order integration at the center of the star. The radius equation is calculated from

$$dm = 4\pi\rho r^2 dr \rightarrow \int_0^{m_1} dm = 4\pi\rho_1 \int_0^{r_1} r^2 dr$$

$$\rightarrow m_1 = \frac{4\pi}{3}\rho_1 r_1^3,$$

assuming that $\rho = \rho_1$ is constant for $0 < m < m_1$. Likewise, the luminosity equation assumes that $(\epsilon - TdS/dt)$ is constant:

$$L_\odot dL = (\epsilon - TdS/dt)dm \rightarrow L_1 = \frac{m_1}{L_\odot}(\epsilon - TdS/dt)_1.$$

A star, including its atmosphere, is divided into three parts by two points (or interfaces) in its one-dimensional model: the surface, which is so defined that the surface temperature equals the effective temperature, and the fitting point, which is so selected that the luminosity above this point can be considered to be constant. Although the surface values mark the surface boundary, the model surface boundary is usually defined at the fitting point. The portion between the fitting point and the surface is called the envelope. We can place the fitting point anywhere in the surface convection zone of the Sun without introducing the magnetic or turbulent perturbations. However, the luminosity in the surface convection zone is no longer constant when these perturbations are present. Therefore, we should eliminate the envelope when including perturbations. In order to obtain the surface pressure and temperature, we need a temperature-optical depth relation for the stellar atmosphere. The Sun's atmosphere is expirically modeled by the Krishna Swamy $T(\tau)$ relation [*Krishna Swamy*, 1966]

$$T^4(\tau) = \frac{3}{4}T_{eff}^4 \left[\tau + 1.39 - 0.815 \exp(-2.54\tau) \right. \qquad (C.1)$$
$$\left. - 0.025 \exp(-30\tau) \right].$$

For this model the surface is specified by $\tau_s = 0.31215633$.

Given the (log L, log T_{eff}) and the total mass M_{tot}, the radius at the base of the atmosphere or the surface radius is determined by $L \cdot L_\odot = 4\pi R^2 \cdot \sigma T_{eff}^4$, and the surface gravity is calculated from $g = GM_{tot}/R^2$. The atmospheric values of P are computed by integrating log P vs. log τ from $\tau \ll 1$ to $\tau = \tau_s$ for a plane parallel atmosphere. Combining $d\tau = -\rho k dr$ and $dp = -\rho g dr$, we obtain

$$d \log P = (g\tau / \kappa P) d \log \tau. \qquad (C.2)$$

The starting values of (P_0, τ_0) are chosen by selecting a small density ρ_0 and then computing

$$P_0 = (a/3)T_0^4 + \rho_0 RT_0,$$

where $T_0 \equiv T(\tau = 0)$. Then (P_0, τ_0) gives ρ_1 which gives $\kappa_0(\rho_1, T_0)$ which gives $\tau_0 = \kappa_0 P_0/g$. This method could be iterated upon by redefining $T_1 = T(\tau_0)$ and so forth. Finally we can obtain the surface values for P and T. Atmosphere integration uses polynomial extrapolation based on the midpoint rule. It is self-starting and automatically readjusts the present step size and estimates the subsequent step size in order to comply with the specified accuracy.

The values of (P, T, R) at the surface for three atmosphere integrations are needed in order to compute the surface boundary coefficients. One solves the following system,

$$\begin{pmatrix} \log P_3 & \log T_3 & 1 \\ \log P_2 & \log T_2 & 1 \\ \log P_1 & \log T_1 & 1 \end{pmatrix} \cdot \begin{pmatrix} a_3 & a_6 \\ a_2 & a_5 \\ a_1 & a_4 \end{pmatrix} = \begin{pmatrix} \log R_3 & \log L_3 \\ \log R_2 & \log L_2 \\ \log R_1 & \log L_1 \end{pmatrix} \qquad (C.3)$$

for the coefficients which are used for the surface boundary conditions,

$$\log R = a_1 \log P + a_2 \log T + a_3 \qquad (C.4)$$

$$\log L = a_4 \log P + a_5 \log T + a_6 \qquad (C.5)$$

where the ($\log P$, $\log T$) refer to the values at the outermost mass point in the model. The effective temperature equals the surface temperature by definition.

The initial model with an estimated ($\log L^*$, $\log T_{eff}^*$) is triangulated in the ($\log L$, $\log T_{eff}$)-plane by constructing three envelopes of the form

E1: $\left(\log L^* - \frac{1}{2} \Delta_L, \log T_{eff}^* + \frac{1}{2} \Delta_T \right)$

E2: $\left(\log L^* - \frac{1}{2} \Delta_L, \log T_{eff}^* + \frac{1}{2} \Delta_T \right)$

E3: $\left(\log L^* - \frac{1}{2} \Delta_L, \log T_{eff}^* \right)$.

If subsequent models or if the model itself during convergence moves significantly out of the triangle, the triangle is flipped until it once again contains the model. The decision as to which point of the triangle should be flipped—if any—can be made by testing

$$c_i = f \Big\{ \big(\log L_{i+1} - \log L_{i+2} \big) \big(\log T_{eff} - \log T_{effi+1} \big)$$
$$+ \big(\log T_{effi+2} - \log T_{effi+1} \big) \big(\log L - \log L_{i+1} \big) \Big\},$$

where $f = \pm 1$ is the orientation of the triangle. The value of c_i is tested against $\epsilon \Delta_L \Delta_T$ where setting $\epsilon = 0$ gives exact triangulation and $\epsilon > 0$ allows the point ($\log L$, $\log T_{eff}$) to be at most ϵ of a triangle outside. Begin testing with $i = 1$ to 3, if $c_i < -\epsilon \Delta_L \Delta_T$ then flip point i,

$$\log L_i \quad \leftarrow \quad \log L_{i+1} + \log L_{i+2} - \log L_i$$
$$\log T_{effi} \quad \leftarrow \quad \log T_{effi+1} + \log T_{effi+2} - \log T_{effi}$$
$$f \quad \leftarrow \quad -f$$

and repeat the testing again starting with i=1 until c_i passes for i=1 to 3. The atmosphere integrations that have been flipped are then recomputed as are all the coefficients a_i.

4. APPENDIX D: INPUT PHYSICS

4.1. The Equations of State

When a magnetic field is present, the equation of state relates the density ρ to the pressure P, temperature T, magnetic energy per unit mass χ, the ratio of specific heats γ, and the chemical composition:

$$\rho = \rho(P, T, \chi, \gamma; X, Z),$$

where $P = P_0 + P_r + P_m$ is the total pressure, P_0 the gas pressure, $P_r = aT^4/3$ the radiative pressure, $P_m = (\gamma - 1)\chi\rho$ the magnetic pressure, X the mass fraction of hydrogen, Z the mass fraction of elements heavier than helium (the so-called metal mass fraction). Its differential form is

$$\frac{dp}{\rho} = \mu \frac{dP}{P} + q \frac{dT}{T} + \upsilon \frac{d\chi}{\chi} + \upsilon' \frac{d\gamma}{\gamma},$$

where

$\mu = (\partial \ln \rho / \partial \ln P)$ at constant T, χ, γ,

$q = (\partial \ln \rho / \partial \ln T)$ at constant P, χ, γ,

$\upsilon = (\partial \ln \rho / \partial \ln \chi)$ at constant P, T, γ,

$\upsilon' = (\partial \ln \rho / \partial \ln \gamma)$ at constant P, T, χ,

here X and Z are assumed to be constant.

Since it is tedious to accurately calculate the equation of state from first principles, the equations of state are usually provided by the numerical tables as functions of (ρ, T, X, Z) for P_0, S (entropy), U (internal energy), $(\partial U/\partial \rho)_T$, $c_v = (\partial U/\partial T)_\rho$, $\chi_\rho = (\partial \ln P_0/\partial \rho)_T$, $\chi_T = (\partial \ln P_0/\partial T)_\rho$, $\Gamma_1 = (\partial \ln P_0/\partial \ln \rho)_S$, $\Gamma_2' = \Gamma_2/(1 - \Gamma_2) = 1/\nabla_{ad}$, and $\Gamma_3' = (\partial \ln T/\partial \ln P)_{P_0} - 1$. The equation of state for the gas is taken from *Rogers et al.* [1996]. In order to take into account a magnetic field based on the EOS tables, one can use the following correction method: (i) Using the total pressure $P = P_0 + P_r + P_m$, the total internal energy $U = U_0 + 3P_r/\rho + \chi$, and the total entropy $S = S_0 + 4P_r/\rho/T + \chi/T$ to replace the gas pressure P_0, the gas internal energy U_0, and the gas entropy S_0 respectively when interpolating to obtain the density for the given P and T; (ii) Using $(P_0 + P_m)/P$ to rescale χ_ρ; (iii) Using P_0/P to rescale χ_T from the EOS tables and add $4P_r/P$; (iv) Adding $12P_r/T$ to c_v from the EOS tables; (v) compute $\Gamma_3' = P\chi_T/c_v \rho T$, $\Gamma_1 = \chi_\rho + \chi_T \Gamma_3'$, and $\Gamma_2' = \Gamma_1/\Gamma_3'$.

Taking these as known, we can calculate $\mu = 1/\chi_\rho$, $q = -\chi_T/\chi_\rho$, $\upsilon = -P_m/P$, $\upsilon' = -[\gamma/(\gamma - 1)]P_m/P$, $\nabla_{ad} = 1/\Gamma_2'$, $c_p = -Pq/\rho T \nabla_{ad}$. These quantities are used in calculating the convective gradient ∇_c.

4.2. Energy Generation

The calculation of the energy generation includes the individual rates for the PP-chain (PPI, PPII, PPIII), the CNO-cycle with a simplified NO approach to equilibrium. The coefficients of all of the reaction rates and the formulae for most of them are taken from *Fowler et al.* [1975].

The reaction rate for the PP-chain is actually that for the $H^1(p, e^+\nu)D^2$ reaction and assumes that all the other reactions in the chain are relatively instantaneous. The burning rate is then

$$(dX/dt)_{PP} = 4.181 \cdot 10^{-15} \rho X^2 T_9^{-2/3} \exp(-3.380/T_9^{1/3})$$
$$\phi(\alpha)(1.0 + 0.123 T_9^{1/3} + 1.09 T_9^{2/3} + 0.938 T_9) \ \mathrm{sec}^{-1},$$

where $T_9 = T/10^9 \,^\circ$K, the screening factor f_s is set equal to 1,

$$\phi(\alpha) = 1 + \alpha\left[(1 + 2/\alpha)^{1/2} - 1\right]$$

$$\alpha = 1.93 \cdot 10^{17} (Y/2X)^2 \exp(-10.0/T_9^{1/3}).$$

The total energy of the PP-chain (subtracting the energy of the neutrinos which are produced) is

$$\epsilon_{PP} = 6.398 \cdot 10^{18} \psi (dX/dt)_{PP} \quad \mathrm{erg/gm/sec},$$

where

$$\psi = 0.979 f_I + 0.960 f_{II} + 0.721 f_{III}$$
$$f_I = \left[(1 + 2/\alpha)^{1/2} - 1\right]/\left[(1 + 2/\alpha)^{1/2} + 3\right]$$
$$f_{II} = (1 - f_I)/(1 + \Gamma)$$
$$f_{III} = 1 - f_I - f_{II}$$
$$\Gamma = 10^{15.6837}\left[X/(1+X)\right]T_9^{-1/6}\exp(-10.262/T_9^{1/3}).$$

The derivatives of ϵ_{PP} can be found directly:

$$(\partial \ln \epsilon_{PP}/\partial \ln \rho)_T = \epsilon_{PP}$$

$$(\partial \ln \epsilon_{PP}/\partial \ln T)_\rho = \epsilon_{PP}\left[-2/3 + 1.1267/T_9^{1/3}\right.$$
$$+ (\partial \ln \phi/\partial \ln T)_\rho + (\partial \ln \psi/\partial \ln T)_\rho$$
$$+ (0.041 T_9^{1/3} + 0.727 T_9^{2/3} + 0.938 T_9)$$
$$\left./(1 + 0.123 T_9^{1/3} + 1.09 T_9^{2/3} + 0.938 T_9)\right]$$

$$(\partial \ln \phi/\partial \ln T)_\rho = (2/\phi - 1)(1 + 2/\alpha)^{-1/2} 3.333/T^{1/3}$$

$$(\partial \ln \psi/\partial \ln T)_\rho = \psi^{-1}\left\{\left[0.258 - 0.239/(1+\Gamma)\right]\right.$$
$$\left.(\partial f_I \partial \ln T)_\rho - 0.239 f_{III}/(1 + \Gamma)(\partial \ln \Gamma/\partial \ln T)_\rho\right\}$$

$$(\partial \ln \Gamma/\partial \ln T)_\rho = -1/6 + 3.4207/T_9^{1/3}$$

$$(\partial f_I/\partial \ln T)_\rho = -4\left\{\alpha(1 + 2/\alpha)^{1/2}\left[(q + 2/\alpha)^{1/2}\right.\right.$$
$$\left.\left.+3\right]^2\right\}^{-1} \cdot 3.333/T_9^{1/3}$$

In the calculation of the CNO bi-cycle, CN equilibrium is assumed and the CN cycle is assumed o be the only source of energy. The hydrogen burning rate due to the CN cycle is then

$$(dX/dt)_{CN} = 1.202 \cdot 10^7 \rho X X_N T_9^{-2/3} \exp\left(-15.228/T_9^{1/3}\right) \ \mathrm{sec}^{-1}$$

and the energy produced is

$$\epsilon_{CN} = 5.977 \cdot 10^{18} (dX/dt)_{CN} \ \mathrm{erg/gm/sec}.$$

The value of X_N (N^{14} abundance by weight) assumes that all the carbon and nitrogen is in the form of N^{14},

$$X_N = Z - Z_m - X_o,$$

where Z is the total metal abundance by weight, Z_m is the weight abundance of all non-CNO metals, and X_o is the weight abundance of O^{16}. The approach to NO equilibrium is taken as a simple burning rate of O^{16} assuming O^{17} equilibrium,

$$(dX_o/dt) = 9.54 \cdot 10^7 \rho X X_o T_9^{-17/21} \exp\left(-16.693/T_9^{1/3}\right)$$
$$-1.6 \cdot 10^{-3} (dX/dt)_{CN} \ \mathrm{sec}^{-1}$$

Between succesive models the value of X_o is decreased at a rate of (dX_o/dt) per second, and thus the value of X_N is correspondingly increased. Here are the derivatives of the CN energy production:

$$(\partial \epsilon_{CN}/\partial \ln \rho)_T = \epsilon_{CN}$$
$$(\partial \epsilon_{CN}/\partial \ln T)_\rho = \epsilon_{CN}\left(-2/3 + 5.076/T_9^{1/3}\right).$$

4.3. Radiative Opacities

An estimate of magnetic effects on the radiative opacities [$\kappa = \kappa(T, \rho, X, Z)$] can be found in *Li and Sofia* [2001]. Since they are small, we use only the OPAL opacities tables [*Iglesias and Rogers*, 1996] together with the low-temperature opacities from *Alexander and Ferguson* [1994]. For X and Z the linear intepolation is used, but for T and ρ the cubic spline interpolation is used. The cubic spline interpolation scheme allows one to obtain the derivatives of κ with respect to T and ρ. These derivatives are needed in the linearization of the equations of energy transport.

Acknowledgments. This work was supported in part by a grant from NASA, and in part by a grant from NSF. We want to thank G. Thuillier for his providing the information about PICARD.

REFERENCES

Alexander, D. R., and J. W. Ferguson, Low-temperature Rosseland opacities, *Astrophys. J.*, 437, 879-891, 1994

Antia, H.M., S. Basu, J. Pintar, and B. Rohl, Solar cycle variation in solar f-mode frequencies and radius, *Sol. Phys.*, 192, 459-468, 2000

Antia, H.M., S. M. Chitre, and M. J. Thompson, The Sun's acoustic asphericity and magneticc fields in the solar convection zone, *Astron. Astrophys.*, 360, 335-344, 2000

Bhatnagar, A., K. Jain and S. C. Tripathy, GONG p-mode frequency changes with solar activity, *Astrophys. J.*, 521, 885-888, 1999

Caccin, B., and Penza, V., Line-depth and T_{eff} variations with the solar cycle, *preprint*, 2002

Deupree, R. G., S. Sofia, and L. H. Li, A hydrodynamic 2-d variability model of the Sun including magnetic fields, *in preparation*, 2002

Dziembowski, W.A., P. R. Goode, and J. Schou, Does the Sun shrink with increasing magnetic activity? *Astrophys. J.*, 553, 897-904, 2001

Egidi, A., B. Caccin, and S. Sofia, Analysis method for SDS data from flights 6-10, *in preparation*, 2002

Emilio, M., J. R. Kuhn, R. I. Bush, and P. Scherrer, On the constancy of the solar diameter, *Astrophys. J.*, 543, 1007-1010, 2000

Endal, A.S., S. Sofia, and L. W. Twigg, Changes of solar luminosity and radius following secular perturbations in the convective envelope, *Astrophys. J.*, 290, 748-757, 1985

Fowler, W. A., G. R. Caughlan, and B. A. Zimmerman, Thermonuclear reaction rates, II, *Ann. Rev. Astro. Astrophys.*, 13, 69-112, 1975

Fox, P., Solar activity and irradiance variations, *this volume*

Fröhlich, C., and J. Lean., Total solar irradiance variations, in *IAU Sympsium 185: New Eyes to See Inside the Sun and Stars*, edited by F.L. Deubner, pp. 89-102, Kluwer Academic Publ., Dortrecht, 1998

Gray, D. F., and W. C. Livingston, Monitoring the solar temperature: spectroscopic temperature variations of the Sun, *Astrophys. J.*, 474, 802-809, 1997

Howe, R., R. Komm, and F. Hill, Solar cycle changes in GONG p-mode frequencies, 1995–1998, *Astrophys. J.*, 524, 1084-1095, 1999

Iglesias, C. A., and F. J. Rogers, Updated opal opacities, *Astrophys. J.*, 464, 943-953, 1996

Krishna Swamy, K. S., Profiles of strong lines in K-Dwarfs, *Astrophys. J.*, 145, 174-194, 1966

Kuhn, J. R., Helioseismological splitting measurements and the nonspherical solar temperature structure, *Astrophys. J.*, 331, L131-134, 1988

Kuhn, J. R., Helioseismic observations of the solar cycle, *Astrophys. J.*, 339, L45-47, 1989

Kuhn, J. R., and K. G. Libbrecht, Nonfacular solar luminosity variations, *Astrophys. J.*, 381, L35-37, 1999

Kuhn, J. R., K. G. Libbrecht and R. H. Dicke, The surface temperature of the sun and changes in the solar constant, *Science*, 242, 908-911, 1988

Kuhn, J. R., H. Lin, and R. Coulter, What can irradiance measurements tell us about the solar magnetic cycle? *Adv. Space Res.*, 24, (2)185-194, 1999

Lean, J. L., J. Cook, W. Marquette, and A. Johannesson, Magnetic sources of the solar irradiance cycle, *Astrophys. J.*, 492, 390-401, 1998

Li, L. H. and S. Sofia, Measurements of solar irradiance and effective temperature as a probe of solar interior magnetic fields, *Astrophys. J.*, 549, 1204-1211, 2001

Li, L. H., F. J. Robinson, P. Demarque, S. Sofia, and D. B. Guenther, Inclusion of turbulence in solar modeling, *Astrophys. J.*, 567, 1192-1201, 2002a

Li, L. H., S. Sofia, and R. G. Deupree, A static 2-d variability model of the Sun I: 2-d effects of a magnetic field, *in preparation*, 2002b

Lydon, T.J. and S. Sofia, A Method for incorporating the effects of large-scale magnetic fields in the study of stellar structure and variability, *Astrophys. J. (Supp.)*, 101, 357-373, 1995

Prather, M.J., The effect of a Brans-Dicke cosmology upon stellar evolution and the evolution pf galaxies, Ph.D. dissertation, Yale University, 1976

Rogers, F. J., F. J. Swenson, and C. A. Iglesias, OPAL equation-of-state tables for astrophysical applications, *Astrophys. J.*, 456, 902-908, 1996

Sofia, S., W. Heaps, and L. W. Twigg, The solar diameter and oblateness measured by the solar disk sextant on the 1992 September 30 balloon flight, *Astrophys. J.*, 427, 1048-1052, 1994

Sofia, S., and L. H. Li, Solar variability and climate, *J. Geophys. Res.*, 106(A7), 12,969-12,974, 2001

Sofia, S., A. Egidi, B. Caccin, and L. Twigg, Solar diameter variations since 1992, *in preparation*

Solanki, S. K., and Y. C. Unruh, A model of the wavelength dependence of solar irradiane variations, *Astron. Astrophys.*, 329, 747-753, 1998

Spruit, H., Theory of solar irradiance variations, *Space Sci. Rev.*, 94, 113-126, 2000

Thuillier, G., private communication, 2002

Willson, R. C., and H. S. Hudson, The sun's luminosity over a complete solar cycle, *Nature*, 351, 42-44, 1991

Woodard, M. F. 1987, Frequencies of low-degree solar acoustic oscillations and the phase of the solar cycle, *Sol. Phys.*, 114, 21-28, 1987

S. Sofia and L. H. Li, Department of Astronomy, Yale University, 260 Whitney Avenue, New Haven, CT 06520-8101 (sofia@astro.yale.edu; li@astro.yale.edu)

Theoretical Models of Solar Magnetic Variability

Manfred Schüssler and Dieter Schmitt

Max-Planck-Institut für Sonnensystemforschung, Katlenburg-Lindau, Germany

Solar variability on all observationally accessible temporal and spatial scales is intimately connected with the variations of the solar magnetic field and its interaction with the non-stationary flow patterns in the convection zone. We briefly review the current status of theoretical models (of conceptual, analytical, and numerical kind) for the various manifestations of solar magnetic variability. These range from magneto-convection on the scale of granulation to the long-term variability of the global solar cycle and its underlying dynamo mechanism. A model aiming at reconstructing the evolution of the total (unsigned) solar magnetic flux since 1610 on the basis of the group sunspot number record is discussed in the final section.

1. MAGNETIC VARIABILITY

The varying solar magnetic field is responsible for the rich variety of phenomena summarized under the term 'solar activity'. Sunspots appear and dissolve in the solar photosphere, where small-scale magnetic flux concentrations flow restlessly like quicksilver through the network of convective downflow lanes, chromosphere and transition region are in a state of permanent unrest driven by the permanent reorganization of the field structure and connectivity, myriads of hot magnetic loops interlace the corona in an ever-changing pattern. Magnetic energy is released in coronal mass ejections and flares, which affect the space environment of the Earth.

The spectrum of temporal and spatial scales on which the solar magnetic field varies covers the whole observationally accessible ranges and most probably extends beyond those. On a time scale of minutes and on length scales of 1 Mm and below, the interaction between magnetic field and granulation leads to the formation of small magnetic flux concentrations and to a variety of dynamical processes, like

reconnection and wave excitation. Local dynamo action driven by granulation may provide a significant level of background magnetic flux, which varies on these scales. Such processes may also be relevant on the scales of meso-granulation (hours and 5 Mm) and supergranulation (one day and 20 Mm). Theoretical approaches to *magneto-convection* are briefly reviewed in Section 2, although no full coverage of this vast field of research is attempted.

The evolution of magnetically active regions and the transport of magnetic flux on the solar surface covers a range of time scales between days and years. While the short-term variability due to active regions dominates the solar-terrestrial relations, from an (astro-)physical point of view the most important timescale of solar magnetic variability is the 11-year activity cycle (or 22-year magnetic cycle). The underlying large-scale magnetic flux systems are believed to be generated by a hydromagnetic dynamo mechanism based upon convective motions and differential rotation, the rotational shear layer ('tachocline') at the bottom of the convection zone probably playing a crucial role. The current status of dynamo models for the solar cycle is reviewed in Section 3.

The variable cycle amplitudes in the record of sunspot numbers and the appearance of 'grand minima' of low solar activity like the Maunder minimum between 1630 and 1710 indicate that the dynamo process is subject to a modulation

Solar Variability and its Effects on Climate
Geophysical Monograph 141
Copyright 2004 by the American Geophysical Union
10.1029/141GM04

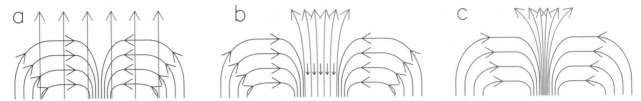

Figure 1. Schematic sketch of the processes that lead to intensification of magnetic field in the solar (sub)photosphere. *a:* the horizontal flows of granular convection sweep magnetic flux towards the downflow regions (lighter vertical lines are magnetic field lines, the other set of lines represents stream lines of the flow). *b:* magnetic forces suppress the convective motions when the magnetic energy density becomes comparable to the kinetic energy density. This throttles the energy supply into the magnetic region, the gas continues to cool owing to radiative losses, and the internal downflow is enhanced. The superadiabatic stratification further amplifies the flow. *c:* the downflow has evacuated the upper layers and the magnetic field in the quenched tube has increased accordingly to maintain the lateral balance of total pressure.

on time scales from decades to centuries. Section 4 gives an overview of existing models for the long-term modulation of the solar dynamo. An attempt to reconstruct the evolution of the solar magnetic flux on the basis of the record of group sunspot numbers is described in Section 5.

2. MAGNETO-CONVECTION

The term *magneto-convection* summarizes the physical processes resulting from the interaction between convectively driven flows and a magnetic field in an electrically conducting fluid. The large length scales of astrophysical systems lead to high (hydrodynamic and magnetic) Reynolds numbers, so that astrophysical magneto-convection typically involves an extended range of spatial scales, nonlinear interactions, and the formation of structures and patterns. Solar magneto-convection is the basis for most physical processes leading to the activity and magnetic variability of the Sun. Such processes involve the *generation* of magnetic flux by a self-excited dynamo mechanism, its spatial *distribution* by flux expulsion and the formation of intermittent structure, the *dynamics* originating from instabilities, wave excitation, and field-line reconnection, and the *energetics* due to non-thermal heating and the interference of the magnetic field with the convective energy transport.

The large magnetic Reynolds numbers corresponding to the various solar convection patterns (i.e., granulation, mesogranulation, supergranulation, possibly also giant convection on a global scale) lead to *magnetic flux freezing*. This means that (with the notable exception of reconnection events) plasma cannot change from one magnetic field line to another: either the field lines have to follow the motion of the plasma (weak field, no significant Lorentz force) or the magnetic field suppresses the fluid motion perpendicular to the field lines (strong field, dominating Lorentz force). The magnetic flux crossing the solar surface is swept by the hor-

izontal motions of the convective flow pattern and accumulates in the convective downflow regions, leading to a hierarchy of network patterns, the most prominent being the supergranular network. This process, known as *flux expulsion* [e.g., *Proctor and Weiss,* 1982; *Hurlburt and Toomre,* 1988], is capable of enhancing the field to about equipartition of the magnetic energy density with the kinetic energy density of the flow. For solar granulation, this corresponds to a field strength of about 500 G.

As the field strength of the expelled flux grows, the convective energy transport is suppressed while radiative cooling of the magnetic structure continues. This drives an enhanced downflow of plasma along the field lines, which is further accelerated by the strongly superadiabatic stratification of the top layers of the convection zone. As a consequence, the upper layers of the magnetic structure become nearly evacuated and the surrounding plasma compresses the magnetic field until an equilibrium of total pressure (i.e., gas pressure plus magnetic pressure) is reached at field strengths in the kG range. This *convective intensification* or *convective collapse* process [*Parker,* 1978; *Spruit and Zweibel,* 1979] is schematically sketched in Figure 1. Observational evidence for the process has been reported by *Zwaan et al.* [1985] and *Bellot Rubio et al.* [2001].

The collapse process loses its efficiency for flux tubes with a diameter smaller than the photon mean free path of about 100 km at the base of the photosphere [*Venkatakrishnan,* 1986]. At that limit, the tube interior is kept at the temperature of its environment and the evacuating downflow is throttled, so that the equipartition limit cannot be exceeded. In fact, observations in the infrared spectral range reveal the existence of small magnetic features with about equipartition field strength in the photosphere [*Lin,* 1995].

Sufficiently strong magnetic fields in large structures, like in sunspots or pores, suppress the convective motions and the associated energy transport, so that they become dark in comparison to the surrounding photosphere. The remaining energy flux (about 20% of the undisturbed flux) is still much too large to be carried by radiation alone. It is assumed that oscillatory convection in a strong magnetic field is the dominant process for energy transport in sunspot umbrae [e.g., *Hurlburt et al.,* 1989; *Weiss et al.,*1996]. On the other hand, small-scale concentrations of strong magnetic field appear brighter than their surroundings because of lateral influx of radiation from the 'hot walls' around the low-density flux concentrations [*Spruit et al.,* 1991] and, presumably, also by dissipation of mechanical energy in its higher atmospheric layers. Note that, for a cylindrical flux tube, the ratio of the heating surface area of its wall and the internal volume to be heated is inversely proportional to the tube radius. The hot walls of small flux tubes become best visible near the limb of the Sun, where the combined effect of many flux tubes gives rise to bright faculae [*Topka et al.,* 1997]. Averaged over the whole Sun, the enhanced brightness of the magnetic elements dominates over the reduced energy flux in the sunspots, so that the total solar output increases with growing amount of magnetic flux in the photosphere [*Fligge et al.,* 2000]. A dissenting view concerning the cyclic variation of solar irradiance is put forward by *Kuhn et al.* [1999].

Numerical simulations have confirmed the basic picture for the formation of the intermittent magnetic structure outlined above (for recent reviews see *Schüssler* [2001] and *Schüssler and Knölker* [2001]). Flux expulsion is nicely demonstrated in Figure 2, which shows a 3D-simulation by *Emonet and Cattaneo* [2001]. These authors consider Boussinesq magneto-convection with high spatial resolution ($512 \times 512 \times 97$ grid points) and a large Rayleigh number of $5 \cdot 10^5$, so that turbulent convection with a wide range of spatial scales develops. Depending on the amount of imposed vertical magnetic flux in the box, the simulations show flux expulsion in network patterns (reminiscent of granulation, 'abnormal' granulation, and mesogranulation on the Sun) or magnetically dominated small-scale convection patterns (possibly relevant for sunspot umbrae).

The convective collapse process has been studied in realistic solar simulations by *Nordlund* [1983], *Nordlund and Stein* [1990], and *Grossmann-Doerth et al.* [1998]. Such simulations include partial ionization and radiative transport. An example is shown in Figure 3. The size of the generated flux concentrations depends on the amount of vertical magnetic flux in the computational box. For larger flux concentrations, the strong downflow may rebound in the high-density layers below the atmosphere. The resulting upflow develops into a shock, which could possibly be related to the spicule phenomenon.

The evolution, dynamics and energetics of already formed small-scale magnetic flux concentrations and their interaction with the convectve flow has been studied on the basis of various approaches. Numerical simulations of two-dimensional magnetic flux sheets with realistic physics (radiative energy transport, partial ionization, compressibility) have been carried out by *Knölker et al.* [1991]; *Grossmann-Doerth et al.* [1994] and *Steiner et al.* [1998]. The simulations show the development of strong downflow jets with velocities up to 6 km s^{-1} next to the magnetic flux sheets. Such downflows lead to asymmetries of the Stokes *V*-profiles by way of the 'canopy effect' [*Grossmann-Doerth et al.,* 1988]. The upper layers of the flux sheet atmosphere become hotter than the environment through radiative illumination by the hot bottom and walls of the flux concentrations, so that small flux concentrations appear bright in comparison to the average photosphere. Flux sheets with diameters in excess of about 500 km appear darker than the average photosphere in continuum radiation if observed near the center of the solar disk (vertical incidence of the line of sight), but become brighter near the limb [*Spruit,* 1976]. External convective motions can significantly bend a magnetic flux sheet, leading to a swaying motion with strong horizontal flows; moreover, they drive field-aligned upward flows that develop into shocks.

On a global scale, the development of magnetic flux on the solar surface has been numerically studied by *Sheeley* [1992]. These authors considered the redistribution of surface flux under the influence of differential rotation, meridional circulation and diffusion (random walk) due to convective flows. Using observed flux emergence events as source terms, they could show that the evolution of the surface flux, including the reversals of the polar fields, can be completely described by flux elements being passively carried by the various surface flows.

Schrijver et al. [1997a, b] have developed a statistical 'magneto-chemistry' approach to study the development of an ensemble of magnetic flux elements in the quiet Sun. The model includes the effects of collisions, coalescence, fragmentation and cancellation of flux elements. More recently, the model has been extended to cover the whole Sun (or other magnetically active stars) with the full range of active regions [*Schrijver,* 2001].

3. THE SOLAR CYCLE

It is generally believed that the cyclic evolution of the large-scale magnetic field of the Sun is the consequence of dynamo action, i.e. the conversion of kinetic into magnetic energy by the inductive effects of fluid motions in an

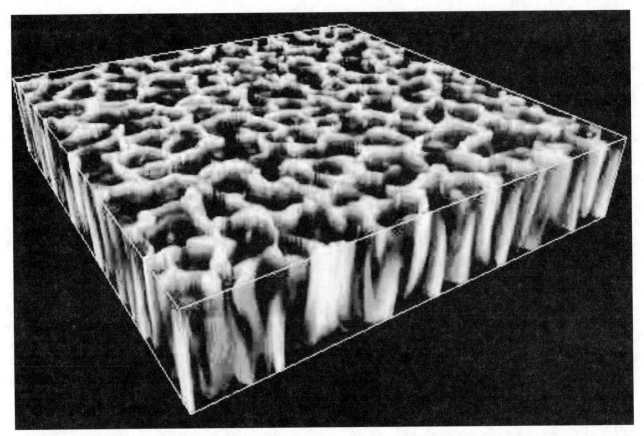

Figure 2. Volume rendering of the magnetic field intensity in a 3D simulation of Boussinesq magneto-convection [*Emonet and Cattaneo*, 2001]. By the process of flux expulsion, the magnetic field becomes concentrated in the network of convective downflow lanes, like in a solar plage area.

electrically conducting fluid. Three processes are thought to play a major role in the solar dynamo process:

- differential rotation generates a toroidal field by winding up a preexisting poloidal field,

- helical turbulence regenerates the poloidal field, and

- meridional circulation transports magnetic flux.

Whereas the first two effects have been considered in many dynamo models in the past, the potential importance of the last process has been realized only recently.

The crucial importance of helical turbulence for a dynamo was first realized by *Parker* [1955]. Rising (sinking) bubbles in the stratified, rotating solar convection zone expand (contract), leading to rotation due to the action of the Coriolis force. Such 'cyclonic' motion bends magnetic field lines into twisted loops (Figure 4), which merge through diffusion and thus form field components perpendicular to the original field. The effect of small-scale motions on the large-scale magnetic field has been systematically investigated within the framework of mean-field electrodynamics, which was established by *Steenbeck et al.* [1966], see also *Krause and Rädler* [1980]. They formally showed that helical motions drive a mean electric current parallel or antiparallel to the mean magnetic field. This current is represented by an additional term in the induction equation for the mean magnetic field, called the α-effect. Furthermore, the mean field is subject to enhanced turbulent diffusion.

Three decades ago, calculations of $\alpha\Omega$-type dynamos, where α-effect and differential rotation (Ω-effect) constitute the induction processes, showed qualitatively good agreement with observations for suitable assumptions about the dynamo effects in the solar convection zone. Periodic solutions with equatorward propagating dynamo waves were obtained, fields were antisymmetric with respect to the equatorial plane, thus reproducing Maunder's butterfly dia-

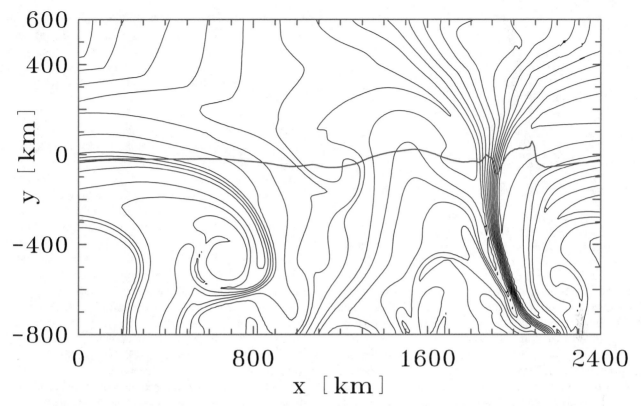

Figure 3. Flux expulsion and convective intensification in a 2D simulation with radiative transfer and ionization [*Grossmann-Doerth et al.*, 1998]. Shown are magnetic field lines; $y = 0$ corresponds to optical depth unity at 500 nm wavelength (the solar 'surface'). A strong magnetic flux concentration has formed in a convective downflow region around $x = 2000$ km.

gram and Hale's polarity rules (e.g., *Steenbeck and Krause* [1969], see reviews by *Stix* [1981] and *Rädler* [1990]). The equatorward propagation, for instance, results from a negative sign of the product of the α-effect and a radial gradient of differential rotation in the northern hemisphere [*Yoshimura*, 1975]. Furthermore, the observed antiphase of toroidal and radial field components requires a positive α-effect and a negative gradient of differential rotation [*Stix*, 1976]. The general agreement of the calculated fields with the observed pattern provided confidence that the basic ideas were correct and that minor disagreements could be resolved with a better knowledge of the solar differential rotation and a realistic turbulence model for the mean electromotive force, i.e. the α-effect and turbulent diffusivity. This picture has, however, repeatedly been questioned during the past three decades.

Most of the solar magnetic flux in the convection zone is concentrated in small-scale intermittent features [*Galloway and Weiss*, 1981; *Proctor and Weiss*, 1982] as observed on the solar surface [*Stenflo*, 1973; *Solanki*, 1993]. These structures are difficult to store in the convection zone for times

comparable to the solar cycle. Magnetic buoyancy and instabilities transport magnetic flux in the convection zone in times of the order of one month, much to short for the dynamo to generate the field [*Parker*, 1975; *Schüssler*, 1979; *Spruit and van Ballegooijen*, 1982].

Helioseismological data have shown that the convection zone rotates like the solar surface, with only a latitudinal but no significant radial gradient, whereas the radiative interior rotates almost rigidly at a rate intermediate between the equatorial and polar rates on the surface [*Brown and Morrow*, 1987; *Libbrecht*, 1988; *Schou et al.*, 1992; *Tomczyk et al.*, 1995; *Schou et al.*, 1998]. Thus a strong radial gradient of angular velocity occurs in a transition region between the base of the convection zone and the top of the interior, the so-called tachocline. On the other hand, dynamo models in the convection zone with only a latitudinal gradient of angular velocity do not show migration towards the equator [*Köhler*, 1973], as is demanded by the observed butterfly diagram.

It has been suggested that the bulk of the solar magnetic flux is stored in the overshoot region, a subadiabatically

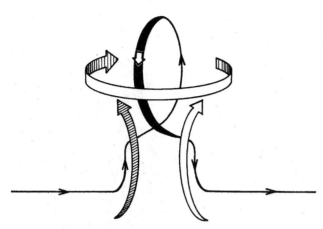

Figure 4. A rising convective cell expands and rotates, twisting a magnetic field line into a loop with components perpendicular to the plane of projection [after *Parker*, 1970].

stratified region below the convection zone, which results from overshooting convection from above [*Spiegel and Weiss*, 1980; *Schüssler*, 1984]. The differential rotation there is able to wind up a strong toroidal field [*Schüssler*, 1987; *Fisher et al.*, 1991]. Magnetic buoyancy is reduced because of the subadiabatic stratification, thus enabling storage of strong magnetic fields [*Moren-Insertis et al.*, 1992; *Ferriz-Mas*, 1996]. Finally, turbulent diffusivity is reduced in the overshoot region.

The field has to be strong enough to emerge at low latitudes as observed and to avoid the poleward slip [*Schüssler et al.*, 1994; *Choudhuri and Gilman*, 1987]. Strong fields that resist tangling by convective motions are also required in order to account for the polarity rules and the tilt angles of sunspot groups. Detailed studies of the dynamics of magnetic flux tubes in the convection zone lead to field strengths of 10^5 G at the bottom of the convection zone [*D'Silva and Choudhuri*, 1993; *Fan et al.*, 1993, 1994; *Caligari et al.*, 1995, 1998]. Such strong fields can only be stored in the subadiabatic overshoot region beneath the convection zone proper [*Moren-Insertis*, 1992; *Ferriz-Mas and Schüssler*, 1993, 1994, 1995].

With the overshoot region as the seat of a strong toroidal magnetic field, which is wound up by differential rotation in the tachocline, three types of dynamo models have been discussed during the last 10 years:

- *overshoot layer dynamos* for which both α-effect and differential rotation are located in the overshoot region,

- *interface dynamos* with the Ω-effect working in the overshoot region and the convective α-effect in the turbulent part of the convection zone just above the overshoot layer, and

- *flux transport dynamos* with the α-effect near the surface, originating from the tilt of active regions, and a meridional circulation connecting the separated dynamo processes, with a poleward flow on the surface and an equatorward return flow at the base of the convection zone.

In the case of a dynamo working completely in the overshoot region, the regeneration of the poloidal field has to be reconsidered because the strong fields resist the turbulent flow and the kinematic α-effect of convective motions hardly works. Magnetic instabilities generating helical waves have been proposed, which give rise to a dynamic α-effect. *Schmitt* [1984, 1985, 2003] considered a magnetic buoyancy instability of a toroidal magnetic layer. This Rayleigh-Taylor-like instability occurs at the top of the layer, where the magnetic field decreases rapidly enough with height. Because of rotation the instability takes the form of growing magnetostrophic waves [*Acheson and Hide*, 1973; *Acheson and Gibbons*, 1978; *Acheson*, 1978, 1979]. These are helical and induce a toroidal electric current which regenerates the poloidal field [*Schmitt*, 1987]. This α-effect is able to work in the presence of a strong field.

For obtaining low-latitude activity belts, the dependence of the α-effect on colatitude θ is important. Conventionally, $\alpha = \alpha_0 \cos\theta$, $\alpha_0 > 0$, is assumed [e.g., *Moffatt*, 1978] because the horizontal component of the Coriolis force is $\propto \cos\theta$. For unstable magnetostrophic waves, a more complicated, nonmonotonous behaviour of $\alpha(\theta)$ results. α is antisymmetric with respect to the equator, where it vanishes. In low northern latitudes α is negative, changes sign around $\theta \approx 60°$ and is positive up to the north pole. The low-latitude contribution dominates. From helioseismology we know that also the radial gradient of angular velocity changes sign at around the same colatitude. It is positive near the equator and negative near the poles, where the gradient is steeper by a factor of approximately 2. Because of the almost simultaneous sign changes of the two induction effects and the negative sign of their product, the dynamo wave migrates equatorward in all latitudes. Moreover, the strong α-effect near the equator compensates for the strong differential rotation near the pole, yielding low-latitude oscillatory dynamo action [*Prautzsch*, 1993, 1997; *Schmitt*, 1993; *Rüdiger and Brandenburg*, 1995].

A problem of this dynamo model is the disagreement with observation of the phase relation between the toroidal and radial field components in the overshoot region. However, this changes when comparing the toroidal field at the bottom with the radial field at the top of the convection zone [*Schlichenmaier and Stix*, 1995]. Another problem is a considerable overlap of the butterfly wings in a thin-layer dynamo. Furthermore, symmetric and antisymmetric solutions with respect to the equator are almost equally excited.

Similar to the magnetic buoyancy instability of a homogeneous magnetic layer, non-axisymmetric instabilities of toroidal magnetic flux tubes under the influence of rotation give rise to an α-effect [*Ferriz-Mas et al.*, 1994]. These instabilities require the field strength to exceed a threshold value. When the field strength falls below the threshold value, dynamo action stops and needs a starting mechanism in order to work again [*Schmitt et al.*, 1996]. This may lead to the occurrence of grand minima. A more detailed discussion follows in the next section.

Linear stability analysis does not yield the actual strength of the induction effect due to the magnetic buoyancy instability. A non-bear numerical simulation confirms the existence of this α-effect [*Brandenburg and Schmitt*, 1998].

In a recent study, *Dikpati and Gilman* [2001a] showed that a global hydrodynamic instability of latitudinal differential rotation in the tachocline gives rise to helical motions and thus provides another source of a non-conventional α-effect. This effect depends on latitude in a complicated fashion with multiple sign changes. It contributes mainly at mid latitudes, where it is positive in the northern hemisphere. Together with the radial differential rotation profile in the tachocline, oscillatory dynamo action takes place mainly at mid and high latitudes, with dynamo waves migrating poleward near the pole and equatorward at mid latitudes. Employing meridional circulation, which transports magnetic flux produced by the dynamo equatorward, *Dikpati and Gilman* [2001b] obtained dominating dynamo waves at low latitudes migrating towards the equator, similar to the butterfly diagram of the large-scale features of the solar cycle (Figure 5). Moreover, this dynamo preferentially excites a toroidal field that is antisymmetric about the equator, as is required for the Sun.

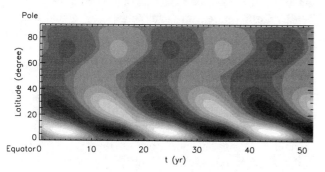

Figure 5. A flux-transport overshoot layer dynamo solution after *Dikpati and Gilman* [2001b]. Shown is the toroidal magnetic field in a time-latitude diagram, which due to an equatorward mean flow exhibits strong butterfly wings near the equator.

The class of *interface dynamos* has been proposed by *Parker* [1993]. Differential rotation builds up a strong toroidal field in the overshoot region, while the regeneration of the poloidal field is achieved by the "classical" convective α-effect in the lower convection zone. Both regions are separated by a jump of the magnetic diffusivity by a factor $\sim 10^{-2\ldots-3}$. As a result, the magnetic field is strong below the interface and weak above. Diffusive transport across the interface is crucial for the operation of this class of dynamo models, while the separation of the dynamo effects determines the dynamo period and the phase relationship between poloidal and toroidal field.

Charbonneau and MacGregor [1997] developed interface dynamo models with a realistic internal differential rotation profile and found various modes of dynamo operation. Hybrid modes rely on the latitudinal shear in the convection zone and overshoot region, while interface modes are driven by the radial shear in the overshoot region. Depending on the assumed latitudinal distribution of the α-effect, the latter show dynamo action at high or low latitudes, similar to the pure overshoot layer dynamo. Which of the modes is preferred depends on details of the model. A similar study by *Markiel and Thomas* [1999] attributes the hybrid mode to an improper boundary condition and emphasizes the difficulty in obtaining solar-like dynamo solutions.

The flux transport dynamos have some similarities to the models of *Babcock* [1961] and *Leighton* [1969] for which reason they are sometimes called Babcock-Leighton dynamos. In these models, the regeneration of the poloidal field from the toroidal field component is assumed to originate with the twist imparted by the Coriolis force on toroidal flux tubes rising through the convective envelope. Thus, the two induction effects, differential rotation and α-effect, are widely separated in space. Dynamo action requires the poloidal field to be transported to the region of strong shear at the bottom of the convection zone. In recent models, large-scale meridional circulation in the convection zone, with a poleward surface flow and an equatorward subsurface return flow, is invoked to achieve this effect [*Choudhuri et al.*, 1995; *Durney*, 1995, 1996, 1997; *Dikpati and Charbonneau*, 1999; *Nandy and Choudhuri*, 2001; *Küker et al.*, 2001]. Meridional circulation is obtained in both analytic theory and numerical simulations. It is observed directly at the solar surface, and recent helioseismic studies have demonstrated this flow to continue well below the surface [*Giles et al.*, 1997; *Braun and Fan*, 1998; *Schou and Bogart*, 1998]. Key properties of the dynamo are determined by the mean meridional flow. It can control the migration of the dynamo wave such that the *Yoshimura*

[1975] sign rule no longer governs the equatorward migration of sunspot belts [*Choudhuri et al.*, 1995], while the time period of the dynamo is determined by the circulation flow speed [*Dikpati and Charbonneau*, 1999].

At present a clear preference to one or another type of the presented models is not possible. The equatorward transport of magnetic flux at the base of the convection zone by meridional circulation, in combination with the various kinds of models, especially also overshoot layer and interface dynamos, offers a promising possibility to avoid some of the problems these models face without this effect.

4. LONG-TERM MODULATION AND GRAND MINIMA

The dominant time scale of variability of solar activity is the 11-year solar cycle, or, if one considers the magnetic polarity reversals, the 22-year magnetic cycle of the Sun. This cycle is modulated by a long-term variability of its amplitude and period on timescales of decades to centuries. Due to insufficient length of the available data records, it is unclear whether this modulation reflects a superposition of quasi-periodic processes or has a chaotic or stochastic character.

The strongest perturbation of the sequence of 11-year cycles is the appearance of 'grand minima' during which solar activity is low for an extend period of time. The best documented example is the Maunder minimum between 1630 and 1710 when only very few sunspots were observed. It has been suggested that the cycle continued at a low level during the Maunder minimum, mainly at the southern hemisphere [*Ribes and Nesme-Ribes*, 1993; *Beer et al.*, 1998].

On the basis of the varying production rate of the cosmogenic isotopes ^{14}C and ^{10}Be, the long-term modulation of solar activity can be followed further back into the past (Figure 6). These isotopes are formed by cosmic rays as spallation products in the upper atmosphere of the Earth, from which they are removed by precipation. The yearly layering in ice cores drilled in Greenland can be used to determine the variation of the incoming flux of galactic cosmic rays, which itself is anticorrelatcd with the level of solar activity [e.g. *Bazilevskaya,* 2000]. Consequently, high concentration levels of cosmogenic isotopes indicate low solar activity, and vice versa. There is some evidence that episodes of reduced activity tend to recur roughly on a 200-year timescale, at least during some time intervals [*Beer,* 2000].

Magnetic activity of cool stars other than the Sun can be detected indirectly by measuring the associated chromospheric excess emission [*Wilson*, 1978]. Many stars show cyclic variations similar to the Sun, while others exhibit either irregular variations on a high activity level or have a flat low activity level [*Baliunas et al.*, 1995]. Typically, stars with a cyclic or flat activity level rotate slowly, while those with a high activity level are rapid rotators [*Montesinos and Jordan*, 1993]. A flat activity level could possibly indicate a grand minimum [*Baliunas and Jastrow*, 1990].

The origin of the long-term modulation of the solar cycle is hardly understood. In terms of dynamo theory, the modulation of the various dynamo effects has been discussed. Among the mechanism are

- a modulation of the differential rotation through the nonlinear back-reaction of the magnetic field,

- a stochastic fluctuation of the α-effect,

- a variation in the meridional circulation, and

- on-off intermittency due to a threshold field strength for dynamo action.

These effects and their consequences are briefly discussed in the following.

The back-reaction of the large-scale magnetic field on the differential rotation in the tachocline leads to complicated nonlinear behaviour. Imposing dipole symmetry of the solutions in a simplified dynamo model, *Tobias* [1996] found significant long-term amplitude modulations when the magnetic Prandtl number Pm, the ratio of kinematic viscosity to magnetic diffusivity, is of order 0.01. The ratio of the period of modulation to the basic cycle period scales as $Pm^{-1/2}$. When the symmetry constraint is relaxed, symmetry-breaking bifurcations can occur. Particularly when the field is weak (during grand minima) the symmetry can be broken and magnetic activity dominates in one hemisphere [*Tobias*, 1997]. Furthermore, the symmetry may actually flip between dipole and quadrupole states in a grand minimum [*Beer et al.*, 1998].

Moss and Brooke [2000] discuss a similar model in spherical symmetry, using a realistic solar rotation profile. For small magnetic Prandtl number they find intermittent and chaotic behaviour. However, they emphasize that the behaviour is strongly dependent on the values of poorly known parameters. A similar conclusion can be drawn from the study of *Küker et al.* [1999].

A different approach to account for the modulation of the solar cycle is the stochastic behaviour of the dynamo itself [*Hoyng*, 1987, 1988]. Defining averages over longitude in a mean-field dynamo model and considering a finite number of

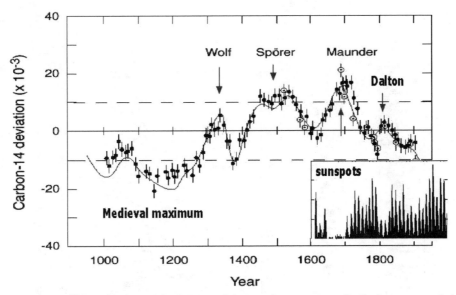

Figure 6. Variation of the ^{14}C production rate in the terrestrial atmosphere as a proxy for the long-term variation of solar activity. Due to the long residence time of ^{14}C of 30–40 years, the 11-year cycle is smoothed out. The Maunder and Dalton minima, as well as earlier grand minima, clearly appear as maxima in the ^{14}C record.

convective cells (giant cells) all mean quantities including the α-effect retain a stochastic component [*Hoyng*, 1993, 1996].

Ossendrijver et al. [1996] study a mean field dynamo model in a spherical shell, where the α-coefficient is a stochastic function of time and latitude. This leads to a variation of the fundamental (oscillatory and dipolar) dynamo mode and to a stochastic excitation of overtones. The phase and the amplitude of the fundamental mode perform a random walk, their fluctuations are anti-correlated as observed in sunspot data. There are long intervals of low activity reminiscent of grand minima. Superposition of the fundamental mode and the overtones leads to an asymmetry of the activity on the two hemispheres. The asymmetry often peaks during the activity minimum and can persist for many dynamo periods.

The effect of stochastic fluctuations on a flux-transport model of the solar cycle is investigated by *Charbonneau and Dikpati* [2000]. They study the consequence of large-amplitude stochastic fluctuations in either or both the meridional flow and a poloidal source term (α-effect). Solar-like oscillatory behaviour persists for high fluctuation amplitudes, indicating the robustness of this class of models. The observed weak anticorrelation between duration and amplitude of the cycle can be recovered if fluctuations in the poloidal source term are considered. The results from the model suggest that the meridional circulation speed, the primary determinant of the cycle period, acts as a clock, so that the cycle periods rarely depart from their average peri-

od for more than a few consecutive periods. The model also exhibits a clear correlation between the toroidal field strength of a given cycle and the strength of the high-latitude surface magnetic field of the preceding cycle, which is in qualitative agreement with observational inferences [*Legrand and Simon*, 1981, 1991]. Producing extended periods of reduced activity, however, turns out to be rather difficult.

The dynamic α-effect due to magnetic buoyancy, one of the candidates for a strong-field dynamo operating in the overshoot layer, only sets in beyond a threshold field strength of several times 10^4 G [*Ferriz-Mas et al.*, 1994]. Therefore, as a starting mechanism, the dynamo requires fluctuating fields, transported by downdrafts from a turbulent convection zone into the overshoot region. On the other hand, such fluctuations, when destructive, can lead to a sequence of low-amplitude cycles or even drive the dynamo subcritical until another, constructive magnetic fluctuation restarts the dynamo again. This leads to on-off intermittent solutions and can be related to the occurrence of grand minima [*Schmitt et al.*, 1996; *Schüssler et al.*, 1997; *Schmitt et al.*, 1998] (Figure 7).

Sufficiently strong fluctuations destroy the cyclic behaviour of the overshoot layer dynamo and lead to irregular activity. Such activity is observed in fast-rotating cool stars. On the other hand, stars with low and non-variable magnetic activity may be in a state with only a turbulent convection zone dynamo active.

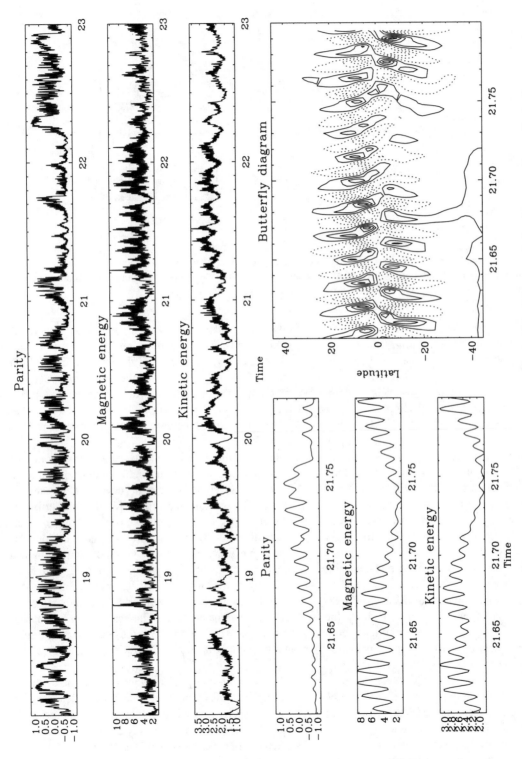

Figure 7. Time evolution of an on-off intermittent dynamo with a lower threshold in field strength for dynamo action, a fluctuating source term [*Schmitt et al.*, 1996] and, as limiting nonlinearity, a back-reaction of the magnetic field on the differential rotation. The three panels in the upper half give the parity, the magnetic energy of the toroidal field, and the kinetic energy of the velocity perturbation, respectively. In the lower half, the same quantities for a small section of the time series, together with the corresponding butterfly diagram (contour plot of the azimuthal field on a latitude-time plane). The dominant dipolar parity during the oscillatory phases gives way to a mixed parity in 'grand minima' phases. Typically, a strong north-south asymmetry develops during such phases.

5. MODELING THE EVOLUTION OF THE SOLAR MAGNETIC FIELD SINCE 1610

The magnetic flux related to the 11/22-year solar cycle emerges at the photosphere in the form of magnetically bipolar regions, which have a continuous spectrum of sizes [Zwaan and Harvey, 1994]. The larger of these are the *active regions*, which usually contain sunspots and are restricted to about ±30 deg latitude. The smaller *ephemeral regions* do not form sunspots and have a broader distribution in latitude [Harvey and Martin, 1973; Harvey, 1992, 1993]. While active regions can have lifetimes up to several months (at least in the form of 'enhanced network'), ephemeral regions typically have a very short decay time of the order of a day. On the other hand, the rate of flux emergence in ephemeral regions is about two orders of magnitude larger than the corresponding rate from active regions [Schrijver et al., 1997b; Hagenaar, 2001].

Active regions and ephemeral regions show a cyclic variation and, therefore, are connected to the basic dynamo process generating the 11/22-year cycle and its long-term modulation. On the other hand, the large amount of 'internetwork' magnetic flux on very small scales is probably produced by a fast dynamo process based on granular convection [Petrovay and Szakaly, 1993; Cattaneo, 1999]. Granulation is practically unaffected by solar rotation, so that such a non-helical 'local' dynamo process generates magnetic field on the spatial scale of the driving velocity pattern, but no large-scale or global field. There is no reason to assume a secular variation of granulation and of the magnetic field it may locally generate. To model the long-term evolution of solar magnetic flux it therefore suffices to consider the flux emerging in active regions and ephemeral regions (the total cycle-related flux), which may be superposed upon a time-independent small-scale background flux of unknown magnitude.

The activity cycle of ephemeral regions is more extended than that of active regions with sunspots and runs ahead of it in phase [Harvey, 1993, 1994]: ephemeral regions belonging to a given cycle (as distinguished by emergence latitude) start emerging 2–3 years before the first sunspots of the corresponding cycle appear. Therefore, around each sunspot minimum we have an overlap of magnetic flux emergence in ephemeral regions from the old and from the new cycle. As shown below, this effect can lead to a secular variation of the total magnetic flux threading the solar surface.

A large part of the emerged flux is rapidly removed from the solar surface by cancellation with opposite-polarity flux within a timescale of days. The remaining flux becomes redistributed over the solar surface through advection by supergranulation (turbulent diffusion), differential rotation, and meridional flow. The corresponding timescales are in the range of months to years. The evolution of the large-scale patterns of the magnetic flux distribution on these scales can be fairly well reproduced by simulations of passive horizontal flux transport by these flow patterns [e.g. Wang and Sheeley, 1994]. In particular, the polarity reversals at the polar caps can be attributed to the dominant transport of opposite-polarity flux from the following parts of the tilted bipolar regions in the activity belts [cf. Petrovay, 2001].

Recently, Solanki et al. [2000, 2002] have developed a simple model which allows a reconstruction of the cycle-related magnetic flux at the solar surface on the basis of sunspot numbers. The model includes emergence of magnetic flux in active regions as well as in ephemeral regions, its decay by cancellation, and its transfer to the 'open' flux component (i.e., the interplanetary or heliospheric field). Here we extend the reconstruction back to 1610, taking the group sunspot number [Hoyt and Schatten, 1998] as a proxy for the rate of flux emergence in active regions. Following the observational results, flux emergence in ephemeral regions is described by a stretched solar cycle starting 2.5 years earlier than the cycle of active regions, so that ephemeral region cycles overlap around sunspot minimum. The amplitude of flux emergence in ephemeral regions is tied to the rate of emergence in active regions of the same cycle, with an amplification factor following from the observations of Hagenaar [2001] near the minimum of cycle 22. The model was validated by comparing its results for the cycles 21 and 22 with the observational determination of the total flux by Harvey [1994].

By connecting all flux emergence with the group sunspot number, the various components of the solar magnetic surface flux can be reconstructed back to the beginning of the record in 1610. Of course, this reconstruction is based upon the assumption that the relation between sunspot numbers and active region flux emergence as well as the relation between the emergence rates in active and in ephemeral regions did not change during the time interval considered. This assumption cannot be proven, but lacking any evidence for the contrary, it seems a reasonable basis for a first attempt, as long as its limitations and uncertainties are kept in mind. Figure 8 shows the reconstructed time series of the active region flux, the ephemeral region flux, and the total flux, the latter also in the form of a 20-year running mean. While the active region flux practically reflects the sunspot numbers, the ephemeral region flux clearly shows a secular variation that is caused by the overlapping of the corresponding activity cycles. As a consequence, a number of

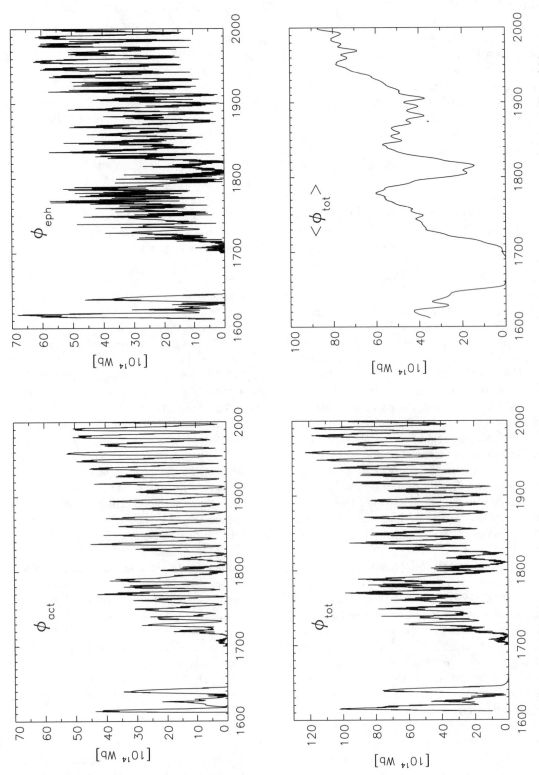

Figure 8. Reconstructed magnetic fluxes since 1610: active regions (upper left), ephemeral regions (upper right), total flux (lower left), and 20-year running mean of total flux (lower right). On average, the total flux has about doubled in the first half of the 20th century.

strong (weak) cycles leads to a secular increase (decrease) of the flux in ephemeral regions. Since this flux represents a major fraction of the total flux, the latter shows similar secular trends. In particular, the running average indicates that the total flux has dropped by a factor of 3 during the Dalton minimum, about doubled in the first half of the 20th century, while staying roughly constant during the second half. Such a variation could have significant consequences for the variability of the total irradiance and the UV flux from the Sun, both quantities of potential importance for terrestrial climate change.

It is of course highly uncertain whether the total surface flux in fact dropped to nearly zero during the Maunder minimum as suggested by Figure 8. There are indications that the 11-year solar cycle continued throughout this grand minimum [*Ribes and Nesme-Ribes*, 1993; *Beer et al.*, 1998], so that probably the assumed relation between flux emergence in ephemeral regions and active regions has to be modified in the case of very low or vanishing sunspot activity.

Lockwood et al. [1999] have reconstructed the evolution of the heliospheric field strength (i.e., the Sun's open magnetic flux) since 1868 on the basis of the *aa*-index of geomagnetic disturbances. They found a secular variation of the open flux superposed upon the 11-year cycle and, more specifically, a doubling of the heliospheric field strength since about 1900, which is in accordance with the general decrease of the ^{10}Be production rate during the same period of time [*Beer*, 2000]. The time evolution of the open flux is well reproduced by the model under the assumption of a long decay time of 3–4 years for the open flux component, owing to its intrinsically large spatial scale (extended unipolar regions and coronal holes). If the decay time is a significant fraction of the solar cycle length, a secular variation of the open flux results from a variation of the cycle length. In particular, a sequence of short cycles leads to an increase of the (cyle-averaged) open flux. The open heliospheric flux affects the flux of cosmic rays in the vicinity of the Earth, which in turn may affect cloud formation with potential consequences for the global climate [*Marsh and Svensmark*, 2000]. This could possibly provide an explanation for the good correlation between the global terrestrial temperature and solar cycle length before ~ 1980 [*Friis-Christensen and Lassen*, 1991; *Fligge et al.*, 1999].

Figure 9 shows the 11-year running average of the reconstructed open flux and, for comparison, the corresponding curve derived for the reconstruction of *Lockwood et al.* [1999]. The assumed decay time for the open flux in the model is 3 years. There is considerable uncertainty about the contribution of ephemeral regions to the open flux [*Harvey*, 1994]. In the reconstruction shown here it has been assumed that they contribute to the open flux with a

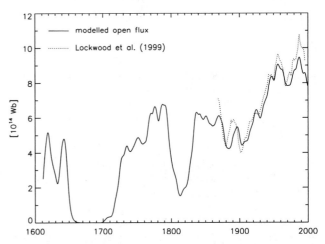

Figure 9. Full line: 11-yr running mean of the reconstructed open solar flux, obtained from the model including ephemeral regions. A decay time of 3 yr for the open flux has been employed. Dotted line: 11-yr running mean of the reconstructed heliospheric flux from *Lockwood et al.* [1999].

factor of 6 smaller efficiency than the active region flux, owing to their intrinsically smaller dipole moments and more randomly distributed orientations. Alternatively, if the ephemeral regions are completely ignored in the determination of the open flux, a similar agreement between the model and the data from Lockwood et al. can be achieved using a decay time of 4 years [*Solanki et al.*, 2000]. Independent evidence for the validity of the reconstructed open flux is provided by comparison with the record of the ^{10}Be data since 1900 [*Beer et al.*, 1990]. Both the cyclic variation and the secular trends in the model are in good agreement with these data.

Recently, *Arge et al.* [2002] argued on the basis of magnetogram data and potential field extrapolations that there was no secular increase of the solar open flux since 1970, in contradiction to the conclusions of *Lockwood et al.* [1999] from satellite measurements made near the ecliptic plane and from *Ulysses* results. On the other hand, *Wang and Sheeley* [1995]; *Wang et al.* [2000a, b] have verified the assumption of a parallel variation of solar open flux and the radial component of the interplanetary magnetic field near Earth, thus supporting the case of Lockwood et al. In any case, the reconstruction of the *total* solar flux (lower panels of Fig. 8) is in agreement with the results of Arge et al. as far as the measured surface fluxes are concerned.

REFERENCES

Acheson, D. J., On the instability of toroidal magnetic fields and differential rotation in stars, *Phil. Trans. Ser. Roy. Soc. London, 289*, 459-500, 1978.

Acheson, D. J., Magnetic buoyancy in the sun, *Nature, 277*, 41, 1979.

Acheson, D. J., and M. P. Gibbons, Magnetic instabilities of a rotating gas, *J. Fluid Mech., 85*, 743-757, 1978.

Acheson, D. J., and R. Hide, Hydromagnetics of rotating fluids, *Rep. Prog. Phys., 36*, 159-221, 1973.

Arge, C. N., E. Hildner, V. J. Pizzo, and J. W. Harvey, Two solar cycles of non-increasing magnetic flux, *J. Geophys. Res.*, in press, 2002.

Babcock, H. W., The topology of the Sun's magnetic field and the 22-year cycle, *Astrophys. J., 133*, 572, 1961.

Baliunas, S., and R. Jastrow, Evidence for long-term brightness changes of solar-type stars, *Nature, 348*, 520-523, 1990.

Baliunas, S. L., et al., Chromospheric variations in main-sequence stars, *Astrophys. J., 438*, 269-287,1995.

Bazilevskaya, G. A., Observations of variability in cosmic rays, *Space Sci. Rev., 94*, 25-38, 2000.

Beer, J., Long term, indirect indices of solar variability, in *Solar Variability and Climate*, edited by E. Friis-Christensen, C. Fröhlich, J. Haigh, M. Schüssler, and R. von Steiger, p. 53, Kluwer Academic Publishers, Dordrecht, 2000.

Beer, J., A. Blinov, G. Bonani, H. J. Hofmann, and R. C. Finkel, Use of Be-10 in polar ice to trace the 11-year cycle of solar activity, *Nature, 347*, 164-166, 1990.

Beer, J., S. Tobias, and N. Weiss, An active sun throughout the maunder minimum, *Solar Phys., 181*, 237-249, 1998.

Bellot Rubio, L. R., I. Rodriguez Hidalgo, M. Collados, E. Khomenko, and B. Ruiz Cobo, Observation of convective collapse and upward-moving shocks in the quiet Sun, *Astrophys. J., 560*, 1010-1019, 2001.

Brandenburg, A., and D. Schmitt, Simulations of an alpha-effect due to magnetic buoyancy, *Astron. Astrophys., 338*, L55-L58, 1998.

Braun, D. C., and Y. Fan, Helioseismic measurements of the subsurface meridional flow, *Astrophys. J., 508*, L105-L108, 1998.

Brown, T. M., and C. A. Morrow, Depth and latitude dependence of solar rotation, *Astrophys. J., 314*, L21-L26, 1987.

Caligari, P., F. Moreno-Insertis, and M. Schüssler, Emerging flux tubes in the solar convection zone. 1: Asymmetry, tilt, and emergence latitude, *Astrophys. J., 441*, 886-902, 1995.

Caligari, P., M. Schüssler, and F. Moreno-Insertis, Emerging flux tubes in the solar convection zone. II. The influence of initial conditions, *Astrophys. J., 502*, 481, 1998.

Cattaneo, F., On the origin of magnetic fields in the quiet photosphere, *Astrophys. J., 515*, L39-L42, 1999.

Charbonneau, P., and M. Dikpati, Stochastic fluctuations in a Babcock-Leighton model of the solar cycle, *Astrophys. J., 543*, 1027-1043, 2000.

Charbonneau, P., and K. B. MacGregor, Solar interface dynamos. II. Linear, kinematic models in spherical geometry, *Astrophys. J., 486*, 502, 1997.

Choudhuri, A. R., and P. A. Gilman, The influence of the Coriolis force on flux tubes rising through the solar convection zone, *Astrophys. J., 316*, 788-800, 1987.

Choudhuri, A. R., M. Schüssler, and M. Dikpati, The solar dynamo with meridional circulation, *Astron. Astrophys., 303*, L29, 1995.

Dikpati, M., and P. Charbonneau, A Babcock-Leighton flux transport dynamo with solar-like differential rotation, *Astrophys. J., 518*, 508-520, 1999.

Dikpati, M., and P. A. Gilman, Analysis of hydrodynamic stability of solar tachocline latitudinal differential rotation using a shallow-water model, *Astrophys. J., 551*, 536-564, 2001a.

Dikpati, M., and P. A. Gilman, Flux-transport dynamos with α-effect from global instability of tachocline differential rotation: A solution for magnetic parity selection in the Sun, *Astrophys. J., 559*, 428-442, 2001b.

D'Silva, S., and A. R. Choudhuri, A theoretical model for tilts of bipolar magnetic regions, *Astron. Astrophys., 272*, 621, 1993.

Durney, B. R., On a Babcock-Leighton dynamo model with a deep-seated generating layer for the toroidal magnetic field, *Solar Phys., 160*, 213-235, 1995.

Durney, B. R., On a Babcock-Leighton dynamo model with a deep-seated generating layer for the toroidal magnetic field, II, *Solar Phys., 166*, 231-260, 1996.

Durney, B. R., On a Babcock-Leighton solar dynamo model with a deep-seated generating layer for the toroidal magnetic field. IV, *Astrophys. J., 486*, 1065, 1997.

Emonet, T., and F. Cattaneo, Small-scale photospheric fields: Observational evidence and numerical simulations, *Astrophys. J., 560*, L197-L200, 2001.

Fan, Y., G. H. Fisher, and E. E. DeLuca, The origin of morphological asymmetries in bipolar active regions, *Astrophys. J., 405*, 390-401, 1993.

Fan, Y., G. H. Fisher, and A. N. McClymont, Dynamics of emerging active region flux loops, *Astrophys. J., 436*, 907-928, 1994.

Ferriz-Mas, A., On the storage of magnetic flux tubes at the base of the solar convection zone, *Astrophys. J., 458*, 802, 1996.

Ferriz-Mas, A., and M. Schüssler, Instabilities of magnetic flux tubes in a stellar convection zone. I. Equatorial flux rings in differentially rotating stars, *Geophys. Astrophys. Fluid Dyn., 72*, 209-247, 1993.

Ferriz-Mas, A., and M. Schüssler, Waves and instabilities of a toroidal magnetic flux tube in a rotating star, *Astrophys. J., 433*, 852-866, 1994.

Ferriz-Mas, A., and M. Schüssler, Instabilities of magnetic flux tubes in a stellar convection zone. II. Flux rings outside the equatorial plane, *Geophys. Astrophys. Fluid Dyn., 81*, 233-265, 1995.

Ferriz-Mas, A., D. Schmitt, and M. Schüssler, A dynamo effect due to instability of magnetic flux tubes, *Astron. Astrophys., 289*, 949-956, 1994.

Fisher, G. H., A. N. McClymont, and D. Chou, The stretching of magnetic flux tubes in the convective overshoot region, *Astrophys. J., 374*, 766-772, 1991.

Fligge, M., S. K. Solanki, and J. Beer, Determination of solar cycle length variations using the continuous wavelet transform, *Astron. Astrophys., 346*, 313-321, 1999.

Fligge, M., S. K. Solanki, and Y. C. Unruh, Modelling irradiance variations from the surface distribution of the solar magnetic field, *Astron. Astrophys., 353*, 380-388, 2000.

Friis-Christensen, E., and K. Lassen, Length of the solar cycle: An indicator of solar activity closely associated with climate, *Science, 254*, 698-700, 1991.

Galloway, D. J., and N. O. Weiss, Convection and magnetic fields in stars, *Astrophys. J., 243*, 945-953, 1981.

Giles, P. M., T. L. Duvall, P. H. Scherrer, and R. S. Bogart, A flow of material from the suns equator to its poles, *Nature, 390*, 52, 1997.

Grossmann-Doerth, U., M. Schüssler, and S. K. Solanki, Unshifted, asymmetric Stokes V-profiles - Possible solution of a riddle, *Astron. Astrophys., 206*, L37-L39, 1988.

Grossmann-Doerth, U., M. Knölker, M. Schüssler, and S. K. Solanki, The deep layers of solar magnetic elements, *Astron. Astrophys., 285*, 648-654, 1994.

Grossmann-Doerth, U., M. Schüssler, and O. Steiner, Convective intensification of solar surface magnetic fields: results of numerical experiments, *Astron. Astrophys., 337*, 928-939, 1998.

Hagenaar, H. J., Ephemeral regions on a sequence of full-disk Michelson Doppler Imager magnetograms, *Astrophys. J., 555*, 448-461, 2001.

Harvey, K. L., The cyclic behavior of solar activity, in *The Solar Cycle*, edited by K. L. Harvey, p. 335, Astronomical Society of the Pacific, ASP Conf. Series Vol. 27, San Francisco, 1992.

Harvey, K. L., Magnetic bipoles on the sun, Ph.D. thesis, University of Utrecht, The Netherlands, 1993.

Harvey, K. L., The solar magnetic cycle, in *Solar Surface Magnetism*, edited by R. J. Rutten and C. J. Schrijver, p. 347, Kluwer, Dordrecht, 1994.

Harvey, K. L., and S. F. Martin, Ephemeral active regions, *Solar Phys., 32*, 389, 1973.

Hoyng, P., Turbulent transport of magnetic fields - Part Two - The role of fluctuations in kinematic theory, *Astron. Astrophys., 171*, 357, 1987.

Hoyng, P., Turbulent transport of magnetic fields. III - Stochastic excitation of global magnetic modes, *Astrophys. J., 332*, 857-871, 1988.

Hoyng, P., Helicity fluctuations in mean field theory: an explanation for the variability of the solar cycle?, *Astron. Astrophys., 272*, 321, 1993.

Hoyng, P., Is the solar cycle timed by a clock?, *Solar Phys., 169*, 253-264, 1996.

Hoyt, D. V., and K. H. Schatten, Group sunspot numbers: A new solar activity reconstruction, *Solar Phys., 179*, 189-219, 1998.

Hurlburt, N. E., and J. Toomre, Magnetic fields interacting with nonlinear compressible convection, *Astrophys. J., 327*, 920-932, 1988.

Hurlburt, N. E., M. R. E. Proctor, N. O. Weiss, and D. P. Brownjohn, Nonlinear compressible magnetoconvection. I - Travelling waves and oscillations, *J. Fluid Mech., 207*, 587-628, 1989.

Köhler, H., The solar dynamo and estimate of the magnetic diffusivity and the α-effect, *Astron. Astrophys., 25*, 467, 1973.

Küker, M., R. Arlt, and G. Rüdiger, The Maunder minimum as due to magnetic lambda-quenching, *Astron. Astrophys., 343*, 977-982, 1999.

Küker, M., G. Rüdiger, and M. Schultz, Circulation-dominated solar shell dynamo models with positive alpha-effect, *Astron. Astrophys., 374*, 301-308, 2001.

Knölker, M., U. Grossmann-Doerth, M. Schüssler, and E. Weisshaar, Some developments in the theory of magnetic flux

concentrations in the solar atmosphere, *Adv. Space Res., 11*, 285-295, 1991.

Krause, F., and K. H. Rädler, *Mean-field magnetohydrodynamics and dynamo theory*, Oxford: Pergamon Press, 1980.

Kuhn, J. R., H. Lin, and R. Coulter, What can irradiance measurements tell us about the solar magnetic cycle?, *Adv. Space Res., 24*, 185-194, 1999.

Legrand, J. P., and P. A. Simon, Ten cycles of solar and geomagnetic activity, *Solar Phys., 70*, 173-195, 1981.

Legrand, J. P., and P. A. Simon, A two-component solar cycle, *Solar Phys., 131*, 187-209, 1991.

Leighton, R. B., A magneto-kinematic model of the solar cycle, *Astrophys. J., 156*, 1, 1969.

Libbrecht, K. G., Solar p-mode frequency splittings, in *Seismology of the Sun and Sun-Like Stars*, pp. 131-136, 1988.

Lin, H., On the distribution of the solar magnetic fields, *Astrophys. J., 446*, 421, 1995.

Lockwood, M., R. Stamper, and M. N. Wild, A doubling of the sun's coronal magnetic field during the past 100 years, *Nature, 399*, 437-439, 1999.

Markiel, J. A., and J. H. Thomas, Solar interface dynamo models with a realistic rotation profile, *Astrophys. J., 523*, 827-837, 1999.

Marsh, N., and H. Svensmark, Long term, indirect indices of solar variability, in *Solar Variability and Climate*, edited by E. Friis-Christensen, C. Fröhlich, J. Haigh, M. Schüssler, and R. von Steiger, p. 215, Kluwer Academic Publishers, Dordrecht, 2000.

Moffatt, H. K., *Magnetic field generation in electrically conducting fluids*, Cambridge University Press, 1978.

Montesinos, B., and C. Jordan, On magnetic fields stellar coronae and dynamo action in late type dwarfs, *Mon. Not. R. Astron. Soc., 264*, 900, 1993.

Moreno-Insertis, F., The motion of magnetic flux tubes in the convection zone and the subsurface origin of active regions, in *NATO ASIC Proc. 375: Sunspots. Theory and Observations*, pp. 385-410, 1992.

Moreno-Insertis, F., M. Schüssler, and A. Ferriz-Mas, Storage of magnetic flux tubes in a convective overshoot region, *Astron. Astrophys., 264*, 686-700, 1992.

Moss, D., and J. Brooke, Towards a model for the solar dynamo, *Mon. Not. R. Astron. Soc., 315*, 521-533, 2000.

Nandy, D., and A. R. Choudhuri, Toward a mean field formulation of the Babcock-Leighton type solar dynamo. I. α-coefficient versus Durney's double-ring approach, *Astrophys. J., 551*, 576-585, 2001.

Nordlund, A., Numerical 3-D simulations of the collapse of photospheric flux tubes, in *IAU Symp. 102: Solar and Stellar Magnetic Fields: Origins and Coronal Effects*, edited by J. O. Stenflo, vol. 102, pp. 79-83, 1983.

Nordlund, A., and R. F. Stein, Solar magnetoconvection, in *IAU Symposium*, vol. 138, pp. 191+, 1990.

Ossendrijver, A. J. H., P. Hoyng, and D. Schmitt, Stochastic excitation and memory of the solar dynamo, *Astron. Astrophys., 313*, 938-948, 1996.

Parker, E. N., Hydromagnetic dynamo models, *Astrophys. J., 122*, 293, 1955.

Parker, E. N., The generation of magnetic fields in astrophysical bodies. I - The dynamo equations, *Astrophys. J., 162*, 665, 1970.

Parker, E. N., The generation of magnetic fields in astrophysical bodies. X - Magnetic buoyancy and the solar dynamo, *Astrophys. J., 198*, 205-209, 1975.

Parker, E. N., Hydraulic concentration of magnetic fields in the solar photosphere. VI - Adiabatic cooling and concentration in downdrafts, *Astrophys. J., 221*, 368-377, 1978.

Parker, E. N., A solar dynamo surface wave at the interface between convection and nonuniform rotation, *Astrophys. J., 408*, 707-719, 1993.

Petrovay, K., What makes the Sun tick? The origin of the solar cycle, in *The Solar Cycle and Terrestrial Climate, ESA SP-463*, pp. 3-14, 2001.

Petrovay, K., and G. Szakaly, The origin of intranetwork fields: a small-scale solar dynamo, *Astron. Astrophys., 274*, 543, 1993.

Prautzsch, T., The dynamo mechanism in the deep convection zone of the Sun, in *Solar and Planetary Dynamos*, pp. 249-256, 1993.

Prautzsch, T., Zum Entstehungsort solarer Magnetfelder, *PhD Thesis, Universität Göttingen*, 1997.

Proctor, M. R. E., and N. O. Weiss, Magnetoconvection, *Rep. Progr. Phys., 45*, 1317-1379, 1982.

Rädler, K., The solar dynamo, in *ASSZ Vol. 159, IAU Colloq. 121, Inside the Sun*, p. 385, 1990.

Ribes, J. C., and E. Nesme-Ribes, The solar sunspot cycle in the Maunder minimum AD 1645 to AD 1715, *Astron. Astrophys., 276*, 549, 1993.

Rüdiger, G., and A. Brandenburg, A solar dynamo in the overshoot layer: cycle period and butterfly diagram, *Astron. Astrophys., 296*, 557, 1995.

Schlichenmaier, R., and M. Stix, The phase of the radial mean field in the solar dynamo, *Astron. Astrophys., 302, 264*, 1995.

Schmitt, D., Dynamo action of magnetostrophic waves, in *The Hydromagnetics of the Sun*, pp. 223-224, 1984.

Schmitt, D., Dynamowirkung magnetostrophischer Wellen, *PhD Thesis, Universität Göttingen*, 1985.

Schmitt, D., An alpha-omega-dynamo with an alpha-effect due to magnetostrophic waves, *Astron. Astrophys., 174*, 281-287, 1987.

Schmitt, D., The solar dynamo, in *IAU Symp. 157: The Cosmic Dynamo*, p. 1, 1993.

Schmitt, D., Dynamo action of magnetostrophics waves, in *Advances in Nonlinear Dynamos*, 2003.

Schmitt, D., M. Schüssler, and A. Ferriz-Mas, Intermittent solar activity by an on-off dynamo., *Astron. Astrophys., 311*, L1, 1996.

Schmitt, D., M. Schüssler, and A. Ferriz-Mas, Variability of solar and stellar activity by two interacting hydromagnetic dynamos, in *ASP Conf. Ser. 154: Cool Stars, Stellar Systems, and the Sun*, vol. 10, p. 1324, 1998.

Schou, J., and R. S. Bogart, Flow and horizontal displacements from ring diagrams, *Astrophys. J., 504*, L131, 1998.

Schou, J., J. Christensen-Dalgaard, and M. J. Thompson, The resolving power of current helioseismic inversions for the sun's internal rotation, *Astrophys. J., 385*, L59-L62, 1992.

Schou, J., et al., Helioseismic studies of differential rotation in the solar envelope by the solar oscillations investigation using the Michelson Doppler Imager, *Astrophys. J., 505*, 390-417, 1998.

Schrijver, C. J., Simulations of the photospheric magnetic activity and outer atmospheric radiative losses of cool stars based on characteristics of the solar magnetic field, *Astrophys. J., 547*, 475-490, 2001.

Schrijver, C. J., A. M. Title, H. J. Hagenaar, and R. A. Shine, Modeling the distribution of magnetic fluxes in field concentrations in a solar active region, *Solar Phys., 175*, 329-340, 1997a.

Schrijver, C. J., A. M. Title, A. A. van Ballegooijen, H. J. Hagenaar, and R. A. Shine, Sustaining the quiet photospheric network: The balance of flux emergence, fragmentation, merging, and cancellation, *Astrophys. J., 487*, 424-436, 1997b.

Schüssler, M., Magnetic buoyancy revisited—Analytical and numerical results for rising flux tubes, *Astron. Astrophys., 71*, 79-91, 1979.

Schüssler, M., On the structure of magnetic fields in the solar convection zone, in *The Hydromagnetics of the Sun*, pp. 67-76, 1984.

Schüssler, M., Magnetic fields and the rotation of the solar convection zone, in *ASSL Vol. 137: The Internal Solar Angular Velocity*, pp. 303-320, 1987.

Schüssler, M., Numerical simulation of solar magneto-convection, in *ASP Conf. Ser. Vol. 236: Advanced Solar Polarimetry— Theory, Observation, and Instrumentation*, edited by M. Sigwarth, p. 343, Astronomical Society of the Pacific, San Francisco, 2001.

Schüssler, M., P. Caligari, A. Ferriz-Mas, and F. Moreno-Insertis, Instability and eruption of magnetic flux tubes in the solar convection zone, *Astron. Astrophys., 281*, L69, 1994.

Schüssler, M., and M. Knölker, Magneto-convection, in *ASP Conf. Ser. 248: Magnetic Fields Across the Hertzsprung-Russell Diagram*, p. 115, 2001.

Schüssler, M., D. Schmitt, and A. Ferriz-Mas, Long-term variation of solar activity by a dynamo based on magnetic flux tubes, in *ASP Conf. Ser. 118: 1st Advances in Solar Physics Euroconference. Advances in Physics of Sunspots*, p. 39, 1997.

Sheeley, N. R., The flux-transport model and its implications, in *The Solar Cycle*, edited by K. L. Harvey, pp. 1-13, Astronomical Society of the Pacific, ASP Conf. Series Vol. 27, San Francisco, 1992.

Solanki, S. K., Small scale solar magnetic fields—an overview, *Space Sci. Rev., 63*, 1-188, 1993.

Solanki, S. K., M. Schüssler, and M. Fligge, Evolution of the Sun's large-scale magnetic field since the Maunder minimum, *Nature, 408*, 445-447, 2000.

Solanki, S. K., M. Schüssler, and M. Fligge, Secular variation of the Sun's magnetic flux, *Astron. Astrophys., 383*, 706-712, 2002.

Spiegel, E. A., and N. O. Weiss, Magnetic activity and variations in solar luminosity, *Nature, 287*, 616, 1980.

Spruit, H. C., Pressure equilibrium and energy balance of small photospheric fluxtubes, *Solar Phys., 50*, 269-295, 1976.

Spruit, H., M. Schüssler, and S. Solanki, Filigree and flux tube physics, in *Solar Interior and Atmosphere*, edited by A. Cox, W.

Livingston, and M. Matthews, p. 890, The University of Arizona Press, Tucson, 1991.

Spruit, H. C., and A. A. van Ballegooijen, Stability of toroidal flux tubes in stars, *Astron. Astrophys., 106*, 58-66, 1982.

Spruit, H. C., and E. G. Zweibel, Convective instability of thin flux tubes, *Solar Phys., 62*, 15-22, 1979.

Steenbeck, M., and F. Krause, On the dynamo theory of stellar and planetary magnetic fields. I. AC dynamos of solar type, *Astronomische Nachrichten, 291*, 49-84, 1969.

Steenbeck, M., F. Krause, and K. H. Rädler, A calculation of the mean electromotive force in an electrically conducting fluid in turbulent motion under the influence of Coriolis forces, *Zeitschrift Naturforschung Teil A, 21*, 369-376, 1966.

Steiner, O., U. Grossmann-Doerth, M. Knölker, and M. Schüssler, Dynamical interaction of solar magnetic elements and granular convection: Results of a numerical simulation, *Astrophys. J., 495*, 468, 1998.

Stenflo, J. O., Magnetic-field structure of the photospheric network, *Solar Phys., 32*, 41, 1973.

Stix, M., Differential rotation and the solar dynamo, *Astron. Astrophys., 47*, 243-254, 1976.

Stix, M., Theory of the solar cycle, *Solar Phys., 74*, 79-101, 1981.

Tobias, S. M., Grand minima in nonlinear dynamos, *Astron. Astrophys., 307*, L21, 1996.

Tobias, S. M., The solar cycle: parity interactions and amplitude modulation, *Astron. Astrophys., 322*, 1007-1017, 1997.

Tomczyk, S., J. Schou, and M. J. Thompson, Measurement of the rotation rate in the deep solar interior, *Astrophys. J., 448*, L57, 1995.

Topka, K. P., T. D. Tarbell, and A. M. Title, Properties of the smallest solar magnetic elements. II. Observations versus hot wall models of faculae, *Astrophys. J., 484*, 479, 1997.

Venkatakrishnan, P., Inhibition of convective collapse of solar magnetic flux tubes by radiative diffusion, *Nature, 322*, 156, 1986.

Wang, Y. M., and N. R. Sheeley, The rotation of photospheric magnetic fields: A random walk transport model, *Astrophys. J., 430*, 399-412, 1994.

Wang, Y. M., and N. R. Sheeley, Solar implications of Ulysses interplanetary field measurements, *Astrophys. J., 447*, L143-146, 1995.

Wang, Y.-M., J. Lean, and N. R. Sheeley, The long-term variation of the sun's open magnetic flux, *Geophys. Res. Lett., 27*, 505-508, 2000a.

Wang, Y.-M., N. R. Sheeley, and J. Lean, Understanding the evolution of the sun's open magnetic flux, *Geophys. Res. Lett.*, 27, 621-624, 2000b.

Weiss, N. O., D. P. Brownjohn, P. C. Matthews, and M. R. E. Proctor, Photospheric convection in strong magnetic fields, *Mon. Not. R. Astron. Soc., 283*, 1153-1164, 1996.

Wilson, O. C., Chromospheric variations in main-sequence stars, *Astrophys. J., 226*, 379-396, 1978.

Yoshimura, H., Solar-cycle dynamo wave propagation, *Astrophys. J., 201*, 740-748, 1975.

Zwaan, C., and K. L. Harvey, Patterns in the solar magnetic field, in *Solar Magnetic Fields*, edited by M. Schüssler and W. Schmidt, p. 27, Cambridge University Press, Cambridge, 1994.

Zwaan, C., J. J. Brants, and L. E. Cram, High-resolution spectroscopy of active regions. I—Observing procedures, *Solar Phys., 95*, 3-14, 1985.

M. Schüssler and D. Schmitt, Max-Planck-Institut für Sonnersystemforschung , Max-Planck-Str. 2, 37191 Katlenburg-Lindau, Germany. (msch@linmpi.mpg.de; schmitt@linmpi.mpg.de)

Global Magnetic-Field Reversal in the Corona

Boon Chye Low and Mei Zhang

High Altitude Observatory, National Center for Atmospheric Research, Boulder, Colorado

This paper addresses the nature of the hydromagnetic processes in the solar corona that bring about the reversal of its global magnetic field, in response to the emergence of new magnetic flux in an 11-year solar cycle. Magnetic reconnection, magnetic fields reconfigured from old and new fluxes, and coronal mass ejections give physical clues for a qualitative theory of this global field reversal phenomenon. We show that twisted magnetic fields may emerge and reorganize into flux ropes in the corona, with implications for the removal of magnetic flux threading across the photosphere, as well as the removal of magnetic flux systems and their helicity out of the corona into interplanetary space in order to make room for the newly emerged flux. This physical picture is constructed out of physical insights and results of idealized calculations from published works.

1. INTRODUCTION

Each 11-year solar cycle begins with the appearance of two sunspot belts on the solar photosphere, at about 30° in latitude on the two sides of the solar equator. As individual sunspots form and decay, they create the appearance of the two belts migrating to meet at the equator when finally no new sunspots appear. The Sun then settles into a period of activity minimum until the next solar cycle commences. Magnetograph observations show that the sunspots in each cycle are of a polarity opposite to that of the previous cycle. For example, this can be inferred from sunspot polarities governed by the Hale-Nicholson law [*Zirin*, 1988]. Full disk magnetograms show that as the sunspot belts drifts to the equator, larger-scale weaker magnetic fields of the new-cycle polarity migrate poleward in the opposite directions [*Harvey*, 1994]. Eventually, within about 2 years from sunspot maximum, the polarity of the global photospheric magnetic field reverses, first one solar pole reversing followed by the other [*Howard*, 1972, *Wilson and Giovannis*, 1994].

Solar Variability and its Effects on Climate
Geophysical Monograph 141
Copyright 2004 by the American Geophysical Union
10.1029/141GM05

Satellite magnetometers at 1 AU observe a corresponding solar-cycle reversal of the interplanetary magnetic field, showing that solar magnetic field reversal is not a strictly photospheric phenomenon but extends through the corona, and out into the heliosphere. It took many years of work to synthesize the global topology of the heliospheric magnetic field from space observations; see the review by *Smith* [2001]. The discovery of the heliospheric current sheet and the observed Parker spiral magnetic fields led to an interpretation of the heliospheric magnetic field that can be related self-consistently to the large-scale photospheric magnetic field [*Wilcox and Scherrer*, 1972, *Levy*, 1976, *Hundhausen*, 1977, *Jokipii and Thomas*, 1981, *Smith*, 1993].

Magnetic fields of astronomical scales cannot be resistively removed in any relevant time scale, certainly not on a time scale of 11 years. This paper addresses the physical question of how the corona and heliosphere rid themselves of the old flux from a previous solar cycle to be replaced with the new flux of the opposite polarity. We carry out below a theoretical analysis of the field-reversal phenomenon in order to identify the basic hydromagnetic processes in terms of which, at least qualitatively, the global field reversal may be understood. In our analysis, we will rely on insights from observations to guide its development and make use of some recent theoretical results [*Zhang and Low*, 2001, 2002, *Low and Zhang*, 2002; hereinafter referred to as *ZL1*, *ZL2*, and *LZ3*, respectively].

The physical picture we will construct suggests that there is an outward transport of magnetic flux and helicity through the corona into the heliosphere in the course of an 11-year solar cycle, with coronal mass ejections playing a significant role [*Low*, 1994, 1996, 1997, 2001].

2. THE HYDROMAGNETIC MIXING OF OLD AND NEW FLUXES

The corona at its million-degree temperature is almost a perfect electrical conductor over scales down to below the limit of telescopic spatial resolution. However, it never means that electrical resistivity is thus unimportant. In the coronal environment characterized with a large magnetic Reynolds number ($> 10^{15-18}$), formation of electric current sheets followed by magnetic reconnection is a ready means of current dissipation despite the high conductivity [*Parker*, 1994]. Under these conditions, magnetic flux bundles can be destroyed only under exceptionally rare circumstances, e.g., if they are aligned anti-parallel in pairs to be annihilated along their *entire* lengths. Under natural circumstances, resistive dissipation by magnetic reconnection takes places in localized regions where two non-parallel flux bundles come into contact. There is no annihilation of entire flux bundles since, outside of the reconnection regions, the plasma is essentially perfectly conducting. In the fluid dominated photosphere and convection zone below, flux may be removed by churning a large scale field into small scale structures for eventual dissipation. This does not happen in the corona where the magnetic field dynamically dominates over the tenuous plasma.

Under conditions of high magnetic Reynolds numbers, the changes of magnetic topology produced by reconnection have consequences for the dynamics of the corona as important as the intensive, localized resistive heating they produce. Changes in magnetic topology enable a magnetic field to seek progressively lower energy states characterized with a simpler field topology. The magnetic energy is released by the Lorentz force doing work the plasma, with certain modes of motion which are otherwise forbidden if magnetic topology absolutely cannot change.

The high electrical conductivity of the solar corona does mean that when a new magnetic flux emerges into the corona, it does not immediately mix with the old flux to reach a potential state. The equilibration between new and old fluxes proceeds through much dynamics and magnetic reconnection and this is the underlying cause of solar activity. The coronal magnetic field is largely anchored to the photosphere. Coronal dynamics proceeds at typically its Alfvenic speeds of the order of 10^3 $km\ s^{-1}$ whereas the photosphere moves slowly at about 0.5 $km\ s^{-1}$. When fresh magnetic flux first emerges at the photosphere, there may be higher speeds at this level, a factor of 2 or 3 [*Schrijver and Zwaan*, 2000]. Hydromagnetic changes in the corona therefore can take place on time scales during which the photosphere hardly moves. Moreover, the astronomical scales also imply that to a first approximation the photosphere is a perfect conductor [*Parker*, 1979]. This means that the identities of old and new fluxes as defined by the photospheric plasma they thread, do survive after magnetic field topological reconnection has taken place in the corona. We shall make these ideas more specific in the rest of this paper, with an analysis of what the relevant hydromagnetic processes and effects are, based on simple constructions. To that end, let us first examine the classical potential-field model to describe the evolution of coronal magnetic fields [*Altschuler and Newkirk*, 1969, *Luhmann et al.*, 1998], and then build up the physics not included in this model.

2.1. A Potential-Field Model for Coronal Field Reversal

Consider the topological evolution of a potential magnetic field outside a theoretical star represented by a unit sphere, stipulated to be changing instantaneously to fit a prescribed changing normal field distribution on the unit sphere. Let us simplify the exercise by taking the system to be axisymmetric and symmetric about the stellar equator. Figure 1 shows an example taken from *ZL1*.

We assume that the stellar surface is a perfect electrical conductor. The left panels in Figure 1 show an evolution of the surface magnetic flux to represent new magnetic flux having emerged at the equator, displacing poleward the two parts of the stellar surface threaded by the old flux. It suits the purpose of illustration in this paper to take the new flux to emerge at the equator which is a simplification from the realistic case of solar flux emergence in two spatially separate sunspot belts. The right panels show the potential fields in the stellar atmosphere as defined by their respective normal field distributions on the stellar surface. The magnetic field in the atmosphere is taken potential by assumption, effectively taking the atmosphere to be a highly resistive medium. We see a quadrupolar stellar magnetic field with three neutral lines in a progressive change as more and more new magnetic flux emerges through the equatorial region.

In this evolution, the field topology in the atmosphere changes with the ratio F_e/F_o of the emerged flux to a fixed amount of preexisting flux. At the stellar surface, we have

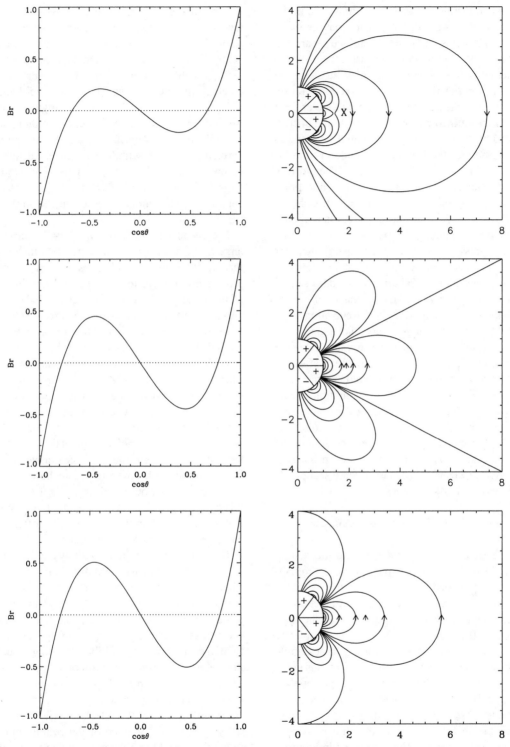

Figure 1. The surface normal fields (left panels) as functions of cos θ and normalized to the magnetic strength at the poles, and their respective matching potential magnetic fields in the $r - \theta$ plane (right panels) in a theoretical, axisymmetric stellar atmosphere with the value of the ratio F_e/F_o of the emerged flux to the preexisting flux varying from 0.67 (top panels), through 2.25 (middle panels) to 2.78 (bottom panels).

assumed perfect conductivity so that the neutral lines in the mid-latitudes on the two sides of the equator may be interpreted to be the boundaries on the stellar surface separating old and new fluxes. As the flux ratio increases, the instantaneous reconnection between the old and new fluxes in the evolving potential field at first produces a reconfiguration of the field confined to the near region. The far dipolar field maintains a polarity defined by the old flux. At the critical stage represented by $F_e/F_o = 2.25$, the potential field due to the emerged flux has for the first time extended to the far coronal region within some limits in latitudes. The old flux still extends to the far corona but is now confined to polar latitudes. Upon crossing this threshold in flux ratio, shown for the case $F_e/F_o = 2.78$, the entire far coronal region becomes occupied by the emerged flux. The old flux is now displaced and compressed inward to occupy two finite polar regions in the atmosphere centered over the stellar poles.

There are two important physical effects not addressed by the potential model in Figure 1. The corona is of course highly conducting rather than resistive. Relaxation to a potential state by reconnection may not take place right away because magnetic fields may be arrested in some intermediate metastable states [ZL2]. The other effect is related to the fact that the identities of the old and new flux are retained at the stellar surface under the condition of perfect electrical conductivity despite magnetic reconnection between the two fluxes in the atmosphere. The old flux pushed to the polar regions need to be removed so that the new flux will dominate everywhere. We take up the first effect in subsection [2.2] and the second effect in section [3] where we treat coronal mass ejections.

2.2. Multipolar Potential Fields with Current Sheets

The metastable states of newly emerged and preexisting fields result from extremely complex plasma and hydromagnetic physics. ZL1 provided an insight into this complex process by considering a particular set of idealized metastable states associated with the potential sequences of the type shown in Figure 1. If magnetic reconnection is completely suppressed in a perfectly conducting corona with negligible pressure and density, the emerged and preexisting fields in the atmosphere may achieve equilibrium by being potential in two spatially distinct regions separated by a current sheet. The total magnetic pressure is continuous across the current sheet in each case so that the current sheet as a macroscopic structure is in force equilibrium. The left panels in Figure 2 show such metastable states satisfying the respective normal field distributions given in Figure 1. The solar wind is completely neglected in this idealized treatment.

Each of these current-sheet fields is at a higher energy level than its corresponding potential state with free energy to be liberated via reconnection. In those cases where the emerged flux is relatively small, reconnection involves a local re-configuration of the field. For the case of the emerged flux being larger than a threshold above the fixed old flux, the $F_e/F_o > 2.25$ case in Figures 1 and 2, reconnection produces a global reconfiguration, involving the emerged flux pushing its way out to occupy the entire far coronal space, and withdrawal of the old flux from the far coronal space to localized regions above the poles. Such large-scale magnetic reconfigurations are suggestive of coronal mass ejections to be discussed in the next section [ZL1].

The reconnection taking an initial metastable state of large F_e/F_o to a potential state involves the novel effect of a reconnection layer which does not stay in one place but propagates outward to reach the old flux in the far reaches of the atmosphere. There is a global balance of magnetic pressure on the two sides of the current sheet in the initial state before reconnection. After reconnection has started, the old flux is reconnected out of the way for the new flux to push its way out into the far reaches of the atmosphere [Antiochos et al., 1999, ZL1, LZ3]. In its dynamical development, the momentum and energy generated may result in the forced opening of a part of the multipolar magnetic field created by reconnection. Opening is taken in the sense that one or more of the bipolar fields making up the global field have the tops of their lines of force dragged to infinity, leaving behind lines of force one end anchored at the stellar surface and the other out to infinity [Aly, 1991]. To understand this hydromagnetic behavior quantitatively requires numerical simulation. For the present, ZL1 offers some insight based on an energy consideration.

The right panels in Figure 2 show equilibrium states compatible with the normal field distributions of Figure 1 identified by the flux-ratio parameter, but with one or more of the bipolar magnetic fields, making up the global field, forced open. Each opened bipolar field carries a current sheet to separate the open fields of opposite polarities. The field in each case is in equilibrium, being potential such that the magnetic pressure is continuous across the current sheet. These partially open fields, being in equilibrium, represent the minimum energy state a field must possess if it has the same boundary flux at the stellar surface and the same partially open topology. Each of the equilibrium states on the right in Figure 2 has a lower magnetic energy than its counterpart on the left in Figure 2, showing that those metastable states in the left panel may energetically transit by partially opening up a part of their reconnected field to attain the

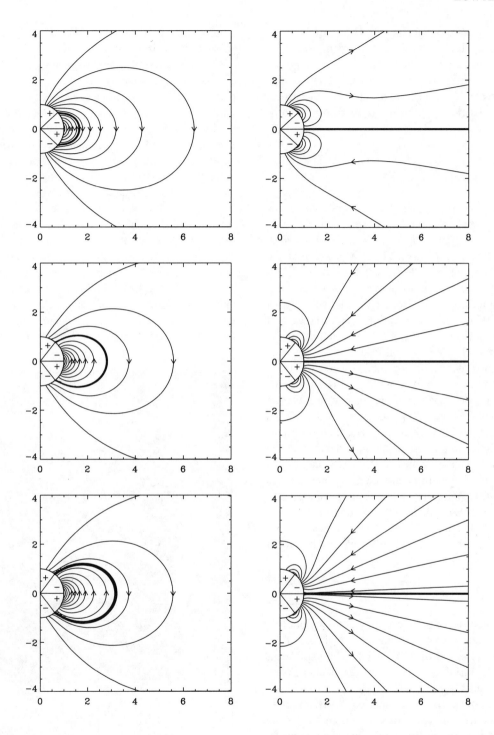

Figure 2. Two sequences of metastable equilibrium states matching the surface normal field distributions in Figures 1. The sequence on the left panels are equilibrium states in which old and new fluxes are suppressed from magnetic reconnection and are kept spatially apart by an electric current sheet. The sequence on the right panels are equilibrium states in which some reconnection between old and new fluxes has occured but a part of the multipolar magnetic field has opended out to infinity. Current sheets across which tangential magnetic fields reverse abruptly are shown as thick curves.

equilibrium states on the right panel. The partially-open states are also possible metastable states from which by magnetic reconnection a transition to the respective potential states in Figure 1 may take place. The relationships between the two classes of metastable states and their common potential end states are relevant to the theory of coronal mass ejections.

3. CORONAL MASS EJECTIONS AND FLUX EMERGENCE

Coronal mass ejections (CMEs) are large-scale expulsions of coronal magnetic structures [*MacQueen*, 1980, *Howard et al.*, 1985, 1997, *Hundhausen*, 1999, *Forbes*, 2000, *Gosling*, 2000, *Low*, 1994, 2001]. Occurring at the rate of 1 to 3 events a day, the typical mass of 10^{15-16} g expelled in a CME contributes to an average mass loss well less than 10% of the mass loss through the solar wind [*Webb and Howard*, 1994]. This observational fact shows that CMEs are not significant as a mass-loss mechanism for the corona, although each CME is significant both as a perturbation in the solar wind and for its influence on space weather [*Thompson et al.*, 1998]. What then is the significance of CMEs for the behavior of the solar corona over an 11-year solar cycle? *Low* [1997] had proposed that CMEs are the basic mechanism by which magnetic flux of a previous cycle is taken out of the corona into interplanetary space to make way for the opposite flux of the new cycle. Let us examine the basis of this proposed idea as the next step in our development. We first discuss the phenomenology of a common kind of CMEs, followed by a theoretical interpretation of the hydromagnetic nature of CMEs, leading to a glimpse of how the corona physically reverses its global magnetic field.

3.1. Three-Part CMEs and Coronal Helmet Streamers

Little change is observed in the slowly evolving photospheric magnetic fields as a CME above it travels out of the corona at a typical speed of 450 $km\ s^{-1}$. The significant magnetic change during a CME expulsion occurs in the corona. As illustrated in Figures 1 and 2, to say something about the circumstance of an impending CME requires knowing the state of the coronal magnetic field. Unfortunately the magnetic field in the corona cannot yet be measured with any useful spatial resolution [*Judge*, 1998, *Lin et al.*, 2000]. This means that we have to rely on inference of coronal magnetic field structures from observations of the morphology of the plasma embedding the magnetic field. Fortunately, this approach has produced

a physical picture of CMEs which seems certain [*Hundhausen*, 1999].

Figure 3 shows a commonly observed kind of CME characterized with a three-part structure: a leading shell of dense plasma enclosing a low-density cavity with a dense core within [*Illing and Hundhausen*, 1986, *Burkepile and St. Cyr*, 1993, *Chen*, 1996, *Wu et al.*, 1997, *Hundhausen*, 1999, *Dere et al.*, 1999, *St. Cyr et al.*, 2000, *Gibson and Low*, 2000]. Many such CMEs originate from the disruption of a coronal helmet-streamer, which may be inferred to be a closed magnetic structure surrounded by open magnetic fields along which flows the more or less steady solar wind. A CME opens up the closed field of the coronal helmet, its three-part structure attributed to the dense helmet plasma forming the CME shell, and the prominence and its cavity often seen at the base of the helmet forming the dense core and cavity of the CME. Following the CME expulsion, the open coronal field reconnects to reform a coronal helmet, with reconnection-heating producing the often observed CME-associated flare. The reformed helmet is without a

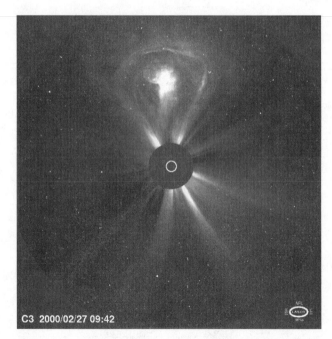

Figure 3. A three-part coronal mass ejection observed by the ESA/NASA Large Angle and Spectrometric Coronagraph Experiment (LASCO) C3 White-light Coronagraph showing its bright leading front containing a mass of about 10^{15-16} g. The field of view extends from the edge of the occulting disk to 32 solar radii; the circle drawn on the occulting disk represents the occulted Sun. The front is a three-dimensional shell of mass draped around a low-density cavity within which a bright core is plasma from an erupted prominence.

cavity or prominence [*Hiei, Hundhausen and Sime*, 1993]. CMEs coming from low down in the strong fields of an active region probably involve a similar process; three-part CMEs also originate from active regions.

3.2. Magnetic Flux Ropes and Magnetic Flux Emergence

This observational picture of three-part CMEs is compatible with the following theoretical explanation of the origin of the three-part helmet streamer [*Low*, 1996, 2001].

Observations suggest that magnetic fields emerge at the photosphere in a significantly twisted state [*Ishii et al.*, 1998, *Kurokawa*, 1987, *Leka et al.*, 1996, *Lites et al.*, 1995, *Fan et al.*, 1999]. Consider the emergence of a twisted magnetic field in the form of a flux rope through the photosphere into the corona depicted in Figure 4. Sketched is a single elemental strand of magnetic flux. Other such flux elements make the rope and they are not sketched in order to keep the figure simple. The intertwining of such flux elements prevents, under the condition of high electrical conductivity, any one of them from straightening out. In Figure 4a, we see bipolar magnetic arches above the photosphere associated with a certain amount of magnetic flux of each sign threading across the photosphere into the atmosphere above. Figure 4b shows the lifting and sinking of gravitational lowest points of the helical flux element as the result of siphon flows and magnetic buoyancy. The lifting of locally U-shape parts of the flux results in the transport of full magnetic twists into the atmosphere and a simultaneous elimination of fluxes of each sign at the photosphere. All the flux elements making up the flux rope are subject to this effect, which implies the transport of a main part of the flux rope into the atmosphere above the photosphere.

Another way of describing this process is to point out that Figure 4a represents a horizontally oriented magnetic flux rope with its lower horizontal part submerged below the photosphere, whereas Figure 4b represents a significant part of that flux rope having risen and arched above the photosphere. The flux rope that has arched into the atmosphere is distinguished from the surrounding bipolar field (not sketched) by its field lines winding once or more times around some common line, the axis of the flux rope. Going from Figure 4a to 4b, there is a removal of equal amounts of magnetic flux across the photosphere. As each U-shaped part of a flux element lifts off, magnetograph observations at the photospheric level would see the flux element's intersections with the photosphere come together to mutually annihilate [*Martin et al.*, 1994].

The scenario in Figure 4 requires a strong field to dominate over the plasma in the photosphere and just below in

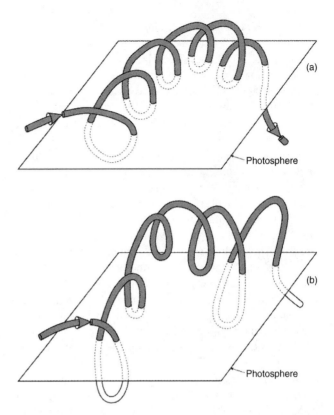

Figure 4. A sketch of the emergence of a helical flux element through the photoshpere treated as a plane. In (a) the flux element threads the photosphere many times to produce inverted-U flux arches each with a pair of footpoints on the photosphere. In (b) siphon flows and magnetic buoyance have resulted in some U-shaped parts of the flux element rising above the photosphere, each rise associated with the elimination of a pair of magnetic footpoints and the cancellation of equal amounts of opposite flux threading across the photosphere. Elsewhere heavy mass accumulated at the bottom of U-shaped parts of the flux resulted in sinking of these parts to deeper layers below the photosphere.

order to maintain its coherence as a flux rope [*Lites et al.*, 1995]. A flux rope of weaker fields would produce the same result in a turbulent fashion. The rising field would be churned by the photosphere into a highly tangled state with the lifting of locally U-shaped field lines taking place on much smaller scales than that considered in Figure 4. Under the condition of high magnetic Reynolds numbers, the twist in the original flux rope is conserved as the net helicity of the tangled magnetic field, even with magnetic reconnection taking place in the photosphere and in the corona [*Berger*, 1984, 1999, *Low*, 1999]. In the photosphere, reconnection is driven by the high-β plasma with two principal effects: draining of heavy plasma from one part of the field to a lower part, and propagation of twist from one part of the field to a higher part [e.g., *Parker*, 1967, *Fan*, 2001, *Manchester*, 2001]. Reconnection in the corona is magnetically driven. The tangled fields dominate over the tenuous coronal plasma to press into each other, producing thin electric current sheets which then resistively dissipate. This may take place in an ongoing manner to heat the quiescent corona non-explosively [*Parker*, 1994]. It could also proceed explosively as flares [*Low and Wolfson*, 1988]. In the end, the large scale field gets into the corona with a complex topology which subsequently simplifies in the transition to a minimum-energy flux-rope state containing the conserved total helicity. Such a flux rope explains the prominence cavity and its manifestation as the filament channel in the chromosphere [*Priest et al.*, 1989, *Low*, 1992, *Low and Hundhausen*, 1995, *Harvey and Gaizauskas*, 1998, *Martin et al.*, 1994, *Amari et al.*, 1999]. Another way to say this is that the flux rope is the natural end state of a twisted magnetic field after it has been bodily transported into the corona, unable to relax to a potential state on account of its conserved helicity [*Low*, 1994].

The important point of Figure 4 and the turbulent version of it is that the formation of magnetic flux rope by its rise into the corona from below *removes* equal amounts of flux of both signs from the photosphere on the two sides of a neutral line. It is a slow process driven relentlessly by magnetic buoyancy acting on a flux-rope magnetic field.

Figure 5 is a sketch in axisymmetric geometry of just such a process over a mid-latitude neutral line on the surface of a theoretical star. For the purpose of making a physical point, we assume that the axisymmetric corona is symmetric about the equator, so that each sketch shows only the northern hemisphere. We consider the case where sufficient new flux has emerged in the equator to correspond to the case of $F_e/F_o = 2.25$ in Figure 1 so that the emerged field dominates in the far region at low latitudes. In contrast to Figure 1, we introduce the solar wind which keeps the magnetic field above about $2R_\odot$ fully open in Figure 5.

In each sub-figure of Figure 5, the corona is depicted to have two helmet streamers in the form of closed bipolar magnetic fields, one over the equatorial neutral line and another over the mid-latitude neutral line. The lines of force are drawn and numbered to represent equal amount of flux between adjacent axisymmetric flux surfaces. In the open lines at the outer boundary of each sketch, we have taken care to sketch them in this region to reflect the distribution of a monopolar potential field. This allows us to render some realism to the field topology associated with the manner a helmet-streamer in the mid-latitude would "bend" towards the solar equator. Suppose, to keep the physics simple, we take the equatorial helmet streamer, representing

newly emerged flux, to be largely unchanged during the evolution described by Figure 5. This evolution is "driven" by the emergence of a magnetic flux rope within the mid-latitudinal helmet streamer which we henceforth refer to as the active helmet.

Figure 5a represents the initial state where the field under the active helmet has a strong azimuthal component. We regard this bipolar field to be the upper part of a flux rope half submerged below the stellar surface. In the assumed axisymmetric geometry, this flux rope runs a closed azimuthal path around the stellar axis. Figure 5b shows the bodily rise of that flux rope into the stellar atmosphere producing closed circulations of poloidal fields about the rope axis that now lay above the stellar surface. Combined with the azimuthal field component, these poloidal fields describe helical field lines winding around the rope axis.

As we have pointed out in connection with Figure 4, the emergence of the flux rope into the corona may not be the simple bodily lifting of a rope into the corona as drawn, but may proceed as the turbulent emergence of a tangled magnetic field that upon reconnection reorganizes itself to form the flux rope.

If we use the number of closed lines of force above the stellar surface in Figure 5b to measure flux, 3 flux units of each sign (numbers 20–22) have been eliminated from the stellar surface. This flux rope forms a low-density cavity within which a prominence may form as sketched [*Low and Hundhausen*, 1995]. Such a helmet streamer by virtue of the free energy associated with the flux rope is capable of breaking up into a CME taking the flux rope and the high-density plasma in the helmet dome out into interplanetary space [*Low and Smith*, 1993].

Figure 5c shows the aftermath of such an event: the CME has left the corona with the mass of the helmet, the cavity flux-rope, and the erupted prominence in the cavity, leaving behind the part of the helmet magnetic field anchored to the stellar surface but combed into an open configuration containing a field reversal layer represented by the dashed line. If we count the field lines open to the outer boundary of Figure 5c, we see that 3 flux units of each sign (numbers 17–19) have joined the open field where the CME has left and taken 3 flux units of each sign (20–22) away.

The development from Figure 5a to 5b is slow, days or weeks, at the photospheric time scale of flux emergence. In the turbulent emergence and formation of the flux rope, small scale fields, well less than 10^5 km, are leaving the photosphere to reconnect and reorganize in the corona to eventually form the flux rope. The flux threading across the photosphere of a given sign is reduced progressively in the topological evolution depicted in Figure 4.

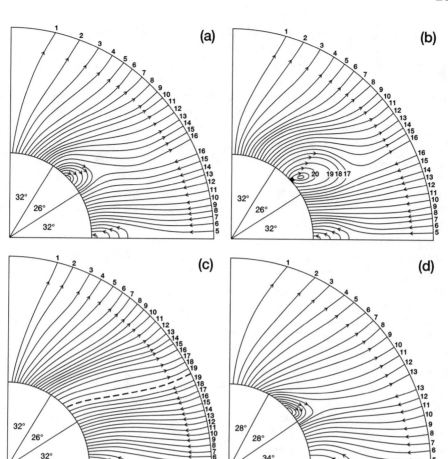

Figure 5. The evolutionary sequence from the formation of a magnetic flux rope under a coronal helmet streamer to a coronal mass ejection and its aftermath. (a) The initial state with a bipolar sheared field under an active helmet at mid-latitude. (b) The appearance of a magnetic flux rope forming a low-density cavity with a prominence condensation in the active helmet. (c) The transient, opened magnetic field shortly after the departure of the CME carrying the cavity and the mass of the erupted helmet and prominence out of the corona. The field reverses across the dashed line. Magnetic reconnection takes this state to a closed-field, new helmet and heats its trapped plasma in a characteristic post-CME flare. (d) The reformed helmet in the mid-latitude containing a bipolar magnetic field with no flux rope. Its bipolar field includes magnetic flux that is open and adjacent to the active helmet in the initial state (a), resulting in a shrinking of the coronal hole and a northward shift of the position of the new helmet relative to the position of the destroyed active helmet. The numbers label axisymmetric flux surfaces with equal flux between adjacent surfaces.

The transition from Figure 5b to 5c is at the rapid hydromagnetic time scale of the corona, a matter of hours for the CME to travel out at the median speed of 450 $km\ s^{-1}$. The dominance of the magnetic field low in the corona results in the collapse of the reversal layer shown in Figure 5c where the CME has left. Magnetic reconnection recloses the opened magnetic field to produce a new helmet streamer where the previous one was destroyed by the CME event. This is shown in Figure 5d, where we have introduced a shrinkage of the base of the polar open field region combined with a poleward shift of the newly

formed helmet streamer. These effects are explained below.

In the initial state, we have an helmet composed of 6 flux units of closed bipolar field. The large-scale corona is controlled by two opposing dynamical drivers. The tendency for heated plasma to expand into the solar wind is opposed by the magnetic tension force of closed bipolar fields with magnetic feet anchored to the base of the corona. If the expansion dominates, the field lines are combed by the solar wind to be open into interplanetary space. Where the field is strong enough, such as found low in the corona over a magnetic polarity inversion line, its magnetic tension force may trap plasma in quasi-static equilibrium against outward expansion. The initial active helmet streamer is in a balance between the two tendencies which dictates dynamically the amount of flux, taken to be 6 flux units, in the closed part of the field (numbers 17–22). A larger amount of flux in the helmet would result in its rising too high into the corona where the solar wind would "peel" off the additional flux to open them up. A smaller amount of flux in the helmet would result in the top of the helmet too low in the corona where open field just above it would dominate over the solar wind to close up by magnetic reconnection.

In Figure 5b, the emerged flux rope creates the three-part structure of the helmet ready for a CME eruption. Despite the reduced flux now threading across the helmet base, the helmet engorged with the flux rope keeps the helmet high in the corona. After the opened field in Figure 5c has reconnected to reform a new helmet, there is no longer the magnetic flux rope that had engorged the previous helmet. Thus, the reformed helmet would grow in flux content strictly with closed fields reconnected from the solar-wind open fields to a size, that we, for simplicity take again to be 6 flux units.

This is an interesting dynamical effect. It means that some of the field initially open and adjacent to the initial active helmet will end up being closed and a part of the new helmet. Thus, the entire process from emergence of flux rope to the relaxation to a new helmet after the dynamical eruption of a CME/flare involves removal of opposite fluxes of equal amounts from the stellar surface, as well as a reduction of magnetic flux from the northern polar coronal hole. Figure 5a and 5d show a reduction of 3 flux units in the polar open field, from 16 to 13 flux units. Combined with the general poleward migration of the large-scale fields on photospheric time scales [*Harvey*, 1994], the reformed helmet naturally would be located further north to the active helmet. This effect could account for the shrinking of the coronal hole observed during global field reversal. From this point of view, CMEs are the culmination of a slow coronal evolutionary process, the final step when the flux and helicity in the flux rope are bodily taken out into the interplanetary solar wind [*Low*, 1997, 2001]. If we postulate that CMEs at rates of 1 to 3 events a day do remove all the magnetic flux of a given sign threading across the photosphere over an 11-year solar cycle, each CME only needs to take out a modest fraction of 10^{-4} of the total photospheric flux. Interpretation of magnetic flux and helicity transport in the solar wind from satellite data taken at 1 AU supports this possibility [*Bieber and Rust*, 1995].

3.3. Magnetic-Field Reversal in the Corona

Figure 1 is based on the assumption that the preexisting flux at the stellar surface is conserved. The high latitude flux between the stellar pole and the mid-latitude neutral line represents this old flux frozen into the stellar surface. If we relax that assumption, the old flux can cancel with the new emerged flux at the mid-latitude neutral line, thereby progressively removing the old flux. This can take place in two ways, either upward or downward transport of a volumetric field, depending on the nature of the field above and below the neutral line.

One way this transport may take place is that depicted in Figure 5, via the passage of a magnetic flux rope into the helmet region above the neutral line, setting the stage for a CME eruption. The alternative is to imagine the converse. Instead of a flux rope emerging at the neutral line, the bipolar inverted U-shaped field above the neutral line in Figure 5a is submerged below the stellar surface. For this to occur, the curvature of the field over the neutral line has to be small since the field counts on its magnetic tension force to pull the atmospheric plasma below the stellar surface [*Rabin et al.*, 1984, *Parker*, 1984]. Although this is possible on the small scales it is intuitively unlikely to be a common event on the large scales because magnetic buoyancy generally causes a field to rise in a stratified atmosphere.

Let us introduce the process depicted in Figure 5 as additional physics operating at the mid-latitude neutral lines in Figure 1. Then, the old flux in the high latitude, instead of being a constant as originally assumed, would be decreasing with successive blowing off of CMEs from helmet-streamers forming repeatedly at the neutral line. This development would not only rid the corona of the old flux by taking it out through the corona into interplanetary space. It would also accelerate the crossing of the threshold for the emergent flux to reverse the field in the far coronal region. Including the effect of the solar wind and polar coronal holes, this process would progress with the shrinking of coronal holes and dominance of the new-cycle flux over an increasingly greater volume of the corona. For example, in Figure 5, the equatorial helmet could also grow with increasing new flux emerging from beneath it. The new flux could emerge as flux ropes so that the equatorial helmet also blows off as CMEs only to reform in a repeated fashion. It is important to appreciate that CMEs do not just get rid of old flux. As a single event, it gets rid of any "excess" flux from any part of the low corona where magnetic flux ropes fail to be confined in stable equilibrium [*Low*, 2001].

4. CONCLUSION

The ultimate cause of the reversal of the Sun's magnetic field is the dynamo in the solar interior. Magnetic buoyancy brings successive waves of new fluxes of alternating polarities first through the photosphere and then into the corona, in 11-year cycles. Flux emergence and field reversal are well observed phenomena at the photosphere, but as our analysis shows, observations confined to the thin photosphere alone are not sufficient to determine what actually happen in the atmosphere above. In our analysis, we provided insights into the complicated hydromagnetic changes in the corona driven by flux emergence. We identified what we regard to be the basic processes and their effects. We conclude with a summary of these processes with comments on how well we understand them physically and on how future work might improve that understanding.

Magnetic reconnection is fundamental to the mixing of new and old fluxes from two adjacent cycles. It is important to appreciate that reconnection rarely can remove two opposite fluxes *in situ*. As its name implies, the process liberates magnetic energy by changing the connection of flux systems without any of the flux systems being removed; the fluxes extending into the plasmas outside of the reconnection region are conserved. In this sense the old and new fluxes retain their identity in terms of where they enter or leave the solar surface. Flux cancellation between fields of the opposite signs on this surface is the only way of removing flux from that surface, but such a process is physically ambiguous without dealing with the actual transport of a volumetric magnetic field across this surface either from above to below, or the other way round.

The mixing of old and new fluxes early in the history of an active region must involve much more complex and richer magnetic metastable states than the idealized ones treated in *ZL1* and *ZL2*, and discussed here. It will be interesting to explore such metastable states in more realistic geometry. There is new physics to discover too. In *LZ3* and references therein, we have found that the mixing of old and new fluxes may be the physical origin of prominences of two magnetic kinds, the normal and inverse configurations, as studied by *Leroy* [1989] and others, as well as the physical reason for two observed dynamical behaviors of CMEs in the corona.

Since magnetic fields are observed to be significantly twisted as they emerge at the photosphere, the formation of a magnetic flux rope in the corona to conserve its intrinsic twist, as magnetic helicity, is a natural consequence of the process. This process is driven principally by magnetic buoyancy and various hydromagnetic instabilities, ideal or resistive, that have the collective effects of draining plasma downward in the magnetic field and transporting magnetic twist upward in the magnetic field. If we observe at the level of the photosphere, inverted-U field lines passing through upward will be seen to produce pairs of additional flux footpoints, whereas rising U-shaped field lines remove them. The converse must also take place of course, that is, sinking inverted-U and U-shaped field lines, but the former is expected to dominate because of the nature of magnetic buoyancy [*Parker*, 1979]. There is much of these complex processes of hydromagnetic turbulence which we do not understand theoretically beyond their qualitative effects, because their nonlinear physics is formidable.

In relation to bodily transport of magnetic flux systems into the corona, we should mention an important observational result. The magnetic helicity of flux ropes, such as inferred from measurements of prominence magnetic fields [*Leroy*, 1989, *van Ballegooijen and Martens*, 1990], for example, has an observed preference of sign in the two solar hemispheres, negative and positive, respectively, in the northern and southern hemispheres; see the reviews of this phenomenon in *Low* [2001] and *Rust* [1994]. Moreover, this hemispherical preference of sign is independent of solar cycle [*Martin et al.*, 1994]. The cycle independence is physically understandable in terms of the topological fact that the handedness of twist in a magnetic field is unchanged if the direction of the field is reversed. This observational result suggests that the new flux of each cycle emerges with helicity of a predominant sign in the respective hemispheres. If not for CMEs, the unceasing injection of magnetic helicity into the corona from one solar cycle to the next, together with helicity conservation, would imply a monotonic, unbounded accumulation of magnetic helicity of a fixed sign in each hemisphere of the corona. By taking flux ropes and their trapped magnetic helicity out with them, CMEs enable the corona to avoid the untenable situation of unbounded accumulation of magnetic helicity in the corona.

The solar atmosphere is thus not just the medium through which the solar luminosity and plasma mass, momentum and energy escape. There is also a continual transport of magnetic flux and helicity through it. The solar atmosphere is thus truly coupled to the solar dynamo. Not only is the solar atmosphere driven by the emergence of magnetic flux and helicity generated by the dynamo, we may need to revise our view of the working of the dynamo to account for the significant amounts of magnetic flux and helicity it generates which escape through the corona into interplanetary space.

Future work should proceed both theoretically and observationally. By working on basic physical calculations we can extend our intuitive command of ideas to interpret phenomena in terms of elementary physical principles. On the other hand, observations in crucial ways can tell us whether we are wrong in an interpretation or point to unsuspected new directions to pursue. Our analysis shows that the global reversal of magnetic field in the corona can be understood in explicit physical terms, even if that understanding is still just a skeleton. This suggests that this area of research is ready for more work and, possibly, exciting progress in the near future.

We conclude with a comment on the topological relationship between the heliospheric and coronal magnetic fields. *Smith et al.* [2001] have pointed out that interplanetary observations agree with speculations that, in effect, the Sun's magnetic dipole rotates from the pole at minimum to the equator at maximum and continues rotating to yield the reversal in polarity [*Saito et al.*, 1978]. Interesting phenomenological ideas have been proposed to relate the behaviors of the large scale heliospheric fields to coronal differential rotation and magnetic reconnection, leading to a suggestion how such a rotation of the solar dipole may occur [*Zurbuchen et al.*, 1997, *Fisk et al.*, 1999, *Fisk and Schwadron*, 2001]. The time seems ripe to reconcile such interplanetary views of the global field with coronal view points such as the one described in this paper. Our view point argues for a significant role for the CME in the global evolution.

Acknowledgements. We thank Randy Jokipii, Phil Judge and Thomas Zurbuchen for helpful comments. This work is supported in parts by fundings from the National Academy of Sciences Arctowski Medal Award to A. J. Hundhausen, NASA, and the NSF National Space Weather Program. The National Center for Atmospheric Research is sponsored by the National Science Foundation.

REFERENCES

Aly, J. J., How much energy can be stored in a three-dimensional force-free magnetic field?, *Astrophys. J., 375*, L61, 1991.

Altschuler, M. D., and G. Newkirk, Jr., Magnetic fields and the structure of the solar corona. I. Method of calculating coronal fields, *Solar Phys., 9*, 131, 1969.

Amari, T., J. F. Luciani, Z. Mikic, and J. Linker, Three-dimensional solutions of MHD equations for prominence magnetic support: twisted magnetic flux rope, *Astrophys. J., 518*, L57, 1999.

Antiochos, S. K., C. R. DeVore, and J. A. Klimchuk, A model of solar mass ejections, *Astrophys. J., 510*, 485, 1999.

Berger, M. A., Rigorous new limits on magnetic helicity dissipation in the corona, *Geophys. Astrophys. Fluid Dyn., 30*, 79, 1984.

Berger, M. A., Magnetic helicity in space physics, *Magnetic helicity in space and laboratory plasmas*, ed. by M. R. Brown, R. Canfield, and A. Pevtsov, AGU, Washington D.C., p.1, 1999.

Bieber, J. W., and D. M. Rust, The escape of magnetic flux from the sun, *Astrophys. J., 453*, 911, 1995.

Burkepile, J. T., and O. C. St. Cyr, A revised and expanded catalogue of coronal mass ejections observed by the Solar Maximum Mission coronagraph, *Tech. Note TN-369+STR*, Nat. Center for Atmos. Res., Boulder, Colorado, 1993.

Chen, J., Theory of prominence eruption and propagation: Interplanetary consequences, *J. Geophys. Res., 101*, 27499, 1996.

Dere, K. P., G. E. Brueckner, R. A. Howard, D. J. Michels, and J. P. Delaboudiniere, LASCO and EIT observations of helical structure in coronal mass ejections, *Astrophys. J., 516*, 465, 1999.

Fan, Y., The emergence of a twisted Ω-tube into the solar atmosphere, *Astrophys. J., 554*, L111, 2001.

Fan, Y., E. G. Zweibel, M. G. Linton, and G. H. Fisher, The rise of kink-unstable magnetic flux tubes and the origin of *d*-configuration sunspot, *Astrophys. J., 521*, 460, 1999.

Fisk, L. A., T. H. Zurbuchen, and N. A. Schwadron, Coronal hole boundaries and their interactions with adjacent regions, *Space Sci. Rev., 87*, 43, 1999.

Fisk, L. A., and N. A. Schwadron, The behavior of the open magnetic field of the Sun, *Astrophys. J., 560*, 425, 2001.

Forbes, T. G., A tutorial review on CME Genesis, *J. Geophys. Res., 105*, 23153, 2000.

Gibson, S. E., and B. C. Low, 3D and twisted: An MHD interpretation of on-disk observational characteristics of CMEs, *J. Geophys. Res., 105*, 18187, 2000.

Gosling, J. T., Coronal mass ejections, in *26th International Cosmic Ray Conference*, ed. by B. L. Dingus et al., AIP, Washington, DC, p.59, 2001.

Harvey, K. L., The solar magnetic cycle, in *Solar Surface Magnetism*, ed. R. J. Rutten and C. J. Schrijver, Kluwer, Dordrecht, p. 347, 1994.

Harvey, K. L., and V. Gaizauskas, Filament channels: Contrasting their structures in H_α and HeI 1083nm, in *New perspectives on solar Prominences*, ed. by D. Webb, D. M. Rust, and B. Schmieder, ASP Publication, p. 269, 1998.

Hiei, E., A. J. Hundhausen, and D. G. Sime, Reformation of a coronal helmet streamer by magnetic reconnection after a coronal mass ejection, *Geophys. Res. Lett., 20*, 2785, 1993.

Howard, R., Polar magnetic fields of the Sun: 1960–1971, *Solar Phys., 25*, 5, 1972.

Howard, R. A., N. R. Sheeley, Jr., M. J. Koomen, and D. J. Michels, Coronal mass ejections: 1979–1981, *J. Geophys. Res., 90*, 8173, 1985.

Howard, R. A. et al., Observations of CMEs from SOHO/LASCO, *Geophys. Mon., 99*, 17, 1997.

Hundhausen, A. J., An interplanetary view of coronal holes in *Coronal holes and high speed streamers*, ed. by J. B. Zirker, p.225, Colo. Assoc. Univ. Press, Boulder, 1977.

Hundhausen, A. J., Coronal mass ejections: A summary of SMM observations from 1980 and 1984–1989, in *The many faces of the Sun*, ed. by K. Strong, J. Saba, B. Haisch, and J. Schmelz, p.143, Springer-Verlag, New York, 1999.

Illing, R. M. E., and A. J. Hundhausen, Disruption of a coronal streamer by an eruptive prominence and coronal mass ejection, *J. Geophys. Res., 91*, 10951, 1986.

Ishii, T. T., H. Kurokawa, and T. T. Takeuchi, Emergence of a twisted magnetic flux bundle as a source of strong flare activity, *Astrophys. J., 499*, 898, 1998.

Jokipii, J. R., and B. T. Thomas, Effects of drifts on the transport of cosmic rays. IV. Modulation by a wavy interplanetary current sheet, *Astrophys. J., 243*, 1115, 1981.

Judge, P. G., Spectral lines for polarization measurements of the coronal magnetic field. I. Theoretical intensities, *Astrophys. J., 500*, 1009, 1998.

Kurokawa, H., Two distinct morphological types of magnetic shear development and their relation to flares, *Solar Phys., 113*, 259, 1987.

Leka, K. D., R. C. Canfield, A. N. McClymont, and L. van Driel-Gesztelyi, Evidence for current-carrying emerging flux, *Astrophys. J., 462*, 547, 1996.

Leroy, J. L., Observation of prominence magnetic fields, in *Dynamics and structure of quiescent prominences*, ed. by E. R. Priest, Kluwer, Dordrecht, p.77, 1989.

Levy, E. H., The interplanetary field structure, *Nature, 261*, 394, 1976.

Lin, H., M. J. Penn, and S. Tomczyk, A new precise measurement of the coronal magnetic field strength, *Astrophys. J., 541*, L83, 2000.

Lites, B. W., B. C. Low, V. Martinez-Pillet, P. Seagrave, A. Skumanich, Z. A. Frank, R. A. Shine, and S. Tsuneta, The possible ascent of a closed magnetic system through the photosphere, *Astrophys. J., 446*, 877, 1995.

Low, B. C., Three dimensional structures of magnetostatic atmospheres. IV. Magnetic structures over solar active regions, *Astrophys. J., 399*, 300, 1992.

Low, B. C., Magnetohydrodynamic processes in the solar corona: Flares, coronal mass ejections, and magnetic helicity, *Phys. Plasma., 1*, 1684, 1994.

Low, B. C., Solar activity and the corona, *Solar Phys., 167*, 217, 1996.

Low, B. C., The role of coronal mass ejections in solar activity, in *Coronal mass ejections*, ed. by N. Crooker, J. Joselyn, and J. Feynman, p.39, AGU Publication, 1997.

Low, B. C., Magnetic energy and helicity in open systems, *Magnetic helicity in space and laboratory plasmas*, ed. by M. R. Brown, R. Canfield, and A. Pevtsov, AGU, Washington D. C., p.25, 1999.

Low, B. C., Coronal mass ejections, magnetic helicity, and solar magnetism, *J. Geophys. Res., 106*, 25141, 2001.

Low, B. C., and J. R. Hundhausen, Magnetostatic structures of the solar corona. II. The Magnetic topology of quiescent prominences, *Astrophys. J., 443*, 818, 1995.

Low, B. C., and D. F. Smith, The free energies of partially open coronal magnetic fields, *Astrophys. J., 410*, 413, 1993.

Low, B. C., and Wolfson, R., Spontaneous formation of current sheets and the origin of solar flares, *Astrophys. J., 324*, 574, 1988.

Low, B. C., and Zhang, M., The hydromagnetic origin of the two dynamical types of coronal mass jections, *Astrophys. J., 564*, L53, 2002; LZ3.

Luhmann, J. G., J. T. Gosling, J. T. Hoksema, and X. Zhao, The relationship between large-scale magnetic field and coronal mass ejections, *J. Geophys. Res., 103*, 6585, 1998.

Manchester, IV., W., The role of nonlinear Alfven waves in shear formation during solar magnetic flux emergence, *Astrophys. J., 547*, 503, 2001.

MacQueen, R. M., Coronal transients: A summary, *Phil. Trans. R. Soc. Lond., A 297*, 605, 1980.

Martin, S. F., R. Bilimoria, and P. W. Tracadas, Magnetic field configurations basic to filament channels and filaments, in *Solar Surface Magnetism*, ed. by R. J. Rutten and C. J. Schrijver, Kluwer, Dordrecht, p. 303, 1994.

Parker, E. N., The dynamical state of interstellar gas and fields. II. Non-linear growth of clouds and forces in three dimensions, *Astrophys. J., 149*, 517, 1967.

Parker, E. N., *Cosmical magnetic fields*, Oxford U. Press, New York, 1979.

Parker, E. N., Depth of origin of solar active regions, *Astrophys. J., 280*, 423, 1984.

Parker, E. N., *Spontaneous current sheets in magnetic fields*, Oxford U. Press, New York, 1994.

Priest, E. R., A. W. Hood, and U. Anzer, A twisted fluxtube model for solar prominences. I. General properties, *Astrophys. J., 344*, 1010, 1989.

Rabin, D., R. Moore, and M. J. Hagyard, A case for submergence of magnetic flux in a solar active region, *Astrophys. J., 287*, 404, 1984.

Rust, D. M., Spawning and shedding helical magnetic fields in the solar atmosphere, *Geophys. Res. Lett., 21*, 241, 1994.

Saito, T., T. Sakurai, and K. Yumoto, The earth's palaeomagnetosphere as the third type of planetary magnetosphere, *Planet. Space Sci., 26*, 413, 1978.

Schrijver, C. J., and C. Zwaan, *Solar and Stellar Magnetic Activity*, Cambridge University Press, New York, 2000.

Smith, E. J., The heliospheric current sheet, *J. Geophys. Res., 106*, 15819, 2001.

Smith, E. J., The Sun and interplanetary magnetic field, in *The Sun in Time*, ed. by C. P. Sonett, M. S. Giampapa, and M. S. Mathews, p. 175, Univ. of Arizona Press, Tucson, 1993.

Smith, E. J., A. Balogh, R. J. Forsyth, and D. J. McComas, Ulysses in the south polar cap at solar maximum: Heliospheric magnetic field, *Geophys. Res. Lett., 28*, 4159, 2001.

Thompson, B. J., S. P. Plunkett, J. B. Gurman, H. S. Hudson, R. A. Howard, and D. J. Michels, SOHO/EIT observations of an Earth-directed coronal mass ejection on May 12, 1997, *Geophys. Res. Lett., 25*, 2465, 1998.

St. Cyr, O. C., et al., Properties of coronal mass ejections: SOHO LASCO observations from January 1996 to June 1998, *J. Geophys. Res., 105*, 18169, 2000.

van Ballegooijen, A. A., and P. C. H. Martens, Magnetic fields in quiescent prominences, *Astrophys. J., 361*, 283, 1990.

Webb, D. F., and R. A. Howard, The solar cycle variation of coronal mass ejections and the solar-wind mass flux, *J. Geophys. Res., 99*, 4201, 1994.

Wilcox, J. M., and P. H. Scherrer, Annual and Solar-magnetic-cycle variations in the interplanetary magnetic field 1926–1971, *J. Geophys. Res., 77*, 5385, 1972.

Wilson, P. R., and J. Giovannis, The reversal of the solar polar magnetic fields. V. The reversal of polar fields in cycle 22, *Solar Phys., 155*, 129, 1994.

Wu. S. T., et al., MHD interpretation of LASCO observations of a coronal mass ejection as a disconnected magnetic structure, *Solar Phys., 175*, 719, 1997.

Zhang, M., and B. C. Low, Magnetic-flux emergence into the solar corona. I. Its role for the reversal of global coronal magnetic field; ZL1, *Astrophys. J., 561*, 406, 2001.

Zhang, M., and B. C. Low, Magnetic-flux emergence into the solar corona. II. Global magnetic fields with current sheets, *Astrophys. J.*, in press, 2002; ZL2.

Zirin, H., *Astrophysics of the Sun*, Cambridge University Press, Cambridge, UK, 1988.

Zurbuchen, T. H., N. A. Schwadron, and L. A. Fisk, Direct observational evidence for a heliospheric magnetic field with large excursions in latitude, *J. Geophys. Res., 102*, 24175, 1997.

Fundamentals of the Earth's Atmosphere and Climate

Joanna D. Haigh

Space and Atmospheric Physics, Blackett Laboratory,
Imperial College London, UK

This tutorial review paper introduces some of the basic physics and chemistry determining the atmosphere's composition and temperature structure. The Earth's radiation budget, factors determining atmospheric heating rates and the "greenhouse effect" are discussed. The concept of the radiative forcing of climate change is introduced and the role of cloud in the radiation balance reviewed. A short section on modes of variability of the atmosphere is included. In each of these areas the potential impact of variations in solar activity are considered.

1. INTRODUCTION

At periods of higher solar activity the Earth is subject to higher solar irradiance, a greater incidence of solar energetic particles and fewer galactic cosmic rays. The absorption of solar radiation determines the Earth's mean temperature and radiation budget, while the latitudinal distribution of the absorbed radiation is the primary driver for atmospheric circulations. Thus changes in incident irradiance are bound to have an effect at least at some level of significance.

At any point in the atmosphere the radiative heating rate is the net effect of solar heating and infrared cooling, the latter being intrinsically related to the atmospheric temperature structure. Because the atmosphere is more transparent to solar than infrared radiation a "greenhouse" effect causes the surface of the Earth to be approximately 30 K warmer than it would be if the atmosphere were not present. Thus any factors which change the radiation budget of the Earth will tend to affect the atmospheric temperature structure in a rather non-linear fashion.

Solar radiation is also key to determining the composition of the atmosphere—and hence its radiative properties. For

Solar Variability and its Effects on Climate
Geophysical Monograph 141
Copyright 2004 by the American Geophysical Union
10.1029/141GM06

example, the presence of stratospheric ozone results from photochemical processes, and the ozone itself modifies the solar radiation reaching the surface. Solar energetic particles can also affect atmospheric composition.

Clouds play a key role in the radiation budget and any changes in their coverage or radiative properties could result in significant changes in climate.

This paper presents some background material on the physics and chemistry of the Earth's lower and middle atmosphere so that the degree to which solar variability may affect the radiation balance, atmospheric composition, cloud cover and atmospheric circulation can be assessed.

2. EARTH RADIATION BUDGET

2.1. Global Average

The global average equilibrium temperature of the Earth is determined by a balance between the energy acquired by the absorption of incoming solar radiation and the energy lost to space by the emission of thermal infrared radiation. The amount of solar energy absorbed depends both on the incoming irradiance and on the Earth's reflective properties. If either of these changes then the temperature structure of the atmosphere-surface system tends to adjust to restore the equilibrium. In order to understand how these processes affect climate it is important to investigate in more detail the solar and infrared radiation streams. Plate 1 shows the com-

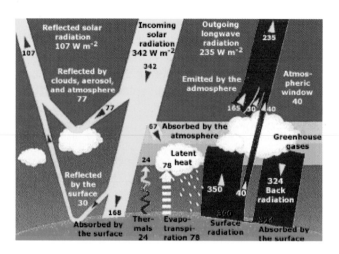

Plate 1. Globally averaged energy budget of the atmosphere. (Figure from http://asd-www.larc.nasa.gov/ceres/ brochure/clouds-and-energy based on data from *Kiehl and Trenberth*, 1997).

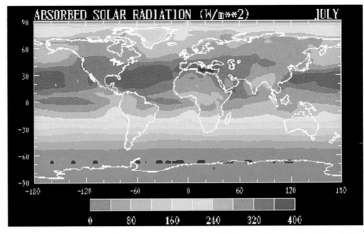

Plate 2a. Absorbed solar radiation in July (W m^{-2}). (Figures from http://rainbow.ldeo.columbia.edu/ees/data/)

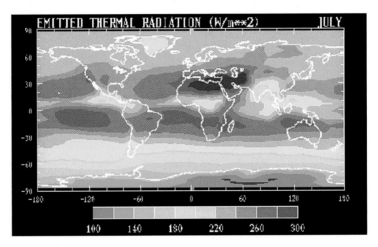

Plate 2b. As Figure 2a but emitted thermal radiation.

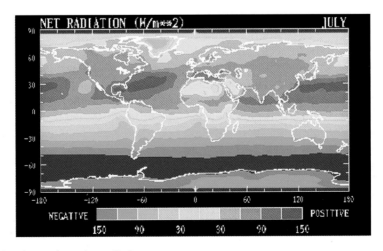

Plate 2c. As Figure 2a but for net incoming radiation.

ponents of the global annual average radiation budget and how much radiation is absorbed, scattered and emitted within the atmosphere and at the surface. The value for the incoming radiation, 342 W m^{-2}, is equivalent to a total solar irradiance at the Earth of 1368 W m^{-2} averaged over the globe. Of this 31% (107 W m^{-2}) is reflected back to space by clouds, aerosols, atmospheric molecules and the surface, with the clouds playing the most important role, so that only 235 W m^{-2} is absorbed by the Earth system. 20% (67 W m^{-2}) of the incident radiation is absorbed within the atmosphere leaving 49% (168 W m^{-2}) to reach and heat the surface.

The temperature and emissivity of the surface are such that 390 W m^{-2} of infrared energy are emitted into the atmosphere. Only 40 W m^{-2} of this, however, escapes to space with the remainder being absorbed by atmospheric gases and cloud. The atmosphere returns 324 W m^{-2} to the surface. The energy balance at the surface is achieved by non-radiative processes such as evaporation and convection. The radiation balance at the top of the atmosphere is achieved by 195 W m^{-2} emitted to space by the atmosphere and clouds.

2.2. Geographical Distribution

The above picture considers only the global annual average situation which gives an indication of how atmospheric and surface properties may affect the vertical temperature structure of the atmosphere. This is discussed further in sections 4.2 and 5.1 below. Hidden within this energy balance, however, there are wide geographic and seasonal variations in the radiation budget components. Plate 2 presents data acquired by the ERBE instrument (on the ERBS and NOAA-9 satellites) which give an indication of the spatial distributions. Plate 2(a) shows the absorbed solar radiation in July. The incident radiation is greatest at the sub-solar point (near 20 °N), and no radiation is incident in the polar night (south of 70 °S), but the pattern of absorbed radiation is complicated by the presence of tropical cloud and reflective surfaces. These enhance the albedo and thus result in patches of reduced absorption. The map of thermal radiation (Plate 2(b)) indicates the temperature of the emitting surface with, in general, more being emitted from the warmer low latitudes and summer hemisphere. However, the presence of cloud near 10 °N and in the Indian summer monsoon provide colder radiating surfaces and thus patches of lower emission. The distribution of net radiation (Plate 2(c)) shows generally positive values (more absorbed than emitted) in the summer hemisphere and negative values south of 10 °S and near the north pole. There are significant exceptions, however, e.g. negative values over the Sahara and the

Arabian peninsular, due to a combination of hot surface temperatures and high albedo. The low latitude excess of energy must be transported, by either the atmosphere or oceans, to make up the deficit at high latitudes. Thus the radiation balance is intrinsically linked with large scale atmospheric and oceanographic circulations.

3. ABSORPTION OF SOLAR RADIATION BY THE ATMOSPHERE

3.1. Solar Irradiance

Absorption by the atmosphere of solar radiation depends on the concentrations and spectral properties of the atmospheric constituents. Figure 1 shows a blackbody spectrum at 5750 K, representing solar irradiance at the top of the atmosphere, and a spectrum of atmospheric absorption. Absorption features due to specific gases are clear with molecular oxygen and ozone being the major absorbers in the ultraviolet and visible regions and water vapor and carbon dioxide more important in the near-infrared.

The solar flux, in the direction of the beam, at wavelength λ and altitude z is given by:

$$F(\lambda, z) = F_0(\lambda) \exp(-\tau(\lambda, z))$$

where F_0 is the flux incident at the top of the atmosphere and the optical depth, τ, depends on the air density, ρ, the mass mixing ratio, c, and extinction coefficient, k, of the absorbing gas and the solar zenith angle, ζ:

$$\tau(\lambda, z) = -\int_z^\infty k(\lambda)c(z')\rho(z')\sec\zeta\, dz'.$$

Clearly the flux at any point depends on the properties and quantity of absorbing gases in the path above. The altitude at which most absorption takes place at each wavelength can be seen in Figure 2 which shows the altitude of unit optical depth for an overhead Sun. At wavelengths shorter than 100 nm most radiation is absorbed at altitudes between 100 and 200 km by atomic and molecular oxygen and nitrogen, mainly resulting in ionized products. Between about 80 and 120 km oxygen is photodissociated as it absorbs in the Schumann-Runge continuum between 130 and 175 nm. The Schumann-Runge bands, 175–200 nm, are associated with electronic plus vibrational transitions of the oxygen molecule and are most significant between 40 and 95 km altitude. The oxygen Herzberg continuum is found in the range 200–242 nm and is overlapped by the ozone Hartley-Huggins bands between 200 and 350 nm which are responsible for the photodissociation of ozone below 50 km.

The ozone Chappuis bands, in the visible and near-infrared, are much weaker than the aforementioned bands but, because they absorb near the peak of the solar spectrum, the energy deposition into the atmosphere is significant. Furthermore, this deposition takes place in the lower atmosphere and so is particularly relevant for climate. The absorption of solar near-infrared by carbon dioxide and water vapor is smaller but makes an important contribution to the heat budget of the lower atmosphere (see section 3.3).

Solar irradiance varies with solar activity, as discussed in Chapter 3 of this volume. The spectral composition of the variation determines which parts of the atmosphere respond most in terms of heating rates (see below). Note, however, that if the composition of the atmosphere were to remain unchanged, then variations in irradiance do not affect the height of unit optical depth shown in Figure 2.

3.2. "Anomalous" Absorption

Observed values of the absorption of solar radiation in the atmosphere almost always exceed theoretical values. This effect has become known as "anomalous absorption" and is the subject of considerable debate (a good review is given by *Ramanathan and Vogelmann*, [1997]). Discussion concerns not only possible physical explanations for its existence but also its magnitude (including whether it actually exists), spectral composition and whether it is a property only of cloudy skies. Some aircraft studies of the visible

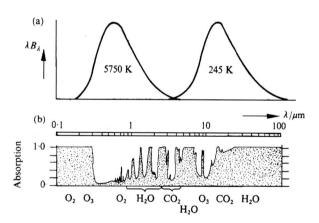

Figure 1. (a) Black body functions at the emitting temperatures of the Sun and the Earth. The functions are scaled to have equal area to represent the Earth's radiation balance. (b) Absorption by a vertical column of atmosphere. The most important gases responsible for absorption are identified below near the appropriate wavelengths.

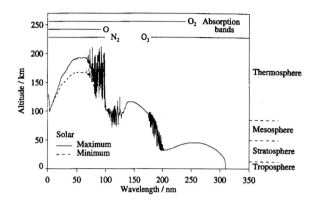

Figure 2. Wavelength dependence of the altitude of one optical depth for absorption of solar radiation with an overhead Sun. After *Andrews* [2000].

radiation field around cloud suggest that the apparent anomaly is an artifact of the imperfect sampling of the 3D structure. However, studies using a combination of satellite and ground-based data have confirmed the existence of a global average anomaly of 25–30 W m^{-2} (i.e., 10–12% of the total solar irradiance absorbed by the Earth), with the near-infrared region appearing to be significant. Some of the studies concluded that cloudy skies were responsible for the excess absorption but others have suggested a significant discrepancy between the results of GCM radiation schemes and observations in clear-sky absorption.

Mechanisms proposed to account for the underestimate of absorption in radiation models include problems with the formulation of the radiative transfer (band models, treatment of scattering, etc.), uncertainties in the radiative properties of water vapor (the spectral database and continuum absorption), the loading, composition and radiative properties of aerosol particles, cloud impurities, cloud drop-size distributions, the inability of models to simulate 3D radiation field in inhomogeneous cloud and the enhancement of the photon mean-free-path in 3D cloud. Work continues in investigating these avenues but, in the context of the effects of solar variability on climate it is worth noting that large uncertainties remain in quantitative estimates of the absorption of solar radiation by the atmosphere.

3.3. Solar Heating Rates

Most of the absorbed solar radiation eventually becomes heat energy so the local solar heating rate, Q (degrees per unit time), can be estimated from the divergence the solar fluxes:

$$Q(\lambda,z) = \frac{1}{\rho(z)C_p \sec\zeta} \frac{dF(\lambda,z)}{dz}$$

$$= \frac{k(\lambda)c(z)}{C_p} F_0(\lambda)\exp(-\tau(\lambda,z))$$

where C_p is the specific heat at constant pressure of air.

Figure 3 presents a vertical profile of diurnally averaged solar heating rates for equatorial equinox conditions, showing the contribution of each of the absorption bands mentioned above. This vertical structure in the absorption of solar radiation is crucial in determining the profile of atmospheric temperatures and plays an important role in atmospheric chemistry and thus composition.

If the solar spectral irradiance varies then, if there is no change in composition, from the equation for heating rate above we can see that the spectral heating rate just varies in proportion to the irradiance. If, however, as is actually the case, the atmospheric composition also responds to solar variability then this will affect both F and Q in a non-linear fashion. For example, an increase in F_0 will tend to increase F and Q. However, an increase in $c(z)$ (of ozone for example) enhances τ tending to reduce F. The sign of the change in F at any altitude depends on the competition between these two factors which is determined by the spectral composition of the change in F_0 and its effects on the photochemistry of the atmosphere. The effect on Q is then a product of the changes in F and $c(z)$.

Plate 3 shows vertical profiles of the solar fluxes and heating rates for solar minimum conditions in the ultravio-

let, visible and near infrared spectral regions. The magnitude of the UV flux is much smaller than in the other two regions but its absorption by ozone causes the largest heating rates in the middle atmosphere. The weaker absorption of visible radiation in the lower stratosphere and of near infrared radiation in the troposphere give much smaller heating rates. Also shown in Plate 3 are the differences in fluxes and heating rates between 11-year solar cycle minimum and maximum conditions. At the top of the atmosphere the increases in incoming radiation in the visible and UV regions are of similar magnitude but the stronger absorption of UV produces a much larger effect. The same data are shown with a linear pressure scale for the ordinate (to emphasise the troposphere) in the last two panels of Plate 3. The spectral changes prescribed for these calculations were such that near-infrared radiation was weaker at solar maximum so decreases in heating rate are shown. This is contentious but the changes are anyway very small being about one part in ten thousand.

3.4. Photochemistry

In the stratosphere the main chemical reactions determining the concentration of ozone are:

$$O_2 + h\nu \rightarrow O + O$$
$$O + O_2 + M \rightarrow O_3 + M$$
$$O_3 + h\nu \rightarrow O_2 + O$$
$$O + O_3 \rightarrow 2O_2$$
$$O_3 + X \rightarrow XO + O_2$$
$$XO + O \rightarrow X + O_2$$

The first of these reactions represents the photodissociation of oxygen at wavelengths less than 242 nm. This process is the key step in ozone formation because the oxygen atoms produced react with oxygen molecules to produce ozone molecules, as depicted in the second reaction (the M represents any other air molecule whose presence is necessary to simultaneously conserve momentum and kinetic energy in the combination reaction). Because the short wavelength ultraviolet radiation gets used up as it passes through the atmosphere, concentrations of atomic oxygen increase with height. This would tend to produce a similar profile for ozone but the effect is counterbalanced by the need for a 3-body collision (reaction 2) which is more likely at higher pressures (lower altitudes). Thus a peak in ozone production occurs at around 50 km. The third reaction is the photodissociation of ozone, mainly by radiation in the Hartley band ($\lambda < 310$ nm), into one atom and one

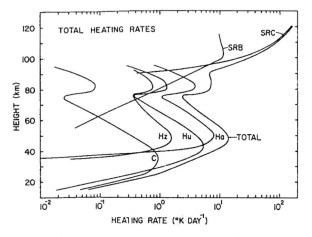

Figure 3. Diurnal average solar heating rate (K d^{-1}, log scale) as a function of altitude for equinoctial conditions at the equator showing contributions by the Schumann-Runge continuum and bands (SRC and SRB), the Herzberg continuum (Hz) and the Hartley (Ha), Huggins (Hu) and Chappuis (Ch) bands. After *Strobel* [1978].

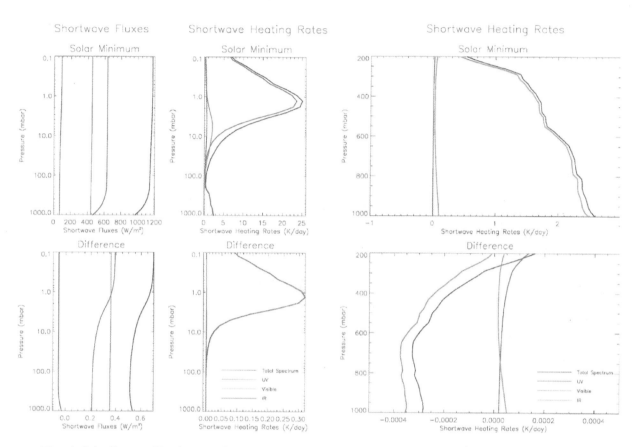

Plate 3. Solar fluxes and heating rates for an overhead Sun as a function of altitude (log pressure). Solar minimum conditions (above) and difference between solar maximum and solar minimum (below) at ultraviolet (blue curves), visible (green curves) and near infrared (red curves) wavelengths. Fluxes (W m^{-2}, left), heating rates (K d^{-1}, center and, on an expanded scale as a function of pressure, right). Spectral data from Judith Lean (personal communication), calculation from *Larkin* [2000].

molecule of oxygen. This does not represent the fundamental destruction of the ozone because the oxygen atom produced can quickly recombine with an oxygen molecule. The fourth reaction represents the destruction of ozone by combination with an oxygen atom. The fifth and sixth reactions represent the destruction by any catalyst X, which may include OH, NO and Cl. The various destruction paths are important at different altitudes but the combined effect is an ozone concentration profile which peaks near 25 km in equatorial regions.

Because photodissociation is an essential component of ozone formation most ozone is produced at low latitudes in the upper stratosphere. Observations show, however, that it is also present in considerable quantities in the mid- and high latitude stratosphere. Indeed, the quantity of ozone above unit area of the Earth's surface (ozone column amount) is usually greater at midlatitudes than the equator, and in the spring rather than mid-summer, as shown in Figure 4. This is a result of the transport of ozone by atmospheric motions. The atmospheric circulations tend to move it away from its source region toward the winter pole and downward. In the lower stratosphere its photochemical lifetime is much longer, because of the reduced penetration of the radiation which destroys it, and its distribution is determined by transport, rather than photochemical, processes. In winter high latitudes photochemical destruction essentially ceases and the ozone accumulates until the spring, as can be seen in Figure 4

As outlined above, ozone is produced by short wavelength solar ultraviolet radiation and destroyed by radiation at somewhat longer wavelengths. The amplitude of solar variability is greater in the far ultraviolet [*Rottman et al*, this volume] so that ozone production is more strongly modulated by solar activity than the destruction mechanisms and thus there are higher concentrations of stratospheric ozone during periods of higher solar activity. Both observational records and model calculations show approximately 2% higher values in ozone columns at 11-year cycle maximum relative to minimum. However, there are some discrepancies between satellite observations and model predictions in the vertical and latitudinal distributions of the response. Figure 5 shows a typical model calculation with largest changes in the middle stratosphere and less above and below. The observational data suggest a larger response near the stratopause and possibly a second maximum in the lower stratosphere [*McCormack and Hood*, 1996]. The data are only available over less than two solar cycles so there remains some doubt about the conclusions drawn from them as well as the validity of the model simulations.

The chemical composition of the atmosphere also responds to the incidence of high energy solar particles.

Precipitating electrons and solar protons affect the nitrogen oxide budget of the middle atmosphere through ionization and dissociation of nitrogen and oxygen molecules. NO catalytically destroys ozone, as discussed above, and reductions in ozone concentration may occur down to the middle stratosphere for a particularly energetic event [*Jackman and McPeters*, this volume]. As the solar particles follow the Earth's magnetic field lines these effects only occur at high latitudes. It is interesting to note that the effect of energetic particle events on ozone is in the opposite sense to that of enhanced ultraviolet irradiance. As particle events are more likely to occur when the Sun is in an active state the combined effect on ozone may be complex in its geographical, altitudinal and temporal distribution.

Solar radiation is also fundamental in determining the composition of the troposphere. The daytime chemistry of the troposphere is dominated by the hydroxyl radical, OH, because its high reactivity leads to the oxidation and chem-

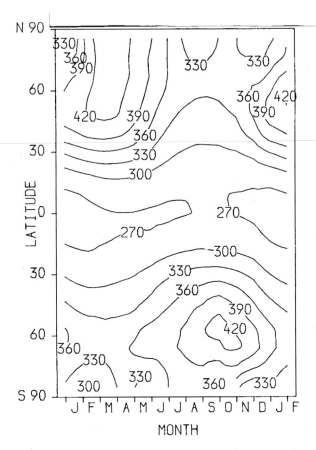

Figure 4. Ozone column amounts (m atm-cm) as a function of latitude and time of year. From a photochemical transport model calculation [*Haigh*, 1994].

ical conversion of most other trace constituents. OH is formed when an excited oxygen atom, O(^{1}D), reacts with water vapor. The source of the O(^{1}D) is the photodissociation (at wavelengths less than about 310 nm) of ozone; thus the presence of ozone is fundamental to the system. A major source of tropospheric ozone is transport from the stratosphere, but it is also formed through the photolysis (at wavelengths less than 400 nm) of nitrogen dioxide, which can be catalytically regenerated. Because OH is photolytically produced its concentration drops at night and the dominant oxidant becomes the nitrate radical, NO$_3$, itself photochemically destroyed during the day.

Thus any variation in the intensity, or spectral composition, of solar radiation may affect the lower atmosphere not only through direct heating but also potentially through modifying its chemical composition.

4. INFRARED RADIATIVE TRANSFER AND THE ATMOSPHERIC TEMPERATURE STRUCTURE

4.1. Infrared Absorbers

The atmosphere absorbs solar radiation, as discussed above, and, while this energy may at first be used in photodissociation, molecule excitation or ionization processes, it essentially ends up as molecular kinetic energy, i.e., in raising atmospheric temperatures. To balance this the atmosphere must lose heat by radiating energy in the thermal infrared. The amount of energy radiated depends on the local temperature and on the infrared spectral properties

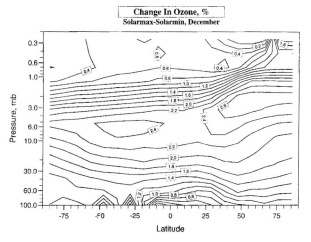

Figure 5. Difference between solar maximum and minimum ozone concentration (%) in December as a function of latitude and height as calculated in a photochemical-transport model [*Haigh*, 1994].

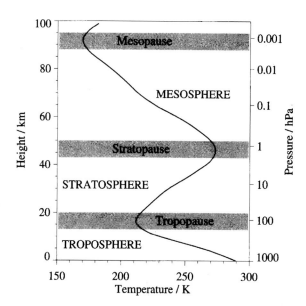

Figure 6. Typical vertical structure of atmospheric temperature (K). After *Andrews* [2000].

(emissivities) of the atmospheric constituents. Figure 1 shows a blackbody spectrum at 245 K, the radiative equilibrium temperature for a planet with albedo 31% at approximately 1AU from the Sun, and the atmospheric absorption spectrum. Far infrared radiation is strongly absorbed by water vapor in its rotation bands and across the thermal infrared there are further water vapor absorption bands as well as features due to other "greenhouse" gases. There are strong carbon dioxide bands at 15 μm, 4.3 μm and 2.7 μm, water vapor bands at 6.3 μm and 2.7 μm, an ozone band at 9.6 μm as well as features due to methane, nitrous oxide and chlorofluorocarbons.

4.2. Atmospheric Temperature Profile

Where the atmosphere is optically thin in the infrared, such as in the stratosphere, radiant heat energy may be transmitted to space, causing local cooling. At lower altitudes where the atmosphere is optically much thicker, however, emitted infrared radiation is absorbed and reemitted by neighboring layers. Thus the atmospheric temperature profile is determined by interactions between levels as well as by solar heating and direct thermal emission.

In the middle atmosphere absorption of solar ultraviolet radiation by oxygen and ozone, as described in the previous section, produces a peak in temperature near 50km called the stratopause as shown in Figure 6. This heating is counteracted by thermal emission, mainly by carbon dioxide in its

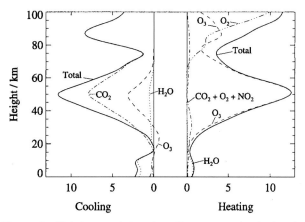

Figure 7. Global mean infrared cooling rates and solar heating rates (K d^{-1}) as a function of altitude showing contributions of the major gases. After *Andrews* [2000].

15 µm band, but also by the ozone 9.6 µm band and the water vapor 6.3 µm band. Typical profiles of infrared cooling and solar heating rates can be seen in Figure 7. The lower stratosphere (between approximately 15 and 25 km) is in approximate radiative equilibrium. Here heating is due to ozone absorption both of visible radiation in its Chappuis bands and also of infrared radiation emanating from lower levels in its 9.6 µm band. Cooling is mainly by carbon dioxide.

In the troposphere radiative transfer is largely accomplished by water vapor, and solar heating is relatively small. However, radiative processes do no not determine the temperature profile in this region. This is because a radiative equilibrium profile would be convectively unstable, i.e., a small upward displacement of an air parcel would result in it remaining hotter than its environment, despite expansion and adiabatic cooling, and thus continuing to rise. The temperature profile of the troposphere is therefore limited by convective processes which result in the adiabatic lapse rate of about –7 K km^{-1} seen in Figure 6. Temperatures are locally warmer than would be the case based on radiative processes alone so that infrared emissions increase. This results in infrared cooling due to tropospheric water vapor, as shown in Figure 7, rather than the warming which would come about from infrared trapping alone. Thus the tropopause marks the region where radiative processes (ozone heating and carbon dioxide cooling) take over from mainly convective processes. Note that an important factor determining the value for the temperature lapse rate is the release of latent heat from the condensation of water vapor into cloud droplets. Thus clouds play an integral part in determining the temperature structure of the lower atmosphere. The accurate representation of clouds and precipitation remains a major challenge in global climate modeling.

Above the ozone layer the effects of ultraviolet absorption by ozone are reduced and there is a minimum in temperature at the mesopause. Higher still, heating due to the absorption by molecular oxygen of far ultraviolet radiation takes over and there is a steep increase in temperature in the lower thermosphere.

The response of atmospheric temperatures to solar variability becomes more marked with altitude. In the thermosphere variations of order 100 K are typical over the 11-year cycle reflecting the large modulation of far ultraviolet radiation. At the stratopause variations of about 2 K are suggested by both satellite measurements and model calculations. In the lower stratosphere the response of temperature to solar variability is not well established but is probably of the order of a few tenths of 1 K. Lower down still, ocean temperatures appear to vary with changes in total solar irradiance [*White et al.,* 1997] while the response of tropospheric air temperatures is probably determined by dynamical factors (i.e., changes in atmospheric circulation) as much, or more than, radiative ones.

5. THE GREENHOUSE EFFECT

5.1. A Simple Model

First it should be noted that the "greenhouse" in this context is a misnomer as garden greenhouses keep warm mainly by inhibiting convection (i.e., trapping the hot air) rather than by limiting infrared emission. Nevertheless, the terminology has achieved such widespread acceptance that it is retained here.

The basic premise of the greenhouse effect is that the atmosphere is relatively transparent to solar radiation, thus allowing it to reach and warm the surface, while being absorptive in the infrared, thus trapping the heat energy at low levels. In a very simple model the atmosphere is assumed to have a single uniform temperature and grey absorption properties (i.e., not varying with wavelength except in as much as they are different for solar and thermal radiation) and to lie above a surface at a different temperature. The whole system is in radiative equilibrium. As discussed in the preceding sections, these assumptions are not accurate but this model does allow us first to observe the principle of greenhouse warming, and second to establish a basis for the concept of the radiative forcing of climate change, discussed below.

Figure 8 shows the fundamentals of the model. Solar irradiance F_s is incident at the top of the atmosphere and a fraction Γ_S of this is transmitted to the surface. The surface, at temperature T_g, emits irradiance F_g and a fraction Γ_L of this reaches the top of the atmosphere. The suffixes $_S$ and $_L$ to the transmittances represent shortwave (solar) and longwave (infrared) properties. The atmosphere, at temperature T_a, emits irradiance F_a in both upward and downward directions.

For radiation balance at the top of the atmosphere and at the surface respectively:

$$F_s = F_g \Gamma_L + F_a \qquad F_s T_S = F_g - F_a$$

from which can be deduced that:

$$F_g = F_s (1 + \Gamma_S) / (1 + \Gamma_L) .$$

If the surface is a black body (i.e., has unit emissivity) then:

$$F_g = \sigma T_g^4 .$$

Furthermore, the solar flux may be related to the equilibrium temperature of the Earth, T_e, by:

$$F_s = \sigma T_e^4 = (1 - \alpha) S / 4$$

where α is albedo, S is total solar irradiance and the factor 4 arises from averaging over the surface of the globe. Thus:

$$T_g^4 = T_e^4 (1 + \Gamma_S) / (1 + \Gamma_L) .$$

With $S = 1368$ W m^{-2} and $\alpha = 0.31$ this expression gives $T_e = 254$ K. If the atmosphere is more transparent to solar

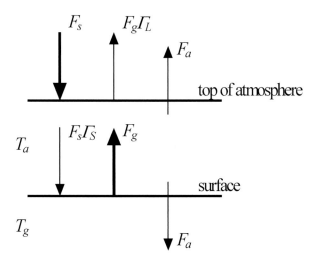

T_a

T_g

Figure 8. Simple model of radiative fluxes used to demonstrate the greenhouse effect. See text for further details.

than thermal radiation then the model predicts that the surface temperature will be greater than the equilibrium temperature. For example, with $\Gamma_S = 0.49$ and $\Gamma_L = 0.10$ (as suggested by Plate 1) $T_g = 1.08 T_e = 274$ K. This is a demonstration of greenhouse warming. In this simple model the atmospheric temperature cools to less than the equilibrium temperature such that the total emitted irradiance remains σT_e^4.

Changes to any of the parameters within the expression for T_g will affect the equilibrium surface temperature. Thus variations in total solar irradiance, planetary albedo, or the concentrations of any of the gases involved in determining atmospheric shortwave or longwave transmittance, will have an impact on climate.

6. RADIATIVE FORCING OF CLIMATE CHANGE

6.1. Radiative Forcing Concept

Continuing for just a little longer with the simple model of the previous section we can express the net downward flux at the top of the atmosphere as:

$$F_N^\downarrow = F_s - F_g T_L - F_a$$
$$= \sigma [T_e^4 - (T_g^4 - T_a^4) \Gamma_L - T_a^4]$$

where it has been assumed that $F_a = (1 - \Gamma_L) T_a^4$, i.e., that the emissivity plus the transmissivity of the atmosphere is unity (Kirchoff's Law). In equilibrium, as discussed above, $F_N^\downarrow = 0$. However, consider a situation in which T_g and T_a have their equilibrium values and some external factor acts to perturb the value of T_e or Γ_L. Then, before T_g and T_a have adjusted and equilibrium is re-established, the instantaneous value of $F_N^\downarrow$ is not zero. The simplest definition of radiative forcing (RF) is just the change in the value of $F_N^\downarrow$. If the RF is positive then there is an increase in energy entering the system (or equivalently a decrease in energy leaving the system) and it will tend to warm until the outgoing energy matches the incoming and the net flux is again zero. The perturbing factors might again be changes in solar irradiance, planetary albedo or the concentrations of radiatively active gases, aerosols or cloud.

The concept of radiative forcing has been found to be a useful tool in analyzing and predicting the response of surface temperature to imposed radiative perturbations. This is because experiments with general circulation models (GCMs) of the coupled atmosphere-ocean system have found that the change in globally averaged equilibrium surface temperature is linearly related to the radiative forcing:

$$\Delta T_g = \lambda \, \Delta F_N^{\downarrow} = \lambda \, RF$$

where λ is the "climate sensitivity parameter" which has been found to be fairly insensitive to the nature of the perturbation and to lie in the range $0.3 < \lambda < 1.0$ K (W m^{-2})$^{-1}$. Thus a calculation of the radiative forcing due to a particular perturbant gives a first-order indication of the potential magnitude of its effect on surface temperature without the need for costly GCM runs.

Note that this relationship between ΔT_g and RF is not consistent with the simple model above (which would suggest that the surface temperature varied with the cube root of the change in flux). This is because of the gross assumptions made in the model: an isothermal, grey atmosphere, neglect of convective adjustment and, most importantly, the role of water vapor in a positive feedback process. This comes about because as the atmosphere warms it can hold more water vapor which acts as a greenhouse gas to increase the warming. Thus the simple model is useful to introduce the fundamentals of the greenhouse effect and the concept of radiative forcing but should not be used in any quantitative assessment of potential temperature change.

6.2. Instantaneous and Adjusted Radiative Forcing

It has been found that the value of λ, and thus the usefulness of the radiative forcing concept, is more robust if, instead of using the instantaneous change in net flux at the top of the atmosphere, RF is defined at the tropopause with the stratosphere first allowed to adjust to the imposed changes. Thus a formal definition of radiative forcing, as used by the Intergovernmental Panel on Climate Change [*Ramaswamy et al., 2001*] is the change in net flux at the tropopause after allowing stratospheric temperatures to adjust to radiative equilibrium but with surface and tropospheric temperatures held fixed. The effects of the stratospheric adjustment are complex as can be illustrated by the case of changes in stratospheric ozone. An increase in ozone masks the lower atmosphere from solar ultraviolet i.e., reducing net flux and thus RF; however, the presence of ozone in the lower stratosphere increases the downward infrared emission (and RF) both directly through the 9.6mm band and also indirectly through the increase in stratospheric temperatures which it produces. Whether the net effect is positive or negative depends on whether the shortwave or longwave effect dominates and this is determined by the vertical distribution of the ozone change.

The direct effect of an increase in total solar irradiance is to increase the radiative forcing; the heating of the stratosphere by the additional irradiance will enhance this by increasing the downward emission of thermal radiation. However, the sign of the radiative forcing due to any solar-induced increases in ozone is not clear—published estimates show both positive and negative values—because of the uncertainties in the distribution of the ozone change.

6.3. Radiative Forcing Since 1750

Plate 4, reproduced from the IPCC 2001 scientific assessment of climate change, shows the RF values deduced for the period 1750 to 2000 for a range of different factors. The largest component, of 2.43 W m^{-2}, is due to the increase in greenhouse gas concentrations. The other components are all of magnitude a few tenths of a W m^{-2}, including the solar contribution which is assessed to be 0.3±0.2 W m^{-2}. The year 1750 is chosen to represent the pre-industrial atmosphere but for a naturally varying factor like the Sun this date is somewhat arbitrary. A choice of later in the 18th century would have given a slightly reduced solar RF but early in the 19th century a significantly larger one.

The value of the climate sensitivity parameter λ is estimated from GCM calculations to be approximately 0.6 K (W m^{-2})$^{-1}$ so that a solar radiative forcing of 0.3 W m^{-2} would indicate that a global average surface warming of only about 0.18 K since 1750 could be ascribed to the Sun. However, the IPCC gives the assignation "very poor" to the level of scientific understanding associated with solar radiative forcing, thereby acknowledging that there may be factors as yet unknown, or not fully understood, which may act to amplify (or even diminish) its effects. These may include additional effects due to solar-induced ozone changes or possibly a direct influence of variations in galactic cosmic rays on cloud cover.

7. CLOUDS

7.1. Radiative Forcing

Clouds have a major impact on both the shortwave and longwave radiation budgets. In the shortwave they reflect solar radiation back to space, thus increasing planetary albedo and reducing the net incoming radiative flux. The magnitude of the reflectance depends on the optical thickness of the cloud, the water phase (liquid or ice), the particle size distribution and possibly particle shape.

In the longwave clouds can trap infrared radiation, acting in a similar way to greenhouse gases. The degree of trapping will depend on the transmissivity of the cloud and also its temperature. High (cold) cloud is more effective because it emits less radiation to space.

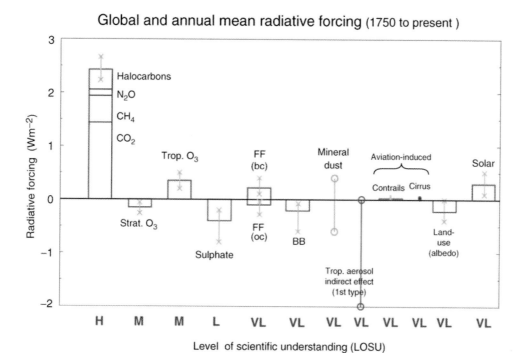

Plate 4. Global, annual average radiative forcing contributions 1750-2000 from the IPCC report [*Ramaswamy et al.*, 2001].

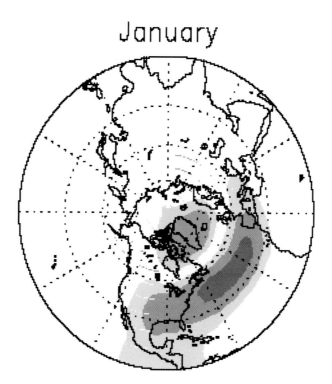

Plate 5. Pattern of sea level pressure modulation due to the North Atlantic Oscillation. The positive phase shows higher than average values in the yellow-red regions and lower in the blue regions.

Thus clouds tend to reduce the incoming solar radiation and reduce the outgoing longwave radiation. Their effect on the net radiation budget depends on whether the shortwave or longwave effect is larger and thus on the location, height and microphysical properties of the cloud. Studies using satellite data suggest that for the global average the net effect of the presence of cloud is to reduce the net absorbed radiation and thus cool the climate (i.e., the shortwave dominates the longwave) but this not true at all locations and times of year. For example, in the winter hemisphere, where solar irradiance is low, or at summer high latitudes, where cloud lies over an icy (high albedo) surface the longwave effect dominates.

In an equilibrium state clouds form part of the overall balance (as shown in Plate 1). However, a factor which induces a change in cloud cover, drop size or altitude will introduce a radiative forcing. If the changes are brought about by another forcing factor then their effects may be viewed as a feedback on the initial forcing. For example, an increase in greenhouse gases might cause a surface warming and this may induce enhanced convection and an increase in cloud cover. The thick convective cloud produced will have a negative radiative forcing and thus reduce the potential greenhouse gas warming. However, such feedback effects are included in the GCMs used to assess the viability of the radiative forcing concept and are thus implicitly included in the value of the climate forcing parameter. Uncertainties and approximations in the representation of cloud formation and cloud radiative properties remain a major cause of uncertainty in current climate prediction models, and are the main reason for the range of uncertainty of approximately a factor two in estimates of λ.

Thus the cloud (or indeed atmospheric humidity) produced by a dynamical response to other forcings can not be viewed as an additional forcing component. Only if changes to cloud properties are induced in situ by chemical or microphysical processes can they produce a radiative forcing, in the climate change sense, rather than a feedback. To investigate how this might occur the processes involved in cloud formation are considered below.

7.2. Cloud Formation

Clouds form when the water vapor in air condenses out. Generally this occurs in rising air masses because these expand due to the reduction in pressure and cool adiabatically. The cooler air has a lower saturated vapor pressure and thus the air mass, with a given humidity, may become saturated. However, this is not a sufficient condition: it would be possible for relative humidities to reach values up to 500% without any condensation occurring. The water vapor requires a suitable surface, called a condensation nucleus, on which to condense. If the condensation nucleus is not a water surface then heterogeneous nucleation is said to take place; if it is then homogeneous nucleation occurs. In the free atmosphere, however, heterogeneous nucleation is the only important process because homogeneous nucleation requires prohibitively high relative humidities. (For a comprehensive discussion of cloud formation see the classic text by *Ludlam*, [1980]).

Particles which act as condensation nuclei include sea salt, sulphates, mineral dust and aerosols produced from biomass burning. The concentration and composition of atmospheric aerosol vary geographically with, for example, sulphate aerosol being more abundant in the northern hemisphere as it is generated in industrial regions. A region with a higher concentration of condensation nuclei will produce a larger number of smaller cloud droplets than a remote area with clean air which will produce fewer larger droplets for the same total water content. Smaller drops are more effective at scattering radiation so that the cloud produced in the air with more aerosol has a higher albedo. This has been demonstrated in satellite images of ship tracks where emissions from a ship's funnel can be seen to modify cloud albedo. This effect has become known as the "first indirect effect of aerosol" in the context of the radiative forcing of climate change by anthropogenic aerosol. Plate 4 shows the direct radiative forcing of various aerosol types since 1750. For example, sulphate aerosol has produced a RF -0.4 W m^{-2}, negative because it enhances the albedo. Also shown in Plate 4 is a bar from 0 to -2 W m^{-2} representing the indirect effect of all aerosol types. The magnitude is very uncertain but probably negative for the reasons outlined above.

Clearly, if solar activity can produce a direct impact on cloud cover then there is scope for a considerable amendment to the solar radiative forcing value based on incident irradiance alone. It has been proposed that variations in cosmic rays could provide such a mechanism by modulating atmospheric ionization resulting in the electrification of aerosol and the effectiveness of this aerosol to act as condensation nuclei [*Dickinson*, 1975]. Other possible processes are discussed by *Tinsley and Yu* [this volume]. That cosmic rays are modulated by solar activity and that this causes a variation in atmospheric ionization are not contentious. The various stages whereby the ionized air may enhance the number of cloud condensation nuclei are the subject of much current research. However, even if such a mechanism proves feasible to produce a measurable effect on cloud cover or properties, the magnitude, and even the sign, of the impact on radiative forcing remains uncertain as it will depend on the cloud location and physical properties, as discussed in the section 7.1.

8. MODES OF VARIABILITY OF THE ATMOSPHERE

8.1. Overview

A description of the general circulation of the atmosphere, and the physical basis for this, is beyond the remit of this paper. It may be useful, however, to briefly consider natural modes of variability of the atmosphere as it has been suggested that the impact of solar variability, as well as other climate forcing factors, may be to affect the frequency of occupation of certain phases of these modes.

Identified modes of variability are several, each with different geographical influences, but all showing specific patterns of response with large regional variations. The El Niño-Southern Oscillation (ENSO) phenomenon is the leading mode in the tropics, although its influence is felt globally. At higher latitudes the leading northern winter modes are the North Atlantic Oscillation (NAO) and the Pacific-North America Oscillation (PNA), which are sometimes viewed as part of the same phenomenon, referred to as the Arctic Oscillation (AO). There is an Antarctic Oscillation (AAO) in the southern hemisphere. In the equatorial lower stratosphere a Quasi-Biennial Oscillation (QBO) modulates winds and temperatures. Here discussion is restricted to the NAO (AO) and the QBO as these modes have been most widely suggested to respond to solar activity.

8.2. The North Atlantic Oscillation

Over the Atlantic in winter the average sea level pressure near 25–45°N is higher than that around 50–70°N. This pressure gradient is associated with the storm-tracks which cross the ocean and determine, to a large extent, the weather and climate of western Europe. Since the 1930s it has been known that variations in the pressure difference are indicative of a large-scale pattern of surface pressure and temperature anomalies from eastern North America to Europe. If the pressure difference is enhanced then stronger than average westerly winds occur across the Atlantic, cold winters are experienced over the north-west Atlantic and warm winters over Europe, Siberia and East Asia, with wetter than average conditions in Scandinavia and drier in the Mediterranean. The fluctuation of this pattern is referred to as the NAO and the pressure difference between, say, Portugal and Iceland can be used as an index of its strength, see Plate 5. Some authors regard the NAO as part of a zonally symmetric mode of variability, the AO, characterized by a barometric seesaw between the north polar region and midlatitudes in both the Atlantic and Pacific.

Using individual station observational records of pressure, temperature and precipitation, values for the NAO index have been reconstructed back to the seventeenth century. The index shows large inter-annual variability but until recently has fluctuated between positive and negative phases with a period of approximately four years. Over the past two decades, however, the NAO has been strongly biased toward its positive, westerly, phase and it has been suggested that this might be a response to global warming. Computer models of the general circulation of the atmosphere (GCMs) are quite successful in simulating NAO-type variability and some GCM studies do show increasing values of the NAO index with increased greenhouse gases. However, this is not true of all models' results and some studies suggest that the NAO pattern itself may be modified in a changing climate so that use of simple indices may not be appropriate. One recent GCM study [Shindell et al., 2001] of the Maunder Minimum period has suggested that the negative phase of the NAO may have been dominant during that period of low solar activity.

Another uncertainty is how significant might be coupling with either sea surface temperatures and/or the state of the middle atmosphere in producing a realistic NAO/AO pattern. Planetary-scale waves, produced in the lower atmosphere by longitudinal variations in topography, propagate upwards in winter high latitudes through the stratosphere and deposit momentum and heat which feeds into the general atmospheric circulation. Where this wave absorption takes place depends on the ambient temperature and wind structure. Thus any changes induced in the mean temperature structure of the stratosphere may result in a feedback effect on lower atmosphere climate. An analysis of zonal wind observations does suggest a downward propagation of AO patterns in many winters [Baldwin and Dunkerton, 2001]. This offers a plausible mechanism for the production of NAO/AO-type signals in tropospheric climate by factors which affect the heat balance of the stratosphere [Kodera, 1995], specifically solar variability. Data and modeling studies have already shown such a response to heating in the lower stratosphere by volcanic eruptions [Robock, 2001].

8.3. The Quasi-Biennial Oscillation

In the equatorial lower stratosphere an oscillation in zonal winds occurs with a period of approximately 28 months. A given phase (east or west) starts in the upper stratosphere and moves downward at a rate of about 1km per month to be replaced by winds of the opposite phase (see Figure 9). The largest amplitude in the zonal wind variation occurs at about 27 km altitude. The QBO comes about because of interactions between vertically propagating waves and the mean flow. When the wind blows from the west (QBO west phase) westward moving waves can propagate freely but

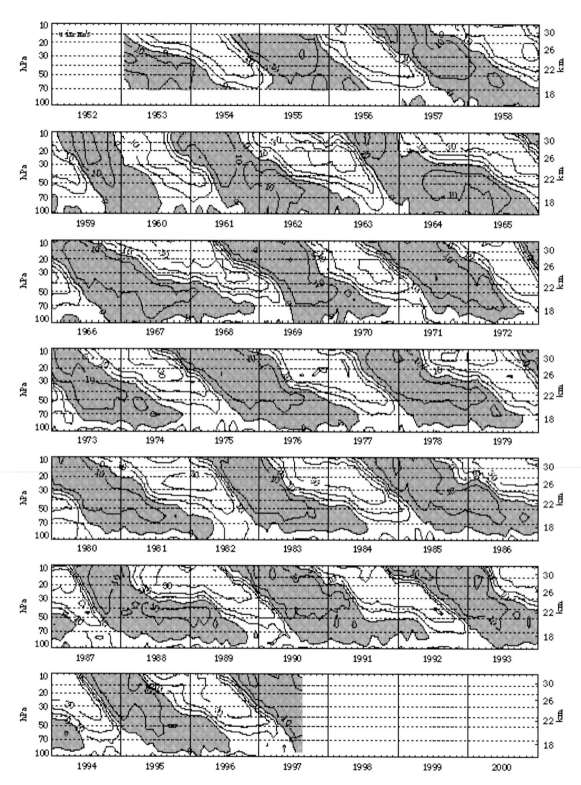

Figure 9. Zonal wind (m s[-1]) at Singapore as a function of height and time showing the Quasi-Biennial Oscillation. Westerly winds shown as shaded, contours at 10 m s[-1]. From the Free University of Berlin.

eastward moving waves are absorbed and deposit their momentum, thus strengthening the existing westerlies and moving the westerly peak downward. Somewhat above this absorption layer the westward moving waves are dissipated and weaken the west wind, eventually changing the direction to easterly. The absorption of the westward waves then starts to propagate downward, reversing the phase of the QBO. *Baldwin et al.* [2001] give a good review of current understanding of the QBO.

The effects of the QBO are not restricted to equatorial regions. The transport of heat, momentum and ozone to high latitudes are all modulated. On average the westerly phase of the QBO is associated with colder winter temperatures at the north pole in the lower stratosphere. This can be explained by an enhanced ability of midlatitude planetary waves to propagate into the equatorial westerlies leaving the cold winter pole undisturbed. However, this relationship only appears to hold when the Sun is at lower levels of activity. Near 11-year cycle maxima the relationship breaks down and possibly reverses. The absorption of solar radiation is greater in the stratosphere than lower down and its modulation by solar activity quite significantly larger (see section 3). The potential for solar modulation of stratospheric temperatures and winds to modify planetary wave propagation, and hence tropospheric climate, is an active area of research [*Hood*, this volume].

Signals with a periodicity of 2–3 years have also been found in records of surface temperature and precipitation and, recently, in the NAO index. This interaction between modes presents a much more complex picture but again suggests a mechanism whereby the response of climate parameters to solar variability could have a marked geographical distribution.

9. CONCLUSIONS

Radiation from the Sun ultimately provides the only energy source for the atmosphere. Changes in solar irradiance undoubtedly impact the Earth's energy balance, thermal structure and composition. Questions remain, however, concerning the detailed mechanisms which determine to what extent, where and when these impacts are felt. It is only by further understanding of the complex interactions between radiative, chemical and dynamical processes in the atmosphere that these questions will be answered.

REFERENCES

Andrews, D. G., An introduction to atmospheric physics, CUP, 2000.

Baldwin, M. P. and T. J. Dunkerton, Stratospheric harbingers of anomalous weather regimes, *Science, 294,* 581-584, 2001.

Baldwin, M. P., L. J. Gray, T. J. Dunkerton, K. Hamilton, P. H. Haynes, W. J. Randel, J. R. Holton, M. J. Alexander, I. Hirota, T. Horinouchi, D. B. A. Jones, J. S. Kinnersley, C. Marquardt, K. Sato and M. Takahashi, The quasi-biennial oscillation, *Rev. Geophys., 39,* 179-229, 2001.

Dickinson, R. E., Solar variability and the lower atmosphere, *Bull. Amer. Meteor. Soc., 56,* 1240-1248, 1975.

Haigh, J. D., The role of stratospheric ozone in modulating the solar radiative forcing of climate, *Nature, 370,* 544-546, 1994.

Hood, L., Effects of solar UV variability on the stratosphere, *this volume.*

Jackman, C. H. and R. D. McPeters, the effect of solar proton events on ozone and other constituents, *this volume.*

Kiehl, J. T. and K. E. Trenberth, Earth's annual global mean energy budget, *Bull. Amer. Meteor. Soc., 78,* 197-208, 1997.

Kodera, K., On the origin and nature of the interannual variability of the winter stratospheric circulation in the northern hemisphere, *J. Geophys. Res., 100,* 14077-14087, 1995.

Larkin, A., Investigation into the effects of solar variability on climate using atmospheric models of the troposphere and stratosphere, Ph.D. thesis, University of London, 2000.

Ludlam, F. H., Clouds and storms: the behavior and effect of water in the atmosphere, Pennsylvania State University Press, 1980.

McCormack, J. P. and L. L. Hood, Apparent solar cycle variations of upper stratospheric ozone and temperature: Latitudinal and seasonal dependences. *J. Geophys. Res., 101,* 20933-20944, 1996.

Ramanathan, V. and A. M. Vogelmann, Greenhouse effect, atmospheric solar absorption and the Earth's radiation budget: from the Arrhenius-Langley era to the 1990s, *Ambio, 26,* 38-46, 1997.

Ramaswamy, V., O. Boucher, J. D. Haigh, D. Hauglustaine, J. Haywood, G. Myhre, T. Nakajima, G. Y. Shi and S. Solomon, Radiative forcing of climate change. In *Climate Change 2001: The Scientific Basis. Contribution of Working Group 1 to the Third Assessment Report of the Intergovernmental Panel on Climate Change*, eds. Houghton, J. T., Y. Ding, D. J. Griggs, M. Noguer, P. J. van der Linden, X. Dai, K. Maskell and C. A. Johnson. (Cambridge University Press, London and New York), 2001.

Robock, A., Stratospheric forcing needed for dynamical seasonal prediction, *Bull. Am. Meteorol. Soc., 82,* 2189-2192, 2001.

Rottman, G., L. Floyd and R. Viereck, Measurement of the solar ultraviolet irradiance, *this volume.*

Shindell, D. T., G. A. Schmidt, M. E. Mann, D. Rind and A. Waple, Solar forcing of regional climate change during the Maunder Minimum, *Science, 294,* 2149-215, 2001.

Strobel, D. F., Parameterization of the atmospheric heating rate from 15 to 120 km due to O_2 and O_3 absorption of solar radiation, *J. Geophys. Res., 83,* 6225-6230, 1978.

Tinsley, B. A. and F. Yu, Atmospheric ionisation and clouds as links between solar activity and climate, *this volume.*

White, W. B., J. Lean, D. R. Cayan and M. D. Dettinger, Response of global upper ocean temperature to changing solar irradiance, *J. Geophys. Res., 102,* 3255-3266, 1997.

Joanna D. Haigh, Space and Atmospheric Physics, Blackett Laboratory, Imperial College London SW7 2AZ, U.K.

Section 2

Solar Energy Flux Variations

Section 2. Solar Energy Flux Variations

Hugh S. Hudson

Space Sciences Laboratory, University of California, Berkeley, California.

The chapters in this section of the monograph deal with the basic raw material of solar variability, namely the measurements themselves. With complete characterization of the spectral components of the irradiance, one might imagine an easy task in putting them all together to determine the energy flux from the Sun. As the length and depth of these chapters shows, however, the simple characterization of a spectral irradiance at the accuracies permitted by the available technology already becomes a sophisticated business. The scope of the problem becomes apparent when one realizes that the observed bolometric variability of the total irradiance does not exceed fractions of a percent, and has well-measured components with amplitudes below one part per million (*Fröhlich*)!

The introductory chapter of this session (*Kuhn and Armstrong*) deals with mechanisms; the subsequent chapters continue with reviews of the total-irradiance observations (*Frölhich*), the ultraviolet (*Rottman, Floyd, and Viereck*), the extreme ultraviolet (*Woods et al.*). The knowledge thus obtained needs a synthesis for further research into causes and effects, and *Thuillier et al.* discuss the current best such synthesis.

Notably in this Section we find no chapters devoted exclusively to infrared and radio variability, an omission justified by the idea that these wavelength ranges are well-behaved and/or trivial as components of the total irradiance. Please see [*Kundu et al.*, 1989] for further information on the longer wavelengths. We do find a chapter on solar energetic particles (*Lario and Simnett*); these also have no real significance in comparison with the total irradiance. But the particles may act as catalysts for important geophysical effects, as reviewed by *Jackman and McPeters* and *Tinsley*

and Yu elsewhere in this monograph. Note also that the "luminosity" associated with the particles of the solar wind might approach one part per million of the total irradiance, not entirely negligible in view of the precision of present total-irradiance measurements now. Thus we must bear non-radiative components in mind as we discuss this subject. We might also note for completeness the neutrino energy loss from the solar core. This subject, including the extremely marginal evidence for variability, properly belongs again with the physicists now that the solar interior models have been vindicated [*SNO Collaboration*, 2002].

Many forms of solar irradiance variability – from the shortest time scales to the longest yet observable – result from solar magnetic activity, as discussed by *Fox*. Nevertheless we note that the power spectrum of total irradiance variability (see *Fröhlich*) contains both line and continuum components attributed to non-magnetic sources. The line features in the spectrum are the global oscillations described as p-modes in helioseismology, and also known as the 5-minute oscillations. The spectral continua show us the solar surface effects of convection, specifically the granulation and supergranulation, as well as the irradiance anisotropies due to magnetic activity. To what extent does our knowledge of stellar convection permit us to understand the non-magnetic variations? *Kuhn and Armstrong* discuss some possible implications of our improving knowledge of convection, but observationally it remains unclear to what extent the simple picture derived from mixing-length theory may not suffice. We may be sure that, if non-mixing-length effects have not been observed yet, they will indeed be observed in the future. The measurement of solar irradiance variability (and relates phenomena such as the shape of

Solar Variability and its Effects on Climate
Geophysical Monograph 141
Copyright 2004 by the American Geophysical Union
10.1029/141GM07

the Sun, as described by *Kuhn and Armstrong*), in fact provides an interesting channel for studies of the solar interior that is independent of helioseismology.

REFERENCES

Kundu, M. R., Woodgate, B. E., and Schmahl, E. J., Energetic phenomena on the Sun, Kluwer, Dordrecht, 1989.

SNO Collaboration, Direct evidence for neutrino flavor transformation from neutral-current interactions in the Sudbury Neutrino Observatory, *Phys. Rev. Lett.*, 89, 011301, 2002.

Hugh S. Hudson, Space Sciences Laboratory, University of California, Berkeley, CA 94720–7450, USA. (hhudson@ssl.berkeley.edu)

Mechanisms of Solar Irradiance Variations

J. R. Kuhn and J. D. Armstrong

Institute for Astronomy, University of Hawaii, 2680 Woodlawn Dr., Honolulu, HI, 96822

This section discusses how measurements of the Sun's surface brightness may be used to determine its total emergent flux. We explore how solar luminosity and irradiance changes are related but distinct phenomena, which are not well treated in one-dimensional or diffusive solar convection zone models. Efforts to improve our knowledge of the solar luminosity are essential, since a refined understanding of the variability of the total solar luminosity, or even of the net emergent energy flux from isolated parts of the photosphere are critical tools for understanding the physics of the convection zone and for probing deeper into the solar interior. In combination with magnetic, helioseismic, and numerical simulation observations and tools, the global solar luminosity and irradiance variability can reveal much about the mechanisms of the solar cycle.

1. INTRODUCTION

One method for obtaining the Sun's luminosity uses knowledge of its radiation field on a spherical surface of radius, r, (perhaps near the photosphere). On this surface we seek knowledge of the specific intensity, $I(\lambda, \theta, \phi, \hat{n}, r, t)$ where θ, and ϕ define a point in solar colatitude and longitude, and $\hat{n}$ defines the field direction from which we measure the surface brightness at wavelength λ at time t. While the detailed wavelength dependence of I contains a wealth of information which is determined by the structure and atomic physics of the Sun's photosphere, chromosphere, and corona, this discussion is focused on the Sun's overall radiative energy balance.

At visible (and near-IR) wavelengths the spectral distribution of the Sun's radiation is dominated by thermal contributions. For example the spatial variation of $I(\theta, \phi)$ due to sunspots and faculae is well represented by small temperature perturbations of a solar reference blackbody spectrum [*Lin and Kuhn* 1992]. Non-thermal and high temperature

UV, EUV, and X-ray contributions are important and the coupling of these higher energy photons to the Earth's upper atmosphere probably has a significant role in determining how solar variations affect terrestrial climate. Nevertheless most of the Sun's energy is radiated with a thermal distribution at temperatures near 5700K.

In practice we must infer the Sun's radiation output from photospheric optical measurements derived from full-disk imagery as projected onto the plane of the sky. We take $F(\lambda, \theta, \phi, t)$ to describe the Sun's surface brightness (or mean intensity) at inferred colatitude and longitude θ and ϕ in the photosphere as it is obtained by integrating I over outward directions $\hat{n}$ corresponding to an earth-bound photometer. We assume a heliocentric spherical coordinate system with $\hat{z}$ in the plane of the sky parallel to the Sun's rotation axis and take the azimuthal angle ϕ to be zero along the line-of-sight from Earth-bound observers to disk center. Broadband optical observations nominally sample the local thermodynamic conditions in the solar photosphere at an optical depth approximately equal to the cosine of the central angle, $\mu = \hat{n} \cdot \hat{r} = \sin\theta \cos\phi$, (the Eddington-Barbier approximation). Here $\hat{n}$ is a unit vector directed toward the observer and $\hat{r}$ is a unit radial vector at a point θ and ϕ in the Sun's atmosphere.

Since observations of F do not determine I at a constant radius or over all angles $\hat{n}$, any determination of the solar

Solar Variability and its Effects on Climate
Geophysical Monograph 141
Copyright 2004 by the American Geophysical Union
10.1029/141GM08

luminosity depends on a model of the mean atmosphere. Practically we account for the limb-darkening effect on the photospheric blackbody spectrum with a function, $l_\lambda(\mu)$. We note that one-dimensional (1-d) atmosphere models, varying only in radius, *fundamentally cannot distinguish* irradiance and luminosity mechanisms, since there is only one outward ray path!

The spatial brightness variations in F obviously require a horizontally inhomogeneous atmosphere to account for the apparently bright and dark magnetic structures seen in the photosphere. Detailed magnetic flux tube models of faculae or sunspots imply an anisotropic radiation distribution [cf. *Spruit* 1976]. Clearly one radiative effect of magnetic fields in a model or observation (that may not even spatially resolve them) is to induce anisotropy in the radiation field. For example, the flux from faculae is enhanced in directions perpendicular to the vertical magnetic field. We use functions $C_f(\mu)$ and $C_s(\mu)$ to describe the facular and spot radiation anisotropy (also called "center-to-limb contrast"). We also assume that these functions account for limb-darkening and any implicit non-thermal dependence with wavelength. If we now take T_p, T_s, and $T_f(\theta, \phi)$ to describe the local temperature of the photosphere, in sunspots, or in faculae then we can write

$$F(\lambda, \theta, \phi, t) = B_\lambda(T_p(\theta, \phi, t)l_\lambda(\mu) + B_\lambda(T_f(\theta, \phi, t) C_f(\mu) + B_\lambda(T_s(\theta, \phi, t)C_s(\mu) \quad (1)$$

where $B_\lambda(T)$ is the Planck function at temperature T. If F is accounted for by this model then it is straightforward to compute its integral over θ, ϕ and outward radiation angles to obtain the irradiance or total solar luminosity. For example, the full-disk spectral irradiance is

$$I_\lambda(t) = \int_{-\pi/2}^{\pi/2} \int_0^\pi F(\lambda, \theta, \phi, t)\mu \sin\theta d\theta d\phi. \quad (2)$$

2. IRRADIANCE AND LUMINOSITY

One may question whether such a simple thermal model of the solar disk adequately describes its surface brightness. Of course if we look in spectral detail at the solar irradiance, or look outside of the optical-IR wave-lengths where most of the Sun's energy is radiated then this description suffers. But, as long as we confine our description of the solar irradiance to broadband or bolometric measurements (as in *Lin and Kuhn* 1992), then this Planckian description is quite adequate to the level of the bolometric observations.

A successful interpretation of solar irradiance observations must match model to data at a fractional level of better than 10^{-3}. This is the change we observe in the Sun's

bolometric irradiance, both on solar rotation timescales (times from a day to a month) and solar cycle timescales (times from a few months to a few years).

One should not expect a single physical mechanism to account for the irradiance perturbations over different timescales. For example, the physical origin of irradiance or luminosity changes on timescales like flares or 5-minute oscillations or the much longer stellar evolutionary brightness timescales obviously involve very different physics. In these cases particle beam heating, adiabatic pressure fluctuations, and thermodynamic changes effectively driven by the varying mean atomic mass as hydrogen is converted to helium, are responsible.

2.1. Outer Atmosphere Changes?

An important question is whether or not changes in the outer atmosphere of the Sun near and above unity optical depth can account for the observed solar variability. In particular, could opacity perturbations in the Sun's outer atmosphere cause the 0.1% bolometric flux changes that are observed at Earth?

2.1.1. One-dimensional? Let's consider a 1-d model for the changing atmosphere, where $T(z)$, $\kappa(z)$, and $S(z) = B(T(z))$ describe a local thermodynamic model, where $T(z)$ is the local temperature at depth z (measured into the Sun), κ is the opacity, and S is the local source function [cf. *Mihalas* 1978]. Let $\kappa(z < 0) = 0$ and optical depth $\tau(z < 0) = 0$ so that $z = 0$ corresponds to a "photosphere" and decreasing z corresponds to the outward direction. We compute the emergent intensity L (which trivially equals the irradiance and luminosity in a 1-d model), from knowledge of $T(z)$ as $L = \int_0^\infty S(\tau) \exp(-\tau)d\tau$. On long timescales we recognize that there is some depth, z_0, below which T is unaffected by changes in the atmospheric opacity above. In other words the energy flux entering the atmosphere from below (the solar interior) is constant. We can ask then what happens to the emergent flux if "sunspot" or "facular" magnetic structures in a one-dimensional solar atmosphere perturb $\kappa(z < z_0)$. Of course, in equilibrium, there can be no change in L in response to changes to $\kappa(z)$ since the temperature in the atmosphere and the source function adjust to conserve energy. The effective temperature of the 1-d photosphere can't be changed by perturbing the atmospheric opacity—it is determined by $T(z_0)$, to the extent that $T(z_0)$ is fixed. In order to change the equilibrium luminosity the energy flux through the reference surface at depth z_0 must be affected.

With these assumptions the only way to affect the luminosity is to make a dynamic change to the atmosphere. For example, a 0.1% luminosity increase requires an average inward gas velocity of more than 10 m/s near unity optical depth (where the gas pressure is 10^5 dyn/cm^2 in the Sun) in order to conserve energy. This isn't a large velocity compared to typical random granulation flows but constraints from observations of the solar gravitational redshift [cf. *Lopresto et al.* 1991] appear to rule out such a systematic (solar cycle) Doppler change. Because the magnetic energy in the solar atmosphere has the wrong phase (it is largest at solar maximum when the irradiance is also largest) it is also unlikely that magnetic fields are the sink/reservoir for a variable (solar cycle) luminosity.

2.1.2. Multidimensional. A two dimensional atmosphere admits anisotropy in the emergent radiation field. Thus a stratified atmosphere perturbed by opacity inhomogeneities due to imbedded magnetic structures can affect both the local photospheric intensity and flux.

In a real Sun we expect that as z_0 gets larger (deeper) the timescale for changing the temperature of the reference surface through which we measure the solar luminosity, $T(z_0)$, grows longer. Outer opacity perturbations should have a smaller effect on the net flux leaving the deeper fiducial reference surface. We will compute the timescale and efficacy of this constant $T(z_0)$ approximation below but note here that if the emergent luminosity at z_0 is constant over a timescale t_0, and the upper layers equilibrate on shorter timescales then even an inhomogeneous opacity distribution can not affect the photospheric luminosity on timescales comparable to t_0. The emergent luminosity will be determined by energy conservation and the flux at z_0.

It should be evident that to understand the origins of solar variability we must know whether flux changes are due to a changing solar luminosity or are due to redistribution of the Sun's radiated energy into or out of the plane of the ecliptic.

Opacity perturbations in a two (or three) dimensional atmosphere may affect the specific intensity but they do not change the mean intensity or the total radiated power (or luminosity) on timescales where the outward energy flux is fixed by the incoming power below the photosphere. We have argued that the alternative, that the observed luminosity change results from storing kinetic or magnetic energy in the atmosphere, is also unlikely.

2.2. Magnetic Fields and Timescales

Magnetic fields dominate any empirical description of the solar dynamics on timescales of decades or shorter. This is especially true for irradiance where fractional bolometric perturbations are of order 10^{-3} but correlated magnetic changes are fractionally of order unity. Because these are so much larger we expect that irradiance changes are directly

or indirectly driven by magnetic fields. Our goal here is to understand the connection between solar magnetic fields, and irradiance and luminosity perturbations over various timescales.

A simple illustration of how magnetic flux affects solar irradiance differently on rotation and solar cycle timescales comes from disk-averaged observations. Figures 1 and 2 show solar irradiance data from ACRIM and Kitt Peak full-disk flux density maps that have been averaged over 30 day intervals. A 150d running boxcar average is also plotted here. The correlation between the irradiance and magnetic data, and the corresponding residuals after removing the 150d running average are shown in Figures 3 and 4.

The appearance of additional magnetic flux on the Sun on short timescales makes it appear *darker* (by about 0.2 W/m^2/G). On timescales longer than about a solar rotation period a 1 Gauss flux density change is associated with a 0.1 W/m^2 *brighter* Sun. Whether the changing solar brightness is due to energy flux distribution or is caused by a genuine change in the solar luminosity is evidently an important question whose answer depends on the timescale of the change.

2.2.1. Evidence of Solar-cycle Luminosity Changes. Although we lack the remote sensing tools needed to routinely measure the solar radiative flux above and below the ecliptic plane there is good indirect evidence that the Sun's luminosity varies on yearly timescales. This comes from both photometry and helioseismic observations.

In principle, sensitive photometry can be used to estimate the functions T_f, T_p, and C_f, C_p defined in Eq. (1). Sunspots are relatively easy to observe and their flux contribution to Eqs. (1) and (2) is obtained by measuring and scaling sunspot areas. In practice the direct measurement of T_f and

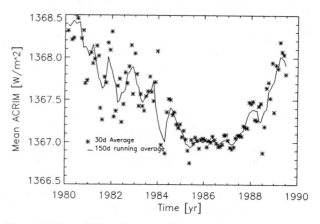

Figure 1. The solar irradiance as measured by the ACRIM satellite is shown here in 30 day averages (points) and the 150 day running average (solid line)

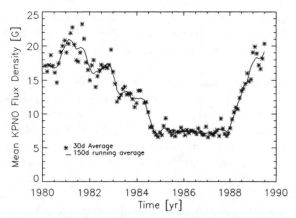

Figure 2. The average flux density on the solar disk as measured by Kitt Peak National Observatory magnetographs is plotted here for the timespan defined by Figure 1. The 30 day average (points) and 150 day running mean (solid line) is shown.

T_p, at a level comparable to observed 0.1% changes in $I(t)$, is difficult. Differential limb photometry obtained during the 80's and 90's had temporal resolution of a few months and enough sensitivity to infer changes in $I(t)$ by directly evaluating Eq. (2) from T_f and T_p [cf. *Kuhn* 1996]. These data imply that the Sun's luminosity varies during the solar cycle in phase with the measured satellite irradiance changes.

During this period (and up to the present) helioseismic normal mode frequency measurements have been obtained that also show solar cycle changes. A straight-forward interpretation of these suggests a changing near-photosphere temperature structure (T_p and T_s). Since the photospheric temperature affects the local sound speed there is a measureable affect on the solar normal-mode acoustic oscillation frequencies [*Kuhn* 1989]. The observed frequency variations are consistent with the photometrically inferred surface temperature functions and support the evidence for a changing solar luminosity. More recently space observations [*Kuhn et al.* 1998] confirm the previous solar cycle T_p structure, but with temporal resolution as good as a few hours.

2.2.2. On Short Timescales the Sun's Luminosity May be Constant. We can estimate functions like $C_f(\mu)$ or $C_s(\mu)$ from a timeseries of flux observations obtained from the rotating Sun. This calculation requires very accurate total flux measurements, $I(t)$, over a long period of time. The only data with sufficient accuracy and stability are obtained from space using instruments which integrate over the full-disk. In this case we can determine only statistical information about the contrast function of photospheric active regions.

One approach for obtaining the angular distribution information is to compare solar irradiance measurements separated in time by a fraction of the 27 day equatorial rotation period. Defining the autocorrelation $A(\tau)$ in terms of the intensity timeseries $I'(t)$, which has had low frequency solar cycle trends removed from $I(t)$, we have $A(\tau) = \Sigma_{ti} \, I'(t_i) \, I'(t_i-\tau)/N_\tau$ with N_τ the number of daily samples at time lag τ. For example, a bright flux change caused by magnetic activity near disk center should be dark about 7 days earlier or later (a time when it is viewed nearly tangentially through the photosphere) if the magnetic feature doesn't affect the Sun's luminosity on rotation timescales. Figure 5 shows $A(\tau)$ evaluated from ACRIM data [*Kuhn* 1996]. The magnitude of the negative dip near a lag of 7 days is consistent with *no change in the solar luminosity* due to sunspots and faculae on timescales shorter than a few months.

3. THE CONVECTIVE INTERIOR

Beneath the photosphere lies a part of the Sun we still can't describe very accurately. The thermal transport properties of this convection zone are in dispute. At one extreme is the view represented by *Spruit* [1991] that the mixing length theory adequately describes a very high conductivity medium which also has a very long thermal reradiation timescale. The implications of this are that solar luminosity changes are very difficult to drive on short timescales unless they are confined to the outer few scale heights below the photosphere. The alternative view represented in *Kuhn* [1996] is that the SCZ conductivity is anisotropic, considerably less than MLT estimates, and that energy reradiation times are controlled by the circulation time between fluid elements and the photosphere.

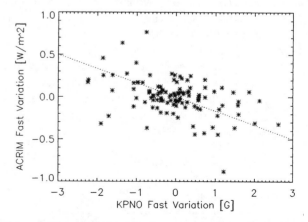

Figure 3. The correlation between magnetic field and irradiance on short timescales is plotted here. The negative correlation is highly significant and was obtained here after subtracting the 150 day running average from each 30 day average point.

3.1. Mixing Length or Plumes

Stellar convection requires a large range of length and timescales to adequately describe the fluid motion and its heat and mass transport properties. Classical mixing length theory [cf. *Bohm – Vitense* 1958] does well to simplify the driving influence of bouyancy and gravity in order to explain the mean convective heat transport and stratification of the solar convection zone (SCZ). Despite such success there is ample evidence that the evolution of heat *perturbations* in the SCZ is *non-diffusive* and requires more than classical mixing length theory (MLT).

Laboratory and numerical experiments [e.g. *Kerr* 1996; *Stein and Nordlund* 1998] explain a new *paradigm* for solar convection – cool dense threads of fluid descend to deeper layers, while broad slow upflows replace the fluid which cools from the surface. In contrast to MLT predictions, circular "eddies" of fluid on the scale of the pressure scale length do not characterize realistic solar CZ simulations. In the vertical direction convective flows or plumes maintain their identity over many pressure scale lengths. Thus while MLT flow exhibits large scale isotropy, the plume motions are anisotropic with larger vertical than horizontal flows.

The success of MLT lies in its ability to account for the mean CZ stratification but it does a poor job of explaining how perturbations in the CZ should evolve. MLT relates the deviation of the actual vertical CZ stratification from the adiabatic gradient. This superadiabaticity then determines an eddy velocity which in turn affects the convective energy transport and the net superadiabaticity.

In the presence of a steady convection zone infinitesmal local entropy (or temperature) perturbations should diffuse in response to local flows. In this case the diffusion of a fluid parcel will have very little to do with its local bouyan-

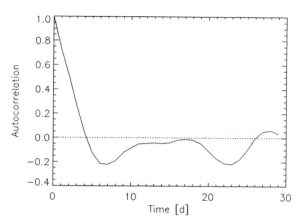

Figure 5. After removing low frequency trends (periods longer than 150 day) the daily sampled ACRIM data autocorrelation function was evaluated versus time lag (measured in days). The negative dip near a lag of 6 days is highly significant for any 1 year period during the solar cycle.

cy but a great deal to do with the effects of nearby turbulent motions. We expect that the evolution of these perturbations within the SCZ can be described by a diffusion equation.

Finite amplitude entropy perturbations, as seen in numerical convection experiments, will generate plumes and non-local (anisotropic) correlated fluid flow over many scale lengths. If such a hot fluid parcel gets advected from the interior to the photosphere, then we should expect rapid radiative equilibrium with the local surroundings to occur. In this picture entropy perturbations introduced into the SCZ are radiated from the photosphere on a timescale over which fluid parcels are recirculated from the interior to the surface.

3.2. Conductivity

To describe the linear evolution of infinitesmal temperature perturbations *Kuhn and Georgobiani* [2000] use an anistropic diffusive model for heat transport

$$\frac{\partial(\kappa_{ij}\partial T)}{\partial x_i \partial x_j} = \rho C_p \frac{\partial T}{\partial t} \tag{3}$$

where κ_{ij} represents an effective thermal conductivity tensor, T is the local temperature, ρ is the density, C_p is the specific heat per mass at constant pressure and summation over repeated indices is assumed.

In the MLT the thermal conductivity is approximately $\kappa_{MLT} = C_p\rho L\upsilon$ where L is the length scale over which an eddy maintains its identity and υ is a characteristic fluid velocity. Computing κ in a realistic plume-like CZ requires a more detailed description of the turbulent flow. The

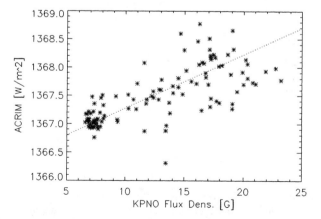

Figure 4. Similar to Figure 3, but here magnetic field and irradiance data is plotted without removing low frequency (150 day period) trends. The positive correlation is highly significant.

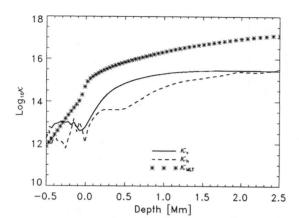

Figure 6. A full 3d realistic numerical simulation of convection was used to derive the effective diffusion conductivity near the top of the solar convection zone. The plotted stars show that the MLT convection is about an order of magnitude larger than the vertical and horizontal diffusive conductivities (solid and dashed lines). The convergence of model calculations at depth is likely to be a boundary effect.

approach *Kuhn and Georgobiani* [2000] took was to solve, in a least-squares sense, for the "coefficients" κ_{ij} using Eq. (3). From a 3-d numerical simulation [*Stein and Nordlund*, 1998] of convection all of the derivatives of T over the simulation volume can be computed and κ_{ij} is directly recovered from Eq. (3). These calculations assumed no external heat source, the intrinsic random convective fluctuations in T were sampled from the simulation to solve for κ.

As expected on physical grounds, the set of tensor coefficients required to describe the diffusion simplifies to only a horizontal and vertical conductivity (κ_h and κ_v. Off-diagonal elements of the tensor are found to be orders of magnitude smaller and insignificant. The fitting technique also allows an evaluation of the significance of the diffusive approximation by considering the goodness-of-fit χ^2 statistic. In fact at all depths from the photosphere downward χ^2 per degree of freedom was very close to unity, supporting the conclusion that diffusion is an accurate representation of the evolution of *small* CZ thermal perturbations.

Figure 6 compares κ_{MLT}, κ_h, and κ_v over the range in depth of the simulation volume (about 3000km). The discussion above anticipates two qualitative features of these curves which are borne out in the numerical calculations: 1) Anisotropy in the plume flow causes horizontal conductivity to be several times smaller than vertical conductivity (the convergence of κ_v and κ_h near the bottom simulation boundary appears to be a numerical boundary effect), and 2) Since turbulent velocities drive the perturbation evolution, this conductivity is significantly smaller (by about an order of magnitude) than the MLT estimate.

3.3. *Diffusion Timescales*

It has been argued [cf. *Spruit* 1991] that the timescale for mixing entropy perturbations within the SCZ is so short that any perturbation caused by magnetic inhomogeneities gets conducted throughout the SCZ before appearing at the photosphere. This would imply that the SCZ effectively "shunts" local fluctuations and distributes their energy throughout the outer envelope of the Sun before the perturbation can be radiated from the photosphere. *Spruit* [1991] further argues that this energy is "lost" and only reradiated on the thermal timescale of the SCZ (about 10^5 yr). In this picture it is difficult to generate a local luminosity variation with position (e.g. $T_p(\theta, \phi)$) at the photosphere. This classical argument is based on a 1-d MLT calculation and must be modified in the anisotropic diffusive approximation.

In the diffusive approximation the timescale for attenuating a perturbation of physical size L is $\tau = L^2 \rho C_p / \kappa$. Beneath the superadiabatic region near the top of the convection zone C_p is nearly constant, the convective velocity v is slowly varying so that $\kappa \propto \rho$ and we expect τ to scale as L^2 considering perturbations of scale length L defined by their vertical depth. Figure 7 plots τ for the diffusive vertical, horizontal and MLT conductivities. Here we have extrapolated results below the depth of the numerical simulation (3000km) using a MLT scaling law.

In the MLT approximation the timescale for convective mixing is less than the 11 year solar cycle period throughout the SCZ. In the case of the anisotropic model both horizontal and vertical timescales are significantly longer than a solar cycle at the SCZ depth (2×10^5km). The rapid timescale decrease near the surface is also independent of L

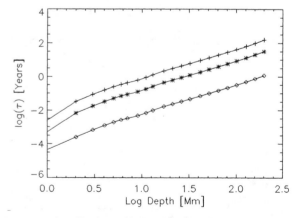

Figure 7. The effective diffusive timescale for the upper part of the CZ down to the depth indicated along the horizontal axis is plotted here. Diamonds correspond to the MLT, stars are the vertical anisotropic conductivity, and crosses are the horizontal anisotropic timescales.

as $C_p\rho/\kappa_{v,h}$ declines by a factor of 4 between 3000km depth and the photosphere. Given these results it is clear why an entropy perturbation due to a sunspot would be "short-circuited" by the convection zone. These short timescales also imply that such perturbations should be spatially mixed with no possibility of residual localized brightness changes associated with the perturbing magnetic field. In the anisotropic model entropy fluctuations are not so readily conducted away.

3.4. Finite Amplitude Reradiation Timescales

The timescale for a perturbed fluid parcel to pass vertically through a convecting volume is not characterized by a diffusion process. *Georgobiani et al.* [1997] found that an average downwelling fluid parcel traveled 2.5Mm in 35 min and it took 66 min for an upwelling fluid element. This is nearly 20 scale heights in a convecting region characterized by vertical fluid velocities of only about 1km/s. These results imply that a high entropy fluid parcel launched from the base of the convection zone could reach the photosphere in approximately one-two weeks. Excess entropy can be radiated by the photosphere over timescales much shorter than the *thermal* time characterized by the solar luminosity and the heat capacity of the convection zone ($t_{thermal} \approx$ 200,000 years).

3.5. Evidence For Non-Diffusive SCZ Coupling

There are several observations that support this view of correlated flows over many scale heights within the SCZ. An important consequence is that heat flow perturbations within the SCZ can be coupled to the radiative photosphere. The diffusion model alone fails to describe the solar photometric observations.

3.5.1. Active Latitude Bands.
What evidence is there that the SCZ is not diffusive? The most striking example may be the mere existence of the magnetic active latitude bands. Since sunspot and active region magnetic fields originate from near the base of the convection zone, we might ask why they are not broadly distributed throughout the SCZ before emerging at the surface. Of course these fields are subject to additional magnetic forces beyond the experience of an entropy perturbation but their latitudinal coherence is still quite surprising if one believed that even the first few scale heights of the convection zone were diffusive.

Consider that we have excellent reasons for expecting 10^5G fields at the base of the SCZ [cf. *Parker* 1996]. Fields this small have an energy density which is comparable to the convective kinetic ($rg y^2$) energy density. A 10 degree latitudinal extent to the active bands is inexplicably narrow if the field migration were to be diffusive through the lower SCZ. Serious attempts to model flux tube migration from the base of a realistic CZ [*Dorch et al.* 2001] depended on vertically correlated (non-diffusive) updrafts to generate realistic active latitude bands.

3.5.2. Sunspot Bright-rings.
Even though the formation and dynamics of sunspots are not well understood, some of their properties are indisputable. For example their strong magnetic fields inhibit local convective motion and must effectively decrease the local thermal conductivity. *Zhao et al.* [2001] have recently measured the vertical structure in a sunspot through helioseismic techniques and found a depth of at least 6000 km. This is actually shallower than many expect but even this spot depth (about equal to the radius of their spot) serves to illustrate the problems with all diffusive bright-ring calculations.

Several have looked at where the convective flux which is blocked by a sunspot should be reradiated. *Spruit* [1982], *Fowler et al.* [1983], and *Chiang and Foukal* [1985] each applied the MLT to estimate how this flux should appear at the photosphere. A feature of all diffusion models is that the conductivity tends to increase exponentially like the density, with depth. Realistic anisotropy and the small perturbation κ calculations exhibit a slower increase with depth, but there is still a strong tendency for most of the blocked flux to diffuse preferentially toward the interior because of this radial dependence in κ.

For such a shallow spot *Fowler et al.* [1983] predicted a bright ring that has a maximum intensity contrast of 0.02% of the photosphere while *Spruit* [1982] claims that only 3% of the total blocked flux should be reradiated in the ring. These predictions are in conflict with the observations.

Coulter and Kuhn [1994] describe a set of precision photometric solar telescopes (PSPT) built to obtain accurate solar surface brightness observations. These telescopes were specifically designed to help solve the solar irradiance problem and to detect small contrast brightness variations like sunspot bright rings. With one of these instruments *Rast et al.* [2001] were able to show that the surface brightness contrast of a sunspot ring is approximately 1% and that at least 10% of the blocked flux is reradiated from the surrounding ring. This estimate should be considered a lower bound because *Rast et al.* did not correct for the angular dependence of the sunspot radiation and could not correct for any large scale excess surface brightness.

The observations are clearly in conflict with diffusion predictions. Too much of the blocked flux is reradiated from the nearby photosphere. The simplest explanation is that correlated flows couple the deeper entropy perturbation to the photosphere on timescales of hours.

3.5.3. Facular Leaks. It has been suggested that small flux tubes or faculae may provide a higher conductivity path to radiate energy from the photosphere, thereby enhancing the luminosity. Since these magnetic flux tubes have a lower internal gas pressure (because of the magnetic pressure contribution) their interior opacity is reduced which could make them more effective radiators.

Unlike the deeper sunspot perturbation these "thermal shunts" are effective in the outer few density scale heights near the photosphere. In a diffusion model the shunts couple to the SCZ interior and tap its thermal energy reservoir on a 10^5 year timescale. In the correlated flow picture the excess heat radiated by the shunt originates from the surrounding plasma which is effectively coupled to the surface by correlated vertical flows over hour timescales. Thus, over short times the facular radiant enhancement and nearby deficit tend to balance with little change in the total luminosity from the local shunt.

3.5.4. Large Space and Time-scale Luminosity Variations. The MLT model for thermal perturbations has at least two additional problems with current observations. This model (e.g. as formulated by *Spruit* 1991) homogenizes perturbations throughout the SCZ and reradiates this energy on the thermal timescale of the entire SCZ. This implies that a solar cycle luminosity perturbation must have an enormous driving amplitude in order to overcome the "impedence" mismatch with the much longer thermal timescale of the SCZ. In this case it is difficult to imagine a mechanism for changing the solar luminosity – in conflict with the observed solar cycle luminosity changes.

SCZ models directly conflict with the diffusive heat transport approximation. *Kuhn* [1996] explored the simple example of a 1% spatial temperature perturbation at a few thousand km depth. In the diffusion approximation this deep perturbation shouldn't result in an observable temperature variation, but what was observed near the photosphere is a significant thermal shadow of several degrees amplitude. Realistic SCZ models provide no evidence of diffusive transport properties.

Finally we note that the observed photospheric brightness temperature, $T_p(\theta, \phi)$, [cf. *Kuhn et al.* 1988] has significant large scale spherical harmonic structure at wavelengths smaller than a solar radius. Direct photometry from above the atmosphere using SOHO/MDI confirms that even at solar minimum, with no measurable magnetic contamina-

tion, the Sun has a latitudinal structure in $T_p(\theta, \phi)$ which could not be sustained by a high MLT conductivity SCZ. In the correlated flows model this temperature structure, like the active latitude bands, can be generated naturally from the base of the convection zone. *Kuhn* [1996] summarizes a mechanism that introduces higher entropy magnetized fluid into the base of the SCZ. The emergence of the magnetized fluid at the photosphere is accompanied by a higher luminosity.

4. OTHER GLOBAL SOLAR CYCLE CHANGES

Knowing about variations in the solar luminosity is critical to diagnosing how the solar cycle works. Just as we depend on the evolution of the largescale magnetic field to offer clues to the dynamo mechanism we must fit the solar luminosity changes into the puzzle of how the SCZ interacts with the cycle's magnetic fluctuations.

4.1. Solar Radius Changes

There is no confirmed detection of solar radius changes. *Ribes et al.* [1991] reviewed the state of the literature and recently *Emilio et al.* [2000] reported a relatively small upper limit to the amplitude of possible radius changes from sensitive satellite measurements. Even though there is no evidence that the Sun's radius changes by more than 9 milliarcsec/year, or 10 milliarcsec amplitude with a solar cycle periodicity, it is interesting to consider the implications of measureable change.

Many physical mechanisms have been proposed to cause solar radius and irradiance changes [cf. *Emilio et al.* 2000]. It is customary to characterize them by their implied dimensionless shape to brightness ratio, $W = (\delta r/r)/(\delta L/L)$. Most of these calculations are one-dimensional and unable to distinguish luminosity and irradiance changes. Therefore they may be of dubious utility for comparison to the real Sum. Nevertheless W ranges between 2×10^{-4} [*Spruit* 1982b] and values as large in magnitude as 0.07 [*Sofia et al.*, 1979] (from an early exploratory calculation).

Determining W is problematic not only because most models are schematic but because the many direct and secondary physical mechanisms are difficult to account for in a realistic 3-dimensional changing SCZ. A conservative interpretation is derived simply from energy conservation. Since we know the Sun's luminosity varies with an amplitude of about 0.05% with a period of 11 years we can find an upper bound to the magnitude of W by equating the total change in radiated energy to the gravitational energy required to expand the outer mass envelope of the Sun. Consider a perturbation acting from 3 locations in the Sun – at the base of the convection zone ($0.7R_\odot$), intermediate within the CZ ($0.8R_\odot$) or near the superadiabatic region ($0.99R_\odot$). One finds that W varies from 8.5×10^{-6}, 6.6×10^{-6} to 0.3. Since the coupling times are short (at least near the surface) compared to the solar cycle we might further expect the radius to be a maximum when the luminosity is smallest ($W < 0$). Although the real Sun must exhibit values of W which are smaller in magnitude, these calculations imply that actual measurements of a changing solar radius could be an excellent discriminant for how and where the solar cycle drives the SCZ.

4.2. Solar Shape Changes

Because the gravitational potential energy required to affect the solar radius is so large its solar cycle dependence is weak. In contrast the gravitational energy required to change the Sun's shape is a small perturbation of its equilibrium state. Consequently there may be more information about the solar cycle contained in the Sun's shape than its radius. Despite the measurement difficulties there is observational evidence that we may already be seeing the solar cycle in limb shape data.

Both *Lydon and Sofia* [1994] and *Armstrong and Kuhn* [1999] have measured the solar hexadecapole shape ($l = 4$) but at different times during the cycle. They also inferred significantly different amplitudes. These are interesting results because, unlike the solar quadrupole, this shape term should be quite sensitive to changes in the toroidal magnetic structure. If measurements like these can be obtained over the cycle and confirmed to be solar they are bound to become another sensitive probe of the solar cycle and the SCZ.

5. CONCLUSIONS CONTROVERSIES AND FUTURE DIRECTIONS

5.1. Faculae and Sunspots Don't Solve the Irradiance Problem

It has been argued that dark sunspots and bright photospheric faculae can account for the solar irradiance changes [cf. *Lean et al.* 1998]. Their model is a multiparametric description of the observed satellite solar irradiance data. It uses scaled sunspot and facular time-series contributions to reproduce the solar irradiance. Similar models based on scaling photospheric magnetic field data to obtain a facular irradiance component have been developed by *Fligge et al.* [2000].

These parametric descriptions can reproduce more than 90% of the variability over most timescales and are useful tools for predicting irradiance changes when they can't be measured. Unfortunately the assumptions they're built on are not physically realistic. For example, even the basic assumption that faculae are bright, thus providing a positive irradiance contribution, is not accurate.

Proxy models of irradiance contributions are often used because of the difficulty of obtaining sufficiently accurate photometry to directly measure non-spot brightness contributions. Fortunately, photometry from the PSPT now allows the direct determination of facular brightness. A robust conclusion from these data, also previously observed by *Topka et al.* [1997], is that faculae are not always bright. Strong faculae (as measured from CaII K observations) are *dark* near disk center. Figure 8 shows how a facular *irradiance* contribution can be negative or positive depending on its CaIIK brightness (which is always positive). A physically accurate proxy model should, evidently, include a continuum of brightness contributions which span the qualitative characteristics between active region faculae and sunspots.

5.2. Diffusion

Observations of large-scale latitudinal structure, sunspot bright rings, and solar cycle luminosity changes argue against a diffusive picture of perturbed heat flow in the SCZ. Realistic numerical SCZ calculations as diverse as the flux tube rise to internal temperature perturbation evolution models all show that large scale correlated flows are essential for describing the transport properties of the SCZ. The high-conductivity diffusion model fails by all measures.

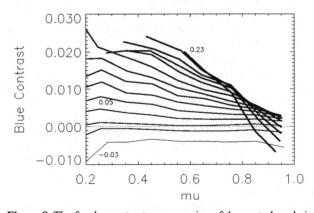

Figure 8. The facular contrast versus cosine of the central angle is plotted here. Disk center is toward the right. Each curve shows the contrast as a function of the CaIIK contrast of the facule. Thus the strongest faculae (contrast near 0.23) have the largest contrast, but are *dark* near disk center.

5.3. Key Questions

Over what timescales then can the solar cycle affect global solar properties like the luminosity? Magnetic fields which enter the SCZ from below are likely to carry excess entropy. When they emerge at the photosphere they will both radiate and redirect the surrounding local photospheric flux. It seems likely that the solar luminosity can be affected on the emergence and evolution timescale of active regions.

Nevertheless, photospheric irradiance/luminosity perturbations are not entirely characterized by sunspots and faculae – understanding their differences may be a new tool for understanding the origin and mechanisms of magnetic field evolution in the photosphere. This is likely to begin with a better understanding of the energy budget associated with such photospheric features.

Acknowledgments. This research has been supported by NASA.

REFERENCES

Armstrong, J.D., Kuhn, J.R., Interpreting the Solar Limb Shape Distortions, *Astrophys. J. 525,* 533 (1999).

Bohm-Vitense, E. Uber die Wasserstoffkonvektionszone in Sternen *Z. Astrophys., 46,* 108 (1958).

Coulter, R. L., Kuhn, J.R., RISE/PSPT as an Experiment to Study Active Region Irradiance and Luminosity Evolution *Solar Active Region Evolution, NSO/SP 14th Summer Workshop,* Ed. K.S. Balasubramaniam, G. W. Simon, PASP, 37 (1994).

Chiang, W. and Foukal, P., The influence of faculae on sunspot heat blocking *Sol. Phys. 97,* 9 (1985).

Dorch, S.B.F., Gudiksen, B.B., Abbett, W.P., Nordlund, A., Flux-loss of buoyant ropes interacting with convective flows *Ast. Astrophys. 380,* 734 (2001).

Emilio, M., Kuhn, J.R., Bush, R.I., Scherrer, P., On the constancy of the solar diameter *Astrophys. J. 543,* 1007 (2000).

Fligge, M., Solanki, S.K., Unruh, Y.C., Modelling short-term spectral irradiance variations *Space Science Reviews 94,* 139 (2000).

Fowler, L.A., Foukal, P.V., Duvall, T.L.Jr Sunspot bright rings and thermal diffusivity of solar convection *Sol. Phys. 84,* 33 (1983).

Georgobiani, D., Kuhn, J.R., and Stein, R.F. Sound speed variations near the photosphere *SCORe'96: Solar Convection and Oscillations and their relationship* eds. Pijpers, J. Christensen-Dalsgaard, C.S. Rosenthal, Kluwer, 127 (1997).

Kerr, R. M., Length scales in turbulent flow *J. Fluid Mech. 310,* 139 (1996).

Kuhn, J., Helioseismic observations of the solar cycle *Ap. J. Letts., 339,* L45 (1989).

Kuhn, J. and Georgobiani, D., A Least-squares Solution for the Effective Conductivity of the Solar Convection Zone *Space Science Reviews 94,* 161 (2000).

Kuhn, J., Libbrecht K.G., and Dicke R.H., The surface temperature of the sun and changes in the solar constant *Science, 242,* 908 (1988).

Kuhn, J. and Libbrecht, K.G., Non-facular solar luminosity variations *Ap. J. Letters, 381,* L35 (1991)

Kuhn, J., Global changes in the sun *The Structure of the Sun: VI Winter School at Instituto d'Astrophysica de Canarias* edited by T. Roca-Cortés, Cambridge Univ. Press, 231, (1996).

Kuhn, J.R., Bush, R., Scheick, X., and Scherrer, P., The sun's shape and brightness *Nature,* 392, 155 (1998).

Lean, J., Cook, J., Marquette, W., Johannesson, A. Magnetic sources of the solar irradiance cycle *Astrophys. J. 492,* 390 (1998).

Lin, H., Kuhn, J.R., Precision IR and visible solar photometry *Solar Phys. 141,* 1 (1992).

Lopresto, J.C., Schrader, C., and Pierce, A. K., Solar gravitational redshift from the infrared oxygen triplet *Astrophys. J., 376,* 757, (1991).

Lydon, T.J. and Sofia, S. Balloon solar oblateness measurements *Phys. Rev. Lett., 76,* 177, (1995).

Mihalas, D., *Stellar Atmospheres,* (Freeman, San Francisco, 1978).

Parker, E.N., in *The Structure of the Sun: VI Winter School at Instituto d'Astrophysica de Canarias* edited by T. Roca-Cortés, Cambridge Univ. Press, Cambridge, 231, 1996.

Rast, M. P. et al., Sunspot bright rings: Evidence from case studies *Astrophys. J., 557,* 864, (2001).

Ribes, E. et al., The variability of the solar diameter *The Sun in Time,* Univ. of Ariz. Press, Tucson, 59, (1991).

Sofia, S., O'keefe, J., Lesh, J. R., & Endal, A. S., Solar constant—Constraints on possible variations derived from solar diameter measurements *Science, 204,* 1306, (1979).

Spruit, H.C., Pressure equilibrium and energy balance of small photospheric fluxtubes *Solar Phys., 50,* 269, (1976).

Spruit, H.C., The flow of heat near a starspot *Ast. Astrophys., 108,* 356, (1982).

Spruit, H.C., Effects of spots on a star's radius and luminosity *Ast. Astrophys., 108,* 348, (1982b).

Spruit, H.C., Theory of luminosity and radius variations *The Sun in Time,* Univ. of Ariz. Press, Tucson, 118, (1991).

Stein, R. H. and Nordlund, A., Simulations of Solar Granulation. I. General Properties *Astrophys. J., 499,* 914, (1998).

Topka, K., Tarbell, T. D., Title, A. M., Properties of the Smallest Solar Magnetic Elements. II. Observations versus Hot Wall Models of Faculae *Astrophys. J., 484,* 479, (1997).

Zhao, J., Kosovichev, A., Duvall, T. Jr., Investigation of Mass Flows beneath a Sunspot by Time-Distance Helioseismology *Astrophys. J., 557,* 384, 2001.

J. R. Kuhn and J. D. Armstrong, Institute for Astronomy University of Hawaii 2680 Woodlawn Dr. Honolulu, HI, 96822. (e-mail: kuhn@ifa.hawaii.edu, armstron@ifa.hawaii.edu)

Solar Irradiance Variability

Claus Fröhlich

Physikalisch-Meteorologisches Observatorium Davos, World Radiation Center, Davos Dorf, Switzerland

Since November 1978 a set of total solar irradiance (TSI) measurements from space is available, yielding a time series of more than 23 years. From measurements made by different space radiometers (HF on NIMBUS 7, ACRIM I on SMM, ACRIM II on UARS and VIRGO on SOHO) a composite record of TSI can be compiled. This leads to a reliable record of TSI with an overall precision of the order of 0.05 Wm^{-2}. This time series is compared to an empirical model based on sunspot darkening and brightening due to faculae and network. Since early 1996 spectral measurements by filter-radiometers of VIRGO provide continuous time series of spectral solar irradiance (SSI) at 402, 500 and 862 nm. These time series are analyzed and compared to TSI yielding information about the redistribution of energy within the spectrum during changes of TSI.

1. INTRODUCTION

Total solar irradiance (TSI) monitoring from space with electrically calibrated radiometers started with the launch of NIMBUS 7 in November 1978. In early 1980 the Solar Maximum Mission Satellite (SMM) was launched and for the last 22 years at least two missions with solar monitors were operating simultaneously (see Plate 1). These more or less continuous measurements are supplemented by point measurements from balloons and rockets, which were very important in the early years when the decrease of TSI during the descending phase of cycle 21 had still to be proven to be of solar origin [*Willson et al.*, 1986]. The radiometric accuracy of irradiance measurements made by individual instruments is of the order of 0.1–0.2%, and is insufficient to determine long-term changes of only about 0.1% observed over the 11-year solar cycle. While the instrument repeatability is adequate to monitor short term changes, the long term behaviour can only be retrieved by careful tracing of one experiment database to the other, incorporating good

knowledge of the degradation of the individual radiometers operating in space. Fortunately several time series exist from different platforms made by different radiometers: HF on NIMBUS 7, ACRIM I on SMM, ERBE on ERBS, ACRIM II on UARS, VIRGO on SOHO and ACRIM III of ACRIMSat. This allows us the construction of a composite time series having improved long-term precision, thus yielding an unbiased estimate of TSI and its variability during the last almost three solar cycles.

In the first section a short introduction to solar radiometry is presented and with the example of the VIRGO measurements it is explained how long-term changes can be investigated and corrected for. Then, most important issues determining the composite TSI are reviewed, and its reliability and precision over the years is discussed. The following section compares the composite with an empirical model of the TSI variations which is based on sunspot darkening and brightening by faculae and network. This comparison allows to calibrate such a model and to give some insight in the possible mechanisms producing the variations. The next section describes the results of the spectral measurements performed with the three-channel filter-radiometers within the VIRGO experiment on SOHO. Although the long-term behaviour of these instruments is not understood well enough to correct the measurements,

Solar Variability and its Effects on Climate
Geophysical Monograph 141
Copyright 2004 by the American Geophysical Union
10.1029/141GM09

the detrended time series allow to analyze their behaviour up to periods of about a year. By comparison of these time series with TSI, the spectral redistribution of the energy during changes of TSI can be determined at the wavelengths of the three channels (402, 500 and 862 nm).

2. SOLAR RADIOMETRY FROM SPACE

Before we discuss the construction of the TSI composite a brief introduction to solar radiometry is presented, followed by a discussion of how the inevitable long-term changes of the space radiometers can be determined and corrected for. The latter will be explained with the VIRGO radiometers as an example.

The presently used radiometers for total solar irradiance observations are based on the measurement of the heat flow produced by the absorbed solar radiation, which is in turn substituted by electrical power to calibrate the heat flow meter. The radiation is collected in cavities which enhance the absorption over the one of a black coating on flat surface by a large amount. Absorptivities of the order of a few 100 parts-per-million (ppm) can be achieved, compared to a paint with typically ≈5%. The simplest cavity is made of a cylindrical wall with a flat bottom and the enhancement for a diffuse paint is roughly proportional to the diameter-to-height ratio. All radiometers which will be discussed here use an organic black paint which can either reflect the radiation diffusively (more or less uniformly distributed over 2π sr) or specularly (similar to a mirror). The following geometries and paints are used: ACRIM [*Willson*, 1979] and ERBE [*Lee III et al.*, 1987] have a conical cavity with specular paint, DIARAD [*Crommelynck and Domingo*, 1984] has a cylindrical cavity with a flat bottom and diffuse paint, PMO6 [*Brusa and Fröhlich*, 1986] and HF [*Hickey et al.*, 1989] have an inverted cone within a cylindrical wall and specular paint.

The thermal heat-flow meter is between the cavity and the heat-sink of the instrument and consists of a thermopile (DIARAD, HF) or a metallic thermal resistor with resistor thermometers (ACRIM, PMO6, ERBE). Normally, the heat flow meter is of differential type to minimize the influence of rate-of-change of the heat-sink temperature and only in the case of DIARAD the arrangement of the two receivers is side by side, so that both can be used for radiation measurements. All other radiometers have their cavities back to back, and the compensating cavity cannot be used for radiation measurements. The electrical substitution heater is in or around the cavity and covers the same area of the cavity as the direct solar radiation. In front of the cavity is a precision aperture which defines the area for the irradiance measurement and a muffler which determines the thermal environment seen by the cavity and absorbs solar stray light. In front of or just behind the view-limiting aperture (constraining the solid angle of the radiometer to ≈0.006 sr) is a shutter which allows to shade the cavity from sunlight.

The radiometers are operated either in active (ACRIM, ERBE, PMO6, DIARAD) or passive (HF) mode: In the *active* mode (hence the designation ACR: active cavity radiometers) the heat flow is maintained at a preset value by controlling the electrical heater power; with periodic shutter-open and closed phases (e.g. 60 sec open and 60 sec closed) the radiant power is the difference between the electrical power during the closed and open phases. In the *passive* mode the output of the heat flow meter (e.g. the temperature difference across the thermal resistor) is read during solar measurements and calibrated from time to time by heating the cavity with known electrical power.

The uncertainty in terms of the SI units (e.g. Wm^{-2}) is determined by the uncertainty of the electrical power measurements, the uncertainty in the area of the precision aperture and the uncertainty of the knowledge of all deviations from ideal behaviour the determination of which is called characterization. Although length measurements can be performed very accurately, the determination of the area of a radiometric aperture is far from being straightforward: measuring several diameters with high accuracy is still insufficient if the aperture has only an approximate circular shape and, moreover, as the land of the aperture has to be small (10–20 μm) the most accurate diameter measurements by physically touching are impossible. Thus, most of the differences between the solar measurements in the seventies and eighties were due to the rather large uncertainty in the area determinations, which were then performed at workshop level and not by metrology institutes. Also the electrical measurements can increase the overall uncertainty if no special precautions are taken [see e.g. *Fröhlich et al.*, 1997]. Nonetheless, the largest contributions to the uncertainty come from the uncertainties in the characterization of the deviations from ideal behaviour. The most important are the non-equivalence between electrical and radiative heating, cavity absorptance, lead heating, stray light, and diffraction. Besides the last effect all are determined by independent experiments and the uncertainty of the determination of the effect adds to the total uncertainty which amounts presently to 0.1 … 0.2 percent [see e.g. *Brusa and Fröhlich*, 1986].

In contrast to well-controlled experiments in the laboratory, radiometers in space are exposed to an environment which depends on the mission and the measurement profile. Mainly the thermal environment is important as the radiometer are very sensitive to temperature changes. The best environment is on SOHO at L1 because it is always in the Sun and the only temperature change is due to the changing distance to the Sun with an annual period. For Earth bound orbits it is important to distinguish between Sun-pointed satellites as SMM (allowing about 60-minute observations each orbit) and Earth-observation satellites as UARS with a pointing platform (allowing for about 15-minute observations during an orbit) or as NIMBUS7 and ERBE with no solar pointing capabilities at all (they rely on the short available time of a few minutes when the Sun crosses the field-of-view of the radiometer at the terminator of the orbit). Earth orbiting satellites exhibit rather large temperatures changes due to the eclipses which do change the thermal environment of the cavity in a complicated way.

Another challenge for space radiometry are the changes of the sensitivity due to the strong solar UV radiation (most importantly Lyman-α) received by the cavities during measurements. A common way to cope with this effect is to have more than one radiometer on the same platform and to determine the exposure dependent changes by comparing the operational radiometer with spare ones which have much lower exposure times. Only the ACRIM experiments with three radiometers and VIRGO with two different types and two radiometers of each type have the possibility to determine the sensitivity changes due to exposure. For the others some assumptions have to be made, as explained later for the HF on NIMBUS. The exposure of ERBE with measurement once every 2 weeks for a few minutes is probably low enough to assume no important degradation (total exposure time of the cavity is 2.3 days for the whole mission up to March 2002). The determination of the sensitivity changes is explained with the example of the VIRGO radiometry, mainly because of the availability of continuous measurements (apart from the interruption during the SOHO vacation) which allow an accurate tracking of the behaviour.

2.1. VIRGO Radiometry

As for all space experiments the evaluation starts with level-1 data (Figure 1) which are transformed from the raw data into physical units using the radiometric constants and electrical calibrations and which are corrected for all a-priori-known effects such as temperature, reduction to 1 AU, correction for radial velocity. Figure 1 illustrates the differences in the long-term behaviour of the four radiometers, the operational PM6V-A and DIARAD-L and the less exposed PMO6V-B and DIARAD-R. In contrast to the observations during SOVA/EURECA with similar radiometers [Crommelynck et al., 1993], DIARAD shows essential-

ly no change, whereas PMO6V-A has with 1.8 ppm/day a slightly smaller degradation as during the 1 year operation of SOVA (note: due to the un-shuttered operation of PMO6V this value has to be reduced by a factor of ≈ 3 to be comparable to the commonly observed ~ 1 ppm/day for a radiometer with 50% shutter duty cycle and 60 minute observations during a 90-minute orbit, as for DIARAD and PMO6 on EURECA or ACRIM I on SMM and ACRIM III). The correction for exposure dependent changes can be determined by comparison of an operational radiometer with a less exposed spare of the same type. The correction can then be determined by interpolating between the comparison points to the observation times of the operational radiometer. In the case of VIRGO a slightly different approach was adopted: In order to understand the behaviour of the radiometer better and to treat the early increase of the PMO6V radiometers properly a functional description of the different effects was developed. A major result is that hyperbolic functions describe the behaviour much better than the usually assumed exponential functions [see e.g. Fröhlich and Finsterle, 2001], mainly because these functions allow to take the actual accumulated UV radiation dose into account which is thought to be the origin of these kinds of changes. Furthermore, with the determined parameters of these functions the responsible wavelength for the changes can be determined, as described in detail in Fröhlich [2003]. Finally, the exposure-dependent corrections for both types of radiometers are determined by the fitted functional parameters and the interpolated residuals which are of the order of ± 80 ppm at most. For DIARAD with its rather small exposure-dependent changes and the fact that DIARAD-R can be assumed not to change, a direct determination from the comparison of L and R would obviously yield the same result. For PMO6V, however, the direct method is not possible as the exposure time of PMO6V-B was important at the beginning of the mission (with ~ 10 days during the first 200 days of the mission, see also Figure 1) and thus it has to be corrected with similar functional parameters as for PMO6V-A in an internally consistent procedure.

After applying the corrections to each of the two types of radiometers individually two independent time series result, which are called level-1.8. These two time series can now be compared and possible non-exposure dependent changes determined, eventually yielding a better estimate for TSI. This will, however, not improve the intrinsic absolute uncertainty of the radiometers, but it allows us to decrease the uncertainty of the long-term changes if it can be shown that there is no remaining systematic trend. The necessary condition for a unique determination of such changes for each radiometer independently is that the temporal behav-

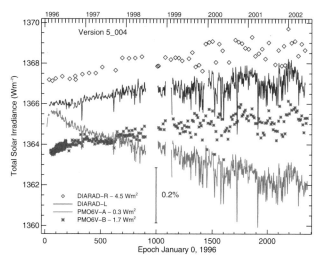

Figure 1. Level-1 data of the two type of radiometers on VIRGO, the DIARAD and PMO6V. Both types have redundant instruments (PMO6V-B) or channels (DIARAD-R) which are used only occasionally in order to assess the radiation dependent changes by comparison with the operational ones. Note the difference in the amount of degradation of PMO6V-A relative to DIARAD-L and the early increase of the PMO6V-A and B which depends obviously on the exposure to solar radiation. The early increase of PMO6V-B is only marginally seen as it reaches 11 days of exposure at around day 190 and a total of 15 days in 2002, which is compared to the increase of PMO6V-A during about 40 days only about one third of the amplitude.

iour of these changes is different enough. Otherwise, it cannot be decided whether the difference has to be allocated to one or with the opposite sign to the other radiometer. In the case of the VIRGO radiometers this condition is met with differences in the relevant time constants of the exponential functions fitted of more than an order of magnitude as shown in Figure 2. The reason for using exponential functions here, is that no dose is involved, just time in space. Most of the functional corrections are allocated to DIARAD and are due to switch-off of this radiometer; the mechanism for this behaviour is still unclear, and most observed sudden changes of radiometers (also after switch-off) have been jumps of the sensitivity not followed by a slow recovery. The only non-exposure-dependent correction for PMO6V is at the very beginning of the measurements, and it may still be due to some incomplete early-increase correction of PMO6V-B[*Fröhlich*, 2003].

Over the gap of the SOHO vacations (late June to early October 1998) a change of ~150 ppm is determined. The allocation of the share of this change to each radiometer is determined from comparison with ACRIM II (corrected as described in Section 3.2). With these corrections applied,

level-2 data are generated for both radiometers, and the VIRGO TSI is the combination of the DIARAD and PMO6V level-2 time series. The actual version of the data used in the following is 5.004 (the VIRGO TSI results are available from ftp://ftp.pmodwrc.ch in the directory data/irradiance/virgo/TSI as daily and hourly values for VIRGO at level 2, and for DIARAD and PMO6V at level 1.8 and 2; VIRGO TSI is a combination of DIARAD and PMO6V level 2 results).

As PMO6V is very similar to all radiometers used to establish the composite, this result is corroborating the assumption that these types of radiometers have no or at most a very small change due to non-exposure-dependent effects. Thus for e.g. the ACRIMs the determination of the degradation by comparison with less exposed spares is a valid approach.

3. COMPOSITE OF TOTAL SOLAR IRRADIANCE

The description of the procedures used to construct the composite from the original data shown in Plate 1 can be

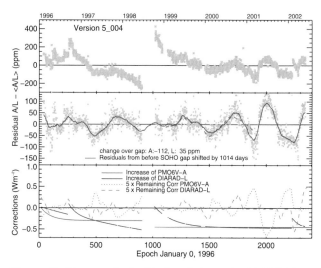

Figure 2. Top panel: Residuals of the ratio PMO6V-A to DIARAD-L, after each has been independently corrected for exposure-dependent changes. Middle panel: Residuals after applying the different exponential functions. An interesting feature is the repetition of the deviations between days 100–900 as indicated by those residuals shifted by 1024 days. The similarity is really striking and has no explanation yet. Bottom panel: The fitted exponential functions for DIARAD are related to switch-offs, included the recovery after the SOHO vacations. The early increase of PMO6V may still be a remainder of the an incomplete corrections for the early increase of PMO6V-B. The allocation of the residuals to each radiometer are determined from short-term comparison with ACRIM and are shown as 5-times amplified corrections in bottom panel.

found in *Fröhlich and Lean* [1998a,b]; *Fröhlich* [2000]. Radiometrically it is based on the ACRIM I and II records; before the start of the ACRIM I measurements in 1980, during the spin mode of SMM, and during the gap between ACRIM I and II, corrected HF data are inserted by shifting the level to fit the corresponding ACRIM data over an overlapping period of 250 days on each side into the ACRIM sets. In early 1996 the VIRGO data take over, again shifted to agree with ACRIM II. Finally the composite record is adjusted via ACRIM II to SARR (Space Absolute Radiometer Reference) which was introduced by *Crommelynck et al.* [1995] and allows the comparison of different space experiments. The data from ERBE and ACRIM III, as well as the empirical model described below (Section 4) are used for comparisons and for internal consistency checks. It is important to note that the model is an independent source of information for comparisons and as long as it is not used over solar cycle time scales it provides a reliable time series for time scales of less than a year. So it will be used in all comparisons as it is available as a daily record which is important for the interpolation between ERBE data with their 14-day sampling.

The following issues will be discussed in the forthcoming sections: the correction for the early measurements of HF on NIMBUS 7 to account for its degradation (as described in *Fröhlich and Lean* [1998b]), the tracking of ACRIM II to ACRIM I by comparison with ERBE, HF and the model (revised version of *Fröhlich* [2000]). A detailed assessment of the influence of the many operational interruptions in the ACRIM II record is presented in Section 3.2.

3.1. Corrections for HF

Because HF has no on-board means to determine irradiation-dependent degradation [*Hoyt et al.*, 1992], an attempt was made to derive corrections by analogy to other experiments. The cavity of HF has the same type of cavity as PMO6-type radiometers and uses the same paint. From Plate 1 it is quite obvious that the early values are rather high with a steep increase at the very beginning which immediately suggests that the behaviour could be similar to the one observed by PMO6V within VIRGO [*Anklin et al.*, 1998; *Fröhlich and Finsterle*, 2001; *Fröhlich*, 2003]. Thus, we apply corrections for the long-term changes prior to 1980 by modelling the early increase from the behaviour of PM06V and by utilizing—after the launch of SMM—data from ACRIM I to determine the exponential function describing the degradation [*Fröhlich and Lean*, 1998a]. Moreover, we adjust HF data prior to the end of 1980 downwards, corresponding to a change in the NIMBUS 7 orienta-

tion relative to the Sun as described in *Fröhlich and Lean* [1998a].

The next important issue is related to possible changes of HF during the gap between ACRIM I and II which is crucial for the determination of the long-term changes of TSI. From comparison with ERBE *Lee III et al.* [1995] have identified two jumps near 1 October, 1989 and 8 May, 1990 of -0.31 and -0.37 Wm^{-2}, respectively, which were confirmed by comparison with models by *Chapman et al.* [1996]. The total change amounts to about 500 ppm. Because it is not so clear that this change happened in two steps and at the proposed dates, a new assessment was made and the total difference determined by comparing the ratio HF/ERBE before and after the jumps [*Fröhlich*, 2000]. This analysis resulted in a reduction of the total change by about 100 ppm. A closer look at Figure 6 of *Fröhlich* [2000] suggests that the two steps could be a continuation of the increase of sensitivity by about 70 ppm/year identified from comparison with ACRIM I between 1980 and '84 [see also Figure 2 *Fröhlich and Lean*, 2002]. A new analysis is performed, the result of which is presented in Plate 2. The shift is determined from the difference between the averages over 81 days before and after day 3560 (29 September 1989, when the data restarted after a 4-day gap) and amounts to 0.417 ± 0.043 Wm^{-2} or 304 ± 31 ppm. Instead of a second slip, which is difficult to locate, a trend of the HF seems more adequate. A linear fit to the ratios yields 0.256 ± 0.076 and 0.226 ± 0.015 ppm/d for HF/ERBE and HF/model respectively. Removal of this trend and the slip corresponds to a total correction over the period between the ACRIMs of -617 ± 50 ppm (the uncertainty corresponds to the formal statistical error). This result is 120 and 220 ppm higher than the total change proposed originally by *Lee III et al.* [1995] and later by *Fröhlich* [2000]. The standard deviation of the comparison is reduced by 21% and 39% for HF/ERBE and HF/model, indicating that the consistency of the three data sets is indeed improved. Interestingly enough, this trend is very close to the linear trend of 0.272 ± 0.006 ppm/d determined over the period up to November 1986 from the comparison of HF with ACRIM I, with the former detrended with an exponential function. Moreover, from Figure 4 it seems that the two fits may indeed be part of the same trend, if the step at the end of September 1989 is accepted. Thus, the statement that HF has changed within the gap between ACRIM I and II is quite robust, it is different from zero at the 12 σ level. Nevertheless, *Willson* [1997] neglects this correction [see also *Kerr*, 1997] and thus concludes that TSI has a long-term upward trend of 0.04% per decade deduced from the difference between the solar minimum in 1996 and the one in 1985/86. The deduced value of 0.04% or 400 ppm equals

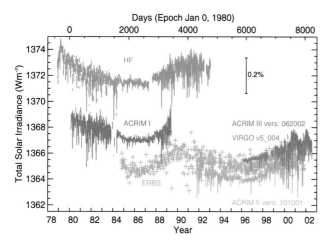

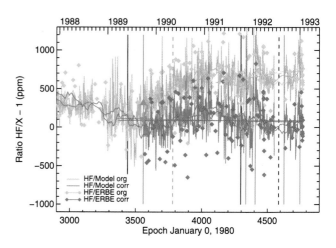

Plate 1. Compared are daily averaged values of the Sun's total irradiance *TSI* from radiometers on different space platforms since November 1978: HF on Nimbus7 [*Hoyt et al.*, 1992], ACRIM I on SMM [*Willson*, 1984], ERBE on ERBS [*Lee III et al.*, 1987], ACRIM II on UARS [*Willson*, 1994], VIRGO on SOHO [*Fröhlich et al.*, 1997], and ACRIM III on ACRIM-Sat [*Willson*, 2001]. The data are plotted as published by the corresponding instrument teams. Note that only the results from the three ACRIMs and VIRGO radiometers have inflight corrections for degradation.

Plate 2. Comparison of the ratio of HF to ERBE and the model for the period 1988–1992. The black vertical lines indicate the end of ACRIM I (1 June 1989) and the beginning of ACRIM II (4 October 1991); the dashed black line is at the time when the data of ACRIM II used for the composite. The green vertical lines indicate interruptions of the operation of HF; the dashed green line is where the 2nd slip was located. The amount of the shift is shown by the green arrow. The linear fits before the correction are plotted as dashed lines, the ones after as full lines.

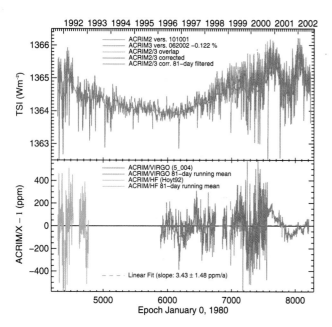

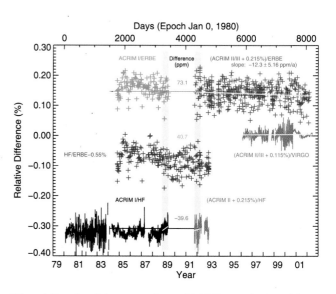

Plate 3. Time series of ACRIM II and III composite with ACRIM II corrected for jumps (the green line corresponds to the original values). ACRIM III is shifted downwards by 0.122% to fit ACRIM II for 83 days after 23 June 2000, otherwise no corrections are needed. The lower panel shows the comparison of the ACRIM II and III record with HF at the beginning and VIRGO after February 1996.

Plate 4. Results of the comparison of ACRIM I and II/III with HF and ERBE after the scaling of ACRIM II/III to ACRIM I by $R_{II\text{-}I}$ = 1.002149 ± 0.000166. The ranges used for the averages are indicated by gray-shaded areas within which the ratio are plotted as linear fits together with the difference in ppm. These differences illustrate the uncertainty of 166 ppm, determined from the weighted average.

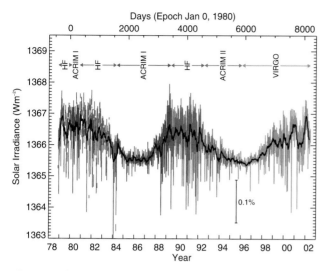

Figure 3. The composite TSI as daily values plotted in different shades for the different originating experiments.

together with the early correction of ACRIM II of -148 ppm (see below) almost exactly the value of the correction of HF over the gap as determined here (see also Figure 4).

3.2. Corrections for ACRIM I, II and III

Corrections for ACRIM I concern only the data in 1980 [*Fröhlich and Lean*, 1998b] due to the fact that *Willson and Hudson* [1991] underestimated the degradation before the spin-mode period. For ACRIM II the situation is complicated by the many interruptions due to satellite and operational problems. Thus, the whole record of ACRIM II (version 3 of 10-10-2002; ACRIM2 and ACRIM3 data are available as daily values from http://www.acrim.com/Data%20 Products.htm) was searched for interruptions with more than 4-day-long gaps. Some of them are too short to be considered separately and others have been merged together, as the differences between them are too small. Finally, we are left with six ranges (5624–5767, 5774–5879, 5888–6048, 6651–6781, 6786–6999, and 7004–7799 in days Epoch January 0, 1980) which are corrected by shifting the values by -54.7, +56.1, +28.8, -183.0, -113.8 and -199.8 ppm respectively. The amount of the shifts is determined from the difference of the weighted mean of the ratios of ACRIM II to ERBE, the model and VIRGO (after Feb 1996) for the adjacent periods. In contrast to the behaviour of DIARAD, these changes seem to persist and do not recover. Similar to these corrections, the early observations of ACRIM II before the 52-day gap starting 4 June 1991 have been shifted by the difference of the weighted averages of the ratios of ACRIM II to HF, ERBE and the model over the periods before and

after; the correction amounts to -148.0 ppm. Although the corrections are rather small, they influence the adjustment of ACRIM II to the level of ACRIM I and the adjustment of VIRGO to ACRIM II. Plate 3 shows the corrected ACRIM II time series on top of the original record. Here, ACRIM III is added to the record by shifting it in order to agree with ACRIM II during a period of 83 days starting on 23 June 2000; after 14 September 2000 ACRIM III takes over. Note the very low relative noise of the ratio to VIRGO during the time of ACRIM III, demonstrating the importance of a large observational duty cycle. Note also that the noise of ACRIM II increased gradually since about mid 1999 and reached a constant value in early 2000, which is about 2 times higher than before. The corrections applied to ACRIM II have substantially improved the consistency between the VIRGO and ACRIM records, especially around the recent solar maximum, and the trend, reported earlier, is removed [e.g. *Fröhlich*, 2003].

In the next step ACRIM II/III has to be adjusted to the scale of ACRIM I. This tracing is based on comparisons with HF and ERBE, the former being corrected as explained

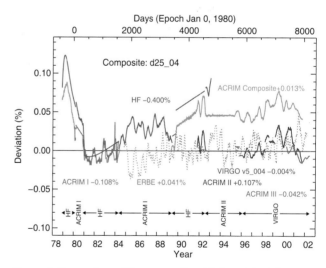

Figure 4. Comparison of the composite with the original data sets. Smooth lines indicate that the radiometer is the basis for the composite and if coincident with zero that no corrections other than an overall shift has been applied (ACRIM I after 1984, ACRIM II and VIRGO). The HF record needed the most important corrections as explained in the text. For the ACRIM I record in 1980 the original corrections as published by *Willson and Hudson* [1991] needed also some revision as described in *Fröhlich and Lean* [1998b]. The ACRIM composite first described in *Willson* [1997] and available from *Willson* [2001] is also shown for comparison. The difference between the two composites is mainly due to the non-acceptance of the correction for HF during the gap between ACRIM I and II by *Willson* [1997].

Table 1. Results from the long-term comparison of ACRIM and ERBE with the composite TSI presented as slopes in ppm per year of a linear fit to the ratio for different periods. The end date is always 20 March 2002, the last day with data from ERBE, the starting dates correspond to the beginning of VIRGO, ACRIM II and ERBE respectively. The data points of ACRIM during the period when it represents the composite are removed. For each comparison also the Spearman rank correlation coefficient ρs is given.

| Start of Comparison (end: 20 March 2002) | ACRIM/TSI | | | | ERBE/TSI | | | |
	slope ppm/a	stddev	number of data points	ρs	slope ppm/a	stddev	number of data points	ρs
7 February 1996	4.13	1.57	2035	0.9559	26.8	10.5	256	0.8213
6 October 1991	2.58	1.11	2257	0.9862	15.6	4.9	449	0.7438
24 October 1984	–	–	–		18.3	2.5	649	0.7702

before. The amount of the shift of the ACRIM II/III record is determined from a weighted average of the mean ratios of ACRIM I to HF and ERBE over the period from November 1988 until June 1989 and the mean ratios of ACRIM II over the period from October 1991 until February 1993. ACRIM II has to be shifted by $+2149 \pm 33$ ppm to be on the same scale as ACRIM I (the uncertainty is a formal error from the statistics and does not include a possible bias). The result is illustrated in Plate 4.

3.3. Construction of the Composite and its Uncertainty

All time series which are part of the composite are now ready to be put together, and it should be stressed that ACRIM I and the shifted ACRIM II/III are the basis and are not changed in the process of producing the composite. Only the HF time series during different periods and the VIRGO record after February 1996 are shifted in order to agree with the adjacent ACRIM series (over a period of 250 days). The result is shown in Figure 3 and a comparison of the composite with the original data in Figure 4.

An estimate for the uncertainty of the long-term behaviour of the composite TSI can be deduced from comparison with ERBE and ACRIM, the result of which is summarized in Table 1. The difference of the standard deviations between ERBE and ACRIM is mainly due to the higher noise of the ERBE data (300 ppm compared to 130 ppm for ACRIM). Moreover, it depends on the number of points and the length of the data set used. With this in mind we may estimate the uncertainty of a possible trend to be about ±3 ppm/a for periods longer than 10–15 years; this implies a possible change of ±70 ppm over the 23 years of the observations. If we add the uncertainties related to the tracing of ACRIM II to I and of the HF correction (±60 ppm) we get a total of ±92 ppm. The observed change of the composite TSI as difference between two successive minima amounts to –52 ppm (–5 ppm/a) which is not significantly different from zero at the 2σ level.

4. COMPARISON OF THE COMPOSITE WITH AN EMPIRICAL MODEL

The sunspots are modelled by the photometric sunspot index (PSI) which was originally developed by *Foukal* [1981] and *Hudson et al.* [1982]. PSI is determined according to *Fröhlich et al.* [1994] from the position on the Sun, the projected area and a size dependent contrast which is deduced from photometric measurements [see e.g. *Steinegger et al.*, 1996]; this determination yields P_s. Current observations of the area and position of spots are available from NGDC Solar Geophysical Data (at ftp.ngdc.noaa.gov/stp/solar_data/sunspot_regions/ in the directories for each data set) where also the data of the Greenwich Observatory back to the eighties of the 19th century are stored. For the facular influence a similar index as for the spots can be developed by taking the area from either plages or from magnetic field strength and their distribution on the photosphere [see e.g. *Fligge et al.*, 1998; *Fontenla et al.*, 1999; *Fligge et al.*, 2000b]; and the long-term changes of irradiance can be explained by the bright magnetic network in and outside active regions [see e.g. *Foukal et al.*, 1991] in a similar way as the facular influence. Indeed, the difference between faculae and network may simply be the density of flux tubes producing the bright points which are at the subarcsecond size and normally not resolved. Both the faculae and the network are produced by magnetic flux tubes. The physical reason for these features to be brighter is that the magnetic field supplies a fraction of the total pressure. Thus, the gas pressure is less and with the lower opacity we see to deeper, hotter, and brighter layers in the vicinity of the flux tube compared to the neighbouring areas. Because a physical representation of such a flux-tube model is very difficult to develop, the influence of the magnetic features are in general calculated from observed temperature distributions in the solar atmosphere above these features. Such models depend on high quality images to determine the fraction of faculae and network on the solar disk which are tedious to analyze and not available on a regular basis.

Another approach is to use some proxy for faculae and network which simulates all these effects in a global way. The MgII index, the core-to-wing ratio of the MgII line at 280 nm, is such a proxy which is available from UV measurements since the start of NIMBUS-7 in November 1978 as described by *Donnelly* [1988]. Composites of the MgII index from several sources have been used in proxy models for irradiance changes since then [see e.g. *Lean et al.*, 1982; *Foukal and Lean*, 1986; *Fröhlich and Lean*, 1998b] and an extension of the one from the last reference to the end of 2001 is used in the following. In order to account for a possible difference between the short-term (rotational 27-day period) and the long-term (solar-cycle period of 11 years) we separate the MgII index into a short- and long-term component, P_{Fs} and P_{Fl}. As a first approximation P_s and the two P_{Fs} and P_{Fl} time-series are linearly regressed against TSI over the period of observations from late 1978 until the end of 2001 and the results in Wm^{-2} are shown in Figure 5.

This procedure is very reasonable for P_S as the center-to-limb variation of sunspots is similar to the quiet Sun. Thus, as a result of the multiple linear regression we use the calibrated $\overline{P_s}$ and correct TSI by calculating TSI$+\overline{P_s}$. From the picture of the faculae and network mentioned before, it becomes clear that the flux contribution of these features to the irradiance has a strong angular dependence such as limb-brightening [see e.g. *Unruh et al.*, 2000] which is not reflected by the MgII proxy. The MgII index represents a sort of projected area of the magnetic fields in the chromosphere producing the faculae on the photosphere and does not mimic the angular distribution of the outgoing radiance. The limb-brightening leads to a double peak modulation of the irradiance during the passage of an active region; even if there is no sunspot in the region, the irradiance is lower at the meridian transit than at 4–5 days before and after the transit. In contrast, the MgII index has its peak influence at the central-meridian transit. To account for this important difference the time series of TSI$+\overline{P_s}$ and the short-term Mg II-index are convolved with a double-peak ('Camel') and a triangular ('Dromedary') filter of 17-day base width as described by *Fröhlich* [2002]. For the 'Camel' filter we have different forms and the best filter is searched by applying different filters to the time series TSI$+\overline{P_s}$; then we calculate the multiple linear regression of corrected TSI$+\overline{P_s}$ with the triangularly filtered P_{Fs} and P_{Fl} (which does not need to be filtered). Then the 'Camel' filter which yields the highest correlation for the regression is selected. With the optimal filter the correlation coefficient is increased to 0.980 from 0.959 without filter, meaning that more than 95% of the variance is formally explained by the model (see Plate 5). An interesting result is also that the factor for P_{Fl} is about 1.7 times larger than the one for P_{Fs}. This may indicate that either the network which is believed to be responsible for the long-term changes is governed by a different physical mechanism, or that there are some other effects (about 40%) which are correlated, but not directly related to the MgII index.

The result of the comparison with the model shows that the present cycle is quite different from the earlier ones, and it is even more pronounced in the model than in TSI: the increase into the cycle is much slower than usual and there may have been a first 'maximum' around the beginning of 1999, indicated by a small shoulder. A major difference is also that there were fewer sunspots in this cycle, even during the present maximum. The overall trend of the residuals amounts to -163 ppm over the 23 years of observation which compares well with the observed trend of -52 ± 92 ppm/a. It should be noted that the residuals depend also on how well the model describes the 3 cycles and from Plate 6 it is obvious that it underestimates the maximum of cycle 21 and overestimates the present one, which could explain the larger trend.

5. SPECTRAL REDISTRIBUTION DURING CHANGES OF TSI

For the first time VIRGO provides reliable spectral measurements together with TSI observations. Former measurements with similar filter-radiometers on SOVA/EURECA

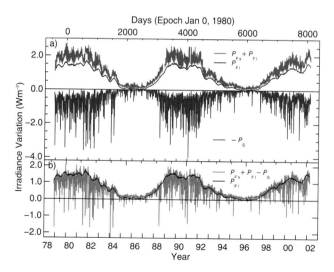

Figure 5. An empirical model of total solar irradiance variability is based on three indices—sunspot darkening by PSI and facular and network brightening by short and long-term MgII index—and calibrated against the TSI composite record yielding the scale in Wm^{-2}.

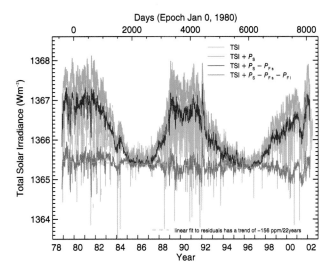

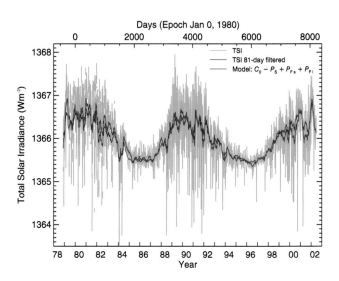

Plate 5. Comparison of the composite TSI with the empirical model based on PSI and the short- and long-term MgII index. The linear fit to the residual yields a slope of -3.8 ± 0.2 ppm/a.

Plate 6. Direct comparison of the model with observed TSI. Note that it underestimates TSI during the maximum of cycle 21 and overestimates the present one.

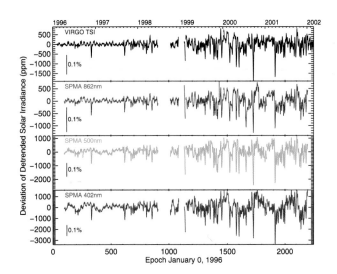

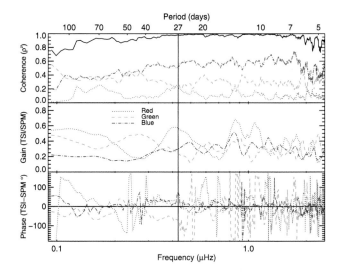

Plate 7. Time series of detrended TSI and the three channels of SPM-A.

Plate 8. Results of the multivariate spectral analysis for the three SPM times series against the TSI time series. The top panel shows the coherence ρ_j^2, the middle the gain and the bottom one the phase of the linear filter.

[*Crommelynck et al.*, 1993] have been hampered by pointing effects, not allowing to asses longer term variations. Even on SOHO, long-term trends of the SPM are still not understood well enough to assess the solar cycle variability of the spectral channels. Thus, a detrending has been devised which allows us to analyze periods shorter than about 10–12 months. The time series of the detrended SPM-A channels as determined by *Fröhlich and Wehrli* [2002] are plotted together with the detrended TSI in Plate 7. The resemblance of the four time series is remarkable. The ratios of the variance of the red, green and blue channels to TSI amount to 0.82, 1.48, and 1.93. But there are also some differences which are most visible in the time series during solar minimum, mainly because of lower solar variability. It is not yet clear whether these differences, e.g. the oscillations in the blue and green channel, are still remnants of instrumental effects or real and of solar origin.

The spectral redistribution can be determined by performing a multi-variate frequency analysis which accounts for the influence of the time-series $X_1(t) \ldots, X_j(t) \ldots, X_m(t)$ on $Y(t)$ by constructing the linear transformation $\mathbf{L}(X_1(t) \ldots, X_j(t) \ldots, X_m(t))$ of these time series which best approximates $Y(t)$. This can be analyzed in the frequency domain and the linear transformations $\mathbf{L}$ become m linear spectral filters $B_j(v)$, determined by comparing, frequency by frequency, the spectra f_{Xj} with f_y. The multiple coherence $\rho_j(v)$ is then a measure of the association between the series and $100 \times \rho_j(v)^2$ gives the amount in percent of the part in $Y(t)$ formally explained by the transformed $X_j(t)$. The total coherence $\rho(v)^2 = \Sigma \rho_j(v)^2$ corresponds to the total amount formally explained by the variance in the combination of all $X_j(t)$. The basic theory can be found in e.g. *Koopmans* [1974, Chapters 5.6f, 8.4f]. This analysis is applied to the three SPM time series as X_R, X_G and X_B and the TSI time series as Y. The results for the coherence, gain and phase are shown in Plate 8 and in Table 2.

The low frequency part of the cross correlations are influenced by the way the detrending is performed. Polynomial fits do not only influence the amplitude, but also the phase. From the plots of the coherence we conclude that the information below about 0.09 μHz (period of ≈130 days) is more and more contaminated by the detrending. At low frequencies the blue and red are contributing to a lesser amount to the variance of TSI than the green channel. It is generally accepted that the slow variations of TSI are mainly due to changes in the UV (up to 30% of the variance) and the observed decrease is somewhat puzzling and may still be due to the data treatment. At periods below about 70 days there are interesting features as the one at the 27-day solar rotation where a wide peak in the red fills in the dip in the blue. Another, even more pronounced sequence is at periods around 20 and 10 days where the green takes over from the blue. Or, the one at 13 days where, as for the 27-day period, the red compensates for the loss in the blue. This is indeed astounding as the UV irradiance has a strong 27-day modulation, which may, however, be mainly due to the strong lines whereas the blue channel at 402 nm may be more representative for a continuum. Variations in the three channels explain 97% of the variance of total solar irradiance, with contributions of 12.2% from the red, 26.4% from the green and 58.3% from the blue, respectively. The spectral redistribution is shared by the red green and blue as 1.0:2.2:4.8. The gain—the factor by which the SPM reading has to be multiplied to get the TSI signal—is in general higher for the red, and the green and blue on the other hand are about the same as listed in Table 2. The phase, the least reliable part of the linear filter due to its ambiguity of $\pm\pi$, indicates with a positive sign a phase lead of the SPM relative to TSI. Up to frequencies of about 0.2 μHz (period of ≈60 days) the green and blue channels are more or less constant whereas the red seems to turn around; at higher frequencies the green changes more than the blue and red. The rather small average values as listed in Table 2 are probably not significantly different from zero. Finally, Figure 6 shows the share of the power of TSI formally explained by the three channels.

6. CONCLUSIONS

The composite has proven to be a very valuable and reliable time series of TSI for both solar physicist and climatologists. As a result of the analysis of the VIRGO radiometry a long-term uncertainty of about 1 ppm/a at the 2σ-level seems to be achievable with present state-of-the-art radiometers. However, from the trends of the ratio of ACRIM II/III and ERBE to the composite TSI the uncertainty amounts to about 6 ppm/a at the 2σ-level. This higher value is mainly due to the fact that radiometers on different platforms are compared and that the environment of Earth-orbiting satellites is less stable than on SOHO. This value together with the uncertainties of the ACRIM II to I tracing and the correction to HF yield a total uncertainty of a possible change over the 23 years of observations of ±92 ppm. Compared to the observed trend determined from the difference between the two minima of -52 ppm it is not significantly different from zero at the 2σ level. Thus, we can conclude that TSI does not show any significant trend during the last 23 years.

The discussion of uncertainties shows that the problem of the lack of sufficient absolute accuracy and the determination of the long-term stability remains as emphasized by *Quinn and Fröhlich* [1999]. The reason that we could still

Table 2. Summary of the coherence, gain and phase as average over the spectral range 0.289 ... 1.653 μHz (periods 7 ... 40 days).

channel	ρ^2	gain	phase (deg)
red	12.2%	0.57	4.7
green	26.4%	0.41	−14.1
blue	58.3%	0.42	2.3

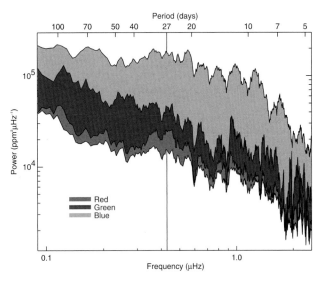

Figure 6. Results of the multivariate spectral analysis for the three SPM times series against the TSI time series displayed as shares of the TSI power spectrum. The explained part covers about one order of magnitude which corresponds to the 97% in the range 0.289 ... 1.653 μHz (periods 7 ... 40 days) listed in Table 1.

produce such a reliable time series of TSI is that we had always at least two independent experiments simultaneously in space. Although the plans of space agencies for the next several years show that we still may have more than one experiment at a time, the plans for the distant future, however, are vague and there may even be a certain reluctance by the agencies to continue such programs. Thus, we need measurements with an accuracy of space radiometry at least a factor of 10 better. Radiometers with such accuracy are available, but have not yet proven to maintain such a level when transferred to space and during longer periods of exposure to solar radiation.

The non-existence of a significant trend is also compatible with the result from the comparison with the empirical model with a trend of -156 ppm over the 23 years. The larger trend is partly due to the fact that the model underestimates the maximum of cycle 21 and overestimates the present one. Such rather simple proxy models are able to explain more than 95% of the TSI variance; this is astounding as the MgII index is only a proxy. More physically based models,

however, confirm these results in general [see e.g. *Fligge et al.*, 2000a] showing, that indeed MgII index is a viable proxy for the irradiance increases due to faculae and network.

The fact that the radiative output of the Sun shows no significant trend over the last 20 years does not mean that the Sun may not change on longer time scales; it may just mean that the observations are coincident with a period of a turnover of the solar output from increasing during the last centuries to decreasing. Thus, the reconstructions of the solar irradiance into the past by e.g. *Hoyt and Schatten* [1993]; *Lean et al.* [1995]; *Solanki and Fligge* [1999] were developed mainly to explain the 'little ice age' during the Maunder Minimum in the 17th century, but they need to replicate the present time series also. More specifically the levels of the last two minima should be about the same; only the most recent reconstruction of *Lean* [2000] satisfies this constraint.

The analysis of the measurements of the 3-channel filter-radiometers on VIRGO and the comparison with TSI show very interesting results. Further investigations will help to better understand the physical mechanisms responsible for the variability of TSI and the related spectral redistribution.

Acknowledgments. The author like to thank the Swiss National Science Foundation for the continuous support of the VIRGO Project at PMOD/WRC. Thanks are extended to Dr. Judith Lean, NRL, Washington DC, for helpful discussions. These results would not be possible without the continuous efforts of the VIRGO and SOHO teams. SOHO is a cooperative ESA/NASA mission.

REFERENCES

Anklin, M., C. Fröhlich, W. Finsterle, D. A. Crommelynck, and S. Dewitte, Assessment of degradation of VIRGO radiometers onboard SOHO, *Metrologia, 35*, 686-688, 1998.

Brusa, R. W., and C. Fröhlich, Absolute radiometers (PMO6) and their experimental characterization, *Appl. Optics, 25*, 4173-4180, 1986.

Chapman, G. A., A. M. Cookson, and J. J. Dobias, Variations in total solar irradiance during solar cycle 22, *J.Geophys.Res., 101*, 13,541-13,548, 1996.

Crommelynck, D., and V. Domingo, Solar irradiance obserations, *Science, 225*, 180-183, 1984.

Crommelynck, D., V. Domingo, A. Fichot, C. Fröhlich, B. Penelle, J. Romero, and C. Wehrli, Preliminary results from the SOVA experiment on board the european retrievable carrier (EURE-CA), *Metrologia, 30*, 375-380, 1993.

Crommelynck, D., A. Fichot, R. B. Lee III, and J. Romero, First realisation of the space absolute radiometric reference (SARR) during the ATLAS 2 flight period, *Adv. Space Res., 16*, (8)17-(8)23, 1995.

Donnelly, R. F., The solar UV MG II core-to-wing ratio from the NOAA9 satellite during the rise of solar cycle 22, *Advances in Space Research, 8*, 77-80, 1988.

Fligge, M., S. K. Solanki, Y. C. Unruh, C. Fröhlich, and C. Wehrli, A model of solar total and spectral irradiance variations, *Astron. Astrophys., 335,* 709-718, 1998.

Fligge, M., S. K. Solanki, and Y. C. Unruh, Modelling irradiance variations from the surface distribution of the solar magnetic field, *Astron. Astrophys., 353,* 380-388, 2000a.

Fligge, M., S. K. Solanki, and Y. C. Unruh, Modelling Short-Term Spectral Irradiance Variations, *Space Science Reviews, 94,* 139-144, 2000b.

Fontenla, J., O. R. White, P. A. Fox, E. H. Avrett, and R. L. Kurucz, Calculation of Solar Irradiances. I. Synthesis of the Solar Spectrum, *Astrophys. J., 518,* 480-499, 1999.

Foukal, P, Sunspots and changes in the global output of the Sun, in *The Physics of Sunspots,* edited by L. E. Cram and J. H. Thomas, pp. 391-398, Sacramento Peak Observatory, New Mexico, 1981.

Foukal, P., K. Harvey, and F. Hill, Do changes in the photospheric magnetic network cause the 11 year variation of total solar irradiance?, *Astrophys. J., 383,* L89-L92, 1991.

Foukal, P. V., and J. Lean, The influence of faculae on total solar irradiance and luminosity, *Astrophys. J., 302,* 826-835, 1986.

Fröhlich, C., Observations of irradiance variability, *Space Science Reviews, 94,* 15-24, 2000.

Fröhlich, C., Long-term behaviour of space radiometers, *Metrologia, 40,* 60-65, 2003.

Fröhlich, C., Total solar irradiance since 1978, *Adv Space Res., 29,* 1409-1416, 2002.

Fröhlich, C., and W. Finsterle, VIRGO radiometry and total solar irradiance 1996-2000 revised, in *Recent Insights Into the Physics of the Sun and Heliosphere: Highlights from SOHO and Other Space Missions,* edited by P. Brekke, B. Fleck, and J. B. Gurman, pp. 105-110, ASP Conference Series, IAU Symposium, Vol. 203, 2001.

Fröhlich, C., and J. Lean, Total solar irradiance variations: The construction of a composite and its comparison with models, in *IAU Symposium. 185: New Eyes to See Inside the Sun and Stars,* edited by F. L. Deubner, J. Christensen-Dalsgaard, and D. Kurtz, pp. 89-102, Kluwer Academic Publ., Dordrecht, The Netherlands, 1998a.

Fröhlich, C., and J. Lean, The Sun's total irradiance: Cycles and trends in the past two decades and associated climate change uncertainties, *Geophys. Res. Let., 25,* 4377-4380, 1998b.

Fröhlich, C., and J. Lean, Solar irradiance variability and climate, *Astron. Nachr., 323,* 203-212, 2002.

Fröhlich, C., and C. Wehrli, Variability of spectral solar irradiance from VIRGO/SPM observations, presented at *SOHO-11: From Solar Minimum to Maximum.*

Fröhlich, C., J. M. Pap, and H. S. Hudson, Improvement of the photometric sunspot index and changes of disk-integrated sunspot contrast with time, *Sol. Phys., 152,* 111-118, 1994.

Fröhlich, C., D. Crommelynck, C. Wehrli, M. Anklin, S. Dewitte, A. Fichot, W Frosterle, A. Jiménez, A. Chevalier, and H. J. Roth, In-flight performances of VIRGO solar irradiance instruments on SOHO, *Sol. Phys, 175,* 267-286, 1997.

Hickey, J. R., B. M. Alton, H. L. Kyle, and D. V. Hoyt, Total solar irradiance measurements by ERB/NIMBUS 7: A review of nine years, *Space Sci. Rev., 48,* 321-342, 1989.

Hoyt, D. V., and K. H. Schatten, A discussion of plausible solar irradiance variations, 1700–1992, *J. Geophys. Res., 98,* 18,895-18,906, 1993.

Hoyt, D. V., H. L. Kyle, J. R. Hickey, and R. H. Maschhoff, The NIMBUS-7 solar total irradiance: A new algorithm for its derivation, *J. Geoph. Res., 97,* 51-63, 1992.

Hudson, H. S., S. Silva, M. Woodard, and R. C. Wilson, The effects of sunspots on solar irradiance, *Sol. Phys., 76,* 211-218, 1982.

Kerr, R., Did satellites spot a brightening Sun?, *Science, 277,* 1923–1924, 1997.

Koopmans, L., *The Spectral Analysis of Time Series,* Academic Press, Inc., London, GB, 1974.

Lean, J., Evolution of the Sun's Spectral Irradiance Since the Maunder Minimum, *Geoph. Res. Let., 27,* 2425-2428, 2000.

Lean, J., J. Beer, and R. Bradley, Reconstruction of solar irradiance since 1610: Implications for climate change, *Geophys. Res. Lett., 22,* 3195-3198, 1995.

Lean, J. L., W. C. Livingston, D. F. Heath, R. F. Donnelly, A. Skumanich, and O. R. White, A three-component model of the variability of the solar ultraviolet flux 145-200 nm, *J. Geophys. Res., 87,* 10,307-10,317, 1982.

Lee III, R. B., B. R. Barkstrom, and R. D. Cess, Characteristics of the Earth radiation budget experiment solar monitors, *Appl. Opt., 26,* 3090-3096, 1987.

Lee III, R. B., M. A. Gibson, R. S. Wilson, and S. Thomas, Long-term total solar irradiance variability during sunspot cycle 22, *J. Geoph. Res., 100,* 1667-1675, 1995.

Quinn, T., and C. Fröhlich, Accurate radiometers should measure the output of the Sun, *Nature, 401,* 841, 1999.

Solanki, S. K., and M. Fligge, A reconstruction of total solar irradiance since 1700, *Geophys. Res. Let., 26,* 2465-2468, 1999.

Steinegger, M., P. Brandt, and H. Haupt, Sunspot irradiance deficit, facular excess, and the energy balance of solar active regions, *Astron. Astrophys., 310,* 635-645, 1996.

Unruh, Y. C., S. K. Solanki, and M. Fligge, Modelling solar irradiance variations: Comparison with observations, including line-ratio variations, *Space Sci. Rev, 94,* 145-152, 2000.

Willson, R. C., Active cavity radiometer type iv, *Appl. Opt., 18,* 179-188, 1979.

Willson, R. C., Measurements of solar total irradiance and its variability, *Space Sci. Rev., 38,* 203-242, 1984.

Willson, R. C., Irradiance observations from SMM, UARS and ATLAS experiments, in *The Sun as a Variable Star, Solar and Stellar Irradiance Variations,* edited by J. Pap, C. Fröhlich, H. S. Hudson, and S. Solanki, pp. 54-62, Cambridge University Press, Cambridge UK, 1994.

Willson, R. C., Total solar irradiance trend during solar cycles 21 and 22, *Science, 277,* 1963-1965, 1997.

Willson, R. C., ACRIM II and ACRIM III data products (version 10/10/01), http://www.acrim.com/ in the directory Data%20 Products.htm, 2001.

Willson, R. C., and H. S. Hudson, The Sun's luminosity over a complete solar cycle, *Nature, 351*, 42-44, 1991.

Willson, R. C., H. S. Hudson, C. Fröhlich, and R. W Brusa, Long-term downward trend in total solar irradiance, *Science, 234*, 1114-1116, 1986.

Claus Fröhlich, Physikalisch-Meteorologisches Observatorium Davos, World Radiation Center, CH-7260 Davos Dorf, Switzerland. (cfrohlich@pmodwrc.ch)

Measurement of the Solar Ultraviolet Irradiance

Gary Rottman

Laboratory for Atmospheric and Space Physics, University of Colorado, Boulder, Colorado

Linton Floyd

Interferometrics, Chantilly, Virginia

Rodney Viereck

Space Environment Center, National Oceanic and Atmospheric Administration, Boulder, Colorado

Solar radiation at wavelengths shorter than 300 nm is almost completely absorbed by the Earth's atmosphere—between 120 and 300 nm it plays a major role in O_2 and O_3 photolysis within the stratosphere and mesosphere. Although this ultraviolet portion of solar radiation comprises less than 1% of the total solar irradiance, it directly influences the physical and chemical processes in the atmosphere, and likely has important indirect influences on climate as well. In the past it has been difficult to measure the solar ultraviolet; not only because instruments must be taken above the atmosphere to make observations, but also because the very energetic photons damage optical systems, causing loss in instrument sensitivity, which leads to measurement uncertainties and ambiguity that exceed solar variations. In the past twenty years instrument capabilities have steadily improved, and current data sets now provide a reasonably clear understanding of solar variability. During the two solar cycles for which measurements are in hand, the Sun has apparently varied by roughly a factor of two at Lyman-α, 5 to 10% out to 200 nm, and only a few percent between 200 to 300 nm. At wavelengths beyond 300 nm the spectral observations set an upper limit of no more than 1% solar variation, with the additional constraint that the variation of total solar irradiance (TSI) is on the order of 0.1%.

1. INTRODUCTION

This chapter reviews our present understanding of solar ultraviolet (UV) spectral irradiance for wavelengths

between 120 and 400 nm, with special attention to wavelengths below 300 nm. Figure 1 shows the entire solar spectrum from the extreme ultraviolet out to 2000 nm (2μm) in the near infrared. The general shape of this spectrum is well represented by a blackbody spectrum for an effective temperature of about 5770 K. The integral over the entire range of spectral irradiance is referred to as the total solar irradiance or TSI. This important climate system variable is considered in detail by *Fröhlich* [this volume]. An

Solar Variability and its Effects on Climate
Geophysical Monograph 141

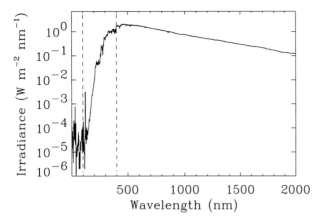

Figure 1. Solar spectrum from 120 to 2000 nm. This comprises about 95% of the total solar irradiance with the remaining 5% in the infrared. The dashed lines define the ultraviolet portion of the spectrum considered in this section.

accepted value of TSI is on the order of 1367 Wm^{-2}, and the portion of TSI represented in Figure 1 includes roughly 95% of this value with the remaining 5% further in the infrared. Note that the ultraviolet spectral irradiance decreases by more than five orders of magnitude over the defined ultraviolet range. The two vertical bars in Figure 1 at 120 and 400 nm limit the UV spectral interval considered in this section, and the portion to the left, the extreme ultraviolet (EUV), is described by *Woods et al.* [this volume]. The ultraviolet ($120 < \lambda < 300$nm), important to our middle atmosphere, is less than 1% of TSI, and even the near UV ($300 < \lambda < 400$nm) is less than 10%. Although their contributions to TSI are relatively small, these UV portions of the solar irradiance are quite variable and extremely important to the Earth's atmosphere.

The visible to near infrared portion of the solar spectrum has yet to be observed with sufficient precision and accuracy to establish solar variability. Since this portion of the solar spectrum ($\lambda > 300$ nm) comprises almost 99% of the solar radiation, the observations of TSI must constrain variability here to less than a few tenths of one percent—perhaps with the exception of a few localized features, e.g., the Fraunhofer absorption lines where variability is larger. Model calculations [*Solanki and Unruh,* 1998; and *Fox,* this volume] confirm this assumption.

The Earth's atmosphere blocks all radiation with wavelengths shorter than about 300 nm—radiation with photon energy greater than 4 eV. This is primarily due to ozone (O_3) in the Earth's stratosphere, which has a strong absorption cross-section for radiation with wavelengths between 200 and 300 nm. A very tenuous amount of ozone is present in the atmosphere as a result of the photolysis of molecular oxygen (O_2), which completely absorbs higher energy photons with wavelengths between 130 to 200 nm. In combination therefore, the O_2 and O_3 in the Earth's atmosphere are an exceptionally effective shield from all photons with wavelengths shorter than 300 nm [*Brasseur and Solomon,* 1986]. This atmospheric protection has been essential to the development and evolution of life as we know it, and its effectiveness is in a vulnerable state (see for example, *Scientific Assessment of Ozone Depletion,* 1998).

Solar radiation absorbed by the atmosphere becomes the dominant direct energy input, and fuels atmospheric photochemistry—establishing atmospheric temperature, composition, and structure. This deposited energy drives the atmosphere's dynamics and circulation. In addition to the direct absorption of solar radiation, the atmosphere, including clouds and aerosols, scatters some fraction of incoming radiation back to space. An additional fraction is scattered at the land and ocean surface, and only the remainder (roughly 50% of the globally averaged value arriving at the top of the atmosphere) contributes to the Earth's radiative balance. The land and ocean surface along with the atmosphere and cloud tops then radiate long wavelength infrared (with mean wavelength near 10 μm) back to space. Greenhouse gases further complicate the balance between incoming shortwave radiation, all of which determine an effective global mean temperature [*Kiehl and Trenberth,* 1997].

To summarize, accurate and precise knowledge of incoming solar radiation is essential to all studies of the Earth's atmosphere, radiative balance, and climate. The ultraviolet portion of solar radiation has a direct impact on the chemistry and dynamics of the middle atmosphere, and may indirectly influence climate change [*Shindell et al.,* 2001].

2. OBSERVATIONS

Because much of the UV is absorbed by the atmosphere, there could be no direct information on the extraterrestrial, or top-of-the-atmosphere, UV until scientific instruments could be placed in space. Some of the very earliest sounding rocket experiments incorporated instrumentation to record solar radiation at ultraviolet wavelengths [e.g. *Baum et al.,* 1946; and *Detwiler et al.,* 1961]. The instruments, calibrations, and observing techniques steadily improved, and by the mid-1970s reasonable confidence in the intensity of UV irradiance and estimates of the solar variability were documented [*White,* 1977; and *Lean,* 1987]. Nevertheless, considerable controversy accompanied these first measurements, and general consensus held that vast improvements in instrument calibration and performance were still required before an understanding of the Sun's influence on the Earth could be realized.

2.1. Long-Term Satellite Observations

Early satellites carrying instruments and providing multi-year data records (OSO-5 [*Vidal-Madjar*, 1977], AE [*Hinteregger et al.*, 1981], Nimbus-7 [*Heath and Schlesinger*, 1986], and SME [*Rottman et al.*, 1982]) provided irradiance time series that clearly showed short- and intermediate-term solar variations, but longer-term solar cycle variations were compromised and obscured by possible long-term changes in the instrument response. These early space missions were planned and intended to operate for only one to two years, and the instruments were of modest design and did not incorporate methods to track changes in the instrument sensitivity. This shortcoming was recognized, and by the mid-1980s new instrument concepts were being tested, including on-board calibration—calibrations that would directly detect changes in instrument responsivity and provide appropriate data correction.

This section concentrates on the evolution and improvements of measurements from early 1980 to the present. The SME and NIMBUS-7 made observations during solar cycle 21, and the NOAA series of SBUV instruments continue the NIMBUS observations today. The GOME instrument on the ERS-2 satellite, launched in 1995 and still flying today, covers an extended spectral range from the ultraviolet through the visible.

Table 1. Major satellite missions carrying solar UV irradiance instruments and providing data for the past two solar cycles.

Instrument	Δλ Range	Time Interval
SME	115 to 300 nm	10/1981 to 4/1989
SBUV		
Nimbus-7	160 to 400 nm	11/1978 to 2/1987
NOAA-9	160 to 405 nm	3/1985 to 2/1998
NOAA-11	160 to 405 nm	2/1989 to 10/1994
NOAA-14	160 to 405 nm	2/1995 to present
NOAA-16	160 to 405 nm	2/2001 to present
SOLSTICE	119 to 420 nm	10/1991 to present
SUSIM	115 to 410 nm	10/1991 to present
GOME	240 to 790 nm	6/1995 to present
ATLAS I, II, III		3/1992, 4/1993,
SSBUV	160 to 405 nm	and 11/1994
SUSIM	115 to 410 nm	
SOLSPEC	200 to 3200 nm	
Other Shuttle		8 flights
SSBUV	160 to 405 nm	10/1989 to 1/1996

Special attention is given to the results from the two UARS instruments, SOLSTICE and SUSIM, whose observations now extend over a period of more than eleven years. These two instruments incorporate internal calibration techniques to monitor changes in instrument responsivity. Additional consideration is given to the SBUV observations that in combination cover by far the longest time interval. Although the SBUV instruments do not have inherent in-flight calibration, the series of Shuttle flights of the SSBUV provides a reliable transfer calibration. These short duration Shuttle observations are limited to only a few days, but nevertheless they provide an important validation of the long-term satellite observations. These "snapshot" type observations have the distinct advantage that the instruments are calibrated prior to flight and again after flight, and as a result, instrument performance during the Sun observation is more accurately specified.

2.1.1. Satellite observing programs 1980 to 1990. The Solar Mesosphere Explorer (SME) was launched in October 1981 and carried five instruments to conduct research on ozone in the mesosphere and upper stratosphere. One of these instruments, the Solar Ultraviolet Irradiance Monitor, was a spectrometer covering the spectral range 115 to 310 nm with 0.75 nm spectral resolution [*Rottman et al.*, 1982]. SME operated successfully until April 1989 and provided more than 7 1/2 years of solar irradiance data—observations spanning the period from near solar maximum in 1981, through solar minimum in 1986, into the ascending phase of solar cycle 22. Unfortunately the SME data set did not continue to the full maximum of the cycle, and most unfortunately the data did not overlap with the UARS observations beginning in late 1991.

The SME mission was designed for a nominal period of only one year, and the instrument did not incorporate a calibration technique for tracking end-to-end changes in instrument responsivity. The instrument used scattering screens to redirect solar irradiance into the spectrometer, and these external elements were acknowledged to be the greatest liability to stable instrument performance. Therefore, multiple screens were incorporated into the instrument design, and the redundant observations helped to determine long-term instrument stability. A series of sounding rocket underflights provided limited in-flight data validation, but the last flight in 1986 did not extend this calibration method into solar cycle 22.

In the final analysis the SME irradiance data set has a relative accuracy of approximately ±4% for the years 1981 to 1989. This level of precision and accuracy is adequate for establishing solar cycle variability for wavelengths shorter than about 200 nm, but is not sufficient at longer wavelengths between 200 and 300 nm [*Rottman,* 1987].

The Solar Backscatter Ultraviolet (SBUV) instrument on NASA's Nimbus-7 satellite was launched in October 1978 with the primary goal of measuring total ozone in the Earth's atmosphere. This double-pass, grating spectrometer provided daily measurements of the solar spectral irradiance in the wavelength region from 160 to 400 nm at 1.1 nm spectral resolution. The instrument did not incorporate a system to monitor long-term changes in instrument responsivity, although an empirical model was developed to account for changes in instrument sensitivity [*Schlesinger and Cebula,* 1992].

SBUV/2 instruments, a second generation of the SBUV, were then flown on NOAA operational weather satellites. These instruments also cover the wavelength range from 160 to 405 nm at 1.1 nm spectral resolution. The first instrument, on the NOAA-9 satellite, was launched in late 1984, followed by NOAA-11 in February 1989, NOAA-14 in February 1995, and NOAA-16 in September 2000. At least two additional SBUV/2 instruments are planned for launch approximately every two to three years throughout the first decade of the twenty-first century. *Cebula and DeLand* [1998] used coincident underflights of the Shuttle SBUV (SSBUV) to derive long-term corrections for the NOAA-11 SBUV/2. The NOAA-9 irradiance data cover all of solar cycle 22, and similar analysis of the data is in progress [*DeLand,* private communication].

NASA's Upper Atmosphere Research Satellite, UARS, was launched in September 1991 with a primary research goal of improving our understanding of the distribution and variation of ozone in the Earth's middle atmosphere. The satellite carries ten scientific instruments to make daily and global observations of the atmosphere's composition, dynamics, and energetics. Three of the ten instruments observe the Sun, one measuring total solar irradiance [see *Fröhlich,* this volume] and two measuring ultraviolet spectral irradiance. The two spectral irradiance instruments, the SOLar STellar Irradiance Comparison Experiment (SOLSTICE) and the Solar Ultraviolet Spectral Irradiance Monitor (SUSIM), cover a similar spectral range of roughly 120 to 400 nm with comparable spectral resolution. They differ significantly in optical design and in their methods of tracking and correcting changes in instrument responsivity.

The SOLSTICE is a three channel grating spectrometer operating over the spectral range 119 to 420 nm. Each channel covers a portion of this spectral range, with optics, detectors, and spectral resolution tailored to varying intensity levels and the energy of the photons [*Rottman et al.,* 1993; *Woods et al.,* 1993; *Rottman et al.,* 1994]. While SOLSTICE is pointed at the Sun, the grating scans to obtain a full irradiance spectrum. Moreover, SOLSTICE has a unique capability of changing its entrance aperture, spectral bandpass (spectrometer exit slit), and integration time to achieve more than eight orders of magnitude enhancement in sensitivity. This transformation allows the instrument to observe bright blue (spectral type O, A, and B) stars employing the same optics and detectors used for the solar observations. The stars are routinely observed during the night portion of each orbit, and they become reference standards to establish changes in the instrument response. Individually the twenty or so stars should vary by only a fraction of one percent. The ensemble average of many stars should display negligible variation, and any changes in stellar signal are interpreted as changes in instrument responsivity. The derived changes in instrument responsivity apply equally to the solar observations, and the solar data are then corrected accordingly.

The SOLSTICE technique has the added feature that the UV output of the Sun is directly established as a ratio to the UV flux from the set of stars. With the assumption that these stars are "standards" and do not vary, future solar observations using the same stars should provide a direct comparison to today's irradiance values.

The UARS SOLSTICE data are still being analyzed, but current estimates demonstrate that any two solar observations can be compared with an accuracy of about $\pm 2\%$ (2σ). Final and complete analysis may improve this estimate slightly. In principle, this technique should be capable of tracking long-term solar variations at a level approaching $\pm 0.5\%$.

The Solar Ultraviolet Spectral Irradiance Monitor (SUSIM) is the other spectral irradiance experiment aboard UARS. This dual-spectrometer is fully redundant, and like SOLSTICE it also measures the ultraviolet irradiance in the spectral range 115 to 410 nm [*Brueckner et al.,* 1993]. SUSIM has three separate spectral resolutions (0.15 nm, 1.1 nm, and 5 nm) available, but its principal solar observing mode is with 1.1 nm resolution. Each spectrometer can be configured using one of four separate grating pairs, with interchangeable detectors, entrance filters, and exit filters as well.

SUSIM carries four stable deuterium lamps that can replace the Sun as the source of illumination. These types of deuterium lamps are routinely used in the laboratory as standard reference sources, and they provide a reliable method of determining changes in instrument responsivity. The "working" optical channel gathers the daily solar UV irradiance measurements. Other optical channels, termed "reference channels," view the Sun less frequently, and accordingly incur less change in responsivity (degradation). Scans of the lamps calibrate the reference channels that are then used to calibrate the working channel. The lamp windows have also experienced degradation in transmission, but intercomparisons of lamp pairs with different run times account for this effect in a self-consistent manner. The SUSIM technique of comparing measurements of the same source (Sun or lamps), using several different and independent optical systems, provides measurements with a precision and long-term relative accuracy of about ±2% (2σ).

Solar data from SOLSTICE and SUSIM have been compared [*Woods et al.*, 1996] and agreement is well within the 2σ uncertainty (6 to 10%) of the individual measurement sets.

The Global Ozone Monitoring Experiment (GOME) was launched on the European Remote Sensing satellite, ERS-2, in April 1995. GOME provides terrestrial radiance and solar irradiance measurements in the wavelength range from 240 to 790 nm at approximate 0.2 nm spectral resolution in the UV and 0.4 nm resolution in the visible. GOME is a double pass spectrometer using a prism and a grating, and the experiment includes an onboard calibration system to track time-dependent instrument changes [*Weber et al.*, 1998; *Weber*, 1999; *Burrows et al.*, 1999].

2.1.2. Shuttle observing programs. The Shuttle Solar Backscatter Ultraviolet (SSBUV) instrument is identical to the SBUV/2 satellite instruments, except that it uses a transmission diffuser rather than a reflection diffuser. SSBUV flew eight Space Shuttle missions between October 1989 and January 1996. SSBUV/2 instruments cover the same wavelength range as the SBUV/2 instruments (160-405 nm), with the same spectral bandpass of 1.1 nm. The instrument's calibration is performed in air, thus the calibrated region of the spectrum extends from 200 to 405 nm. The estimated 2σ uncertainty in the SSBUV solar irradiance ranges from ±2.4% near 400 nm to ±6% near 200 nm [*Cebula et al.*, 1991, 1992, 1996b].

The SUSIM on UARS is an improved version of an earlier design that first flew on the Space Shuttle in 1982 (OSS-1). This earlier model first successfully measured the UV solar irradiance in August 1985 (Spacelab II). Three additional Shuttle flights (ATLAS I, II, and III) during the UARS time period have provided important comparisons and validations. Because of the relatively short duration of these missions, the Shuttle instrument carries only a single standard deuterium lamp and does not employ the redundant filters and gratings. In a manner similar to the SSBUV and the SOLSPEC (discussed below) instruments, the Shuttle SUSIM takes advantage of pre- and post-flight calibrations. These calibrations provide reliable knowledge of the instrument responsivity during the solar observation sequence. The inherent limitation of the Shuttle observations is the restricted viewing time—providing only a "snapshot" of the Sun.

The Solar Spectrum (SOLSPEC) instrument first flew in 1983 on the Spacelab-I Mission [*Labs et al.*, 1987], and subsequently on the three ATLAS missions. SOLSPEC uses three double pass spectrometers to cover the extended spectral range from 200 to 3000 nm. The UV spectrometer (200 to 350 nm) has a spectral bandpass of 1.1 nm. This short-wavelength channel incorporates hollow-cathode lamps to monitor the wavelength calibration and two deuterium lamps to verify instrument responsivity. The calibration of this instrument derives from the irradiance blackbody standard of the Heidelberg Observatory [*Thuillier et al.*, 1997]. In addition to the UV band, SOLSPEC also covered visible (180 – 370) and IR (800 – 3000nm) bands with a spectral resolution of 1 nm (see details by *Thuillier et al.*, this volume).

3. DESCRIPTION OF UV IRRADIANCE

3.1. The Solar Ultraviolet Spectrum

Figure 2 is an expanded view of the ultraviolet portion of the solar spectrum shown in Figure 1. Referring to both figures, the visible (approximately 500 nm) radiation in the solar spectrum originates low in the photosphere, and proceeding to shorter wavelengths, the radiation generally originates in progressively higher layers of the solar atmosphere. Near 200 nm radiation originates near the top of the photosphere, and near 160 nm it originates slightly above the photosphere where the temperature of the solar atmosphere has decreased to its minimum value. Shorter wavelengths then originate from the higher layers of the chromosphere and transition region where the temperature of the solar atmosphere abruptly rises. At wavelengths below about 180 nm the general character of the solar spectrum changes from a moderately strong continuum superposed with absorption features to a much weaker continuum superposed with emission features. The EUV chapter of this monograph [*Woods et al.*, this volume] concentrates on these shorter wavelengths where the solar spectrum consists predominantly of emission lines originating in the transition region and in the higher temperature corona. Structure within the solar spectrum, both emission and absorption, results from complex, non-local thermodynamic equilibrium (non-LTE) effects in the radiative environment of the Sun.

Although the spectrum of Figure 2 was measured by SOLSTICE, one of the two UARS instruments, irradiance data from any of the other instruments discussed above would give nearly the same result. Noting that *y axis* plot covers six orders of magnitude and that the general agreement in recent measurement programs is typically better than 10%, the substitution of data from any or all of the other experiments described above would fit tightly to the curve shown here.

The spectral resolution of the displayed spectrum is approximately 0.2 nm, determined by the resolution of the SOLSTICE instrument. The solar spectrum itself has inherent line widths that are 1/10 of those provided by SOL-STICE. Irradiance observations at higher spectral resolution would also show greater line to continuum contrast. Likewise, a lower resolution irradiance observation, for example from the 1.1 nm spectral mode of SUSIM and SBUV, would appear with proportionally lower contrast of line to continuum, but with the same overall spectral shape.

3.2. Solar Variations

The Earth's orbit about the Sun is eccentric and the Earth is roughly 3% closer in January at perihelion than in July at aphelion. This change in distance causes an approximate 7% irradiance variation throughout the calendar year, greatest in early January and smallest in early July. Irradiance data are typically corrected to a mean distance of one astronomical unit (1 AU). The spectral irradiance data are then reported as W/m^2 for TSI and W/m^3 (alternately in W/m^2nm or mW/m^2nm, or the generally accepted astronomical unit of photons/cm^2 sec Å).

Spectra of the type shown in Figure 2 are obtained on a daily basis by the two UARS instruments, the SBUV (160 < λ < 400 nm), and by the GOME (240 < λ < 400 nm). Time series of irradiance data are then constructed by selecting a given wavelength range, either containing a specific feature or spanning a given range of wavelengths, and plotting the data as a function of time. Figure 3 displays such an example for the strong Lyman-α emission line (at 121.6 nm) found in Figure 2. This line results from the resonance transition of atomic hydrogen. Lyman-α plus the other lines of the Lyman series (all occurring at shorter EUV wavelengths) are emitted from a fairly broad altitude/temperature range of the solar atmosphere [*Woods et al.*, 2000b].

The time series in Figure 3 displays several interesting characteristics: a low-frequency component having approximately one cycle over the time period, a higher-frequency component having a period of roughly one month, and perhaps other intermediate frequencies of smaller amplitude—including one with a period of approximately one year.

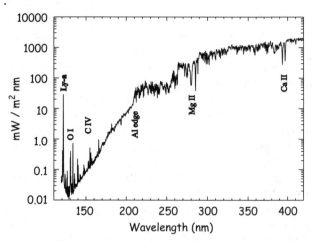

Figure 2. The ultraviolet spectral irradiance obtained by SOL-STICE, one of the two UARS irradiance instruments. Certain spectral features, both emission and absorption features, are identified.

Changes in the Sun's radiative output are primarily related to bright, irregular patches seen on the Sun's surface. The general consensus is that these active regions are the direct result of internal magnetic disturbances that erupt through the visible photosphere of the Sun [*Spruit*, 1994; *Cook et al,* 1980; and *Schüssler and Schmitt*, this volume]. The character of these active regions changes with height in the solar atmosphere. Extended regions of brightening are referred to as *faculae* as seen in the photosphere (visible and near infrared). These features exhibit only small brightness enhancement when seen on the disk of the Sun, but greater enhancement near the limb. Within the *faculae* the much smaller sunspots have even more intense magnetic field strength and lower brightness compared to emission from the neighboring photosphere. At photospheric wavelengths (λ > 300 nm) a competition thus exists between the enhanced emission from *faculae* and deficit emission from the sunspots (see for example, *Fröhlich*, this volume). For higher layers of the solar atmosphere, the chromosphere and above, the bright active regions have even higher contrast and are referred to as *plage,* but at these levels the sunspot irradiance deficit is not present. The high levels of irradiance seen in Figure 3, first in late 1991 and then again in 2001, are the result of the dramatic increase in number, size, and intensity of the *plage* on the solar disk. Additional contributions to UV irradiance variability may come from the active network, but its quantitative role is far less certain [*Lean et al.,* 1982].

Measurements related to the global properties of the Sun demonstrate that there is a strong and consistent cycle from conditions of an active, disturbed Sun to more quiescent conditions. On average, the cycle period is about 11 years, but it can vary by ±2 years. When images of the Sun are examined, the number, size and complexity of recognizable features distinguish the "active" Sun from the "quiet" Sun. If the measurement is an integrated global parameter, as for example the irradiance, then the signal proceeds from maximum level to minimum level in a period of roughly 5 to 6 years. The single cycle in Figure 3 is the manifestation of the most recent solar cycle—maximum in 1991 to minimum in 1996 and approaching solar maximum again in 2001 to 2002. Other solar records that extend much further back than the UV irradiance data, have been assigned a numbering convention that counts solar maxima beginning in the eighteenth century. Using this convention, the cycle that peaked in 1991 is 22 and the current cycle is 23. Solar irradiance variations on various timescales are discussed in the sections to follow.

3.2.1. Short-term variations. As seen from Earth, the Sun rotates with a period of about 27 days (this is the synodic period, while the sidereal rotation period is closer to 25 days). It does not rotate as a rigid body, and low latitude regions rotate somewhat faster than the polar regions. New active regions appear on the solar disk and old regions disperse in a fairly random, non-uniform fashion. This means that at any time the longitudinal distribution of solar activity is irregular, and as the Sun rotates the distant irradiance is modulated. This aspect of solar variability is analogous to signals from a distant lighthouse—as the light rotates the signal observed at a distance is modulated with the well-defined rotation period of the source. The high frequency of Figure 3 is therefore this 27-day rotation of the Sun slowly modified in amplitude by the distribution and intensity of plage present on the solar disk. Near solar minimum the

level of activity is low and the amplitude of the rotational signal is correspondingly small. Likewise, near solar maximum the active regions are much stronger and the signal is strong. Note that the strength of the "27-day" signal is determined as much by the distribution of activity on the solar disk as it is by the combined strength of all active regions.

The rotational signal in solar irradiance is considered in more detail later. From the standpoint of atmospheric and climate studies, it is important to consider that the rotation of the Sun provides a useful modulation of the irradiance, similar to what could be achieved if the Sun was under rheostat control. For studies of the Earth's atmosphere, especially those incorporating photochemical response, the Sun's 27-day variation provides a strong and deterministic change in the forcing over a period of only a few days. The amplitude and phase of the solar variation provides a test and a constraint of a photochemical model's capability and validity (see for example, the 27-day ozone response in the tropics discussed by *Hood*, this volume).

Another feature conspicuous in Figure 3 has a period of some few months, perhaps as long as a year (for example, the envelope with a peak in 1994). The cause of this modulation of the rotational amplitude is somewhat speculative, but it is likely related to the eruption and development of a new active region(s) on the solar disk, for which the typical lifetime is five to six months [*Bouwer*, 1992]. During this time the active regions change in size, shape, and appearance, but stay generally anchored in solar longitude, and therefore maintain the phase of the rotational signal. Meanwhile, the amplitude of the variation grows and weakens with the number, strength, and location of the active regions present.

Figure 3 displays the irradiance time series for one selected wavelength, namely Lyman-α. Time series generated for different spectral features in Figure 2 will have source regions in the solar atmosphere somewhat different from those for the Lyman-α line. The detailed variations in their individual time series will be produced from slightly different emission within the active features (size, shape, intensity, and variation of emission as features moves across the visible disk of the Sun). The resulting time series will be qualitatively similar, and not surprisingly, may exhibit quantitative differences as well.

Figure 4 shows an irradiance time series at 160 nm and covering only the early part of 1992, thereby giving an expanded view of 27-day variation seen in Figure 3. In particular, attention is drawn to the single rotation period denoted "A" in both Figures 3 and 4. We calculate the magnitude of the variation as the maximum value divided by the value at an adjacent minimum (maximum/minimum). This

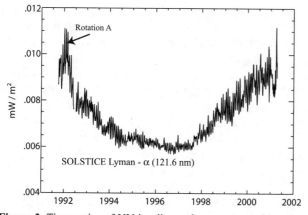

Figure 3. Time series of UV irradiance data constructed by plotting daily values of a particular spectral feature, in this case the Lyman-α line identified in Figure 2, against time.

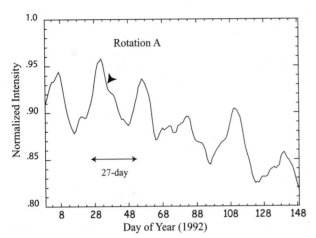

Figure 4. Time series of irradiance at 160 nm, with the time scale expanded from that of Figure 3. This shows the approximate 27-day variation directly related to the rotation of the Sun.

can also be expressed as a percent variation by subtracting the value one (mathematically this becomes (maximum - minimum)/minimum). If the minimum reference value for rotation "A" is taken as the one preceding the peak, then the magnitude of the variation at 160 nm appears to be about 1.06, while the variation of Lyman-α from Figure 3 is closer to 1.2. As anticipated, the two wavelengths show quantitatively different variation for the rotation marked "A". This analysis has been carried out wavelength-by-wavelength, and Figure 5 shows the wavelength dependence of the magnitude of variation for the rotation identified as "A", illustrating the striking wavelength dependence of the variation.

One important characteristic of Figure 5 is that the amount of rotational variation has structure that corresponds directly to the spectral features of Figure 2. The strong emission features show the most dramatic variation. For longer wavelengths (which is equivalent to observing lower layers of the solar atmosphere), the amount of variation decreases. Across absorption edges, for example the aluminum edge at 208 nm, the variation also decreases. The precision and quality of this measurement is quite good and most features have significance. Near 290 nm the phase of the variation changes sign, a phenomenon possibly related to the presence of large sunspot groups on the solar disk at this particular time. (These data are presently being analyzed in conjunction with sunspot and TSI observations.) Note also the relatively high variation for the 1 nm sample including the magnesium II (i.e., singly ionized magnesium) feature at 280 nm. This very strong Fraunhofer absorption line displays variation that far exceeds the variation of the neighboring spectrum. Consequences of this important spectral feature are considered in a discussion to follow.

Beyond 300 nm, analyses of individual rotation periods indicate variation of only 0.1% or less—about the noise level of the individual SUSIM, SOLSTICE, and SBUV measurements. Nevertheless, time series analysis of these long duration data sets (e.g., Fourier analysis) can take statistical advantage of numerous cycles of rotation to return a significant 27-day signal for even longer wavelengths [*Lean et al.*, 1997].

How does the analysis of one individual rotation, for example the rotation "A", apply to other periods of rotational modulation? Each rotational variation is a manifestation of the active regions present on the solar disk at a specific time—a function of their number, location, intensity, and most important their non-uniform distribution across longitude. No two rotational variations are identical, but their similarities statistically outweigh differences, so that scaling is appropriate. For example, the rotations in early 1992 (rotation "A") seen in Figure 3 are approximately five times larger than rotations at solar minimum in mid-1996. Therefore, the wavelength dependence seen in Figure 5 might be scaled down by this "average" factor of five, and deemed an appropriate estimate for one of the solar minimum rotations. There is no substitute for actual measurements of solar irradiance; however, in circumstances where observations are not available, scaling of data in hand may provide an adequate proxy.

3.2.2. Long-term solar variations. Long-term variation, or solar cycle variation, is also easily seen in Figure 3. Using a similar approach to the rotational variations, the ratio of maximum value to minimum value is constructed. Inherent in both analyses is the assumption that all data have first been properly corrected for changes in the instrument responsivity. When a ratio of two irradiance values is taken, there is an inherent ratio of the responsivity at the two times under consideration. The uncertainty in the calculation of solar variation is then directly related to knowledge of the instrument responsivity at the two times. If the two samples under consideration are closely spaced in time, and if the responsivity has changed relatively little, confidence in the ratio of the two values is high. In the preceding discussion of rotational variations two samples of irradiance data separated by only 10 to 15 days are usually compared, and consequently the confidence in the ratio of instrument responsivity at these two times is usually quite high. Therefore, the resulting uncertainty in the solar variation as provided in Figure 5 can be small, perhaps on the order of 0.1%. On the other hand, determining solar cycle variation will require comparing two irradiance observations separated in time by several years. The corresponding uncertainty from the

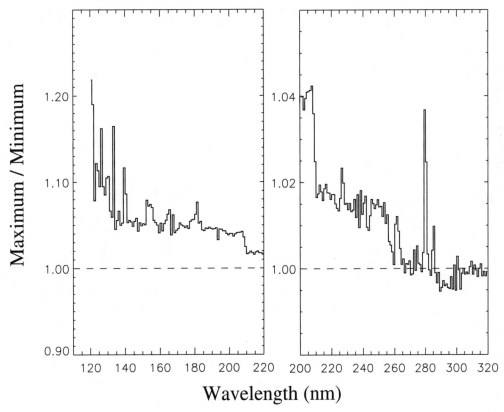

Figure 5. The amount of solar variability at each wavelength 120 to 320 nm (note the slight overlap of the two panels from 200 to 220 nm) for the single rotation noted "A" in Figures 3 and 4. Variation of the irradiance is defined as maximum/minimum.

ratio of two instrument responsivity values will increase significantly.

As discussed above, the high frequency, solar rotation signal in Figure 3 is a manifestation of the distribution of activity on the solar disk, and not the global activity level per se. To accurately describe the mean activity of the Sun, we have applied an 81-day running mean to the time series data. The low-pass filter samples the entire solar surface roughly three times, and thereby establishes a reliable mean value. The resulting "maximum" value is of course not the very highest value recorded, and the "minimum" value is not the very lowest—although it is close since the 27-day variation is quite small at solar minimum. Selecting the very highest value of the unfiltered time series would provide somewhat higher solar variability, but this extreme value is not truly representative of the global mean brightness of the Sun at solar maximum.

After applying the 81-day averaging to the UARS data, the maximum activity level occurs in early 1992 and the minimum in mid-1996 [*Harvey and White*, 1999]. UARS did not observe the entirety of solar cycle 22, which reached

its highest levels earlier in 1989 and continued high until the beginning of 1992. In fact, UARS only captured the final outburst of solar activity for cycle 22 [*Floyd et al.*, 1998]. Based on other indicators of solar activity (sunspot number, 10.7 cm radio flux, and the Mg II index discussed below), the UV irradiance values in early 1992 are within 10% of the most extreme levels reached at the solar cycle 22 maximum.

Considering the Lyman-α time series of Figure 3, the ratio of maximum in early 1992 to minimum in 1996 is about 1.8. That is, the solar irradiance is roughly 80% higher during solar maximum than during solar minimum. Figure 6 shows the wavelength-by-wavelength composite solar cycle variation for cycle 22. The SOLSTICE and SUSIM teams presently estimate uncertainties on the order of ±2% for ratios of this type, with hope that additional analyses may still yield some improvement. The uncertainty inherent in this curve is considerably larger than the uncertainty in Figure 5. Nevertheless, note the close similarity of the wavelength dependence of the rotational variability and the solar cycle variability. The structure in both figures corre-

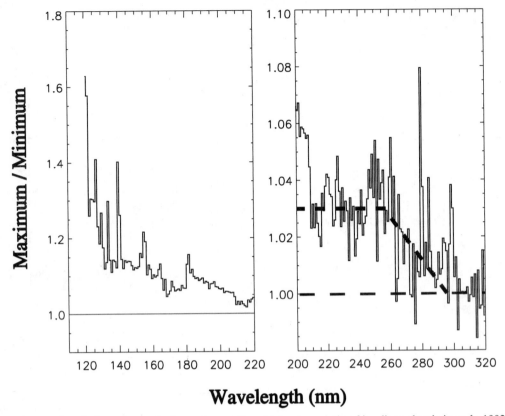

Figure 6. Wavelength dependence of solar cycle variation obtained as a ratio of irradiance levels in early 1992 to values in 1996. The character of the variation is similar to the rotational variation of Figure 5, but perhaps three to four times larger.

sponds closely with the emission and absorption lines and edges present in the solar spectrum (Figure 2).

The magnitude of the solar cycle 22 variation is roughly three times the magnitude of the rotational variation of Figure 5, but again with the caveat that Figure 5 represents the variation of only one selected rotation period, and an alternate rotation would likely display larger or smaller variation. In addition, the calculated rotational variation is a lower limit, for it compares an active hemisphere of the Sun with the opposite hemisphere—usually filled with some residual activity and not necessarily a completely "quiet" hemisphere.

The SME observations from solar cycle 21 provided a similar result for solar cycle variation [*London and Rottman*, 1990]. That observing program lacked a reliable method of tracking long-term changes in instrument responsivity, and the resulting calculation of solar cycle variability had uncertainty at least a factor of two larger than the uncertainty pertaining to the UARS estimates considered here. Nevertheless, the close agreement of observations over two solar cycles is encouraging as other indicators of

solar activity imply that solar cycles 21 and 22 were closely matched.

Figures 5 and 6 have the same general appearance, with stronger variation at shorter wavelengths originating higher in the solar atmosphere and progressively less variation as the source of the radiation becomes the lower levels near the photosphere. As discussed earlier the measurement of the rotational variations is so precise and reliable that even small structures can be identified with corresponding solar emission. This is especially obvious for dominant features, as for example the strong chromospheric emission lines, the aluminum absorption edge at 208 nm, and the Mg II emission at 280 nm (discussed next); but it is also apparent in much smaller structures. The enhanced variability near 260 nm is related to numerous Fe emission lines (see the modeling discussion by *Fox,* this volume).

For wavelengths longer than 300 nm, extending to the visible and infrared, the solar variation is small, and observing programs have not met the challenge of determining true solar variability. The VIRGO instrument on SOHO included a three-channel photometer in addition to two independent TSI devices [*Fröhlich*, this volume]. Time series of the photometer data at 402 nm, 500 nm and 862 nm, each with a bandwidth of 5 nm, provided important information on short-term solar variations related to the passage of faculae and sunspots across the solar disk [*Fligge et al.*, 2001]. These data were first detrended to remove the ambiguous mixture of possible solar cycle variation and change in instrument responsivity.

4. THE MAGNESIUM INDEX

As seen in Figures 5 and 6, the Mg II emission at 280 nm is one of the highly variable wavelengths in the UV. *Heath and Schlesinger* [1986] first recognized the potential value of the strong variation of the chromospheric Mg II emission lines at 280 nm adjacent to the small variations in the neighboring photospheric emissions (e.g. 275 nm and 285 nm). They postulated that if an index were calculated by dividing the irradiance value at 280 nm by the irradiance at the neighboring point(s), then most, if not all, instrumental effects would cancel. Therefore the index would be free of changes in instrument responsivity, and resulting time series of the index would represent true solar variation. Their analysis asserted that with some ancillary knowledge of the wavelength dependence of solar variation, derived for example from the rotational variation (e.g. Figure 5), the time series of the Mg II index could be scaled to provide approximate time series at other UV wavelengths [*Cebula et al.*, 1992; *Cebula and DeLand*, 1998].

Figure 7 provides detail on the solar spectrum near the Mg II line. The thin line is from very high spectral resolution (0.01 nm) balloon data of *Hall and Anderson* [1991]. It provides a fine example of the very rich detail in the solar spectrum, where each feature can be identified with an atomic transition. The heavy solid line is a SUSIM spectrum at its highest instrument resolution (0.15 nm), and for solar conditions similar to solar activity levels during the balloon flight. (The balloon flight was in 1983 during solar cycle 21 and the SUSIM data are from solar cycle 22.) Finally, the thick dashed line is the medium resolution (1.10 nm) SUSIM data. These medium resolution SUSIM data are quite similar to those of the SBUV instruments that currently provide the longest combined record of the Mg II data. Details of the absorption/emission within the Mg II features are complex and beyond the scope of this discussion (see for example *DeToma et al.*, 1998; *Cebula and DeLand*, 1998; and *Viereck and Puga*, 1999). Suffice it to say that in the core of the strong Mg II Fraunhofer absorption line(s) that are formed in the upper photosphere, is a small re-emission (h and k lines) from the overlying chromosphere. This emis-

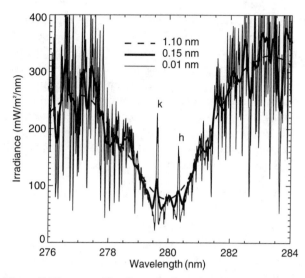

Figure 7. The spectral irradiance in the region of the Mg II absorption feature displayed at three wavelength resolutions. The 1.1 nm and the 0.15 nm data are from the SUSIM UARS experiment, Jan. 11, 1993. The 0.01 nm are from a balloon-borne spectrometer for April 20, 1983 [*Hall and Anderson*, 1991] adjusted to the SUSIM level. The two dates have comparable levels of solar activity and represent similar phases of solar cycles 22 and 21, respectively.

sion in the core is controlled by conditions in the solar chromosphere, and is more intense when the chromosphere is active.

The Mg II index has proven to be an excellent proxy for solar irradiances at many other wavelengths. It is a reliable proxy for the solar EUV wavelengths important to thermosphere and ionosphere heating [*Viereck et al.*, 2001; *Thullier and Bruinsma*, 2001], and it is one of the primary inputs to the empirical model of solar EUV spectral irradiance [*Tobiska et al.*, 2000]. The Mg II index has also provided input to models of total solar irradiance [*Pap et al.*, 1994; *Fröhlich*, this volume].

There have been several compilations of the Mg II index combining data from different observers into a single time series [*Donnelly*, 1988; *DeLand and Cebula*, 1993; *DeToma et al.*, 1997; and *Viereck and Puga*, 1999]. In general, these analyses have shown that Mg II ratios from instruments with different spectral resolution can be linearly scaled to form a single time series. Figure 8 is an extension of the earlier analysis of *Viereck and Puga* [1999]. Based on the quality of the contributing data sets, the relative uncertainty of long-term trends in this data set is less than 1%.

A second dominant Fraunhofer absorption feature in the solar spectrum (Figure 2) is the ionized calcium K line at 393.2 nm. The emission core of this Ca II K line originates from similar, but slightly lower altitudes of the chromosphere than does the Mg II K line [*Avrett,* 1992]. The Mg II and Ca II enhancements on the solar disk involve essentially the same source regions, and the temporal variations of the two emission cores are highly correlated [*Donnelly, White and Livingston,* 1994]. The Ca II emission has, of course, the additional advantage that it is easily observed from the ground and a long-term data record exists back to 1974 [*White and Livingston,* 1981]. This ground based observing program includes the He I 10830, C I 5380, Fe I 5394, and Ca II 8542 lines in addition to the Ca II K line.

5. FUTURE OBSERVING PROGRAMS

The UARS program has completed more than eleven years of continuous solar observations, and NASA is extending the mission year-by-year. The exceptional value of the long-term irradiance data sets being returned by SUSIM and SOLSTICE cannot be overemphasized, along with other extremely valuable atmospheric measurements collected by the UARS instruments. Likewise, the GOME instrument on the ERS-2 mission continues to gather solar irradiance data as well. In combination, the SBUV instruments on the NOAA satellites now provide a very long-duration data set [*Cebula et al.,* 1996a]—their data made even more valuable since all instruments are similar in design and function.

There are two new irradiance programs that are now in operation. The Thermosphere Ionosphere Mesosphere Energetics and Dynamics (TIMED) satellite was launched in December 2001 and carries the Solar EUV Experiment (SEE), primarily intended to make EUV observations [*Woods et al.,* 1998]. However, its spectral measurements also extend to 200 nm and overlap the short wavelength observations of SOLSTICE and SUSIM.

The Solar Radiation and Climate Experiment (*SORCE*) was launched in January, 2003. This mission is one asset of NASA's Earth Observing System (*EOS*) and is dedicated to the measurement of solar irradiance [*Woods et al.,* 2000a]. The *SORCE* satellite carries four different instruments— one that measures total solar irradiance (TSI) and three that measure spectral irradiance. One of these spectral instruments is a second generation SOLSTICE, a new and improved version of the instrument on UARS. Another of the spectral instruments, the Spectral Irradiance Monitor (SIM), is an entirely new concept to measure spectral irradiance from 300 nm to 2 μm.

The UARS instruments have a limiting capability of about ± 1% when it comes to establishing solar cycle variability. Although this capability has proven adequate for wavelengths shorter than 300 nm, the longer wavelengths will require vast improvement over UARS of at least one order of magnitude in precision and accuracy. The SIM instrument on *SORCE* is specifically designed to have absolute accuracy of 300 ppm (0.03%: 1σ), and precision and relative accuracy a factor of 10 better. This high level of accuracy is achieved by characterizing the instrument sensitivity, and not by calibration [*Harder et al.,* 2000]. The distinction is that SIM has each element of its measurement equation directly established in SI (or SI derived) units (e.g.,

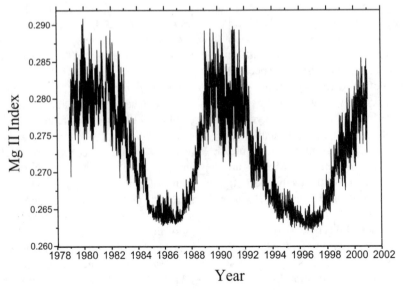

Figure 8. Mg II index as a continuation of the work of *Viereck and Puga* [1999]. This time series uses data combined from several observational programs.

the aperture is directly measured in m², the power in W or J/sec, etc.). This method is the standard approach for TSI devices, but it differs from the approach typically used for spectral instruments (e.g., UARS, SBUV, and GOME), which are calibrated by direct comparison to primary or secondary irradiance standards. The SIM observations should provide the first true measurement of solar variation from the near ultraviolet to the visible, or at least it will improve understanding of the upper limit of solar variability to levels well below 0.1%.

6. SUMMARY

For more than fifty years instruments have made observations of the Sun from the vantage point of space. This has allowed ultraviolet irradiance to be measured at wavelengths below 300 nm. The process has proven to be difficult, and techniques have struggled to achieve sufficient accuracy and instrument stability to make true observations of solar variability. Instrumentation has steadily evolved and improved, and now it is generally accepted that measurements, at least for the past ten to fifteen years, do indeed provide a reliable determination of solar variation. These spectral measurement techniques are still only accurate to about 1%, and in cases where solar variability is less only upper limits of solar variation are established.

Our view of long-term solar cycle variation is that for the past two solar cycles, 21 and 22, the ultraviolet has varied by 10 to 20% for most of the wavelength range 120 to 200 nm, with three to four times larger variation in the strong chromospheric lines. Figure 6 illustrates this wavelength dependent variability. Proceeding to longer wavelengths, 200 to 300 nm, the solar variability steadily decreases down to the order of 1%—the detection limit of present techniques. Wavelengths longer that 300 nm evidently vary less then 1%, perhaps as little as 0.1%, which would be consistent with TSI observations, but such small values have yet to be truly measured. Observational programs now being developed are actively pursuing the goal of making these spectral irradiance observations with sufficient accuracy to establish solar cycle variability from the near ultraviolet through the visible and into the infrared.

Although reliable observations now extend over two solar cycles, these results must not be considered a final and universal determination. All long-term records of the Sun's activity indicate that cycle-to-cycle differences are large, even to the extent that the solar cycle can diminish to a quiescent mode, as for example seen in the Maunder Minimum in the middle of the seventeenth century. Experimental techniques steadily improve, and for the foreseeable future there can be no substitute for continuing the very best observational programs possible. These new irradiance data will continue to provide important insight to the nature of the Sun's variability. The new measurements are not only essential to ongoing atmospheric and climate studies, but today's irradiance data will form the basis for even longer-term studies conducted by future generations of scientists.

Acknowledgments. The support for the UARS SOLSTICE at the University of Colorado is from NASA grants NAG5-6850 and NAG5-11028, and NASA contract NAS5-97145. Support for the SORCE project at the University of Colorado is from NASA contract NAS5-97045. The support for the UARS SUSIM at the Naval Research Laboratory came from NASA under S-14798-D and S-44772-G and the Office of Naval Research. The authors would like to thank T. Woods, B. Knapp, and L. Herring, and extend a special thanks to the UARS Project Scientists, C. Jackman and A. Douglass.

REFERENCES

Avrett, E. H., Temporal Variations of the Near-UV, Visible, And Infrared Spectral Irradiance from a Theoretical Viewpoint, *Proc. Of the Workshop on Solar Electromagnetic Radiation Study for Solar Cycle 22* (ed. R. F. Donnelly) Space Environment Lab., NOAA ERL, Boulder, Colorado, 20-42, 1992.

Baum, W. A., F. S. Johnson, J. J. Oberly, C. C. Rockwood, C. V. Strain, and R. Tousey, Solar Ultraviolet Spectrum at 88 km, *Phys. Rev.,* 70, pp. 781, 1946.

Bouwer, S. D., Periodicities of Solar Irradiance Activity Indices, II, *Solar Physics,* 142, pp. 365-389, 1992.

Brasseur, G. P. and S. C. Solomon, *Aeronomy of the Middle Atmosphere*, D. Reidel Pub. Co., 1986.

Brueckner, G. E., K. L. Edlow, L. E. Floyd, J. L. Lean, and M. E. VanHoosier, The Solar Ultraviolet Spectral Irradiance Monitor (SUSIM) Experiment on Board the Upper Atmospheric Research Satellite (UARS), *J. Geophys. Res.*, 98, pp. 10695-10711, 1993.

Burrows, J. P., et al., The Global Ozone Monitoring Experiment (GOME): Mission Concept and First Scientific Results, *J. Atmos. Sci.,* 56, pp. 151-175. 1999.

Cebula, R. P., M. T. DeLand, E. Hilsenrath, B. M. Schlesinger, R. D. Hudson, and D. F. Heath, Intercomparisons of the solar irradiance measurements from the Nimbus-7 SBUV, the NOAA-9 and NOAA-11 SBUV/2, and the STS-34 SSBUV instruments: a preliminary study, *J. Atmos. Sol-Terr. Physics*, 53, pp. 993-997, 1991.

Cebula, R. P., M. T. DeLand, and B. M. Schlesinger, Estimates of solar variability using the Solar Backscatter Ultraviolet (SBUV) 2 Mg II index from the NOAA 9 satellite, *J. Geophys. Res.*, 97, pp. 11613-11620, 1992.

Cebula, R. P., E. Hilsenrath, and M. T. DeLand, Middle ultraviolet solar spectral irradiance measurements, 1985-1992, from the SBUV/2 and the SSBUV instruments, *The Sun as a Variable*

Star, (ed. J. M. Pap, C. Fröhlich, H. S. Hudson, and S. K. Solanki), Cambridge Univ. Press, pp. 81-88, 1996a.

Cebula, R. P., G. O. Thuillier, M. E. VanHoosier, E. Hilsernath, M. Herse, G. E. Brueckner, and P. C. Simon, Observations of the solar irradiance in the 200–350 nm interval during the ATLAS-1 mission: a comparison among three sets of measurements—SSBUV, SOLSPEC, and SUSIM, *Geophys. Res. Lett.*, 23, pp. 2289-2292, 1996b.

Cebula, R. P., and M. T. DeLand, Comparisons of the NOAA-11 SBUV/2, UARS SOLSTICE, and UARS SUSIM Mg II solar activity proxy indexes, *Solar Physics*, 177, pp. 117-132, 1998.

Cook, J. W., G. E. Brueckner, and M. E. VanHoosier, Variability of the solar flux in the far ultraviolet 1175-2100 AA. *J. Geophys. Res.*, 85, pp. 2257-2268, 1980.

DeLand, M. T., and R. P. Cebula, Composite Mg II solar activity index for solar cycles 21 and 22, *J. Geophys. Res.*, 98, pp. 12809-12823, 1993.

DeToma, G., O. R. White, B. G. Knapp, G. J. Rottman, and T. N. Woods, Solar Mg II core-to-wing index: comparison of SBUV2 and SOSLTICE time series, *J. Geophys. Res.*, 102, pp. 2597-2610, 1997.

DeToma, G., O. R. White, B. G. Knapp, G. J. Rottman, and T. N. Woods, Effect of spectral resolution on the Mg II index as a measure of solar variability, *Solar Physics,* 177, pp. 117-132, 1998.

Detwiler, C. R., D. L. Garret, J. D. Purcell, and R. Tousey, The Intensity Distribution in the Ultraviolet Spectrum, *Ann. Geophys.*, 17, pp. 9-18, 1961.

Donnelly, R. F., The solar UV Mg II core-to-wing ratio from the NOAA9 satellite during the rise of solar cycle 22, *Adv. Space Res.*, 8, pp. 7(77)-7(80), 1988.

Donnelly, R. F., O. R. White, and W. C. Livingston, The Solar Ca II K Index and the Mg II Core-to-Wing Ration, *Solar Physics,* 152, pp. 69-76, 1994.

Fligge, M., S. K. Solanki, J. M. Pap, C. Fröhlich, and Ch. Wehrli, Variations of Solar Spectral Irradiance from Near UV to the Infrared, *J. Atmos. Sol-Terr. Physic*, 63, pp. 1479-1487, 2001.

Floyd, L. E., P. A. Reiser, P. C. Crane, L. C. Herring, D. K. Prinz, and G. E. Brueckner, Solar Cycle 22 UV Spectral Irradiance Variability: Current Measurements by SUSIM UARS, *Solar Physics*, 177, pp. 79-87, 1998.

Fox, P., Solar Activity and Irradiance Variations, *this volume.*

Fröhlich, C. Solar Irradiance Variability, *this volume.*

Hall, L. A., and G. P. Anderson, High-Resolution Solar Spectrum Between 2000 and 3100 A, *J. Geophys. Res.*, 96, 12927, 1991.

Harder, J., G. Lawrence, G. Rottman, and T. Woods, The Spectral Irradiance Mionitor (SIM) for the SORCE Mission, *SPIE Proceedings*, 4135, pp. 204-214, 2000.

Harvey, K. L., and O. R. White, What is Solar Cycle Minimum?, *J. Geophys. Res.*, 104, pp. 19759-19764, 1999.

Heath, D. F., and B. M. Schlesinger, The Mg 280-nm doublet as a monitor of changes in the solar ultraviolet irradiance, *J. Geophys. Res.*, 91, pp. 8672-8682, 1986.

Hinteregger, H. E., K. Fukui, and B. R. Gilson, Observational, reference and model data on solar EUV, from measurements on AE-E, *Geophys. Res. Lett.*, 8, pp. 1147-1150, 1981.

Hood, L. L., Effects of Solar UV Variability on the Stratosphere and Troposphere, *this volume.*

Kiehl, J., and K. Trenberth, Earth's Annual Global Mean Energy Budget, *Bull. Amer. Meteor. Soc.*, 78, pp. 197-208, 1997.

Labs, D., H. Neckel, P. Simon, and G. Thuillier, Ultraviolet solar irradiance measurement from 200 to 358 nm during the Spacelab 1 Mission, *Solar Physics,* 107, pp. 203, 1987.

Lean, J., O. R. White, W. C. Livingston, D. F. Heath, R. F. Donnelly, and A. Skumanich, A three-component model of the variability of the solar ultraviolet flux: 145-200 nm, *J. Geophys. Res.*, 87, pp. 10307-10317, 1982.

Lean, J., Solar ultraviolet irradiance variations: A review, *J. Geophys. Res.*, 92, pp. 839, 1987.

Lean, J., G. Rottman, L. Kyle, T. Woods, J. Hickey, and L. Puga, Detection and Parameterization of Variations in Solar Mid- and Near-ultraviolet radiation (200–400 nm), *J. Geophys. Res.*, 102, pp. 29,939-29,956, 1997.

London, J., and G. J. Rottman, Wavelength Dependence of Solar Rotation and Solar Cycle UV Irradiance Variations, *Climate Impact of Solar Variability,* NASA Conf. Pub. 3086, 1990.

Pap, J.M., R. C. Willson, C. Fröhlich, R. F. Donnelly, and L. Puga, Long-term Variations in Total Solar Irradiance, *The Sun as a Variable Star: Solar and Stellar Irradiance Variations,* (ed. J. M. Pap, C. Fröhlich, H. S. Hudson and W. K. Tobiska), Kluwer Academic Publs., pp. 73-80, 1994.

Rottman, G. J., C. A. Barth, R. J. Thomas, G. H. Mount, G. M. Lawrence, D. W. Rusch, R. W. Sanders, G. E. Thomas, and J. London, Solar Spectral Irradiance, 1200 to 1900Å, October 13, 1981 to January 3, 1982, *Geophys. Res. Lett.*, 9, pp. 587, 1982.

Rottman, G. J., Results from space measurements of solar UV and EUV flux, in *Solar Radiative Output and its Variation,* edited by P. Foukal, pp. 71-86, Cambridge Research and Instrumentation Inc., Boulder, Colorado, 1987.

Rottman, G. J., T. N. Woods, and T. P. Sparn, Solar Stellar Irradiance Comparison Experiment I: 1 Instrument Design and Operation, *J. Geophys. Res.*, 98, pp. 10667-10677, 1993.

Rottman, G. J., T. N. Woods, O. R. White, and J. London, Irradiance Observations from the UARS SOLSTICE Experiment, *The Sun as a Variable Star: Solar and Stellar Irradiance Variations,* (ed. J. M. Pap, C. Fröhlich, H. S. Hudson, and S. K. Solanki), Cambridge Univ. Press, pp. 73-80, 1994.

Schlesinger, B. M., and R. P. Cebula, Solar variation 1979–1987 estimated from an empirical model for changes with time in the sensitivity of the Solar Backscatter Ultraviolet instrument, *J. Geophys. Res.*, 97, pp. 10119-10134, 1992.

Schüssler, M., and D. Schmitt, Theoretical Models of Solar magnetic Variability, *this volume.*

Scientific Assessment of Ozone Depletion: 1998, *Global Ozone Research and Monitoring Project—Report #44*, WMO, Geneva, Switzerland, 1998.

Shindell, D. T., G. A. Schmidt, M. E. Mann, D. Rind, and A. Waple, Solar Forcing of Regional Climate Change During the Maunder Minimum, *Science*, 294, pp. 2149-2151, 2001.

Solanki, S. K., and Y. C. Unruh, A Model of the Wavelength Dependence of Solar Irradiance Variations, *Astron. Astrophys.*, 329, pp. 747-753, 1998.

Spruit, H. C., Theoretical Interpretation of Solar and Stellar Irradiance Variations, *The Sun as a Variable Star,* (ed. J. M. Pap, C. Fröhlich, H. S. Hudson, and S. K. Solanki), Cambridge Univ. Press, pp. 270-279, 1994.

Thuillier, G., M. Herse, P. Simon, D. Labs, H. Mandel, and D. Gillotay, Observation of the UV Solar Spectral Irradiance between 200 and 350 nm during the ATLAS-1 Mission by the SOLSPEC Spectrometer, *Solar Physics,* 171, pp. 283-302, 1997.

Thuillier, G., and S. Bruinsma, The MgII index for upper atmosphere modeling, *Ann. Geophys.*, 19, pp. 219-228, 2001.

Thuillier G., L. Floyd, T. N. Woods, R. Cebula, E. Hilsenrath, M. Hersé, and D. Labs, Sun Irradiance Spectra, *this volume*.

Tobiska, K., T. Woods, F. Eparvier, R. Viereck, L. Floyd, D. Bouwer, G. Rottman, and O. White, The SOLAR2000 Empirical Solar Irradiance Model and Forecast Tool, *J. Atmos. Sol-Terr. Physic*, 62, pp. 1233-1250, 2000.

Vidal-Madjar, A., The solar spectrum at Lyman-alpha 1216 Å, in *The Solar Output and Its Variation*, ed. O. R. White, Colorado Assoc. Univ. Press, Boulder, pp. 213-236, 1977.

Viereck, R. A., and L. C. Puga, The NOAA Mg II core-to-wing solar index; Construction of 20-year time series of chromospheric variability from multiple satellites, *J. Geophys. Res.,* 104, pp. 9995, 1999.

Viereck , R., L. Puga, D. McMullin, D. Judge, M. Weber, and W. K. Tobiska, The Mg II Index: A Proxy for Solar EUV, *Geophys. Res. Lett.,* 28, pp. 1343, 2001.

Weber, M., J. P. Burrows, and R. P. Cebula, Solar UV and visible spectral irradiance measurements from GOME in 1995 and 1997, *Solar Physics*, 177, pp. 63-77, 1998.

Weber, M., Solar activity during solar cycle 23 monitored by GOME, *Proc. European Symposium on Atmospheric Measurements from Space (ESAMS'99)*, pp. 611-616, European Space Agency, 1999.

White, O. R. (editor), *The Solar Output and Its Variation*, Colorado Assoc. Univ. Press, Boulder, 1977.

White, O. R., and W. C. Livingston, Solar Luminosity Variations III: Calcium K Variations from Solar Minimum to Maximum in Cycle 21, *Astrophys. J.*, 249, pp. 798-816, 1981.

Woods, T. N., G. J. Rottman, and G. Ucker, Solar Stellar Irradiance Comparison Experiment I: 2 Instrument Calibration, *J. Geophys. Res.*, 98, pp. 10679-10694, 1993.

Woods, T. N., D. K. Prinz, G. J. Rottman, J. London, P. C. Crane, R. P. Cebula, E. Hilsenrath, G. E. Brueckner, M. D. Andrews, O. R. White, M. E. VanHoosier, L. E. Floyd, L. C. Herring, B. G. Knapp, C. K. Pankratz, and P. A. Reiser, Validation of the UARS Solar Ultraviolet Irradiances: Comparison with the ATLAS - 1, - 2 Measurements, *J. Geophys. Res.,* 101, pp. 9541-9569, 1996.

Woods, T. N., F. G. Eparvier, S. M. Bailey, S. C. Solomon, G. J. Rottman, G. M. Lawrence, R. G. Roble, O. R. White, J. Lean, and W. K. Tobiska, TIMED Solar EUV Experiment, *SPIE Proceedings*, 3442, pp. 180, 1998.

Woods, T., G. Rottman, J. Harder, G. Lawrence, B. McClintock, G. Kopp, and C. Pankratz, Overview of the EOS SORCE mission, *SPIE Proceedings*, 4135, pp. 192, 2000a.

Woods, T. N., W. K. Tobiska, G. J. Rottman, and J. R. Worden, Improved solar Lyman α irradiance modeling from 1947 through 1999 based on UARS observations, *J. Geophys. Res.*, 105, 27,195, 2000b.

Woods, T. N., L. W. Acton, S. Bailey, F. Eparvier, H. Garcia, D. Judge. J. Lean, J. T. Mariska, D. McMullin, G. Schmidtke, S. C. Solomon, W. K. Tobiska, H. P. Warren, and R. Viereck, Solar Extreme Ultraviolet and X-ray Irradiance Variations, *this volume*.

Gary Rottman, Laboratory for Atmospheric and Space Physics, University of Colorado, 1234 Innovation Dr., Boulder, CO 80303

Linton Floyd, Code 7660, Navel Research Laboratory, 4555 Overlook Ave. SW, Washington, DC 20375

Rodney Viereck, National Oceanic and Atmospheric Administration, Space Environment Center, 325 Broadway, Boulder, CO 80303

Solar Extreme Ultraviolet and X-ray Irradiance Variations

Tom Woods[1], Loren W. Acton[2], Scott Bailey[3], Frank Eparvier[1], Howard Garcia[4], Darrell Judge[5], Judith Lean[6], John T. Mariska[6], Don McMullin[5], Gerhard Schmidtke[7], Stanley C. Solomon[8], W. Kent Tobiska[9], Harry P. Warren[10] and Rodney Viereck[4]

The solar extreme ultraviolet (EUV) radiation at wavelengths shortward of 120 nm is a primary energy source for planetary atmospheres and is also a tool for remote sensing of the planets. For such aeronomic studies, accurate values of the solar EUV irradiance are needed over time periods of minutes to decades. There has been a variety of solar EUV irradiance measurements since the 1960s, but most of the recent observations have been broadband measurements in the X-ray ultraviolet (XUV) at wavelengths shortward of 35 nm. A summary of the solar EUV irradiance measurements and their variability during the last decade is presented. One of the most significant new solar irradiance results is the possibility that the irradiance below 20 nm is as much as a factor of 4 higher than the reference Atmospheric Explorer E (AE-E) spectra established in the 1970s and 1980s. The primary short-term irradiance variability is caused by the solar rotation, which has a mean period of 27 days. The primary long-term variability is related to the solar dynamo and is known best by the 11-year sunspot cycle. The solar cycle variability as a function of wavelength can be characterized as 20% to 70% between 120 and 65 nm and as a factor of 1.5 to 10 between 65 and 1 nm. The variability of the total solar EUV irradiance, integrated from 0 to 120 nm, is estimated to be 30–40% for a large 27-day rotational period and a factor of about 2 for the 11-year solar cycle during the recent, rather active, solar cycles.

[1]Laboratory for Atmospheric and Space Physics, University of Colorado, Boulder, Colorado.

[2]Physics Department, Montana State University, Bozeman, Montana.

[3]Geophysics Institute, University of Alaska, Fairbanks, Alaska.

[4]Space Environment Laboratory, National Oceanic & Atmospheric Administration, Boulder, Colorado.

[5]Space Sciences Center, University of Southern California, Los Angeles, California.

[6]E.O. Hulburt Center for Space Research, Naval Research Laboratory, Washington, DC.

[7]Fraunhofer-Institut für Physikalische Messtechnik, Freiburg, Germany.

[8]High Altitude Observatory, National Center for Atmospheric Research, Boulder, Colorado.

[9]Space Environment Technologies, SpaceWx Division, Pacific Palisades, California.

[10]Harvard-Smithsonian Center for Astrophysics, Boston, Massachusetts.

Solar Variability and its Effects on Climate
Geophysical Monograph 141
Copyright 2004 by the American Geophysical Union
10.1029/141GM11

1. INTRODUCTION

The extreme ultraviolet (EUV) region of the solar spectrum, defined here for wavelengths shorter than 120 nm, is only about 10 ppm of the total solar energy. None of this energy penetrates to the surface of the Earth, or even to the troposphere. What, then, motivates the study of the EUV region of the solar spectrum?

The most energetic portion of the Sun's photon flux is worthy of study because it has significant effects on planetary upper atmospheres and because it is so variable. The solar EUV radiation creates planetary ionospheres through photoionization and energizes the photoelectron and neutral components of the thermosphere/ionosphere system. Chemical processes following ionization lead to dissociation of atmospheric species and the creation of radiatively active compounds of odd-nitrogen and odd-oxygen. The changes in upper atmospheric heating caused by changes in the solar photon output cause the thermosphere/ionosphere system to be the most variable region of the Earth's atmosphere. For example, the density at altitudes populated by numerous orbiting artificial satellites changes by an order of magnitude with solar activity, thus significant changes occur for the rate at which the spacecraft orbits decay. To understand these solar influences fully and to construct theoretical models of the thermosphere and ionosphere, accurate specification of the key energetic input to the region, namely the solar EUV irradiance, is needed. Additionally, the solar measurements and/or reliable proxy models are needed for the correct interpretation of airglow observations, which are used to monitor the density and composition of the atmosphere.

Emission lines and ionization continua from the solar upper atmosphere dominate the solar EUV spectrum. The emission lines arise from the dominant species, H and He, and the many minor species in the higher layers of the solar atmosphere as a non-local thermodynamic equilibrium (non-LTE) effect and are strongly sensitive to the magnetic activity on the Sun. The solar EUV emissions also include ionization continua, such as the bright H ionization continuum shortward of 91 nm. These general characteristics of the solar UV spectrum are evident in Figure 1, which displays a solar spectrum at 0.1 nm spectral resolution [*Woods et al.,* 1998a]. This spectrum and others discussed here are the solar spectral irradiance, which is the spectral radiance (or intensity) at a single wavelength integrated over the full disk of the Sun and observed at a distance of 1 AU. The distribution of the emission features with wavelength is caused by the complex atomic energy levels of the source gases, so there is not a general relation of irradiance or its variability

with wavelength. However, there are better relations between the irradiance variability and the different layers of the solar atmosphere, being the photosphere, chromosphere, transition region, and corona. With the density decreasing with altitude and the temperature increasing at higher layers of the solar atmosphere, the radiation from the higher layers is, in general, more variable. In other words, the coronal emissions vary more than the transition region emissions, which in turn vary more than the chromospheric emissions.

The measured amount of the solar EUV irradiance variability is the primary focus for this paper. First the historical background for solar EUV spectral irradiance measurements is briefly discussed, and then summaries of the current measurement programs are presented. The recent spacecraft measurements of the solar EUV irradiance are concentrated in the X-ray ultraviolet (XUV) shorter than 34 nm, and there are only limited measurements, mostly from suborbital rockets, during the last 20 years for wavelengths between 34 and 115 nm. Then two models of the solar EUV irradiance are introduced. Finally, the measurements and models are compared to estimate the amount of the solar EUV variability.

2. HISTORICAL MEASUREMENTS

The earliest spacecraft measurements of the solar EUV and soft X-ray irradiance came in the 1960s from the SOL-RAD [*Kreplin,* 1970; *Kreplin and Horan,* 1992], the Orbiting Solar Observatory (OSO), and the Atmospheric

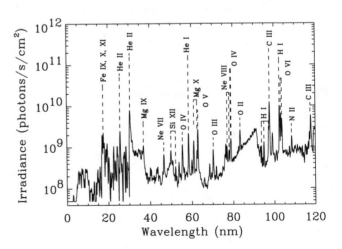

Figure 1. The solar EUV irradiance shortward of 120 nm is dominated by emissions from the chromosphere, transition region, and corona layers of the solar atmosphere. A few of the brighter emissions are labeled in the spectrum.

Explorer (AE) series of satellites [*Gibson and Van Allen*, 1970]. Table 1 summarizes the spacecraft observations of the solar irradiance including their spectral coverage and time period of observations. In the 1970s, there followed the AEROS series of satellites [*Schmidtke et al.*, 1974, 1977]. For many years the most comprehensive knowledge of the EUV and solar soft X-ray irradiance came from measurements on the AE spacecraft [*Hinteregger et al.*, 1977, 1981; *Torr and Torr*, 1985]. Sounding rockets were used for solar irradiance measurements and for in-flight calibrations of spacecraft experiments throughout this period. Many of the rocket results are summarized by *Hall et al.* [1969], *Heroux and Higgins* [1977], *Manson* [1976], *Feng et al.* [1989], and references therein. The AE and sounding rocket data have served as the basis for several empirical models of the solar irradiance at soft X-ray and EUV wavelengths. For more detailed information about the historical background and previous results of the solar UV irradiance, one should examine reviews, for example, by *White* [1977], *Schmidtke* [1984], *Rottman* [1987], *Lean*, [1987 and 1991], *Tobiska* [1993], and *Pap et al.* [1994].

The quoted uncertainties for many of the historical measurements lie in the range of 20 to 40%. Although the AEROS instruments were calibrated by a synchrotron source on the ground and contained an on-board radioactive source, most of the other experiments were calibrated by comparison to tungsten and aluminum oxide photodiodes calibrated by the National Bureau of Standards (now the National Institute for Standards and Technology). Recent measurements, as well as analyses of geophysical observations, suggest that these historical measurements may be uncertain by as high as factors of 4 at the shorter wavelengths [see *Bailey et al.*; 2000, 2001]. The larger uncertainties may be the result of in-flight degradation of the responsivity of the instruments, which can be very strong at soft X-ray and EUV wavelengths.

3. RECENT MEASUREMENTS

The recent measurements of the solar EUV irradiance are mostly concentrated at the short wavelengths with fairly broad bandpasses. The recent spacecraft measurements in the X-ray and XUV region include the NOAA GOES 0.05–0.8 nm, Yohkoh SXT 0.2–3 nm, SNOE SXP 2–20 nm, and SOHO CELIAS/SEM 26–34 nm. There are also some recent rocket measurements of the solar EUV irradiance between 30 and 120 nm with higher spectral resolution, but these are limited to a few measurements during the last solar cycle. There are also full Sun images in the EUV range, such as from the SOHO Extreme ultraviolet Imaging

Telescope (EIT) on a daily basis [*Newmark et al.*, 2002] and from SOHO Coronal Diagnostic Spectrometer (CDS) on a monthly basis [*Thompson and Carter*, 1998], and these solar images, when integrated over the solar disk, provide irradiance values. The measurements of the solar EUV irradiance during the current solar cycle are introduced in this section and then the solar variability from these measurements are discussed in a later section. The irradiances derived from SOHO images are not discussed here as it is work in progress.

3.1. GOES XRS

The NOAA GOES program has been obtaining X-ray measurements of the Sun since 1974 with the X-ray Sensor (XRS), but these X-ray data from ionization cells are readily available only back to 1986. Typically, two GOES spacecraft are operational, and each has an XRS instrument that operates continuously at 0.5 sec resolution (the distributed data is at 3 sec resolution). The overlapping wavelength bands, spanning roughly 0.05 to 0.8 nm (25 to 1 keV), were originally selected because full-disk measurements at these energies provide optimal sensitivity for the early detection of solar flares and have additional application to D-region ionospheric disturbances. For these purposes it is usually sufficient to monitor only the softer X-ray channel from 0.1 to 0.8 nm. The second harder X-ray channel from 0.05 to 0.4 nm provides additional information about the state of the coronal plasma [*Garcia*, 1994a].

The GOES X-ray sensors operate on the ionization-chamber principle: measured electric current is proportional to the net ionization rate caused by incident X-ray flux on encapsulated noble gases. The ratio of the output in electric current of the two channels is uniquely a function of the electron temperature of the emitting plasma and the magnitude of each of these currents is proportional to a quantity, known as the emission measure, which convolves the volume and the density of the plasma. During quiet times the emission is spread more or less evenly over the full solar disk; however, during flares, particularly large events, nearly all of the emission originates from a very small, discrete location.

3.2. Yohkoh SXT

The joint Japan-US-UK solar mission Yohkoh was launched August 30, 1991 for the study of high-energy processes on the Sun. The Soft X-ray Telescope (SXT) experiment on Yohkoh is a well-calibrated imaging instrument sensitive in the 0.2–3 nm band [*Tsuneta et al.*, 1991].

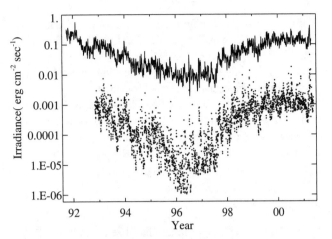

Figure 2. The time series of the solar X-ray irradiance is shown for the last 10 years. The upper data are the 0.2–3 nm irradiances from the Yohkoh SXT instrument, and the lower data are the 0.1–0.8 nm irradiances from the GOES XRS instrument.

The SXT collects about 50 full-Sun soft X-ray images per day. Issues connected with converting the total signal in these images to X-ray irradiance in specified spectral bands have been treated in detail by *Acton, Bruner, and Weston* [1999]. They show that taking the electron temperature of the X-rays emissions into account, through 2-filter photometry, greatly increases the accuracy of the calculated irradiance. Figure 2 displays the daily-averaged, non-flare, X-ray spectral irradiance from SXT computed in this manner. Also displayed in the figure (lower curve) are the daily irradiance values published by NOAA for the GOES 0.1–0.8 nm channel.

3.3. SNOE SXP

The Solar X-ray Photometer (SXP) on the Student Nitric Oxide Explorer (SNOE) spacecraft performs photometric measurements of the solar soft X-ray irradiance in three wavelength channels, 2–7, 6–19, and 17–20 nm. The channels consist of X-ray sensitive Si photodiodes with thin films deposited directly onto the active areas. Irradiances are determined by comparing measured broadband photodiode currents to currents predicted using a high-resolution empirical model of the solar spectrum. The instrument, observations, calibrations, data reduction techniques, and analysis of 1.5 years of data are presented by *Bailey et al.* [2000, 2001]. Calibrations of the coated photodiodes were performed prior to launch using the Synchrotron Ultraviolet Radiation Facility (SURF) at the National Institute for Standards and Technology (NIST) at wavelengths longer than 5 nm. In order to enhance the accuracy of the calibrations and extend the wavelength coverage below 5 nm the

sensitivity is modeled based on the known sensitivity of a bare photodiode and the calculated transmission of the thin-film filters. The modeled sensitivity is used in the data processing. The absolute uncertainties of the SNOE measurements, including uncertainties due to the removal of long wavelength background signals, are approximately 20%. The SNOE irradiance results from *Bailey et al.* [2000, 2001] are presented in Figure 3 both in units of absolute energy flux, mW m^{-2}, and normalized energy flux where the results are normalized to the SC#21REFW [*Hinteregger et al.*, 1981] solar minimum irradiances for the SNOE bandpasses.

One of the significant results of the SNOE solar XUV measurements is that the SNOE solar XUV irradiance is about a factor of 4 higher than the reference AE-E spectra. The AE measurements only extended to 20 nm, so the *Hinteregger et al.* [1981] empirical model shortward of there is based on a few rocket flights with uncertain calibrations. There has been considerable evidence from diverse sources that the empirical model underestimates the XUV solar irradiance, such as comparisons of models with

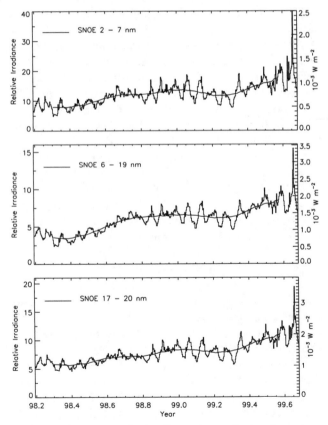

Figure 3. The SNOE solar irradiance time series are shown normalized to the *Hinteregger et al.* [1981] solar minimum values as well as in units of energy flux. The solid line is the 81-day average of the data.

measurements of the photoelectron flux [e.g., *Richards and Torr*, 1984; *Winningham et al.*, 1989], electron density [e.g., *Buonsanto et al.*, 1993, 1995], airglow emissions [e.g., *Link et al.*, 1988; *Solomon and Abreu*, 1989], and odd-nitrogen densities [e.g., *Siskind et al.*, 1990]. This evidence remained unquantified until measurements from the SNOE spacecraft showed XUV fluxes a factor of 4 to 5 above the Hinteregger model at all levels of solar activity [*Bailey et al.*, 2000; *Solomon et al.*, 2001]. This finding is critical for many aspects of thermosphere/ionosphere modeling and is one of the most significant, recent developments in solar irradiance studies.

3.4. SOHO CELIAS/SEM

Since January 1, 1996, the Solar EUV Monitor (SEM) on the Solar and Heliospheric Observatory (SOHO) has continuously measured the solar EUV irradiance with a 15 second time cadence from the Sun-Earth L1 Lagrange point. The SEM instrument is a solar EUV spectrometer [*Judge et al.*, 1998], and the SEM is an integrated part of the SOHO Charge, Element, and Isotope Analysis System (CELIAS) [*Hovestadt et al.*, 1995]. It consists of a high-density transmission grating to disperse the incident radiation, with an aluminum filter directly in front of it to limit the radiation bandpass of the spectrometer. The dispersed radiation is detected by three highly efficient, aluminum coated, silicon photodiodes positioned 200 mm behind the grating at zero order and at both first orders at 30.4 nm. The zero order detector monitors the full-disk solar irradiance from 0.1 to 50 nm. The first order detectors monitor the full-disk solar irradiance within an 8 nm bandpass centered at 30.4 nm. In both cases (zero and first orders), the detector output is a full-disk solar irradiance measurement of the radiation within the bandpasses of the detectors.

The SEM data show evidence of persistent solar active regions throughout solar cycle 23, which give rise to long-term (months to years), 27-day rotation, and short-term (minutes to hours) variations as evident in Figure 4. In order to assure the reliability of the data reported throughout the mission, rocket underflights of the SOHO SEM [*Judge et al.*, 1999], using an identical SEM instrument calibrated at NIST SURF, have provided periodic checks on the absolute calibration of the SOHO SEM. The uncertainty of the SEM irradiance data is 10% (1σ). From the Figure 4 daily averaged data, the solar minimum occurred around April 1996 and solar maximum occurred, so far, around July 2000. It is interesting to note the disappearance of the 27-day modulation at the time of solar minimum. Only very low amplitude "high frequency" variability is evident during the time of minimum solar irradiance. It is also to be noted that the per-

sistent 27-day modulation evident throughout the ascending part of the solar cycle 23 is characterized by relatively low amplitude variability. As solar maximum is approached the absolute flux as well as the amplitude of its modulation increase significantly. However, while the absolute value of both the solar irradiance and its variability increase from solar minimum to solar maximum the percentage change of the 27-day modulation scales approximately linearly with the irradiance. Recent analysis of Mg-II core-to-wing ratio data [*Viereck et al.*, 2001] has shown that these data are highly correlated with the SEM EUV data values throughout the SOHO mission to date. Such data may provide a valuable guide to EUV variability in the absence of the preferred direct EUV measurements. In addition to providing a standard against which the validity of proxies can be measured the SEM data may also be quite important in providing early warning of potentially destructive space weather events [*Judge et al.*, 2002]. The SEM data have been recently incorporated into operational models for space environment modeling and forecasting [*Tobiska et al.*, 2000].

3.5. LASP Rocket Instruments

The Laboratory for Atmospheric and Space Physics (LASP) at the University of Colorado has a sounding rocket payload to measure the solar spectral irradiance in the EUV and XUV wavelength range. The LASP solar irradiance rocket has flown seven times since 1988. It will be flown again once per year for the duration of the TIMED mission for underflight calibration purposes starting in early 2002. While the instrumentation on the payload has evolved over the years, the current incarnation (since 1997) includes a EUV Grating Spectrograph (EGS) that covers the wavelength range from 26 to 195 nm and an XUV Photometer

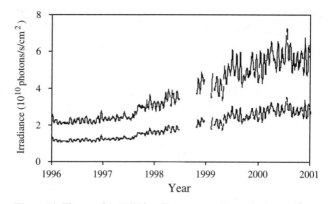

Figure 4. These solar EUV irradiance measurements are the daily averaged values from SOHO CELIAS/SEM for the zero order (upper curve) and first order (lower curve) channels. The data gaps occur during spacecraft operational problems.

Table 1. Solar EUV Irradiance Spacecraft Measurements.

Spacecraft / Instrument	λ Range (nm)	Δλ Res. (nm)	Time Period
SOLRAD-1-11	1–10	1	1960–1976
OSO-3-6 / SES	2–40	0.1	1967–1970
OSO-3-6 / EUVS	27–131	0.2	1967–1970
AEROS-A-B	20–104	1	1973–1975
AE-C-E / EUVS	14–185	0.2–1	1974–1981
AE-C-E/ ESUM	27–122	1-30	1974–1981
GOES / XRS	0.1–0.8	1	1974–present
Phobos 1–2	121.6	10	1988–1989
San Marco / ASSI	30–400	1	1988
Yohkoh / SXT	0.2–3	3	1992–present
CORONAS-I	1,7,28–50,121.6	2	1994
SOHO / SEM	26–34	8	1996–present
SNOE / SXP	0.2–20	4–7	1998–present

System (XPS) that covers 1 to 35 nm. These are prototype versions of the instruments for the TIMED Solar EUV Experiment (SEE). The EGS is a normal-incidence, 1/4 meter Rowland circle spectrograph with a resolution of about 0.4 nm. The XPS consists of an array of twelve silicon photodiodes with metal filters deposited directly on the surface of the diodes. These filters provide different band-passes, each about 7 nm wide, across the XUV wavelength range plus at H I Lyman α (121.6 nm). The EGS as a unit and the individual XPS photodiodes are calibrated at the NIST Synchrotron Ultraviolet Radiation Facility in Gaithersburg, Maryland. This calibration to a standard and repeatable EUV source is essential for determining the absolute solar spectral irradiance from flight to flight of the sounding rocket instruments.

These rocket measurements, having uncertainties of 10–30%, have been useful for SOHO underflight calibrations [*Brekke et al.*, 2000] and for establishing reference spectra [*Woods and Rottman*, 2002]. However, these rocket measurements have provided limited results on the solar variability [*Woods et al.*, 1998a] because they have very limited time coverage and because their absolute uncertainties are similar to the amount of solar variability expected at most wavelengths. Daily measurements from the historical spacecraft, such as from AE-E, still provide the best estimate of the irradiance variability in the EUV range.

4. IRRADIANCE VARIABILITY TIME SCALES

The EUV and X-ray irradiance observations illustrate extensive variability over a wide range of time scales, with strong wavelength dependence. Among the causes of these observed irradiance variations are flares (minutes), solar rotation (27 days), active region evolution (months), and the

solar sunspot cycle (11 years). Very short-term variations, lasting from minutes to hours, are related to eruptive phenomena on the Sun, short-term and intermediate-term variations, modulated by the 27-day rotation period of the Sun, are related to the appearance and disappearance of active regions on the solar disk, and the prominent long-term solar cycle variability is related to the 22-year magnetic field cycle of the Sun caused presumably by the internal solar dynamo. The solar variability, being the peak-to-peak variations for discussions here, is presented starting in the X-ray and moving towards longer wavelengths.

4.1. X-ray Flares

The GOES X-ray measurements have been most valuable to study solar flare events as the X-ray radiation is most sensitive to flare activities. The space weather forecasters use the GOES XRS as the first indicator of a solar flare event and as the standard in defining the magnitude of the X-ray flares. In a recent survey of approximately 1100 flares during solar cycles 21 and 22, *Garcia* [2000] tabulated the number of moderate to intense flares, i.e., M1 to >X5 in the NOAA classification nomenclature, as shown in Table 2. The average number of flares per typical 11-year solar cycle in each of these categories may be approximated by one half of the values shown in Table 2.

The *Garcia* [2000] flare study investigated the major thermodynamic parameters of flares by means of a semi-empirical model that analyzed flares individually, utilizing the observed temperature, emission measure, and duration to determine the flare's time dependent pressure, loop-top temperature, average density, total thermal energy, volume, and mass. These parameters were found to vary systematically with respect to the flare's intensity class as well as its physical size. Therefore, the data were organized around these two main driving parameters: the peak X-ray intensity, representing the external excitation, and the loop length, representing the preexisting magnetic morphology. These relationships are shown in compact form in Figure 5, collecting all (odd numbered) X-ray classes of each parameter with respect to loop length in a single plot. Only temporal maximum temperature and total thermal energy at the time of maximum emission measure are depicted here. The temperature plot (Fig. 5, left panel) demonstrates the finding that

Table 2. Distribution of Flare Types for 1976 to 1995.

M1	M2	M3	M4	M5	M6	M7	M8	M9
2789	790	410	244	133	94	70	62	56

X1	X2	X3	X4	X5	> X5
181	61	31	14	11	29

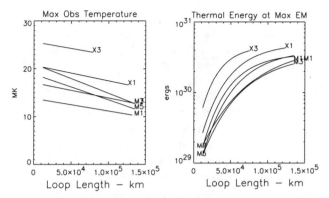

Figure 5. The left panel shows the summary of maximum flare temperatures of odd-numbered NOAA class flares versus loop length. Linear curves were obtained by least squares fits of observed flare temperatures in each intensity class during solar cycles 21 and 22. Loop lengths were computed from flare duration formulation [*Garcia*, 2000]. The right panel, similar to left panel, shows the total thermal energy modeled at the epoch of maximum emission measure and fitted to formulation of a rising exponential.

maximum temperature increases systematically with maximum X-ray intensity (*Garcia and McIntosh*, 1992; *Garcia*, 1994b; *Feldman et al.*, 1996; *Garcia, Greer, and Viereck*, 1999); it also shows the unexpected result that maximum temperature decreases systematically with increasing loop length. The total thermal energy plot (Fig. 5 right panel) shows that internal energy increases with loop length despite smaller average densities for large loops. The largest thermal energies for X3–X4 flares was approximately 7 x 10^{30} ergs, substantially smaller than the highest putative energies of 10^{32} ergs for the most intense flares. However, thermal energies for >X4 flares could not be determined owing to saturation of the GOES X-ray detector at these intensity levels.

4.2. XUV Variations

The X-ray data from GOES XRS and Yohkoh SXT show the most variability over the XUV range. The 11-year variability, smoothed over several days, is a factor of about 10 for the SXT 0.2–3 nm irradiance and a factor of about 100 for the XRS 0.1–0.8 nm irradiance. This variation is almost an order of magnitude larger than the variation at all other longer wavelengths.

The SNOE SXP data at 2 to 20 nm clearly show the 27-day periodicity due to solar rotation and the longer-term variability due to the 11-year solar cycle. The strength of the 27-day variability ranges from 25% to a factor of 2 from peak to trough. For the 2–7 nm channel the irradiances for the levels of activity during the observation period (81-day average of $F_{10.7}$ varied between 100 and 175 x 10^{-22} W m^{-2}

Hz^{-1}) vary between 4.8 and 40 times the *Hinteregger et al.* [1981] solar minimum value, demonstrating variability of a factor of 8.3. This result suggests that the variability of the 2–7 nm irradiance may be larger than one order of magnitude over the full range of solar activity. The variability decreases with longer wavelength with a maximum to minimum ratio of 6.4 for the 6–19 nm daily irradiances and a ratio of 4.2 for the 17–20 nm daily irradiances. The 27-day averages show smaller variability with closer agreement among the channels. The 27-day average of the 2–7 nm irradiances varies by a factor 2.7, the 6–19 nm irradiances by 3.1, and the 17–20 nm irradiances by 2.3. Because the SNOE measurements started after the last solar minimum, the solar cycle variations for the SNOE measurements are estimated from a linear fit of the 27-day smoothed SNOE data to the 27-day smoothed F10.7. Using this method the solar cycle 23 variations are factors of 3.7, 4.1, and 3.2 for the SNOE channels at 2–7 nm, 6–19 nm, and 17–20 nm, respectively.

The SOHO SEM data shows similar amount of variability as the SNOE data. The smoothed SEM data have long-term (11-year solar cycle) variability by a factor of 2.5 for both channels. The 27-day variability from SEM is about 25% and 35% for the 26–34 nm channel and 0.1–50 nm channel, respectively. The higher variability for the 0.1–50 nm channel suggests that coronal X-ray variations, which are larger by almost an order of magnitude than other EUV emissions, affect the 0.1–50 nm channel data the most. With SEM's high time cadence of 15 sec, flares are often observed by SEM, and the full-disk irradiance changes at 30.4 nm for these flares are about 5–10% during M-class flares and as much as 40% during larger X-class flares, which last up to 30 minutes.

4.3. EUV Variations

Because there have been only a handful of solar EUV irradiance measurements during the past solar cycle (1986–1996), it is difficult to characterize the variability in the 30 to 120 nm range. The last, more complete observation of the solar EUV irradiance by the AE-E spacecraft is often used for the reference EUV spectra [*Hinteregger et al.*, 1981; *Torr and Torr*, 1985]. Comparisons of the AE-E results to the more recent rocket measurements show differences by as much as a factor of 2 at some wavelengths [*Woods et al.*, 1998a]. However, it appears that the amount of variability recorded by AE-E is consistent with the limited set of recent measurements. The best characterization of the solar EUV is probably derived from combining the absolute levels of the recent, better calibrated, rocket instruments with the relative variability derived from the longer

AE-E time series. Adopting this approach, the solar EUV irradiance measurement from a sounding rocket in 1994 [*Woods et al.*, 1998a] is used as the solar minimum reference spectrum, and the relative variability established by the AE-E measurements [*Hinteregger et al.*, 1981] is used to scale the solar minimum reference spectrum to a solar maximum spectrum [*Woods and Rottman*, 2002]. From this approach, the 27-day variability ranges from 10% to 70% for the chromospheric and coronal emissions, respectively, and the 11-year solar cycle variations, after smoothed over 27 days, range from 1.2 to 7.

5. MODELS

Because of the limited amount of solar EUV irradiance data on multiple time scales and because of the low spectral resolution for most of these measurements, models of the solar EUV variability are routinely employed by aeronomic studies. One type of solar irradiance model is an empirical model, frequently called a proxy model, that is derived using linear relationships between a proxy of solar activity and direct observations of the solar UV irradiance. These models typically use a commonly available solar measurement, such as the ground-based 10.7 cm radio solar flux (F10.7), that serves as the proxy for the solar irradiance at other wavelengths. The first widely used proxy model was the *Hinteregger et al.* [1981] model, which is based on the AE-E observations and several sounding rocket measurements. The original proxies for this model were the chromospheric H I Lyman-β (102.6 nm) and the coronal Fe XVI (33.5 nm) emissions. As measurements of these emissions are not generally available, they are constructed from correlations with the daily F10.7 and its 81-day average, which have been available on a daily basis since 1947. This Hinteregger model is also referred to as SERF 1 by the Solar Electromagnetic Radiation Flux (SERF) subgroup of the World Ionosphere-Thermosphere Study. At about the same time, *Nusinov* [1984] developed a model with an empirically-determined active region background component, which incorporated physical solar features in the model. More recently, *Richards et al.* [1994] developed a F10.7 proxy model called EUVAC that is similar to SERF 1 but increased the solar soft X-ray irradiances by a factor of 2 to 3 as compared to the SERF 1 model. In addition, Tobiska has developed several proxy models of the solar EUV irradiance: SERF 2 by *Tobiska and Barth* [1990], EUV91 by *Tobiska* [1991], EUV97 by *Tobiska and Eparvier* [1998], and the latest version, SOLAR2000, by *Tobiska et al.* [2000].

The lack of a suitable database of EUV irradiance observations covering the needed range of wavelengths and time scale ultimately precludes the proper constraint and validation of empirical models and has motivated, therefore, the development of more physics-based approaches for quantifying UV irradiance variability [e.g., *Warren et al.*, 1998a, 1998b, 2001; *Fontenla et al.*, 1999; *Lean et al.*, 1982; *Cook et al.*, 1980]. The NRLEUV model, for example, constructs the EUV spectral irradiance as a function of time from information about temperature-dependent emission levels characteristic of the solar magnetic features that are sources of EUV radiation, combined with time-dependent determinations of the fractional disk areas covered by these features. The SOLAR2000 empirical model and the NRLEUV physical model are discussed in more detail, and then the predictions from these models are compared.

5.1. SOLAR2000 Model

The SOLAR2000 model is a collaborative project for characterizing the daily solar irradiance spectral variability from X-ray to infrared wavelengths [*Tobiska et al.*, 2000; *Tobiska*, 2000, 2002a, 2002b]. SOLAR2000 fulfills two primary modeling purposes: one function is to provide a simple, easy-to-use, model of the solar irradiances for research and for space weather operations, and a second purpose is to provide a linkage from the historical measurements, in this case from 1947, to the present. SOLAR2000 is derived using measurements, weighted by their uncertainty, from multiple instruments, captured across many spacecraft and rockets, spectral bands, and periods of time. The proxy parameters for the SOLAR2000 model are the short-term (<81 day) and long-term (>81 day) components of the composite Lyman-α time series [*Woods et al.*, 2000b] and the 10.7 cm radio flux (F10.7). The Lyman-α time series is only updated every few months, so the Mg II core-to-wing index (C/W) can also be used as the proxy through a conversion to units of Lyman-α based on a fit between the Mg II C/W and Lyman-α.

Upgrades to the operational and research grade SOLAR2000 version 1.17 include the definition of the summed spectrum, the incorporation of the revised composite Lyman-α time series, incorporation of NOAA 16 Solar Backscatter UltraViolet (SBUV) Mg II data for nowcast and forecast irradiances, nowcast and 3-hourly forecasts for the next 72-hours, and improved solar irradiance bulletins for spacecraft operators. Upgrades that are currently in progress for the SOLAR2000 model include improved algorithms on six time scales, incorporation of SNOE, SOHO SUMER, UARS, and TIMED SEE data, FUV irradiance variability, development of a callable SOLAR2000 routine for research applications, and addition of wavelength formats for TIME-GCM model compatibility.

A unique part of the SOLAR2000 model is the generation of the "E10.7" index, which is derived by integrating the solar energy flux between 1 and 105 nm and scaled to the F10.7. Because the F10.7 is primarily a coronal emission, the E10.7 index provides a better substitute for F10.7 in atmospheric modeling, which requires as input the information about the solar EUV irradiance emitted from the chromosphere and transition region as well as from the corona.

5.2. NRLEUV Model

The NRLEUV irradiance variability model is based on the physical properties of the solar atmosphere in the quiet Sun, bright active regions and network, and dark coronal holes [Warren et al., 2001] whose changing proportions modulate the net emitted EUV radiation. Central to the NRLEUV model is the differential emission measures (DEMs), which provide information about electron density in these features as a function of temperature. The DEMs are constructed from intensity measurements of 19 emission lines observed at 0.16 nm resolution by the Harvard College Observatory Spectroheliometer instrument on Skylab [Vernazza and Reeves, 1978]. The coronal abundances of Meyer [1985], ionization balance of Arnaud and Rothenflug [1985], and the CHIANTI atomic databases of Dere et al., [1997] and Mewe et al. [1985, 1995] provide additional information about the gases that emit EUV lines and continua, and about their temperature-dependent atomic properties per unit electron density. Images of the solar chromosphere made by the Big Bear Solar Observatory (BBSO) in the Ca K line [Johannesson et al., 1998] and of the corona made by Yohkoh SXT [Acton et al., 1999] provide information about the daily fractional coverage of active regions, active network, and coronal holes. Applying center-to-limb functions permits the derivation of the irradiance of optically thin EUV emission lines and five continua by integration of the radiance derived from the emission measures over the full solar disk. For optically thick lines, direct contrasts of the active regions and coronal holes are used instead of emission measures.

The absolute irradiance scale of the NRLEUV model is that of the Skylab measurements used to construct the DEMs. The solar minimum spectral irradiance is calculated by integrating the quiet-Sun DEM over the entire solar atmosphere, in the absence of any magnetic features [Warren et al., 1998a, 1998b]. At other times, NRLEUV estimates the EUV irradiance by adjusting the adopted quiet solar spectrum for the presence of active regions, active network, and coronal holes, which modify the disk-integrated irradiance by altering the local emission [Warren et al., 2001].

A recent version of NRLEUV includes proxy indicators of the coronal and chromospheric fluxes to parameterize the EUV irradiance. The F10.7 is used as the proxy for coronal emissions ($T > 0.8 \times 10^6$ K), and the Mg II C/W is the proxy to represent the chromosphere and transition region emissions ($T \leq 0.8 \times 10^6$ K). The Mg II C/W is the one defined by the NOAA Space Environment Center [Viereck and Puga, 1999], extended with the Mg index from the GOME instrument [Weber, 1999] and independently validated with Ca II K chromospheric indices [Lean et al., 2001]. The relations between the proxies and the NRLEUV model, which incorporates the Ca K and X-ray images, were parameterized during the descending phase of solar cycle 22 (1992–1996). As presently formulated, the NRLEUV model estimates the EUV spectral irradiance for 1474 emission lines and 5 continua at wavelengths from 5 to 120 nm, daily from 1947 to the present.

Recently, the approach of the NRLEUV model has been utilized to estimate the increase in the EUV spectrum during the Bastille Day flare (14 July 2000) [Meier et al., 2002]. Because the temperature distribution observed in a flare is significantly different than that of the three solar regions currently included in the model, it was necessary to derive a new emission measure from GOES and TRACE observations. According to initial model estimates, the EUV spectrum can increase during a flare by an amount equivalent to its solar cycle variation, and include, as well, emissions at wavelengths not typically present in non-flaring conditions.

5.3. Comparisons of Measurements and Models

The Woods and Rottman [2002] reference solar irradiance spectra are used as a common set of spectra for comparing the different measurements and models. These reference spectra include spectra for solar minimum and maximum conditions and an example of 27-day rotational variability and are representative of conditions during solar cycle 22 (1986–1996). These reference spectra are primarily established from actual observations; however, results from selected solar irradiance proxy models are used in creating and validating the reference solar spectra in the XUV and EUV range. These reference spectra are given in 1 nm intervals on 0.5 nm centers. The solar minimum and maximum reference spectra are chosen centered on 1996/108 and 1992/032, respectively. These solar cycle spectra are obtained by averaging over 27-days to represent the mean value during a solar rotation. The variability of the total solar EUV irradiance, integrated from 0 to 120 nm for these reference spectra, is estimated to be 25% for a large 27-day rotational period and about 85% for the 11-year solar cycle.

A crucial unresolved aspect of EUV and X-ray irradiance is the differences in absolute values among the various measurements and models. The *Woods and Rottman* [2002] reference spectrum for solar cycle minimum condition is compared to the NRLEUV and SOLAR2000 models in Figure 6. While the differences are larger at higher spectral resolution, the differences at 5 nm resolution are mostly less than 30% and thus are consistent with the ±30% uncertainty of the reference spectrum irradiance. The largest difference between the NRLEUV model and the *Woods and Rottman* [2002] reference spectrum is below 15 nm where the NRLEUV model irradiance is about a factor of 3 lower than the reference spectrum. The largest difference between the SOLAR2000 model and the *Woods and Rottman* [2002] reference spectrum is above 100 nm where the SOLAR2000 model irradiance is a factor of 2–10 higher than the reference spectrum. Resolving these differences likely requires both improvements in the models and more accurate measurements of the solar EUV irradiance.

Bailey et al. [2000, 2001] have compared the predictions of empirical models with the SNOE XUV observations shortward of 20 nm. They found that the SNOE results are as much as a factor of four times larger than the predictions of the empirical model of *Hinteregger et al.* [1981] in terms of absolute irradiance. *Solomon et al.* [2001] have shown that the SNOE results resolve historical discrepancies between geophysical observations and models that are based on solar soft X-ray energetic inputs.

The other aspect of comparisons is the solar variability, which is the relative value of the irradiances from each data set or model. The solar cycle variability from the *Woods and Rottman* [2002] reference spectra, recent measurements, and models are compared in Figure 7. Each spectrum is smoothed over 5 nm before deriving the solar cycle variation, being the ratio of solar maximum irradiance to solar minimum irradiance. The variability for these reference spectra is based primarily on the variability from the *Hinteregger et al.* [1981] empirical model. So the fact that the SNOE SXP and SOHO SEM results show reasonable agreement for the variability indicates that the variability from the AE-E measurements are reasonable values to use for the latest solar cycle, at least for the SNOE and SEM wavelength bands. An exception for this SNOE comparison is that the SNOE 6–19 nm channel shows more variability than expected from the reference spectra or models.

The spectral shapes of the solar variability are similar for the reference spectra and the NRLEUV and SOLAR2000 models, but they differ in the amount of variability. The SOLAR2000 model predicts similar levels of solar cycle variability as the reference spectra longward of 50 nm but

diverge shortward of 50 nm. For the shorter wavelengths, the SOLAR2000 model predicts more solar cycle variability than the reference spectra. The NRLEUV model shows similar solar cycle variability as the reference spectra at wavelengths shortward of 20 nm, but the NRLEUV model indicates much less variability at the longer wavelengths. *Lean et al.* [2002] give more detailed comparisons of the NRLEUV model to data and other models, including the SOLAR2000 model.

The solar maximum to minimum ratio below 2 nm is off scale in Figure 7 and is about a factor of 10 at 1–2 nm from Yohkoh SXT results and about a factor of 100 at 0–1 nm from GOES XRS results. The *Hinteregger et al.* [1981] empirical model did not have values below 1.8 nm, so these reference spectra are based on the recent Yohkoh and GOES results.

6. FUTURE MEASUREMENTS

Improved solar EUV irradiance measurements are needed to resolve the differences in both the absolute values of the solar irradiance and the levels of solar variability and to provide a reliable database for the development of improved variability models. There are several missions being planned by NASA, NOAA, ESA, and Russia that will address this need.

The new NASA observations of the solar EUV irradiance during solar cycle 23 include the Thermosphere Ionosphere Mesosphere Energetics and Dynamics (TIMED) and Solar

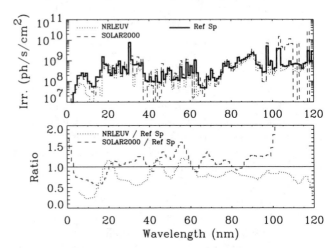

Figure 6. The solar EUV irradiances for solar cycle minimum condition are shown in the top panel in 1 nm intervals for the *Woods and Rottman* [2002] reference spectrum (solid line) and for the NRLEUV and SOLAR2000 models. The ratio of the models to the reference spectrum is shown in the bottom panel. The reference and model spectra are smoothed over 5 nm before taking the ratio.

Radiation and Climate Experiment (SORCE) spacecraft. The TIMED Solar EUV Experiment (SEE) will provide solar irradiances from 1 to 195 nm with ~10% accuracy [*Woods et al.*, 1998b]. The TIMED launch is currently planned for December 2001. The SORCE spacecraft contains 4 solar irradiance instruments that cover the spectral range from 1 to 2000 nm, but with no EUV observations between 35 and 115 nm [*Woods et al.*, 2000a]. The SORCE XUV measurements shortward of 35 nm will be similar to the TIMED XUV measurements. The SORCE launch is currently planned for the fall of 2002. The solar EUV and XUV irradiance instruments on the TIMED and SORCE spacecraft have pre-flight calibrations with accuracies of 5-20% traceable to NIST standards and in-flight calibrations based on redundant channels and on-board reference detectors. In addition to these existing NASA programs, the NASA Solar Dynamics Observatory (SDO) is planning solar EUV irradiance measurements below 120 nm with 0.1 nm resolution and at 10 sec cadence. These SDO measurements, perhaps beginning in 2007, will provide faster measurement cadence and higher spectral resolution to improve the understanding of flare events and the sources of the solar variability.

There are also new NOAA missions being planned for solar cycle 24 that will include solar EUV irradiance observations. For example, the XRS instruments, discussed earlier, are planned for the next series of GOES spacecraft. NOAA will also begin to monitor the solar EUV flux between 10 and 124 nm at a ten-second sample rate starting with the GOES-13 spacecraft, whose launch is expected to

be in late 2002 or 2003. The GOES EUV Sensor (EUVS) will have five broadband channels centered at 15 nm, 30.4 nm, 60 nm, 90 nm, and 121.6 nm; at this time there is some technical uncertainty regarding the 90 nm channel. The EUVS incorporates silicon photodiode detectors with metallic thin-film filters and transmission gratings for wavelength isolation. There will be EUVS instruments on each of the GOES spacecraft so there will be two instruments operating simultaneously. The instrument stability will be monitored by launching freshly calibrated instruments on each GOES spacecraft. In the past, GOES spacecraft have been launched at about two year intervals. The EUVS will provide a measure of the solar EUV irradiance for estimation of thermosphere and ionosphere densities. These operational parameters are important for radio communication and navigation, as well as for the specification and prediction of spacecraft orbits. Updates and improvements to the EUVS are planned for the future GOES spacecraft to be launch-ready around 2010.

There are also Russian and ESA missions being prepared that include solar EUV irradiance instruments. The Russian Solar Patrol mission is planned to fly in 2004 aboard the Russian part of the International Space Station (ISS) [*Avahyan and Kuvaldin*, 2000]. It consists of three instruments. One instrument is a set of X-ray and EUV radiometers in the spectral range of 0.14 to 135 nm with a filter wheel and radioactive sources to trace efficiency changes. A second instrument is an X-ray grating spectrometer that will measure the solar emissions from 2 to 60 nm with a spectral resolution of 0.3 nm. The other instrument is a EUV grating spectrometer to cover the wavelength range from 57 to 153 nm and from 16 to 63 nm with a spectral resolution of 1 nm. The ESA Solar Auto-Calibrating EUV/UV Spectrometers (SOL-ACES) is planning to fly in 2005–2006 [*Wienhold et al.*, 2000]. It will measure the solar radiation from 17 to 220 nm with a spectral resolution from 0.3 to 2 nm with four grazing incidence planar grating spectrometers. To obtain high radiometric accuracy of <5%, a double ionization chamber with an additional silicon detector is assigned to each of the spectrometers as a primary detector standard. Optical bandpass filters are mounted on a filter wheel in front of the entrance apertures of the spectrometers and ionization chambers and thereby will establish the radiometric link among these devices.

These future missions will surely extend and improve our understanding of the solar irradiance over a longer timescale and with a higher accuracy as technology and calibration standards continue to improve. In the meantime, several of the past measurements and models are available for solar and atmospheric research (see Table 3).

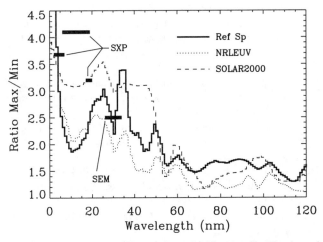

Figure 7. The 11-year solar cycle variability for the *Woods and Rottman* [2002] reference spectra (solid line) is compared to the recent measurements by SNOE SXP and SOHO SEM (thick lines) and to the NRLEUV and SOLAR2000 models The reference and model spectra are smoothed over 5 nm before taking the ratio of solar maximum to solar minimum.

Table 3. Web Sites for Solar EUV Irradiances.

Measurement / Model	Web Site (http:// prefix) [NSSDC=nssdc.gsfc.nasa.gov/space]
OSO	NSSDC/space_phys/add_spcraft.html
SOLRAD	NSSDC/space_phys/add_spcraft.html
AEROS	NSSDC/space_phys/haeros.html
AE	NSSDC/space_phys/hae.html
San Marco / ASSI	NSSDC/space_phys/hsan.html
GOES	spdir.ngdc.noaa.gov/spdir/
SOHO/SEM	www.usc.edu/go/spacescience/
SNOE / SXP	lasp.colorado.edu/snoedata/
Woods Rockets	lasp.colorado.edu/rocket/
EUV81 (SERF 1) Model	lasp.colorado.edu/rocket/
SERF 2 Model	NSSDC/model/Sun/
EUVAC Model	lasp.colorado.edu/rocket/
SOLAR2000 Model	SpaceWx.com/
Reference Spectra [*Woods and Rottman*, 2002]	ftp://laspftp.colorado.edu/pub/solstice/ ref_min_27day_11yr.dat

Acknowledgments. The support for Tom Woods and Frank Eparvier is through JHU/APL contract JH774017. Support for Kent Tobiska has been provided by the NASA TIMED contract NAS5-97179 (APL Contract 774017, UCB SPO BS0059849), by the NASA SOHO contract NAS5-98172, by the NATO EST.CLG 975301 collaborative travel grant, and by the NOAA/SEC-SET CRADA. The development of the NRLEUV model by J. Lean, J. Mariska, and H. Warren was supported by NASA and ONR.

REFERENCES

Acton, L. W., D. C. Weston, and M. E. Bruner, Deriving solar X ray irradiance from Yohkoh observations, *J. Geophys. Res., 104*, 14,827, 1999.

Arnaud, M., and R. Rothenflug, An update evaluation of recombination and ionization rates, *Astron. Astrophys., 60*, 425, 1985.

Avahyan, S. V. and E. V. Kuvaldin, The results of work creating XUV instrumentation for the Solar Patrol mission, *Physics and Chemistry of the Earth, 25*, 441, 2000.

Bailey, S. M, T. N. Woods, C. A. Barth, and S. C. Solomon, R. Korde, L. R. Canfield. Measurements of the Solar Soft X-ray Irradiance by the Student Nitric Oxide Explorer: First Analysis and Underflight Calibrations, *J. Geophys. Res., 105*, A12, 27179, 2000.

Bailey, S. M, T. N. Woods, C. A. Barth, and S. C. Solomon, R. Korde, L. R. Canfield. Correction to "Measurements of the Solar Soft X-ray Irradiance by the Student Nitric Oxide Explorer: First Analysis and Underflight Calibrations", *J. Geophys. Res., 106*, A8, 15791, 2001.

Brekke, P., W. T. Thompson, T. N. Woods, and F. G. Eparvier, The extreme-ultraviolet solar irradiance spectrum observed with the Coronal Diagnostic Spectrometer (CDS) on SOHO, *Astrophys. J., 536*, 959, 2000.

Buonsanto, M. J., M. E. Hagan, J. E. Salah, and B. G. Fejer, Solar cycle and seasonal variations in *F* region electrodynamics at Millstone Hill, *J. Geophys. Res., 98*, 15, 677, 1993.

Buonsanto, M. J., P. G. Richards, W. K. Tobiska, S. C. Solomon, Y. K. Tung, and J. A. Fennelly, Ionospheric electron densities calculated using different EUV flux models and cross sections: Comparison with radar data, *J. Geophys. Res., 100*, 14569, 1995.

Cook, J. W., G. E. Brueckner, and M. E. VanHoosier, Variability of the solar flux in the far ultraviolet 1175-2100 Å, *J. Geophys. Res., 85*, 2257, 1980.

Dere, K. P., E. Landi, H. E. Mason, B. C. Mosignori Fossi, and P. R. Young, CHIANTI – An atomic database for emission lines. Paper I: Wavelengths greater than 50 Å, *Astron. Astrophys. Suppl. Ser., 125*, 149, 1997.

Feldman, U., Doschek, G. A., Behring, W. E. and Phillips, K. J., Electron temperature, emission measure, and X-ray flux in A2 to X2 X-ray class solar flares, *Ap. J., 460*, 1034, 1996.

Feng, W., H. S. Ogawa, and D. L. Judge, The absolute solar Soft X-ray Flux in the 20 - 100Å region, *J. Geophys. Res., 94*, 9125, 1989.

Fontenla, J. M., O. R. White, P. A. Fox, E. H. Avrett, and R. L. Kurucz, Calculation of solar irradiances, I, Synthesis of the solar spectrum, *Astrophys. J., 518*, 480, 1999.

Garcia, H., Temperature and emission measure from GOES soft X-ray measurements, *Solar Phys., 154*, 275, 1994a.

Garcia, H., Temperature and hard X-ray signatures for energetic proton events, *Ap. J., 420*, 422, 1994b.

Garcia, H., Thermal-spatial analysis of medium and large solar flares, 1976 to 1996, *Ap. J. Suppl., 127*, 189, 2000.

Garcia, H. and P. S. McIntosh, High-temperature flares observed in broadband soft X-rays, *Sol. Phys., 141*, 109, 1992.

Garcia, H., S. Greer, and R. Viereck, Predicting solar energetic proton events from low temperature soft X-ray flares, in *9th European Meeting on Solar Physics*, ed. A. Wilson, p. 939, European Space Agency, SP-448, 1999.

Gibson, Sister Jean and J. A. Van Allen, Correlation of X-ray radiation (2-12A) with microwave radiation (10.7 centimeters) from the nonflaring Sun, *Astrophys. J., 161*, 1135, 1970.

Hall, L. A., J. E. Higgins, C. W. Chagnon, and H. E. Hinteregger, Solar cycle variation of extreme ultraviolet radiation, *J. Geophys. Res., 74*, 4181, 1969.

Heroux, L. and J. E. Higgins, Summary of full-disk solar fluxes between 250 and 1940 Å, *J. Geophys. Res., 82*, 3307, 1977.

Hinteregger, H. E., EUV flux variation during end of solar cycle 20 and beginning cycle 21, observed from AE-C satellite, *Geophys. Res. Lett., 4*, 231, 1977.

Hinteregger, H. E., K. Fukui, and G. R. Gilson, Observational, reference and model data on solar EUV from measurements on AE-E, *Geophys. Res. Lett., 8*, 1147, 1981.

Hovestadt, D., et al., CELIAS – Charge, element, and isotope analysis system for SOHO, *Solar Physics 162*, 441, 1995.

Johannesson, A., W. H. Marquette, and H. Zirin, A 10-year set of CaII K-line filtergrams, *Solar Phys., 177*, 265, 1998.

Judge, D.L., et al., First solar EUV irradiances obtained from SOHO by the SEM, *Solar Physics, 177*, 161, 1998.

Judge, D. L., D. R. McMullin, and H. S. Ogawa, Absolute solar 30.4 nm flux from sounding rocket observations during the solar cycle 23 minimum, *J. Geophys. Res., 104*, 28,321, 1999.

Judge, D.L., et al., Space Weather Observations Using the SOHO CELIAS Complement of Instruments, *J. Geophys. Res., 106,* (A12), 29, 963, 2002.

Kreplin, R. W., The solar cycle variation of soft X-ray emission, *Ann. Geophys., 26*, 567, 1970.

Kreplin, R. W. and D. M. Horan, Variability of X-ray and EUV solar radiation in solar cycles 20 and 21, *Proceedings of the Workshop on the Solar Electromagnetic Radiation Study for Solar Cycle 22*, edited by R. F. Donnelly, p. 405, NOAA, Boulder, 1992.

Lean, J., Solar ultraviolet irradiance variations: A review, *J. Geophys. Res., 92*, 839, 1987.

Lean, J., Variations in the Sun's radiative output, *Reviews of Geophysics, 29*, 505, 1991.

Lean, J. L., W. C. Livingston, D. F. Heath, R. F. Donnelly, A. Skumanich, and O. R. White, A three-component model of the variability of the solar ultraviolet flux 145-200 nm, *J. Geophys. Res., 87*, 10,307, 1982.

Lean, J. L., O. R. White, W. C. Livingston and J. M. Picone, Variability of a composite chromospheric irradiance index during the 11-year activity cycle and over longer time periods, *J. Geophys. Res., 106*, 10,645, 2001.

Link, R., S. Chakrabarti, G. R. Gladstone, and J. C. McConnell, An analysis of satellite observations of the OI EUV dayglow, *J. Geophys. Res., 93*, 2693, 1988.

Manson, J. E., The solar extreme ultraviolet between 30 and 205Å on November 9, 1971, compared with previous measurements in this spectral region, *J. Geophys. Res., 81*, 1629, 1976.

Meier, R. R., et al., Ionospheric and dayglow responses to the radiative phase of the Bastille Day flare, *Geophys. Res. Lett., 29*(10), 1461, doi:10.1029/2001GL013956, 2002.

Mewe, R., E. H. B. M. Groenschild, and G. H. J. van den Oord, Calculated X-radiation from optically thin plasmas: V, *Astrophys. J. Suppl. Ser., 62*, 197, 1985.

Mewe, R., J. S. Kaastra, and D. A. Liedahl, Update of MEKA: MEKAL, *Legacy, 6*, 16, 1995.

Meyer, J. P., Solar-stellar atmospheres and energetic particles, and galactic cosmic rays, *Astrophys. J. Suppl. Ser., 57*, 173, 1985.

Newmark, J. S., J. W. Cook, J. D., Moses, F. Auchere, and F. Clette, Solar EUV variability as measured by SOHO EIT, *Astrophys. J.*, in preparation, 2002.

Nusinov, A. A., Dependence of the intensity of lines of short-wave solar emission on the activity level, *Geomagn. Aeron., 24*, 529, 1984.

Pap, J. M., C. Fröhlich, H. S. Hudson, and S. K. Solanki (editors), *The Sun as a Variable Star: Solar and Stellar Irradiance Variations*, Cambridge University Press, Cambridge, 1994.

Richards, P. G., and D. G. Torr, An investigation of the consistency of the ionospheric measurements of the photoelectron flux and solar EUV flux, *J. Geophys. Res., 89*, 5625, 1984.

Richards, P. G., J. A. Fennelly, and D. G. Torr, EUVAC: A solar EUV flux model for aeronomic calculations, *J. Geophys. Res. 99*, 8981, 1994.

Rottman, G. J., Results from space measurements of solar UV and EUV flux, in *Solar Radiative Output Variation*, edited by P. Foukal, pp. 71-86, Cambridge Research and Instrumentation Inc, Boulder,1987.

Schmidtke, G., Modeling of the solar extreme ultraviolet irradiance for aeronomic applications, in *Encyclopedia of Physics*, Vol. XLIX/7, Geophys. III, Pt .VII, pp.1-55, 1984.

Schmidtke, G., W. Schweizer, and M. Knothe, The AEROS-EUV-Spectrometer, Z, *Geophys., 40*, 577-584, 1974.

Schmidtke, G., K. Rawer, H. Botzek, D. Norbert, and K. Holzer, Solar EUV photon fluxes measured aboard AEROS A, *J. Geophys. Res., 82*, 2423, 1977.

Siskind, D. E., C. A. Barth and D. D. Cleary, The possible effect of solar soft X-rays on thermospheric nitric oxide, *J. Geophys. Res., 95*, 4311, 1990.

Solomon, S. C., and V. J. Abreu, The 630 nm dayglow, *J. Geophys. Res., 94*, 6817, 1989.

Solomon, S. C., S. M. Bailey, and T. N. Woods, Effect of Solar Soft X-rays in the Lower Ionosphere, *Geophys, Res., Lett., 28*, 2149, 2001.

Thompson, W., and M. Carter, EUV full-Sun imaging and pointing calibration of the SOHO Coronal Diagnostic Spectrometer, *Solar Phys., 178*, 509, 1998.

Tobiska, W. K., Revised solar extreme ultraviolet flux model, *J. Atmos. Terr. Phys., 53*, 1005, 1991.

Tobiska, W. K., Recent solar extreme ultraviolet irradiance observations and modeling: A review, *J. Geophys. Res., 98,* 18,879, 1993.

Tobiska, W.K., Status of the SOLAR2000 solar irradiance model, *Phys. Chem. Earth,* Vol. 25, 383, 2000.

Tobiska, W.K., Variability in the TSI from Irradiances Shortward of Lyman-alpha, *Adv. Space Research,* in press, 2002a.

Tobiska, W. K., Validating the Solar EUV Proxy, E10.7, *J. Geophys. Res., 106* (A12), 29, 969, 2001.

Tobiska, W. K. and C. A. Barth, A solar EUV flux model, *J. Geophys. Res., 95*, 8243, 1990.

Tobiska, W. K., and F. G. Eparvier, EUV97: Improvements to EUV irradiance modeling in the soft X-rays and FUV, *Solar Phys., 177*, 147, 1998.

Tobiska, W. K., T. N. Woods, F. G. Eparvier, R. Viereck, L. Floyd, D. Bouwer, G. J. Rottman, and O. R. White, The SOLAR2000 empirical solar irradiance model and forecast tool, *J. Atmos. and Sol. Terr. Phys., 62*, 1233, 2000.

Torr, M. R., and D. G. Torr, Ionization frequencies for solar cycle 21: Revised, *J. Geophys. Res., 90*, 6675, 1985.

Tsuneta, S., and 12 colleagues, The Soft X-ray Telescope for the SOLAR-A Mission, *Solar Phys., 136*, 37, 1991.

Vernazza, J. E., and E. M. Reeves, Extreme ultraviolet composite spectra of representative solar features, *Astrophys. J. Suppl. Ser., 37*, 485, 1978.

Viereck, R. A. and L. C. Puga, The NOAA Mg II core-to-wing solar index; Construction of 20-year time series of chromospheric variability from multiple satellites, *J. Geophys. Res., 104*, 9995, 1999.

Viereck , R., L. Puga, D. McMullin, D. Judge, M. Weber, and W. K. Tobiska, The Mg II Index: A Proxy for Solar EUV, *Geophys. Res. Lett., 28*, 1343, 2001.

Warren, H. P., J. T. Mariska, and J. Lean, A new reference spectrum for the EUV irradiance of the quiet Sun: 1. Emission measure formulation, *J. Geophys. Res., 103*, 12,077, 1998a.

Warren, H. P., J. T. Mariska, and J. Lean, A new reference spectrum for the EUV irradiance of the quiet Sun: 2. Comparisons with observations and previous models, *J. Geophys. Res., 103*, 12,091, 1998b.

Warren, H.P., J.T. Mariska, and J. Lean, A new model of solar EUV irradiance variability. 1. Model Formulation, *J. Geophys. Res., 106*, 15,745, 2001.

Weber, M., Solar activity during solar cycle 23 monitored by GOME, in *Proc. European Symposium on Atmospheric Measurements from Space (ESAMS'99)*, pp. 611-616, European Space Agency, 1999.

White, O. R. (editor), *The Solar Output and Its Variation*, Colorado Ass. Univ. Press, Boulder, 1977.

Wienhold, F. G., J. Anders, B. Galuska, U. Klocke, M. Knothe, W. J. Riedel, G. Schmidtke, R. Singler, U. Ulmer and H. Wolf, The Solar Package on ISS: SOL-ACES, *Physics and Chemistry of the Earth, 25*, 473, 2000.

Winningham, J. D., D. T. Decker, J. U. Kozyra, J. R. Jasperse, and A. F. Nagy, Energetic (>60 eV) atmospheric photoelectrons, *J. Geophys. Res., 94*, 15,335, 1989.

Woods, T. N. and G. J. Rottman, Solar ultraviolet variability over time periods of aeronomic interest, in *Comparative Aeronomy in the Solar System*, eds. M. Mendillo, A. Nagy, and J. Hunter Waite, Jr., Geophys. Monograph Series, Wash. DC, pp. 221-234, 2002.

Woods, T. N., G. J. Rottman, S. M. Bailey, S. C. Solomon, and J. Worden, Solar extreme ultraviolet irradiance measurements during solar cycle 22, *Solar Physics, 177*, 133, 1998a.

Woods, T. N., F. G. Eparvier, S. M. Bailey, S. C. Solomon, G. J. Rottman, G. M. Lawrence, R. G. Roble, O. R. White, J. Lean, and W. K. Tobiska, TIMED Solar EUV Experiment, *SPIE Proceedings*, 3442, 180, 1998b.

Woods, T., G. Rottman, J. Harder, G. Lawrence, B. McClintock, G. Kopp, and C. Pankratz, Overview of the EOS SORCE mission, *SPIE Proceedings, 4135*, 192, 2000a.

Woods, T. N., W. K. Tobiska, G. J. Rottman, and J. R. Worden, Improved solar Lyman a irradiance modeling from 1947 through 1999 based on UARS observations, *J. Geophys. Res., 105*, 27,195, 2000b.

Tom Woods and Frank Eparvier, Laboratory for Atmospheric and Space Physics, University of Colorado, 1234 Innovation Dr., Boulder, CO 8030.

Loren W. Acton, Physics Department, Montana State University, Bozeman, MT 59717.

Scott Bailey, Geophysics Institute, University of Alaska, Fairbanks, AK 99775.

Howard Garcia and Rodney Viereck, Space Environment Laboratory SEL/R/E/SE, National Oceanic & Atmospheric Administration, 325 Broadway, Boulder, CO 80303.

Darrell Judge and Don McMullin, Space Sciences Center, University of Southern California, Los Angeles, CA 90007.

Judith Lean and John Mariska, Code 7673, Naval Research Laboratory, Washington, DC 20375.

Gerhard Schmidtke, Fraunhofer-Institut für Physikalische Messtechnik, Heidenhofstraße 8, D-7800 Freiburg, Germany.

Stanley C. Solomon, High Altitude Observatory, National Center for Atmospheric Research, PO Box 3000, Boulder, CO 80307-3000.

W. Kent Tobiska, Space Environment Technologies, SpaceWx Division, 1676 Palisades Dr., Pacific Palisades, CA 90272.

Harry P. Warren, Code 7673, Naval Research Laboratory, Washington, DC 20375.

Solar Activity and Irradiance Variations

Peter Fox

High Altitude Observatory, National Center for Atmospheric Research, Boulder, CO

This chapter explores the relation between solar activity and solar irradiance variability and reviews the current state of models of the total and spectral irradiance and variability, including uncertainties in both the theory, assumptions, free parameters, and inputs. Over the past decade, regular measurement programs for parts of the solar reference spectrum have been established and the physical understanding of these measurements is rapidly improving. We discuss all of the present modeling approaches which range from simple time series proxy models for the total solar irradiance to the spectral irradiance which may use a combination of semi-empirical models and empirical image analysis with the theory for line-by-line calculation of emission, absorption, and transfer of radiation in the solar atmosphere. This chapter starts with some definitions and proceeds to a recent history of irradiance models. Next, attention is given to similarities and differences between the models and the common assumptions related to solar activity influences and then the different types of models are presented along with some recent results and their comparison to extant observations. The chapter concludes with an assessment of the current state of models, what the present level of accuracy and precision they can achieve, their uncertainties and what advances are in the near future.

1. HISTORY OF IRRADIANCE MODELS

The context for models of the solar irradiance (both total and spectral) was set by early measurement programs and focused on accounting for the actual presence of a variation of the radiation from the Sun and the recognition that those variations were measurable and possibly of significance to the terrestrial system. Other chapters in this volume set the stage for observations, interpretations, and the myriad impacts of the radiative variability of the Sun.

The initial context for models suggested that the explanations should only take into account sources or mechanisms from the photospheric layers and above, i.e. where meas-

urements originate. This context has shifted and sources for contributions to the variability from inside the Sun, of both magnetic and thermal origin, are now being considered. This internal variability, including the theoretical underpinnings and models are discussed elsewhere. This chapter focuses on assessing the state of models within the original context but does not necessarily presuppose that they are the only explanation for the observed solar variability.

1.1. Definitions

Since the solar irradiance is a primary input to the terrestrial system and it varies, it is useful to define some fundamental physical quantities related to the Sun and this emerging radiation, the geometry with respect to the Earth, etc.

The solar luminosity ($L_\odot$) is the amount of energy per unit time emerging from the solar surface. It is usually expressed

Solar Variability and its Effects on Climate
Geophysical Monograph 141
Copyright 2004 by the American Geophysical Union
10.1029/141GM12

in units of ergs/sec. A typical value from solar structure theory is 3.845×10^{33} [Cox, 2000].

The solar radius ($R_\odot$) is the nominal radius of the Sun, obtained from a solar structure model corresponding to the unit optical depth $\tau_{5000} = 1$ or 2/3 (depending on the atmosphere model assumed) surface and is usually expressed in units of cm. The presently quoted value is 6.96×10^{10}.

The solar flux ($F_\odot = L_\odot / (4\pi R_\odot^2)$), is the amount of energy per unit area per unit time, over all directions. It is usually expressed in units of ergs/cm^2/sec. Using the above-mentioned value for $L_\odot$, its value is 6.316×10^{10}.

The average disk intensity (I), is the sum over the visible solar disk at any given time of the emergent intensities weighted by the apparent emitting area divided by the total disk area. It is measured or expressed in ergs/cm^2/sec/sr or ergs/cm^2/sec/sr/Å (spectral intensity).

The solar irradiance $S_\odot$ is the incident power per unit area at 1AU from the Sun to the top of earth's atmosphere, viz. it is the flux at 1AU through a surface normal to the Sun-Earth line of sight. The total irradiance is integrated over the entire spectrum. It is usually measured or expressed in W/m^2 (or W/m^2/nm for spectral irradiance).

The mean Sun-Earth distance (D), defined as 1 Astronomical Unit (AU), is expressed in cm and its value is 1.496×10^{13} cm [Cox, 2000].

Thus, the solar irradiance is related to the average disk intensity by $S_\odot = I(\pi R_\odot^2/D^2)$. The factor, $\pi R_\odot^2/D^2 = 6.8 \times 10^{-5}$, is the angle subtended by the solar disk at 1AU given in units of steradians.

Converting this expression from CGS units to the metric units used for solar spectral irradiance and solar irradiance (two forms) we get: $S_\odot$ (W/m^2/nm) $= I$ (ergs/cm^2/sec/sr/Å) $\times (\pi R^2/D^2)/100 = I \times 6.8 \times 10^{-7}$, $S_\odot$ (W/m^2) $= I$ (ergs/cm^2/sec/sr) $\times (\pi R^2/D^2)/1000 = I \times 6.8 \times 10^{-8}$, and $S_\odot$ (mW/m^2) $= I$ (ergs/cm^2/sec/sr) $\times (\pi R^2/D^2) \times 1000/1000 = I \times 6.8 \times 10^{-5}$.

Lastly, the radiance is the specific emergent intensity of a particular solar feature at a particular location on the solar disk. For example, when a plage spectrum is measured at disk center, this radiance is part of the contribution to the overall irradiance, which is the sum over the visible solar disk.

Putting this all together gives, for a Sun free of solar features over all wavelengths, a conversion between solar flux and specific intensity using the luminosity from above: I_{free} (ergs/cm^2/sec/sr)$=F_\odot/\pi = 2.011 \times 10^{10}$.

The total solar irradiance, free of solar activity features is then: $S_\odot$ (free) (W/m^2)$= 6.8 \times 10^{10}\, I_{\text{free}} = 6.8 \times 10^{-8} \times 2.011 \times 10^{10} = 1367 W/m^2$.

In practice, the total solar irradiance $S_\odot$ is a combination of the $S_\odot$ (free) and any variations introduced by surface (or internal) features or processes. The solar spectral irradiance is wavelength dependent and is also referred to as the solar spectrum. In turn, the specific intensity is the emergent radiation per wavelength per unit direction from the solar atmosphere, and similarly the spectral flux is the emergent flux.

A model in the present context is defined as a mathematical expression for the solar irradiance as a function of time in some units, the solar spectral irradiance as a function of wavelength at some resolution and time in some units (often at a specific time or state of the Sun). Expressions that go into models range from direct observations, to phenomenological representations, all the way to theoretical terms. This aspect will be discussed in a later section.

1.2. Total Solar Irradiance

Early models were motivated by the existence of ground-based and later space-based observations such those discussed in the other chapters. In the case of the total solar irradiance, it was not until the Nimbus 7/ERB and SMM/ACRIM satellites experiments that a serious effort was put into model construction due to the detection of a variation in the measured radiation coming from the Sun, i.e. not much effort was expended into modeling a quantity that was thought to be *constant*. The measurement of ground-based solar reference spectra, especially those extending over substantial wavelength ranges equivalently generated interest in models of the solar atmosphere but relatively little attention was given to their variability.

In its simplest form the Sun can be modeled as a blackbody with an effective temperature, T_{eff} of $\approx 5770K$. However, as has been known for a long time [Planck, 1901], the 'blanketing' by many lines in the spectrum becomes very important, especially shortward of 430 nm, i.e. the shape is determined by the line overlap. This is represented in Figure 1 which shows the solar spectral irradiance versus wavelength, both in log units and overlaid with the above-mentioned blackbody curve.

Integration under the spectral irradiance curve then gives the total solar irradiance. The blackbody models give $\approx 1680\ Wm^{-2}$ which is an overestimate due to missing absorption features. Thus, the simple model was not sufficient.

The first and most obvious component to a time series model of the solar irradiance is due to sunspots [Willson and Hudson, 1981]. To incorporate the effect of sunspots into a model for solar irradiance in turn required a model of the intensity/radiance/irradiance properties of sunspots – thus the advent of models of specific contributions to models of the solar irradiance.

In the case of sunspots which, to first order, seemed to be of fairly uniform (assumed) contrast relative to the quiet Sun, it was deemed that the position and area of any feature on the solar surface that looked like a spot would be required to develop a model of the amount of blocked emergent radiation. The term Photometric Sunspot Index [PSI; *Hudson et al.*, 1982] is used to characterize this contribution and there have been many efforts to refine this quantity [e.g. *Fröhlich, Pap and Hudson*, 1994; *Brandt, Stix and Weinhardt*, 1994].

Figure 2 [from *Hudson et al.*, 1982] shows one of the first efforts at accounting for what was termed the Sunspot Flux Deficit in a ≈ 130 day timeseries in 1980. The result indicated that even though there was a strong correlation between the sunspot deficit and the dips in the irradiance data that not all of the variations were accounted for.

For the most part, this first order estimate was quite successful. Then it was realized that when the sunspot signal was subtracted from the observed signal, the residual also suggested that another feature was contributing excess radiation. *Chapman* [1987] correctly pointed out that bright facular and plage could account for this excess and *Lawrence et al.* [1985] demonstrated a 'model' of the effect of a passage of a sunspot and accompanying plage across the solar disk in August 1982. The observed irradiance timeseries is shown in Figure 3 along with sunspot, faculae and net (sum) contributions. The term Photometric Facular Index [PFI, or some derivative of this; *Chapman et al.*, 1984; *Chapman and Meyer*, 1986; *Foukal and Lean*, 1986)

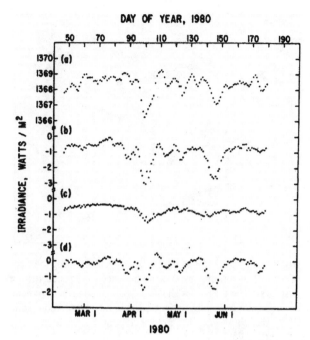

Figure 2. a) daily averages of total irradiance during 1980 from ACRIM, b) the sunspot deficit contribution, c) global deficit after correction, and d) a simple model for reradiation of the spot deficit to account for the residuals.

is now used to characterize the contribution of facular elements.

In the case of faculae and plage, accurate measurements were not routinely available to derive the properties required to form a model. Thus proxies (also known as surrogates or indices) were adopted. In this context, a proxy is a measured or inferred value or values representing the component to be modeled. Common examples are the Calcium 1 Å index [*White and Livingston*, 1981] as a proxy for the plage component (hence its name Calcium Plage Index), the Photometric Facular Index (PFI; derived from 1 nm Ca II K images at San Fernando Observatory) and the Magnesium II core-to-wing ratio also known as the Mg II index [*Heath and Schlesinger*, 1986; *Donnelly et al.*, 1994].

At this stage of irradiance modeling, the explanations were appealingly simple and prompted the question: Could the solar irradiance time series be explained by just (accurately) accounting for spots and faculae/plage – i.e. two well known surface features?

In the course of further modeling and the pursuit of accurately accounting for dark and bright features [in work such as: *Schatten et al.*, 1982; *Oster et al.*, 1982; *Sofia et al.*, 1982] in fact indicated that while the balance seemed promising at first it did not match the observations in a consistent manner over the first full solar cycles worth of observations–

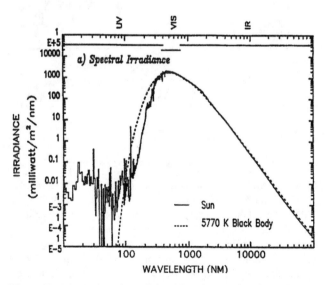

Figure 1. A representation of the solar spectral irradiance versus wavelength along with the blackbody curve with an effective temperature of 5770K, from *Lean* [1991].

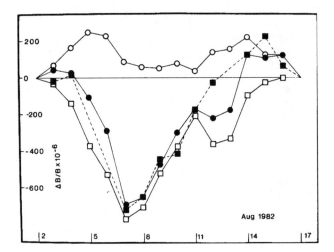

Figure 3. A simple model of total solar irradiance during the transit of a combination of sunspot and plage across the solar disk in August 1982. The timeseries shows the facular contribution (open circles), sunspot deficit (open squares), and net contribution (solid circles). Also shown is a net irradiance contribution from proxy data (solid squares) from *Lawrence et al.* [1985].

from the maximum of cycle 21 (1980) to the maximum of cycle 22 (1991). Around this time, other contributions (based on known solar activity features) were beginning to be considered. *Foukal and Lean* [1988] suggested that the so-called magnetic network may be important. This suggestion spurred research into measuring and characterizing the intensity component of the active magnetic network which continues to the present day [see *Ermolli et al.*, 1998].

Starting with *Fröhlich* [1984] multivariate spectral analysis based on Fourier transform and power spectral distribution methods appeared while *Foukal and Lean* [1988] used multiple regression techniques. Around the maximum of solar cycle 22, i.e. with a little over one cycle worth of satellite irradiance data, a series of models based on proxies was published using these analyses by *Fröhlich and Pap* [1989] for the total solar irradiance. Figure 4 [from *Fröhlich and Pap*, 1989] shows an example of how the contributions of multivariate spectral components in a model of an observed irradiance timeseries can be accounted for in a statistically quantitative way. The analysis also gives information about which component may be most representative on a particular timescale, e.g. the effects of rotational modulation of active regions clearly show a signal at 9 days in such analyses. The majority of these analyses quantified what was already known from simpler models; that at this time, no combination of proxy components could account for the entire irradiance signal, i.e. something was still missing. The summary by *Pap and Wehrli* [1992] of Working Group

1 of the SOLERS22 (SOLar Electromagnetic Radiation study for solar cycle 22) and follow-up papers by *Chapman, Cookson and Dobias* [1996; 1997], and *Pap* [1997] represent the state of understanding at the time and indicate that accurately accounting for sunspots with PSI, and a facular component with, for example, the Mg II index, produces a total irradiance model that accounts for nearly all of the variability (filtered with an 81-day running mean which averages over the rotational modulation and active region signals) during cycle 22. In these follow up studies a series of proxies appeared which started to separate the known solar activity timescales and thus construct proxies which either represented, or excluded, those timescales.

Recently, wavelet techniques have been compared to the usual Fourier methods and show some promise for further identifying physical couplings between different layers in the solar atmosphere based on phase shifts identified in the timeseries [*Vigouroux, Pap and Delache*, 1997] in an even more quantitative way that previous regression analyses.

Some of the most promising results in total irradiance

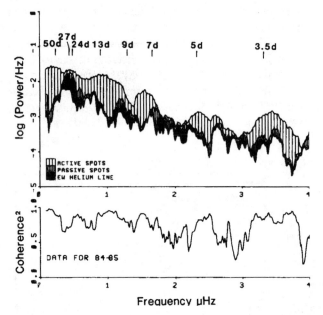

Figure 4. Multivariate spectral analysis from *Fröhlich and Pap* [1989] for a timeseries during solar maximum around 1980 of the SMM/ACRIM data. Upper panel shows the log of the power spectra with frequency in μHz (period in days marked on the top axis) for active spots (broad vertical line shading), passive spots (narrow vertical line shading), and the He I 1083 nm equivalent width. The lower panel shows the square coherence which, for each frequency (period) indicates how much of the observed power spectrum is accounted for by these components based on how close to 1.0 they rise.

modeling are due to efforts at the San Fernando Observatory. The first step in improving these models came from *Walton and Preminger* [1999] which combined some of the early simplicity of an 'effective-temperature' model discussed earlier with spatially derived temperature measurements from the San Fernando Observatory Cartesian Full Disk Telescope (CFDT) red continuum images. This model known as TIRR is given by

$$TIRR = \sum_{allpixels} \left[\frac{T_i}{T(\mu_i)} \right]^4 - 1, \tag{1}$$

where T_i is the temperature of the i-th pixel from the red continuum images (wide band centered on 672.3 nm) and $T(\mu_i)$ is the quiet Sun photospheric temperature as a function of the cosine of the heliocentric viewing angle, μ_i also for the i-th pixel. *Walton and Preminger* [1999] have applied this model to a set of restored CFDT images for particular time series with good results. Following this work, TIRR was replaced in favor of direct indices based on bright and dark regions in the SFO red band images and their 1 nm Ca II K images. The total irradiance model is derived from summing these over the full disk [*Walton, Chapman and Preminger*, private communication].

Given the relative success for proxy modeling of the total solar irradiance for solar cycle 22, solar cycle 23 is providing a test of those models. *DeToma et al.* [2001] suggest that the application of existing models and proxies (e.g. Mg II index and TIRR) to the ascending phase of cycle 23 does not reproduce the observed time series as well and that adjustments are suggested some of the coefficients in the proxy model (e.g. the facular term as shown in Figure 5). The main deductions from these studies that impact models are the presence of a phase shift between the rise of solar activity (either from analysis of magnetograms or indices) and that of the irradiance time series. In addition discrepancies around the solar maxima where solar activity tends to plateau (e.g. in the F10.7 radio flux) and the irradiance time series does not [*Pap*, 1997] point to missing 'physics' or rather in the proxy models, the inability of existing proxies to represent a physical process.

In summary, models of the total solar irradiance became possible as consistent time-series observations came along, and were improved or discarded as errors and uncertainties in the representation of their components went down. At present, no single proxy model accounts for all, or even most, of the existing timeseries on timescales longer than one rotation. The best model fits give about ±0.5 Wm^{-2} peak to peak, or a standard deviation of ~ 0.15 Wm^{-2}.

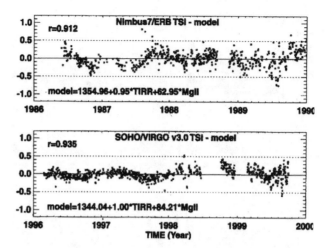

Figure 5. Empirical models of the total solar irradiance for the ascending phases of solar cycle 22 (upper panel) and 23 (lower panel) using the TIRR and Mg II index from *DeToma et al.* [2001].

1.3. EUV/UV

In the same way that satellite observations of the total solar irradiance prompted research into models, the Atmospheric Explorer series, particularly E (AE-E) satellite motivated a substantial number of empirical models and for many years remained the only source of EUV irradiance time series.

While the work on the total irradiance proceeded there were parallel model efforts for specific spectral regions, those in the EUV/UV in particular were motivated by their importance in studies of the terrestrial upper atmosphere and fueled by satellite missions and reference spectra measurements, both discussed in other chapters.

Based on the early work of *Hinteregger et al.* [1981] that used different proxies for parts of the EUV spectrum, in particular Lyman-β and the F10.7cm radio flux [*Tapping and Harvey*, 1994], the models that followed all used some combination of F10.7, Lyman-α and observationally determined EUV fluxes [*Nusinov*, 1984; *Fennelly et al.*, 1994; *Richards, Fennelly and Torr*, 1994; *Worden*, 1997; *Tobiska and Eparvier*, 1998; *Worden et al.*, 1999] for characterizing binned EUV fluxes. Figure 6 shows an example of the spectrally dependent spectrum with activity from the EUV97 model of *Tobiska and Eparvier* [1998].

This succession of models continues to the present day and the reader is referred to further details of these models and intercomparisons reviewed by *Lean* [1990], *Tobiska* [1993], *Tobiska and Eparvier* [1998] and *Tobiska et al.* [2000].

Due to the increased magnitude of variability in the EUV/UV spectrum with solar activity a separation between

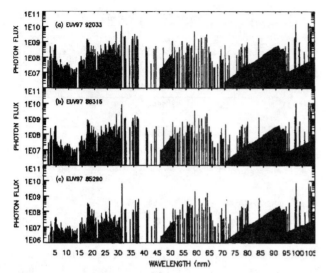

Figure 6. EUV flux spectrum (photons cm^{-2}s$^{-1}\Delta\lambda^{-1}$) from 0 to 110 nm on the SC#21REFW wavelength grid for three levels of solar activity (high: upper panel, moderate: middle panel, low: lower panel).

timescales of proxy components in the models is often required. For example, several activity proxies are sampled on a daily or daily-averaged cadence, or smoothed with 27-day or 81-day running means to filter out shorter term activity and even rotational modulation effects. The consequence of this is that some models may contain two terms based on the same proxy but with different time sampling. While some of this is based on observed phenomenology of solar activity, the remainder is often based on trial and error.

Lean et al. [1997] presented a set of multiple regression-based proxy models for the mid and near UV (200–400 nm). In this work they account for both sunspot darkening and faculae brightening (based on the He I 1083 nm EW) and Figure 7 shows a set of 16 year reconstructed time series for UV irradiance in 50 nm bands between 200 and 400 nm (top to second from bottom) as well as the total irradiance (bottom). Figure 8 shows the modeled spectral irradiance in the 200–300 nm (left top) and 300–400 nm (right top) wavelength regions (with specific spectral features marked) along with an estimate of the solar activity maximum to solar minimum variability (lower panels). These variability estimates indicate, at the 1 nm wavelength bin size, what the range of irradiance change is expected, i.e. ranging from 8–9% around 200 nm to 0.5–1.0% at longer wavelengths.

Figure 9 [*Floyd et al.*, 1998] shows a comparison of measured and modeled UV spectral irradiance solar cycle 22 (minimum to maximum) variability in three wavelength regions: 120 to 160 nm over which the variability varies from about 100% down to ~ 20%, 140 to 220 nm over

which the variability goes from ~ 20% down to ~ 5% and 220 to 400 nm, indicating that the variability drops for ~ 5% down to almost zero, and even some negative (of order 1–2 %) variability is apparent. The models use the He I 1083 nm EW and Mg II index proxies, and give comparable predictions. They also provide a short discussion of the relationship of parts of the UV spectrum to empirical solar atmosphere models.

A presentation of the current status of spectral indices/proxies from existing datasets and other time series incorporated into these models can be found, along with specific discussion and comparisons of proxies, particularly the Mg II index for UV irradiances can be found in *Cebula and Deland* [1998], *Floyd et al.* [1998], *Parker, Ulrich and Pap* [1998] and *Viereck et al.* [2001].

Multiple regression models for Ly-α started to appear with the work of *Pap et al.* [1990; 1991] which set the stage for later models. *Woods et al.* [2000] present the latest efforts on combining proxies and different time series in

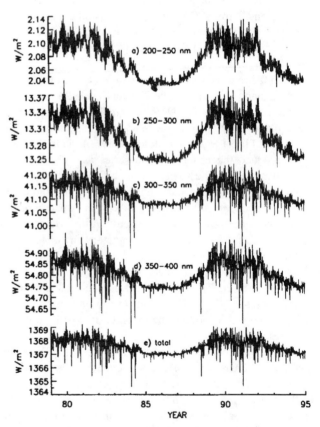

Figure 7. 16 year reconstructed time series for UV irradiance in 50 nm bands between 200 and 400 nm (top to second from bottom) as well as the total irradiance (bottom) from *Lean et al.* [1997].

building a model for Ly-α irradiance spanning over 50 years (see Figure 10). This model makes use of proxies (He I 1083 nm equivalent width, Mg II index, and F10.7 cm radio flux) *both* in their daily time series and smoothed with an 81-day running mean. Their equation (1) and ensuing discussion of the variation of the determined coefficients highlight the present difficulties in proxy-based models. However, in the case of Lyman-α they determine that a combined proxy model with an uncertainty of 10% is sufficient for modeling long term solar variability, i.e. when compared to the cycle-to-cycle measured Lyman-α variations. They conclude that the nature of the formation of the Lyman-α in the upper chromosphere and transition region is not well represented by either coronal or strictly chromospheric proxies (for example Mg II index) and that a proxy in the transition region itself is required. They also provide a discussion of the complexities of identifying both the areas and contrasts of surface structures used in models from at least two different methods. This result highlights the basic problem of the increasing inhomogeneity of the solar atmosphere with height.

The culmination of a series of community research efforts, together with motivation from international efforts to coordinate advancements in the measurement and understanding of solar irradiance and variability (such as SOLERS22 and ISCS) has led *Tobiska et al.* [2000] to present the SOLAR2000 empirical solar irradiance model and forecast tool. This paper also includes some history of the development of EUV/UV/FUV empirical models as noted above.

SOLAR2000 provides a comparative historical database of both the solar reference spectra and its variability on a variety of timescales incorporating over two decades of spaced-based and ground-based measurements as well as a variety of evolving empirical and first principles models (to be described in this section).

SOLAR2000 in principle covers 1 nm to 1 mm in sufficient spectral and time resolutions to be used in the commercial-government-military application areas on time-scales of minutes to weeks as well as for education/outreach and research applications on timescales of days to multiple solar cycles. The accuracy of this model varies with wavelength with the most attention being given to the EUV and UV, and less attention to the visible and infrared. To date there have been several model releases and a series of updates and upgrades to improve the accuracy of the model are planned well into the next decade. Of particular note is the alignment of this effort with the developing International Standards Organization (ISO) solar irradiance process standard [ISO TC20/SC14/WG4; *Tobiska and Nusinov*, private communication].

Despite the widespread use of empirical and proxy-based models there has been a clear need to understand the proxy results, identify the physical processes at work and further to develop, i.e. not guess, new proxies for parts of the spectrum such as the Al edge, and for entire 200–300 nm [*Cebula*, 1992].

In response to this motivation and community need, a new proxy, E10.7, has emerged. This is a synthetic proxy in the sense that it is derived from a combination of integrated solar EUV energy fluxes from the EUV97 [*Tobiska and Eparvier*, 1998] empirical model which was found to be highly correlated with the EUV energy heating at the top of the Earth's atmosphere [*Tobiska*, 1988] and much easier to calculate. Since it uses the same units as the F10.7 cm proxy, it is very straightforward to include it into existing thermospheric and ionospheric models. Further details and some observational comparisons are presented in *Tobiska et al.* [2000] and Figure 11 shows a comparison of the E10.7 and F10.7 timeseries in 1982. The emergence of such a proxy sets the stage for similar constructions in other parts of the solar spectrum.

Another source of uncertainty in models that utilize proxies based in spectrally resolved lines is that of the measuring spectral resolution. *White et al.* [1998] discuss this topic for the Mg II index for time series from two instruments (SS-BUV and UARS/SOLSTICE) with 1.1 to 1.15 nm compared to 0.2 to 0.25 nm resolution. They find that the measurement with higher resolution leads to a Mg II index (normalized to the solar minimum) that is $\approx$ 2 times the one derived from the measurement with lower resolution. This result is found over both the half solar rotation, solar rotational period, and solar cycle timescales.

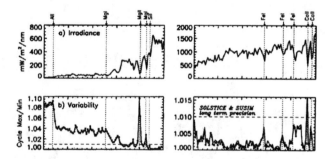

Figure 8. Upper panels: proxy models for UV spectral irradiance in the 200–300 nm (left) and 300–400 nm (right) wavelength regions in 1 nm bins. Specific spectral features are marked on the top axis. Lower panels: solar activity maximum (Nov. 1989) to solar minimum (Sep. 1986) irradiance variability spectrum for same wavelength ranges. From *Lean et al.* [1997].

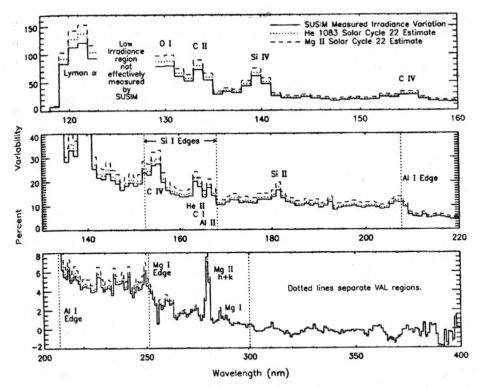

Figure 9. Percentage variability UV spectrum from 120 to 400 nm showing a combination of measurements and empirical proxy models (using He 1083 nm EW and Mg II index) for solar cycle 22 from *Floyd et al.* [1998].

Another set of proxies that may depend on spectral resolution are those based on equivalent width of a spectral line; the most commonly reported are those for the He I 1083 nm and C I 538 nm. The Helium line has been used as a proxy for upper chromospheric activity for some time but to date, no studies of the effects of spectral resolution or other measurement uncertainties have been reported.

Filter widths have similar considerations, and may identify different surface structures due to the sampling of different combination of heights in the solar atmosphere and may lead to different areas and contrasts. This effect is most well known in the comparison of filter-based CCD images in the Ca II K line with filter widths between 0.1 and 1.0 nm (many being around 0.3 nm) compared to Ca II K spectroheliograms (NSO/SP) which sample at 0.05 nm. These matters are discussed further in *Harvey and White* [1999a] for example who highlighted the relation of solar activity to irradiance proxies using the NSO/KP spectromagnetograph data and NSO/SP Ca K spectroheliograms and BBSO Ca K filtergrams.

Present proxies—both generated and used—for the EUV and UV spectrum include: Fe XIII, He I (58.4 nm) and He II (30.4 nm), which are presently measured using SOHO/EIT, He I (1083 nm) equivalent width which has been measured at the National Solar Observatory [Kitt Peak; *Harvey*, 1984], Ly-α (121.5 nm) measured from a variety of instruments including AE-E, SME, and UARS [*Woods et al.*, 2000], various measures using Ca II K (393.3

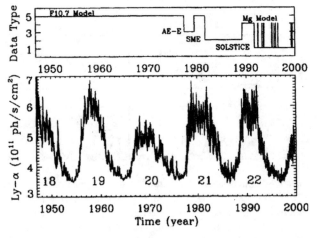

Figure 10. Composite Lyman-α timeseries (lower panel) and proxy and model coverage (upper panel) from *Woods et al.* [2000].

nm; both in narrow band – 0.1 nm, broad band – 0.2– 0.3 nm up to 1 nm, and as a 0.1 nm index) have been measured at a variety of observatories: e.g. Mt Wilson Solar Observatory, *Foukal* [1996]; Arcetri Solar Tower, *Godoli* [1961]; NSO/Sacramento Peak, *White and Livingston* [1981]; Big Bear Solar Observatory, *Johannesson, Marquette and Zirin* [1998]; the San Fernando Observatory, *Chapman, Cookson and Dobias* [1997]; the Precision Solar Photometric Telescope in Rome, *Ermolli et al.* [1998], and Hawaii, *Rast et al.* [1999], F10.7 cm radio flux, *Tapping and Charrois* [1994], E10.7 synthetic/ semi-empirical proxy, *Tobiska et al.* [2000], and the Mg II index, *Heath and Schlesinger* [1986]; *Donnelly et al.* [1994]; *Viereck et al.* [2001]. For reference, the latest composite timeseries for irradiances and a number of these proxies is shown in Figure 12 and 13.

Efforts at developing more theoretical models, based on the emission measure formulation started to appear around 1980 from NRL [*Cook, Brueckner and VanHoosier* 1980] for the optically thin far EUV (117.5–210.0 nm) spectral lines that constructed a model of the emergent flux in a two component model (plage and quiet sun and ignoring sunspots). This work introduced many of the concepts that were to become part of later models; determination of areas of the components, determination of contrasts and the assumption that those individual contrasts did not vary over the solar cycle, or other timescales of interest. Even though this work did not address the spatial distribution of the components (except in restricting the flux computation in its latitude range) it did note the possible importance of center-to-

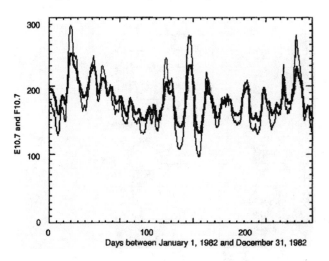

Figure 11. E10.7 (thick line) and F10.7 (thin line) timeseries measured in units of $10^{-22} Wm^2 Hz^{-1}$ during 1982 from *Tobiska et al.* [2000]. Note how F10.7 is more variable (i.e. over and underestimates) than E10.7, which is computed to represent the total EUV energy deposition for the Earth's atmosphere.

limb variations in intensities at different wavelengths and the ability of activity proxies (like the Ca K plage index) to properly represent areas in the mid to upper chromosphere where much of the far UV spectrum is formed.

Cook, et al. [1994] extended this work by including the spatial distribution of solar features using full disk magnetograms and compared the model results to EUV irradiance (117.5–210.0 nm) measurements from UARS/SUSIM.

This work has most recently been continued and expanded by *Warren, Mariska and Lean* [1996; 1998a; 1998b] especially to include spectrally *and* spatially resolved EUV radiance observations to calculate quiet Sun irradiances at wavelengths of interest. *Warren, Mariska and Lean* [2001] compute the variability for solar EUV irradiance also using the emission measure formulation and empirical relations (proxies) for longer wavelengths greater than 120 nm. Further details and recent results of this work appear later in this Chapter.

1.4. Visible And Infrared

In the late 1980's the work on stellar atmospheres (e.g. Kurucz's ATLAS code; *Kurucz*, 1970] started to be applied to model the solar visible and infrared spectrum. For this application more realistic solar atmosphere models were required (i.e. beyond radiative equilibrium) but were still restricted to the quiet Sun (VAL model C and later FAL model C). Around the same time, Avrett and colleagues [VAL series; *Vernazza, Avrett and Loeser*, 1981; and FAL series; *Fontenla, Avrett and Loeser*, 1990; 1991; 1993] were continuing to develop and publish semi-empirical atmosphere models for different solar features (quiet Sun, network, faculae, and plage). In conjunction with this work was the development of the PANDORA program and similar ones, which gave a small set of detailed multi-level, non-LTE spectral line calculations [*Avrett*, 1990; *Avrett and Loeser*, 1992].

Figure 14 [*Avrett*, 1998] shows the temperature and density profiles corresponding to the quiet Sun empirical atmosphere model from the photosphere to the low corona. The formation heights are shown for a variety of important solar spectral features (Ca II K, Mg II, Ly-α, He I 1083 nm, etc.), many of which are used as irradiance proxies, and the others are important irradiance indicators.

As in the case of total irradiance and EUV/UV irradiance studies, the existence of observational datasets to provide the context for the development of models is equally true for the visible and IR parts of the spectrum. It wasn't until the mid-1980's that reliable and well calibrated spectra became available for this part of the spectrum [e.g. *Neckel and Labs*,

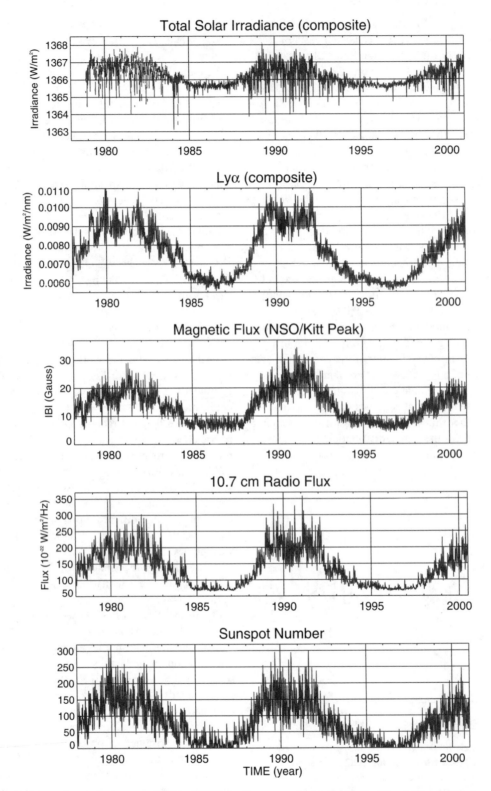

Figure 12. Proxy timeseries between 1978 and 2001 for the composite total solar irradiance (upper), Ly-α composite, NSO/KP magnetic flux, F10.7 cm radio flux, and the sunspot number (lower). Courtesy of G. DeToma.

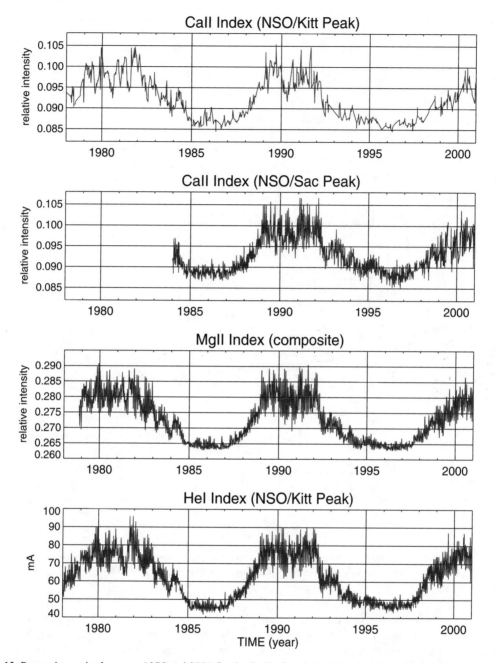

Figure 13. Proxy timeseries between 1975 and 2001 for the Ca K plage index from NSO/Kitt Peak (upper), Ca K plage index from NSO/Sacramento Peak, the composite Mg II index and the He I index from NSO/Kitt Peak (lower). Courtesy of G. DeToma.

1984; *Kurucz, Furenlid, Brault and Testerman*, 1984; *Beckers et al.*, 1976]. It was also not until the early 1990's that regular ground based irradiance observing programs were established (e.g. at the San Fernando Observatory) which measured banded irradiance of parts of the visible continuum. By the mid-1990's programs such the ground-based NSF RISE sponsored Precision Solar Photometric Telescope and the space-based VIRGO/SPM [*Fröhlich et al.*, 1997] mission started to collect more precise visible and near infrared irradiances on a regular basis. The availability of these datasets has provided an invaluable motivation for modeling this part of the solar spectrum.

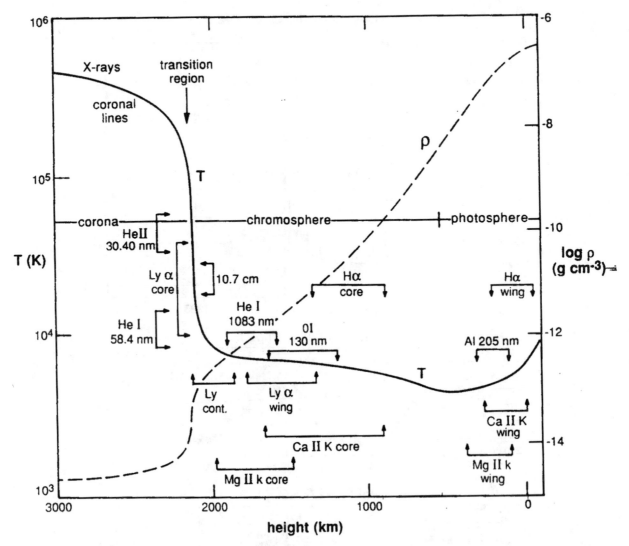

Figure 14. Temperature and density with height relative to the unit optical depth at 500 nm for the quiet Sun. The formation heights are shown for a variety of important solar spectral features, many of which are used as irradiance proxies, and the others are important irradiance indicators from *Avrett* [1998].

Around 1991, Kurucz performed the first LTE calculation of the full spectrum using a quiet sun model normalized so that the integrated area matched the then measured value of the total irradiance (1370 Wm^{-2}). The model was compared to the *Neckel and Labs* [1984] spectrum and demonstrated very good general agreement with the shape and for many features in the spectrum. At the level of detail of the observed total and spectral irradiance however, the model still had many improvements ahead [*Pap and Wehrli*, 1992]. What was absent at this time was any systematic theoretical investigation of the irradiance variability in the visible and IR.

These two pieces of work led to the realization in the early 1990's that it would now be possible to compute the

solar variability (intensity) spectrum much more realistically by combining the stellar atmospheres approach with some improvements for non-LTE, the best and most complete available atomic data, a set of semi-empirical solar atmosphere models, and a way to estimate the coverage of specific features on the solar surface. Early results of this work were given by *Fox et al.* [1994], *Fontenla et al.* [1995] and *Fox et al.* [1996; 1998]. This effort was funded by the NSF RISE program and originated at the SOLERS22 meeting in 1991 [*Pap and Wehrli*, 1992].

Avrett [1998] summarized the state of modeling around the mid-1990's noting that the calculations could be used to fill in the gaps between well calibrated observed parts of the spectrum and highlighted the importance of efforts to improve the radiances for specific solar features using very high resolution (spatial and spectral where possible) observations and concluded that the stage was set for improved irradiance determinations relevant to the terrestrial atmosphere.

The present state of this effort, the SunRISE Spectral Irradiance Synthesis model, is presented by *Fontenla et al.* [1999] and *Fox et al.* [2002]. Further details and discussion of recent results are presented later in this Chapter.

In the mid to late 1990's, another effort was started by Solanki and colleagues [*Solanki and Unruh*, 1998; *Solanki and Fligge*, 1998] aimed at covering a broad range of wavelengths with a simpler synthesis model which utilizes the Kurucz flux spectra calculation and a limited set of solar features to model the solar spectrum and its variability. This work has covered reconstructions of irradiance time series back to the 1700's [*Solanki and Fligge*, 1999; *Fligge and Solanki*, 2000], simulation of VIRGO/SPM irradiance time series [*Fligge et al.*, 1998a; 1998b], studies of facular contrast [*Unruh, Solanki and Fligge*, 1999], studies of line ratios [*Unruh, Solanki and Fligge*, 2000], and the use of magnetograms to identify regions of irradiance variability [*Fligge, Solanki and Unruh*, 2000]. Much of the present work is reviewed by *Solanki* [2001], *Unruh and Solanki* [2001] and *Solanki, Fligge, and Unruh* [2001]. Further details and discussion of recent results are presented later in this Chapter.

The development of these physics-based models is still hampered by the lack of a constraining and consistent set of spectral variability measurements. With about 20 years of modeling experience in this area of solar physics, it is safe to say that a lot is known about the general character of the solar spectrum and even some specific parts are quite well modeled. However, the details of the variability with wavelength and particularly the physical understanding of the nature and origins of those changes is only partially formed. The remainder of this Chapter explores some of the key areas in irradiance modeling, identifies similarities and differences in approaches, gives examples of current results and lays out some suggestions for future work.

2. RELATIONSHIP TO SOLAR ACTIVITY

Driven by the success of a variety of models of both total and spectral irradiance, many manifestations of solar activity have been explored as contributors to solar irradiance variations. As previously mentioned, these contributions themselves require 'models'–sometimes these models are proxy measurements and in other cases they are physical

approximations to the feature observed on the solar surface. This section briefly discusses the present state of understanding of how surface solar activity and solar irradiance are linked (the issue of interior and other effects of solar activity on the radiative output of the Sun is discussed in [*Kuhn and Armstrong*, and *Sofia and Li*, this volume]).

The surface manifestations of solar activity that may contribute to radiative variability include active regions and their remnants as well as those more globally distributed such as network, faculae, coronal holes, and so on.

To assess the effect that different surface structures may have on the continuum contribution to the irradiance Figure 15 [from *Fontenla et al.*, 1999] shows the radiation temperatures for the quiet Sun, plage and sunspot with wavelength. Of note are the departures from the blackbody effective temperature of 5778K as well as the changing nature of the sign and magnitude between these different structures for different parts of the spectrum. It is to be noted that the opacity of the many absorption lines changes the shape of the spectrum below 450 nm (log(wavelength)=−0.35) and reduces the brightness temperatures and their relative contrasts below the values shown here.

Contributions of the major components of sunspots, plage and the magnetic network and quiet Sun differ depending on their location on the solar disk–sunspots over most wavelengths have maximum effect near disk center and minimum effect near the limb, plage tend to have the opposite behavior but the wavelength dependence of this effect, including almost zero contrast in some places, is more complicated, while the active network tends to concentrate in active solar latitudes and the quiet Sun is assumed only to vary with the particular limb darkening function appropriate for the wavelength/bandpass of interest.

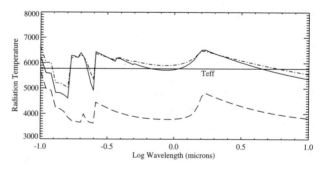

Figure 15. Radiation temperatures from continuum sources for the quiet Sun (solid line), plage (short dash line) and sunspot (long dash line) model atmospheres versus log wavelength (100 nm to 10 μm). The effective temperature (5778K) of the Sun treated as a blackbody is also labeled.

Observationally, each of these features has a size and intensity (contrast) distribution although it is common in irradiance modeling to represent these features by a single contrast and not account for the details of the projected area coverage. The level of both of these assumptions has an impact in the precision and accuracy to which models can reproduce detailed irradiance measurements. One of the features of recent empirical proxy models is that the actual size and contrast distributions are incorporated as distinct from using a PSI or PFI [*Walton and Preminger*, 1999].

One of the important conclusions from this discussion is that the effects of these solar activity features vary significantly across the solar spectrum and the conventional representation of some activity proxies by values such as a linear relationship to sunspot numbers, etc. (basically any monotonic relationship) can lead to inaccurate, possibly in sign or magnitude or both, irradiance contribution estimates. It is in this area that physics-based models hold promise for a more correct account of these structures.

Turning now to the inter-relation between solar activity and irradiance *DeToma et al.* [2001], building on the work of *Harvey* [1994], *White* [1994] and *Harvey and White* [1999b], report a greater increase in total solar irradiance for the rise of solar cycle 23 and at a comparable level to that of cycle 22, than the present calibrations for proxy models based on solar activity would predict. Further, the presence of a time delay between increases in the irradiance and the total magnetic flux prompts for a physical explanation which is presently beyond the capability of the proxy models and suggests that the spatial distribution of the magnetic flux is more important for variability than its magnitude. The stage is set for those models with other direct, or better, representations of the radiance of particular features of solar activity, i.e. spots, plage, quiet and enhanced network to address these observations.

The connection of activity to irradiance variability on timescales that are common, and thus of interest, to both highlights the relatively primitive state of theoretical understanding of their interrelation. On one hand, while the specific structures representing solar activity have been well studied, when we consider timescales longer than active region emergence, or their nature over any substantial fraction of the solar disk the Sun, it is only the solar activity indices that come close to capturing the nature of the solar cycle. On the other hand, it is apparent that using activity indices or proxies to represent the radiative variation also fails to capture sufficient reality of the radiative variations. In particular, when a model requires the decomposition of the many forms of solar activity into discrete types, with modeled radiances which do not vary over the cycle, again some degree of accuracy is also lost.

A good example of this is the comparison of the total solar irradiance between successive solar minimum conditions: a recent topic of interest [see the discussions in *Fröhlich and Lean*, 1998; *Fröhlich*, 1998; *Pap and Fröhlich*, 1999; *Willson*, 1999; *Willson and Mordinov*, 1999; *DeToma et al.*, 2001]. In the very specific case where there is 'no' solar activity which one may assume may occur at each solar minima, ≈ 11 years apart, then the irradiance predicted by both proxy and physical models would be the same for each minima. While this specific case has not been studied in any detail as yet, it does open the question of how models are developed and for what purposes they may and may not be used. A very similar question arises for successive solar maxima [*DeToma et al.*, 2000; *Pap et al.*, 2002].

Another example occurs once particular solar activity features are identified. Each of the models assumes that the location, coverage and distribution from that identification will also apply at the actual height, or range of heights, of formation of the spectral line or irradiance of interest being modeled. This example occurs frequently in the UV and EUV where proxies are identified in the low chromosphere and applied in the mid to upper chromosphere or transition region (see the discussion in *Woods et al.* [2000]). It also emphasizes that consideration of or, better yet, an understanding of the underlying magnetic-thermal structure present in different layers of the solar atmosphere (especially the transition from high-β to low-β) is very important in developing variability models based on surface activity.

At least one desirable use for models within this context is to assist in isolating contributions to changes in solar irradiance (total and perhaps even in spectral bands) between strictly surface activity features, and those that have been postulated in the quiet sun and in the solar interior, at a variety of depths. For example, it is plausible that the topic of relative timescale differences between activity and radiative variability timeseries and onset of either of them at cycle minima is best posed not just as one of surface variability [see *Sofia and Li*, and *Kuhn and Armstrong*, this volume].

3. COMMON METHODOLOGY IN MODELS

Building on the discussion from the previous section, it is useful to characterize how solar irradiance models adopt the same or similar concepts of the influence of solar activity before discussing their specific differences.

All models separate the solar irradiance (total or spectral) signal into components (also known as structures or terms). Since the early 1980's these came from two places: available time series and correlative observations, e.g. white light continuum, magnetograms, etc. There is a basic acceptance that a contribution (ranging all the way from very significant to modest) to the irradiance variability is due to magnetic fields. Thus, the typical solar surface features that models include and may include more than one type are: average quiet sun, network component for quiet and active sun, facular component, plage component for the active sun, sunspots, and even coronal holes, filaments, loops, and network bright points.

These features are determined or derived from a variety of sources: intensity images (e.g. BBSO CaIIK, SFO CaIIK/Red, EIT) are most often used to identify sunspots, faculae/plage and bright network, magnetograms (e.g. NSO/KP, SOHO/MDI) provide complementary information on the overall magnetic network, active regions in general and their decaying remnants. To date resolved spectroscopic observations, especially over large fractions of the solar surface, have not been used to identify features for irradiance models. Some features are identified by proxies, two notable examples are the Ca K plage index, and the Mg II index. It is often necessary to use a combination of techniques to correctly identify some features. For example, while magnetograms measure active region magnetic fields they do not necessarily accurately identify a sunspot well enough for use in irradiance studies [see the discussion in *Harvey and White*, 1999a]. Finally, there is no general consensus on the precise definition of these features within the context of irradiance models although several authors have proposed working definitions (see for example: *Harvey and White* [1999a], and *Woods et al.* [2000]). A full discussion of these issues is beyond the scope of this Chapter.

Thus, all models use some scheme (ranging from simple to sophisticated) to identify surface components that contribute differently to irradiance/radiance for that component. Once identified, a model may then associate different proxies with them, or use more physical (often semi-empirical) models of the solar atmosphere. Some models treat the Sun as a star, or account for the detailed center to limb variations. Finally, the method to determine the weighting of these components may use time series (correlation, power spectrum, etc.) analyses to calibrate the coefficients, or use detailed spectral synthesis where the calculations may be in absolute units.

Each of these model considerations introduces specific limitations to their efficacy. While the origins of a general distribution of solar surface components go back to the Harvard classification, which represents bins of increasing intensity in a distribution of an intensity image using letters, originally labeled L, M and S in *Vernazza, Avrett and Loeser* [1976], and A through F in *Vernazza, Avrett and Loeser* [1981], but more recently A, C, E, F, H, P, and S discussed

in *Fontenla, Avrett and Loeser* [1993; 2002], and *Skumanich, Smythe, and Frazier* [1975], efforts to fully quantify the location and coverage of even the best known features (sunspots and plage) are still subjective. For example, the way to account for sunspot umbra and penumbra is not settled and plage may be identified quite differently in continuum/white light compared to specific parts of the spectrum, e.g. infrared. While the determinations themselves can be internally consistent, it is in their application to combining radiance or irradiance calculations that the uncertainty arises. For example, how well does the mapping of 'network elements' from a photospheric magnetogram allow a particular model to accurately predict the Lyman-α irradiance contribution which is formed much higher in the solar atmosphere? These are extremely important research topics related to irradiance modeling.

Fortunately, there is much work underway in this area. The reader is referred to recent work by *Harvey and White* [1996; 1999a] in the area of magnetograms and calcium images, *Turmon, Pap and Mukhtar* [2002] for analysis of SOHO images, *Preminger, Walton and Chapman* [2001] for analysis of San Fernando Observatory images, and *Ermolli et al.* [1998] and *Berrilli et al.* [1999] on identifying the details of intensity network contributions to the irradiance and from *Worden, White and Woods* [1998] in relation to the challenges in modeling the EUV irradiance.

An important study of the correspondence of magnetic fields and UV radiance formed near or within the temperature minimum-UV continuum at 160 nm was performed by *Cook and Ewing* [1990]. They developed a linear relation between the field strength (0–200 Gauss) and the brightness temperature (4300–5800 K) in fine-scale bright elements (about 2% of the total area though still significant) and discuss possible physical mechanisms at work in this region of the solar atmosphere. Further work on the magnetic origin of significant brightness fluctuations in the solar chromosphere are needed. All of these are examples of where the research challenges are.

In the case of models which incorporate some physical representation of the surface features there is widespread use of a set of semi-empirical solar atmosphere models that originated from the work of *Vernazza, Avrett and Loeser* [1981], *Fontenla, Avrett and Loeser* [1990; 1991; 1993; 2002], and *Fontenla et al.* [1999]. These atmosphere models provide height dependent values for temperature, hydrogen and electron density, turbulent broadening parameters, abundances, and departure coefficients. Figure 16 [from *Fontenla et al.*, 1999] shows a set of the latest semi-empirical atmosphere models which are constructed to reproduce observed emergent intensities and profiles at wavelengths from the UV to

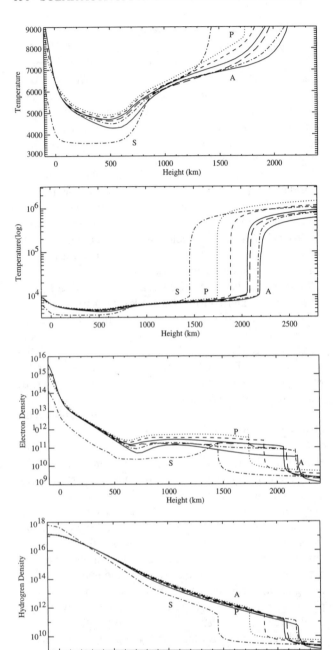

Figure 16. Height dependence of temperature (upper: photosphere and mid-chromosphere), and log temperature (through the transition region and low corona), hydrogen and electron (lower) density for the semi-empirical models of *Fontenla et al.* [1999].

radio wavelengths. Thus, the intensities computed from these models are expected give reasonable absolute intensities and good irradiance estimates. Further details on how the models are constructed and what radiance observations they incorpo-

rated are provided in *Fontenla et al.* [1999] and *Fontenla, Avrett and Loeser* [2002].

4. TYPES OF MODELS

In this section we present details on the types of models, where they may and may not apply, and their characteristics.

4.1. Proxy Models

The first and earliest class of models may be categorized as proxy models and appear in a few different forms.

The single parameter models (activity or irradiance measures) take the form:

$$I(\lambda,t) = I_0 + b \times X(\lambda_0,t), \qquad (2)$$

where $I(\lambda)$ is intensity (or irradiance) as a function of, or at a particular, wavelength λ, I_0 is a constant (e.g. continuum without activity), $X(\lambda_0)$ is the proxy at, or representing, another wavelength λ_0 and b is the proxy coefficient, i.e. a free parameter. Note that these proxies can be related to structures, other measured quantities or just combinations of proxies that seem to give good statistical fits.

Multiple parameter models (also known as multiple regression) in general take the form:

$$I(\lambda) = I_0 + b_0 \times X_0(\lambda_0) + b_1 \times X_1(\lambda_1) + b_2 \times X_2(\lambda_2), \quad (3)$$

where the terms have similar definitions as in the single proxy model and the subscripts 1, 2, etc. denote the different components, wavelengths, proxies, etc. each with its own coefficient.

As work on these models progressed it was realized that these expressions may not be strictly linear but that, in the above equation for example, b_2 could be a function of activity X_2 (or even X_1) [e.g. *Vrsnak et al.*, 1991; *Pap*, 1992] or that a particular term may have an exponent.

As an example, the SOLAR2000 models [given by *Tobiska et al.*, 2000; their Eqs. 1 and 2] includes a representation of two different timescales of the proxies:

$$I(\lambda,k,t)/I_0(\lambda,k,t) = F(\lambda,k=2,t)_{\text{corona}} + \\ F(\lambda,k=1,t)_{\text{chrom.}} + F(\lambda,k=0,t)_{\text{phot.}}, \qquad (4)$$

$$I(\lambda,k,t) = \sum_{k=1}^{3}[a_0(\lambda,k) + a_1(\lambda,k)P_{81}(k,t)^z + \\ a_2(\lambda,k)(P(k,t) - P_{81}(k,t))], \qquad (5)$$

where a_0, a_1, and a_2 are the proxy coefficients, P is the daily proxy value and P_{81} is the 81-day running mean of the proxy.

4.2. Physical Models

In the early 1990's models which included some physical representation of the components and/or some theoretical basis started to appear. These models take a few different forms but all aim to calculate the solar spectrum in detail using radiative transfer calculations.

These calculations became possible due to a number of research efforts: The development of the emission measure formulation by *Cook et al.* [1980; et seq.], the first full spectrum calculation in LTE by *Kurucz* [1991] and the publication of atomic data on CDROM [*Kurucz*, 1993]. This was accompanied by detailed non-LTE calculations from *Avrett* [1990] and *Avrett and Loeser* [1992], *Anderson and Athay* [1989], and the publication of significantly better semi-empirical solar atmosphere models [*Fontenla, Avrett and Loeser*, 1993].

The overall theme of this class of models is directed toward detailed spectral synthesis. Particular choices that are made in building these models are: LTE or NLTE radiative transfer, radiative equilibrium or empirical atmosphere models, two-level and/or multi-level atoms, LTE or NLTE ionization, binned opacities or continuum and line-by-line opacity contributions, fixed or variable spectral resolution, calculation of flux or intensity. Each of the models that follow however addresses these choices in a different manner.

Two of the notable choices that are made are that some models choose to account for the detailed center to limb variations while other just represent the contributing structures by an area weighting factor. The other distinguishing feature is that some models produce calculations that are in absolute units and thus ready for direct comparison to observations while other require a calibration step for the coefficients in the models when combining the components together.

4.2.1. Flux Spectra.
Using the analogy with stellar atmosphere models, the flux spectra at a particular wavelength and time may be expressed as:

$$F_\lambda(t) = (1 - f_F - f_s)F_\lambda^{qs} + f_F F_\lambda^{F} + f_s F_\lambda^{s}, \qquad (6)$$

where the superscripts *qs*, *F* and *s* refer to quiet Sun, faculae and sunspots respectively and the quantities *f* and *F* are the fractional area of each of the three components and their flux spectral contribution.

In the calculation of *F*, the *Kurucz* [1992] spectral synthesis model ATLAS9 is used. This code calculates the spectrum in LTE and either uses binned opacities (as used for stellar atmosphere calculations) or more recently, opac-

ity distribution functions (ODFs) derived from the Kurucz atomic line data. Each *F* is calculated using the FAL model C, a modified FAL model P and a rescaled Kurucz model atmosphere of effective temperature 5150 K for sunspots.

Early work utilized a disk averaged filling factor for the components while recent work has explored the use of magnetograms and continuum images together with a construction of the center-to-limb variation of the facular contrast to account for more detailed disk position of features [*Fligge, Solanki and Unruh*, 2000].

Figure 17 [*Unruh, Solanki and Fligge*, 2000] shows the results of their 3-component model of solar spectral irradiance variability assuming facular and spot filling factors of 0.027 and 0.0023 respectively. Of note is the dashed curve

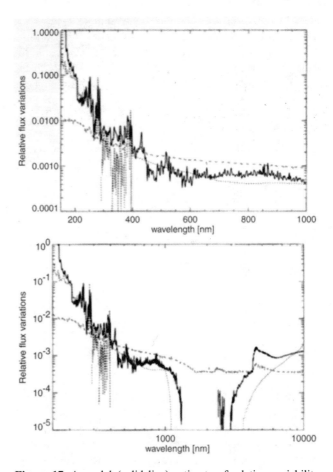

Figure 17. A model (solid line) estimate of relative variability between of solar maximum and minimum of the solar irradiance with wavelength (150 nm to 10 µm) and a compilation of data (dotted line, *Lean* [1997]). For comparison (dashed line), the relative model variability for an activity maximum where the effective temperature of the Sun is increased by 1 K is also shown [from *Unruh, Solanki and Fligge*, 2000].

for a 1 K hotter effective temperature Sun at solar maximum which departs markedly from both the other model and the collected data.

Figure 18 [*Fligge, Solanki and Unruh*, 2000] shows a detailed comparison of a modeled and observed banded irradiance times series. This comparison shows irradiances between Nov. 6, 1996 and Jan. 6, 1997 of the SOHO/VIRGO measurements of total, blue, green and red continuum during the passage of small sunspots early in solar cycle 22.

4.2.2. Intensity/Emission Measure Spectra. Based on early work by *Cook et al.* [1980], using the differential emission measure (DEM) for optically thin emission lines this model (NRLEUV) has been further developed for EUV and other parts of the spectrum. The method is similar to the other models in that radiances are used to compute the emission measure and derive temperature structures of features in the solar atmosphere; ones that are likely to contribute to the spectral irradiance wavelengths of interest. This model has also evolved from using full disk proxies, or filling factors, to now using more detailed distributions of features derived from full disk images.

The flux in this formulation [see *Warren et al.*, 2001] is:

$$F(\lambda,t) = F_{QS}(\lambda)\left(1 + \sum_{structures} \frac{\Delta F_{structure}(\lambda)}{F_{QS}}\right), \quad (7)$$

where

$$F_{QS}(\lambda) = 2\pi \frac{R_\odot^2}{D_\odot^2} I_{QS}(\lambda) \int_0^1 R_\lambda(\mu)\mu d\mu, \quad (8)$$

where I_{QS} is the quiet Sun intensity, and R_λ is the center-to-limb variation and

$$\Delta F_k(\lambda) = \frac{\mu_k A_k}{R^2} I_{QS}(\lambda) R_\lambda(\mu_k)(C_k - 1), \quad (9)$$

$\mu_k A_k$ is the projected area of the k-th structure and $C_k = I_k (\lambda)/I_{QS}(\lambda)$ is its intensity contrast. Thus, the computation of EUV irradiances requires reference spectra for each of the structures (computed or observed) and estimates of their fractional contribution and center-to-limb variation ($R_\lambda(\mu)$).

The intensity of a structure on the solar disk with wavelength λ_{ul} is

$$I_{structure}(\lambda_{ul}) = \int_T G_{ul}(T)\xi(T)dT, \quad (10)$$

where $\xi(T)$ is the differential emission measure ($n_e^2 ds/T$, n_e

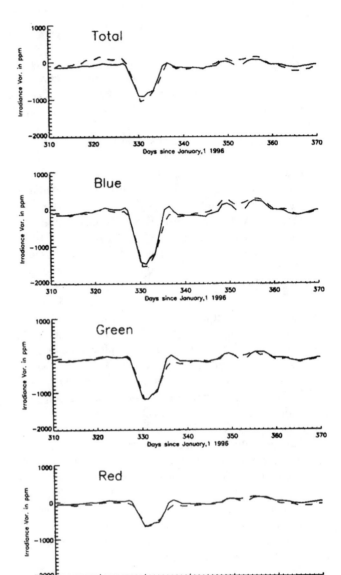

Figure 18. Modeled and measured irradiance times series between Nov. 6, 1996 and Jan. 6, 1997. The irradiance measurements (solid line) are from SOHO/VIRGO of total (upper), blue, green and red (lower) continuum. The model (dashed line) incorporates three components based on detailed solar disk distributions, from *Fligge, Solanki and Unruh* [2000].

is the electron density, ds is the differential distance along the line of sight, and T is the electron temperature), and G_{ul} (T) is a (known) contribution function and is the emissivity of the atomic transition divided by n_e^2 [details can be found in *Warren et al.*, 1998a; 1998b].

Figure 19 [*Warren et al.*, 2001] shows the differential emission measure (cm^{-5}K^{-1}) as a function of temperature for the quiet Sun and active region components of the NRLEUV model.

For basic atomic physics quantities such as oscillator strengths, NRLEUV uses the CHIANTI [*Dere et al.*, 1997; *Mewe et al.*, 1995] databases, ionization balances from *Arnaud and Rothenflug* [1985], coronal abundances from *Meyer* [1985] and emission data from Skylab and SOHO. The NRLEUV model is considered to have 3 components accounting for three structures–quiet Sun, active regions and coronal holes, while the center-to-limb variations are modeled or assumed with functional forms. In the most recent work, full disk images of Ca II K from Big Bear Solar Observatory and Yohkoh/SXT are used to identify contributions from the chromosphere and corona respectively.

Warren, Mariska and Lean [1998a; 1998b] developed reference EUV spectra for quiet Sun using this model and obtained good agreement with a variety of measurements of the spectrum [*Donnelly and Pope*, 1973; *Heroux and Hinteregger*, 1978; *Hinteregger et al.*, 1981; *Woods et al.*, 1998] and comparisons to EUV97 [*Tobiska and Eparvier*, 1998].

Warren, Mariska and Lean [2001] use computed intensities and assume a particular form of R_λ for optically thin lines, and use observed intensities and $R_\lambda = 1$ for optically thick lines.

An example of how well the NRLEUV model reproduced short term effects such as rotational modulation is shown in Figure 20 [from *Warren et al.*, 2001] which compares that model with AE-E measurements of He II 30.4 nm during the peak of solar cycle 21. *Warren et al.* [2001] note that the model does not capture the full range over variability (short and solar cycle timescales) that is evident in the data time-series.

4.2.3. Intensity/Radiance Spectra. In contrast to the flux spectra approach, the emergent spectral intensity may be calculated (using basic radiative transfer in an optically thick medium) as a sum over all contributing structures of the individual intensities of features or their radiance, i.e.

$$I(\lambda,t) = \sum_{structures} \sum_\mu I_{structure}(\lambda,\mu), \quad (11)$$

where μ is the heliocentric viewing angle (center to limb) and

$$I_{structure}(\lambda,\mu) \equiv \int S_\lambda(\tau/\mu)\exp(-\tau/\mu)d\tau, \quad (12)$$

where S_λ is the source function, τ is the monochromatic optical depth, and λ is the wavelength.

The SunRISE model uses three different approaches for obtaining realistic, non-LTE, line source functions. For strong resonance lines: (H Lyα and Mg II h and k), it uses non-LTE source functions from the PANDORA code [*Avrett and Loeser*, 1992] using Partial frequency Re-Distribution (PRD). When PRD is not important it uses non-LTE populations from PANDORA, and assumes Complete Re-Distribution (CRD) for opacity and emissivity computations (Balmer, Paschen, and higher Lyman, etc.). For all other cases it computes an approximate source function based on a "net radiative bracket" formulation in conjunction with the Planck function (medium and small strength lines [*Fontenla et al.*, 1999]).

The opacity sources in these line-by-line calculations use the elements from H to Zn, from the >60 million atomic and molecular lines primarily from *Kurucz and Bell* [1993] and the NIST archive [*NIST Atomic Spectra Database*]. In addition, all known continuum sources are included: H⁻: bound-free and free-free, H2⁺: bound-free and free-free, electrons: free-free, H: 91.2 nm, 364 nm, and higher level ionization,

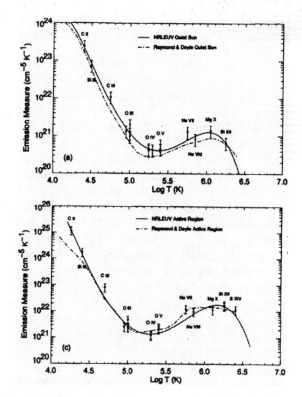

Figure 19. Differential emission measure (cm^{-5}K^{-1}) as a function of temperature for the quiet Sun (upper panel) and active region (lower panel) from the NRLEUV model (solid line), *Raymond and Doyle* [1981; dash-dot line] along with the emission data points and their identifications [from *Warren et al.*, 2001].

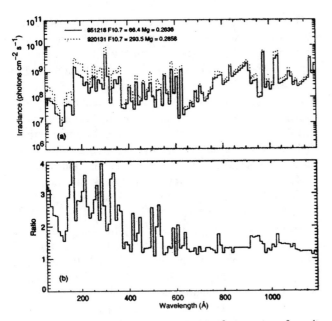

Figure 20. Detailed irradiance (in units of 10^9 photons cm^{-2} sec^{-1}) time series comparison of He II 30.4 nm modeled by NRLEUV (solid line) and the AE-E measurements (dots) between 1979 to 1981, from *Warren et al.* [2001].

C: edge 110 nm, Al: edge 207.6 nm, Si: edges 152.6 nm, 168.2 nm, 388 nm, and Mg: edges 162 nm, 251.5 nm, 365 nm as well as Thomson (electron) and Rayleigh (Hydrogen) scattering processes.

The SunRISE model has 70 components (i.e. 10 disk positions and 7 structures).

Figure 21 [Fig. 5a from *Fontenla et al.*, 1999] shows the part of the visible spectrum between 500 and 508 nm, comparing the synthesis model with a reference spectrum. Note that despite the very close correspondence there are several observed lines missing from the computation and some of the line strengths do not match–this can result from inaccuracies in the atomic data or different levels of solar activity in the measurements, while the model is quiet Sun only.

Figure 22 [*from Fox et al.*, 2002] show profiles for Lyman-α compared to observations [*Lemaire et al.*, 1998]. The top three panels refer to the quiet, moderate and active levels and the lower panel shows the relative difference between the active and quiet sun computations, i.e. the expected variability across the profile.

This figure shows the clear presence of chromospheric variability in the cores of strong lines of abundant elements due to solar activity, i.e. spots and plage. For Lyman-α the line core is formed at about 2100km above the solar photosphere and the wings at 1300–1800km (estimated from Figure 14). The computed variability (lower panel) between

active and quiet Sun indicates variability of ~80% in the line core and the extended wings of the line clearly indicate the sensitivity to activity level. In contrast however, are the peaks around the line core which suggest much lower variability (~55%).

Table 1 presents a comparison of observed and computed solar irradiances, and an example of solar cycle variability estimates from the SunRISE model [subset of Table 6 of *Fontenla et al.*, 1999].

Table 2 presents a comparison of Lya irradiance calculations from the SunRISE model [*Fox et al.*, 2002] and observations from UARS/SOLSTICE [version 9; *Rottman et al.*, 1993]. The two periods are for high and moderate (declining) stages of solar cycle 22.

The UV spectrum between 200 and 400 nm features about 167 strong lines displaying broad or 'winged' characteristics. Some strong lines of FeI between 260 nm and 264 nm show increases between 0.4× and 3× in their solar activity maximum to minimum ratio [*White, Fontenla and Fox*, 2000].

Using the measure of variability in intensity of $V_\lambda = I_{active}/I_{quiet} -1$, they found that in this interval, the computed irradiance at 0.01 nm resolution gave seven lines with V_λ values greater than 0.5, i.e. irradiance increases of 1.5× from solar minimum to maximum in cores of these lines.

However, at 1 nm resolution, the variability spectrum loses almost all of its spectral structure and gives V_λ ~0.02 between 260 nm and 263 nm. Above 263 nm, negative variability is seen which may still be due to missing opacity in the computations.

The band from 306–312 nm has similar results while the band from 259.8 to 260.0 nm contains three strong lines of FeI, and V_λ is greater than 0.25 in bands 0.025 nm wide in the line cores. The maximum variability in the line core

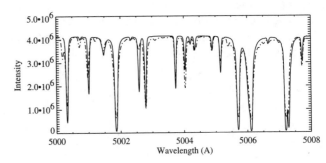

Figure 21. Computed synthetic spectrum (*solid line*) for the quiet Sun (model C) at disk center between 5000 and 5008 Å. Dash line shows the observed spectrum [*Delbouille et al.*, 1981] normalized to match the computed continuum intensity. Intensity is in erg cm^{-2}s^{-1}Å^{-1}sr^{-1}.

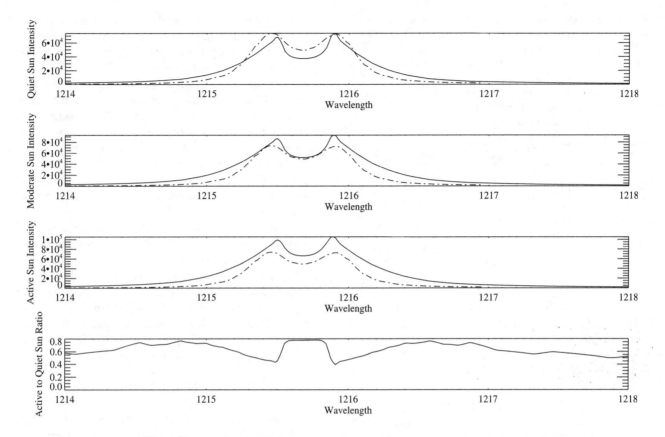

Figure 22. Calculated intensity spectrum (in ergs cm^{-2} sec^{-1} Å^{-1} sr^{-1}) for Ly-α (solid line) and an observed spectrum from Lemaire et al. (1998: dashed line) as a function of wavelength (in Å), for a) quiet Sun, b) moderate acitity Sun, c) active Sun, and d) relative ratio of active to quite Sun calculations. No adjustment of scale has been made to match the profiles; they are a direct comparison in absolute units.

intensities is 2× to 3×. The average value of variability, $< V_\lambda >$, is ~0.5% for the 260 nm and 300 nm bands at 1 nm resolution. For higher resolutions of 0.01 nm and 0.1 nm, V_λ values are larger than 25% near 260 nm and larger than 16% near 300 nm.

This theoretical study suggests measurements of the solar cycle variation in a small number of well-chosen bands with sufficient precision and spectral resolution to quantify this extreme radiative variability.

Figures 23 and 24 [*White et al.*, 2000] demonstrates one of the advantages of synthesis models by comparing the effects of spectral resolution on predictions for solar cycle timescale UV variability with wavelength between 260–264 nm. The high (0.01 nm) and low (1 nm; characteristic of ongoing space-based measurements) resolution spectrum are different in magnitude and form. Thus, the model is an important tool in understanding the physical basis for such variability.

5. PRESENT STATUS OF MODELS

As indicated earlier in this Chapter, one goal of models is to reproduce observations, but then what? The answer to this is most often limited by the ability of a model to capture physical effects or processes that can be identified with the Sun, and logically solar activity. Note that 'capturing physical effect or processes' may be different from directly modeling those effects or processes in a model and thus some rather simple models have lead to significant progress in understanding solar irradiance variability. Proxy models have been used to verify and correct total solar irradiance data, e.g. corrections to ERB in 1989 and 1990 from comparisons by *Lee* [1990] and *Chapman et al.* [1986] which removed an ~ 0.6 Wm^{-2} change in the solar minimum total irradiance. All of the present models appear to be capable of calculating relative or absolute synthetic intensities based on an approximation of the state of solar activity on the Sun.

Table 1. Quiet Sun computed and observed solar irradiances (in mW/m²/nm), and percentage irradiance solar cycle changes for a moderately active Sun (small plage area) and very active Sun (large plage area) for different wavelength bands in nm. Observed values are from *Neckel and Labs* [1984], except for Lyα (121.5 ± 0.45 nm) which is from *Rottman et al.* [1993]. The computation is the SunRISE Spectral Irradiance Synthesis (SIS) model. *This value is the ratio of the irradiance in the band to the total irradiance and here the units do not apply.

Band	Comp	Obs	Mod	Very
Lyα	59.2	60.	34.0	49.0
409–410	2032.	1708.	0.002	0.006
410–411	1642.	1504.	0.06	0.10
411–412	2084.	1822.	−0.04	−0.06
852–854	941.0	974.	0.18	0.26
854–856	854.8	878.	0.32	0.46
856–858	996.9	1012.	0.11	0.14
1.3–2.2 μm	0.14132*	uncertain	−0.13	−0.17

An example of what models could provide is to give insight in the reliability of existing proxies to estimate total and spectral irradiances in the absence of direct measurement. Or, to directly calculate spectral irradiances in important parts of the spectrum which may not be observed regularly enough, or with enough spectral or temporal resolution.

Models may also allow the exploitation of the natural redundancy and coherence in the solar spectrum, i.e. contiguous wavelength regions of the spectrum whose intensity or radiance respond in the same sense and magnitude to specific solar activity (or other) features. This, in principle, would lead to more efficient measurement and analysis programs.

The intersection of these two examples is to give physical understanding of the accuracy of the calculations and of empirical scaling rules, and perhaps to improve their application by specifying which bands are physically well coupled. A current example of this question is posed by the extremely high correlation between the SEM/304 channel and the Mg II index (with some timescale averaging [*Viereck et al.*, 2001], see Figure 25).

Given the different dependence of irradiance variability for different solar activity features across the spectrum, i.e. variations in the size and sign of the relative contrasts that do not lend themselves to models with a simple proxy, it should soon be possible to conceive new 'banded' proxies, developed from the physics-based models now in hand (and discussed earlier). This suggestion parallels the one behind the E10.7 synthetic proxy—developed for effective use in terrestrial applications while more correctly representing the EUV heating term. To be more specific, using the synthetic models for a range of solar activity conditions it will be possible to identify wavelength bands for which well defined and well behaved irradiance variability can be predicted with suitable proxies—with the proxies themselves perhaps also identified by models.

One bridge between proxy models and those with more of a physical basis was through the development of the TIRR model [*Walton and Preminger*, 1999] which used a T_{eff} and some spectral shape at a given wavelength to model irradiance and when combined with the Mg II index (for example) has led to a good reproduction of the observed time series. Although this model has now been discarded, this success prompts for research into how further hybrid models may be constructed.

In addition to focusing on parts of the solar spectrum which vary with solar activity, there is also value in determining parts of the spectrum which have little or no variability and why—something that a model containing physics should be able to provide. Some work on this by *Unruh, Solanki and Fligge* [2000] and *Fox et al.* [2002] show promise in also understanding either the indirect effects of solar activity on the surrounding atmosphere, or other thermal origins of spectral line changes.

6. UNCERTAINTIES AND CONSTRAINTS

Empirical or proxy-based models tend to have better defined uncertainties and constraints. The development of their coefficients is usually statistical and based on regression analysis (simple or sophisticated), and the uncertainties are based on the statistical scatter (standard deviation if reported). To a large extent the actual choice of proxies is based on the development of these statistical relationship and an available timeseries or database and to a lesser extent on any assumed or implied physical relationship between the proxy and the measure it is contributing to. Thus, the

Table 2. Lyα irradiance (mWm^{-2}) summaries for three days in 1992 and four days in 1993 constructed using the SunRISE SIS model with the NSO/SP CaIIK spectroheliograms as inputs for the surface structures. The model is compared to the observed irradiances from UARS/SOLSTICE [version 9; *Rottman et al.*, 1993].

Date	Irrad(I)	Obs.	I-O/O
92/01/16	6.66	8.52	−21%
92/01/31	7.41	10.00	−26%
92/02/19	7.36	8.87	−17%
93/06/03	7.09	6.62	7%
93/06/16	5.79	5.66	2%
93/06/28	6.78	6.84	−.9%
93/07/08	5.66	5.86	−3%

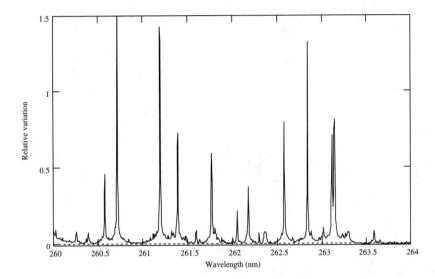

Figure 23. High resolution (0.01 nm) synthetic spectrum variability from minimum to maximum in the solar cycle (upper panel) between 260 and 264 nm. Note extreme variability in the cores of the broad lines in this spectral band.

choice of proxies is more of a constraint and one that should be revisited from time to time to make sure that it is still valid. On the positive side, experience from a few decades of proxy models across the solar spectrum suggests that proxies developed over the full range of solar activity levels and phases of a solar cycle tend to be the most robust.

In contrast, physics-based models have uncertainties in a number of areas. The assumption of plane-parallel radiative transfer through an atmosphere is an uncertainty but one that is likely not the largest source. For detailed synthesis it is the atomic data those presents the greatest uncertainty [as noted for many decades now by *Kurucz*, 1992]. Current tabulations are still missing lines, and those that are may have incorrect oscillator strengths, broadening parameters, or energy level information. The missing lines lead to missing opacity sources, for example between 260–270 nm the inclusion of free-bound continua of Fe, Ca, K, and Na solved a significant discrepancy between computations and

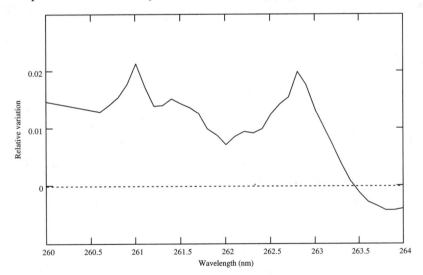

Figure 24. Low resolution (1 nm) synthetic spectrum variability between 260 and 264 nm. Note the loss of spectral detail at this resolution. Only the variability due to individual lines near 261 nm and 263 nm is detectable. Negative variability is also evident.

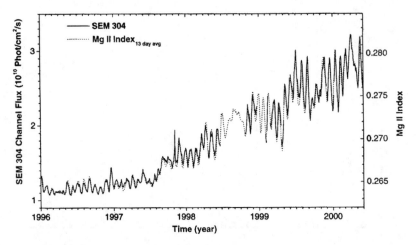

Figure 25. Timeseries comparison of 13-day average Mg II index and the SOHO/SEM He II 30.4 nm flux during the ascending phase of solar cycle 23 from *Viereck et al.* [2001].

observations (see Figure 26 from *White, Fontenla and Fox* [2000]) which shows the clear improvement in matching observations between 260 and 264 nm but also indicates room for further improvement between 264 and 270 nm). Molecules are not included mainly due to the absence of complete electronic, rotational and vibrational parameters for the ones of interest for the Sun.

Other approximations include: line broadening includes only micro-turbulent velocities tabulated from the semi-empirical model solar atmospheres. The models of *Fontenla et al.* [1999] estimate the lower population level by assuming LTE and the elemental ionization is also computed in LTE. Their approximation to the non-LTE line source function relies on a two-level atom

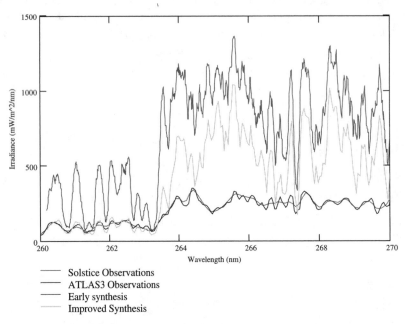

Figure 26. Comparison of observed and synthetic UV synthetic spectrum for the 260–270 nm band at 0.1 nm resolution, early calculation (upper curve), improved spectrum (intermediate curve) together with observed irradiance spectra from ATLAS3 and SOLSTICE (lower curves).

approximation (does not include recombination) for most lines.

Based on the previous section on the contribution and identification of decomposing the Sun into surface structures is one of the main sources of uncertainties in all models. Simply put: what components are to be included and how are they identified and accounted for in the model?

Despite the presence of these uncertainties in all models, much progress has and can still be made in understanding solar irradiance and its variability. Thus depending on the particular aspect of solar irradiance of interest, there are choices that can be made in which model, or models, that may be applied.

The major considerations are: wavelength (with some spectral resolution or integrated), time (particular day or time, phase of solar cycle, or timescale) physics (complexity based on past modeling or interpretation efforts), calibration (whether this is possible or required), and the solar activity state (the trend, ascending or descending phase, or the specific state).

7. PREVIEW—WHAT DO MODELS NEED

7.1. Reference Spectra

Solar reference spectra which are discussed in detail in the Chapter by Thuillier et al. have always played a critical role in the development of models of the solar spectrum. Their role in solar variability was emphasized by Hinteregger et al. [1981] with SC#21REFW and further elevated during the successful international SCOSTEP program SOLERS22 which produced a number of reference spectra, for solar minimum conditions. Other reference spectra have been developed for the relative variation from maximum to minimum. On the solar cycle time scale, these spectra are useful for gauging the correspondence between solar variability and solar activity on those time scales. However they are not as applicable to constraining models which are able to represent the dependence of the solar irradiance (total or spectral) on a particular state of the Sun. Nor can they follow variations over relatively short periods of time, such as the time for sunspot passage across the disk, active region emergence or decay, or the appearance of plage at the solar limbs. Thus, a new need is for accurate reference spectra over shorter time intervals with, at the very least, very accurate relative intensity measurements and preferably at high spectral resolution (0.1 nm or better below 1 μm and 1 nm above 1 μm).

Reference spectra that have been used in some of the model developments previously discussed, both for prediction and

validation include: over broad spectral regions—Neckel and Labs [1984], Kurucz, Furenlid, Brault and Testerman [1984], the UVAFGL atlas from Hall and Anderson [1991], the AFGL atlas from Beckers et al. [1976], Delbouille et al. [1981], and Wallace, Hinkle, and Livingston [1998]. For more details, see Thuillier et al. [this volume].

Radiance spectra continue to be an important element in empirical solar atmosphere models, and thus the irradiance models that associate these models with the underlying solar surface features that contribute to the variability. Some of these include: Lyman series from SOHO/CDS [Warren, Mariska and Wilhelm, 1998], the HRTS-9 quiet sun and active region in the mid-UV [Morrill et al., 2001], sunspot umbra—Wallace, Hinkle, and Livingston [2000], sunspot and photosphere: Wallace, Livingston, and Bernath [1994] near-infrared: Livingston and Wallace [1991], SOLSTICE/UV: [Rottman and Woods, 1994], and SUSIM/UV [Lean et al., 1992; Brueckner et al., 1993].

The need for 'variability' reference spectra is extremely important in the UV, between 200 and 400 nm, especially for studies of the extreme variability of the winged/broad lines [White et al., 2000].

Finally, these spectra need to be readily available to the modeling community and updated as new spectra become available.

7.2. Summary and Immediate Suggestions to Help Modelers

Solar irradiance modeling has significantly changed over the past 20 years and especially in the last ~ 8 years. Proxy models have found their place in modeling the solar irradiance and much work remains in this area. The application of these models should be in two areas: directed toward understanding the Sun and as increasingly quantitative and meaningful inputs to models of the terrestrial upper atmosphere. There has been progress toward true photometric indices based on decompositions of full disk images, i.e. the SFO approach, which is remarkable considering the telescope apertures. The incorporation of information from high precision, full disk photometric images into models is a logical next step.

The emergence or re-emergence of physics-based models holds the most promise for a deep understanding of solar irradiance variability and in particular toward the nature of the relation to solar (surface and perhaps sub-surface) activity on a variety of timescales.

For these detailed synthesis models, many of the important lines in the solar spectrum are accounted for and have correct atomic physics parameters, but there are still many errors and omissions to be corrected and this is a difficult and painstaking task.

In terms of the common element for solar irradiance models, i.e. that of solar surface activity, a substantial amount of work on decomposing the Sun into the important structures and quantifying their uncertainty is required, i.e. more images of the Sun in more wavelengths. This remains a solar physics problem since the study of relations between magnetic flux density distribution and topology, and intensity of structures (radiance) in the broad sense is still an evolving research area.

In the near future, it will be possible to compute emergent intensities (radiances) based on vector magnetic fields measured in the solar atmosphere and coupled to the local thermodynamics: these will arise from projects such as SOLIS, ATST, and new efforts to measure magnetic fields in the solar corona.

To support the trend of improvements and advance the state of the art in physical models of the solar total and spectral irradiance, members of the solar physics and physics communities could collect a set of databases (model and observed) for the entire solar spectrum as a tool for investigating solar irradiance variability and make these available to the community. It would also be very useful to accumulate a set of reference solar conditions (decompositions) ranging from solar activity minimum to maximum, and including activity effects on shorter timescales.

To aid in model validation and intercomparison, it is suggested that the community identify a set of spectral lines and continua and a set or sets of time series so that quantitative estimates of model uncertainties and reliability can be made.

Acknowledgments. In preparing this Chapter, the author would like to acknowledge the contributions of the following people both directly and through collaborations, their published work and meeting presentations over many years: Oran R. White, Juan Fontenla, Gene Avrett, Bob Kurucz, Karen Harvey, Jack Harvey, Bill Livingston, Sabatino Sofia, Judith Lean, Kent Tobiska, Sami Solanki, Marcel Fligge, Giuliana DeToma, Gary Chapman, Steve Walton, Tom Woods, Gary Rottman. The author is also grateful to the reviewers and editors of this volume whose comments improved the chapter.

REFERENCES

Anderson, L. S. and R. G. Athay, Model solar chromosphere with prescribed heating, *Astrophys. J.*, *346*, 1010, 1989.

Arnaud, M. and R. Rothenflug, An updated evaluation of recombination and ionization rates, *Astron. Astrophys.*, *60*, 425-457, 1985.

Avrett, E. H., Models of the solar outer photosphere, in *Solar Photosphere: Structure, Convection, and Magnetic Fields*, ed. by J. O. Stenflo, Kluwer, Dordrecht, p.3, 1990.

Avrett, E. H., Modeling solar variability—synthetic models, in *Solar Electromagnetic Radiation Study for Solar Cycle 22, Proceedings of the SOLERS22 Workshop held at the National Solar Observatory, Sacramento Peak, Sunspot, New Mexico, June 17–21, 1996*, ed. by J. M. Pap, C. Fröhlich, and R. K. Ulrich, Kluwer Academic Publishers, Dordrecht, p.449-469, 1998.

Avrett, E. H. and R. Loeser, The PANDORA atmosphere program, in *Cool Stars, Stellar Systems, and the Sun, ASP Conf. Series 26*, ed. by M. S. Giampapa and J. A. Bookbinder, ASP, San Francisco, p.489, 1992.

Beckers, J.M., C. A. Bridges, and L. B. Gilliam, A high resolution spectral atlas of the solar irradiance from 380 to 700 nanometers, AFGL-TR-76-0126 (II), AFGL, Sunspot, 1976.

Berrilli, F., I. Ermolli, A. Florio, and E. Pietropaolo, Average properties and temporal variations of the geometry of solar network cells, *Astron. Astrophys.*, *344*, 965, 1999.

Brandt, P. N., M. Stix, and H. Weinhardt, Modeling solar irradiance variations with an area dependent photometric sunspot index, *Solar Phys.*, *152*, 119-124, 1994.

Brueckner, G. E., K. L. Edlow, L. E. Floyd, J. Lean, and M. E. Van Hoosier, The solar ultraviolet spectral irradiance monitor (SUSIM) experiment on board the Upper Atmosphere Research Satellite (UARS), *J. Geophys. Res.*, *98*, 10,695-10,711, 1993.

Cebula, R. P., SOLERS22 working group 2 workshop report, in *Proceedings of the Workshop on Solar Electromagnetic Radiation Study for Solar Cycle 22*, ed. by R. Donnelly, NOAA, Boulder, p.286, 1992.

Cebula, R. P. and M. T. Deland, Comparisons of the NOAA-11 SBUV/2, UARS SOLSTICE, and UARS SUSIM MG II solar activity proxy indexes, *Solar Physics*, *177*, 117, 1998.

Chapman, G. A., Solar variability due to sunspots and faculae, *J. Geophys. Res.*, *92*, 809, 1987.

Chapman, G. A., A. M. Cookson, and J. J. Dobias, Solar variability and the relation of faculae to sunspot areas during cycle 22, *Astrophys. J.*, *482*, 541-545, 1997.

Chapman, G. A., A. D. Herzog, and J. K. Lawrence, Time-integrated energy budget of a solar activity complex, *Nature*, *319*, 654, 1986.

Chapman, G. A., A. D. Herzog, J. K. Lawrence, and J. C. Shelton, Solar luminosity fluctuations and active region photometry, *Astrophys. J.*, *282*, L99, 1984.

Chapman, G. A. and A. D. Meyer, Solar irradiance variations from photometry of active regions, *Solar Phys.*, *103*, 21, 1986.

Cook, J. W., G. E. Brueckner, D. K. Prinz, L. E. Floyd, and P. A. Lund, Model for variability of the solar far ultraviolet flux using full disk magnetogram, *BAAS*, *185*, #123.07, 1994.

Cook, J. W., G. E. Brueckner, and M. E. Vanhoosier, Variability of the solar flux in the far ultraviolet 1175-2100 Å, *J. Geophys. Res.*, *85*, 2257-2268, 1980.

Cook, J. W. and J. A. Ewing, Relationship of magnetic field strength and brightness of fine-structure elements in the solar temperature minimum region, *Astrophys. J.*, *355*, 719-725, 1990.

Cox, A. N., Allen's astrophysical quantities, 4th ed., Springer, Berlin, 2000.

Delbouille, L., G. Roland, J. Brault, and L. Testerman, Photometric atlas of the solar spectrum from 1850 to 10000 cm^{-1}, Kitt Peak National Observatory, Tucson, 1981.

Dere, K. P., E. Landi, H. E. Mason, B. C. Monsignori Fossi, and P. R. Young, CHIANTI—an atomic database for emission lines, *Astron. Astrophys. Suppl. Ser.*, *125*, 149-173, 1997.

DeToma, G., O. R. White, G. A. Chapman, S. R. Walton, D. G. Preminger, A. M. Cookson, and K. L. Harvey, Differences in the Sun's radiative output in cycles 22 and 23, *Astrophys. J.*, *549*, L131-134, 2001.

DeToma, G., O. R. White, and K. L. Harvey, A picture of solar minimum and the onset of solar cycle 23. I. Global magnetic field evolution, *Astrophys. J.*, *529*, 1101-1114, 2000.

Donnelly, R. F. and J. H. Pope, Tech. Rpt. ERL 276-SEL 25, NOAA, Boulder, 1973.

Donnelly, R. F., O. R. White, and W. C. Livingston, The solar Ca II K index and the Mg II core-to-wing ratio, *Solar Phys.*, *152*, 69-76, 1994.

Ermolli, I., M. Fofi, C. Bernacchia, F. Berrilli, B. Caccin, E. Egidi, and A. Florio, The prototype RISE-PSPT instrument operating in Rome, *Solar Phys.*, *177*, 1, 1998.

Fennelly, J. A., D. G. Torr, P. G. Richards, and M. R. Torr, Simultaneous retrieval of the solar EUV flux and neutral thermospheric O, O2, N2, and temperature from twilight airglow, *J. Geophys. Res.*, *99*, 6483, 1994.

Fligge, M. and S. K. Solanki, The solar spectral irradiance since 1700, *Geophys. Res. Lett.*, *27*, 2157, 2000.

Fligge, M., S. K. Solanki, and Y. C. Unruh, Modelling irradiance variations from the surface distribution of the solar magnetic field, *Astron. Astrophys.*, *353*, 380, 2000.

Fligge, M., S. K. Solanki, Y. C. Unruh, C. Fröhlich, and C. Wehrli, Modelling spectral irradiance variations obtained by VIRGO, *ASP Conf. Ser. 140*: Synoptic Solar Physics, 311, 1998a.

Fligge, M., S. K. Solanki, Y. C. Unruh, C. Fröhlich, and C. Wehrli, A model of solar total and spectral irradiance variations, *Astron. Astrophys.*, *335*, 709, 1998b.

Floyd, L. E., P. A. Reiser, P. C. Crane, L. C. Herring, D. K. Prinz, and G. E. Brueckner, Solar cycle 22 UV spectral irradiance variability: Current measurements of SUSIM UARS, *Solar Phys.*, *177*, 79-87, 1998.

Fontenla, J. M., E. H. Avrett, and R. Loeser, Energy balance in the solar transition region. I—Hydrostatic thermal models with ambipolar diffusion, *Astrophys. J.*, *355*, 700, 1990.

Fontenla, J. M., E. H. Avrett, and R. Loeser, Energy balance in the solar transition region. II—Effects of pressure and energy input on hydrostatic models, *Astrophys. J.*, *377*, 712, 1991.

Fontenla, J. M., E. H. Avrett, and R. Loeser, Energy balance in the solar transition region. III. Helium emission in hydrostatic, constant-abundance models with diffusion, *Astrophys. J.*, *406*, 319, 1993.

Fontenla, J. M., E. H. Avrett, and R. Loeser, Energy balance in the solar transition region. IV. Hydrogen and helium mass flows with diffusion, *Astrophys. J.*, *572*, 636, 2002.

Fontenla, J., O. R. White, P. A. Fox, E. H. Avrett, and K. L. Harvey, Calculation of the time variation of the solar spectrum, *EOS*, *76*, S234, 1995.

Fontenla, J. M., O. R. White, P. A. Fox, E. H. Avrett, and R. L. Kurucz, Calculation of solar irradiances I. Synthesis of the solar spectrum, *Astrophys. J.*, *518*, 480, 1999.

Foukal, P., The behavior of solar magnetic plages measured from Mt. Wilson observations between 1915–1984, *Geophys. Res. Lett.*, *23*, 2169, 1996.

Foukal, P. and J. Lean, The influence of faculae on total solar irradiance and luminosity, *Astrophys. J.*, *302*, 826, 1986.

Foukal, P. and J. Lean, Magnetic modulation of solar luminosity by photospheric activity, *Astrophys. J.*, *328*, 347, 1988.

Fox, P. A., O. R. White, R. Meisner, M. Knölker, and M. P. Rast, Precision solar images from PSPT as input for spectrum synthesis and variability studies, *EOS*, *79*, F9, 1998.

Fox, P. A., O. R. White, J. Fontenla, E. H. Avrett, and K. L. Harvey, Calculation of solar irradiances II: Synthesis of spectra, irradiances and images, *Astrophys. J.*, to appear, 2002.

Fox, P. A., O. R. White, J. Fontenla, E. H. Avrett, and R. L. Kurucz, Application of spectral synthesis to the study of improve surrogates for solar radiation in the visible, UV and EUV, *EOS*, *75*, 291, 1994.

Fox, P. A., O. R. White, J. Fontenla, E. H. Avrett, and K. L. Harvey, Calculation of solar spectral irradiances from .1 to 100 microns, SOLERS22, Sacramento Peak, NSO, Sunspot, 1996.

Fröhlich, C., Solar variability for periods of days to months, *Adv. Space Res.*, *4*, 117-120, 1984.

Fröhlich, C., Total solar irradiance monitoring programs, in *Solar Electromagnetic Radiation Study for Solar Cycle 22, Proceedings of the SOLERS22 Workshop held at the National Solar Observatory, Sacramento Peak, Sunspot, New Mexico, June 17–21, 1996*, ed. by J. M. Pap, C. Fröhlich, and R. K. Ulrich, Kluwer Academic Publishers, Dordrecht, p.391, 1998.

Fröhlich, C., B. Andersen, T. Appourchaux, *et al.*, First results from VIRGO, the experiment for helioseismology and solar irradiance modeling on SOHO, *Solar Phys.*, *170*, 1-25, 1997.

Fröhlich, C. and J. Lean, The Sun's total irradiance: Cycles, trends, and related climate change uncertainties since 1976, *Geophys. Res. Lett.*, *25*, 4377, 1998.

Fröhlich, C. and J. M. Pap, Multi-spectral analysis of total solar irradiance variations, *Astron. Astrophys.*, *220*, 272-280, 1989.

Fröhlich, C., J. M. Pap, and H. S. Hudson, Improvement of the photometric sunspot index and changes of the disk integrated sunspot contrast with time, *Solar Phys.*, *152*, 111-118, 1994.

Godoli, G., Area e posizione dei flocculi di calcio per il 1960 secondo le osservazioni eseguite alla Torre Solare di Arcetri, *Memorie della Societa Astronomica Italiana*, *32*, 181, 1961.

Hall, L. A. and G. P. Anderson, High-resolution solar spectrum between 2000 and 3100 Å, *J. Geophys. Res.*, *96*, 12,927-12,931, 1991.

Harvey, J. W., Helium 10830 angstrom irradiance: 1975–1983, in *Solar Irradiance Variations on Active Region Time Scales (SEE N84–27635 17-92)*, NASA, Washington D.C., p. 197-211, 1984.

Harvey, K. L., Irradiance models based on solar magnetic fields, in *IAU Colloquium 143: The Sun as a Variable Star: Solar and Stellar Irradiance Variations*, ed. by J. M. Pap, C. Frolich, H. S. Hudson and S. Solanki, Cambridge Univ. Press, Cambridge, p.217-225, 1994.

Harvey, K. L. and O. R. White, Technical Report, Solar Physics Research Corp., Tucson, AZ, 1996.

Harvey, K. L. and O. R. White, Variability of solar surface structures, *Astrophys. J.*, *515*, 812, 1999a.

Harvey, K. L. and O. R. White, What is Solar Cycle Minimum *J. Geophys. Res.*, *104*, 19759-19764, 1999b.

Heath, D. F. and B. M. Schlesinger, The Mg 280-nm doublet as a monitor of changes in solar ultraviolet irradiance, *J. Geophys. Res.*, *91*, 8672, 1986.

Heroux, L. and H. E. Hinteregger, Aeronomical reference spectrum for solar UV below 2000 Å, *J. Geophys. Res.*, *83*, 5305-5308, 1978.

Hinteregger, H. E., K. Fukui, and B. R. Gilson, Observational, reference and model data on solar EUV, from measurements on AE-E, *Geophys. Res. Lett.*, *8*, 1147, 1981.

Hudson, H. S., S. Silva, M. Woodard, and R. C. Willson, The effect of sunspots on solar irradiance, *Solar Phys.*, *76*, 211, 1982.

Johannesson, A., W. H. Marquette, and H. Zirin, A 10-year set of Ca II K-line filtergrams, *Solar Phys.*, *177*, 265, 1998.

Kuhn, J. R. and J. D. Armstrong, Mechanisms of solar irradiance variations, this volume, 2004.

Kurucz, R. L., Atlas: a computer program for calculating model stellar atmospheres, SAO Spec. Rept. No. 309, SAO, Cambridge, p.291, 1970.

Kurucz, R. L., New opacity calculations, in *Stellar Atmospheres; Beyond Classical Models*, ed. by L. Crivellari, I. Hubeny and D. G. Hummer, Kluwer, Dordrecht, p.440, 1991.

Kurucz, R. L., Remaining line opacity problems for the solar spectrum, *Rev. Mex. de Astron. y Astrof.*, *23*, 187, 1992.

Kurucz, R. L. and B. Bell, Atomic line list, CDROM-23, SAO, Cambridge, 1993.

Kurucz, R. L., I. Furenlid, J. Brault, and L. Testerman, Solar flux atlas from 296 to 1300 nm, NSO, Sunspot, p.240, 1984.

Lawrence, J. K., G. A. Chapman, A. D. Herzog, and J. C. Shelton, Solar luminosity fluctuations during the disk transit of an active region, *Astrophys. J.*, *292*, 297, 1985.

Lean, J., A comparison of models of the sun's extreme ultraviolet irradiance variations, *J. Geophys. Res.*, *95*, 11933, 1990.

Lean, J., Variations in the Sun's radiative output, *Rev. of Geophys.*, *29*, 505-535, 1991.

Lean, J., The Sun's variable radiation and its relevance for earth, *Ann. Rev. Astron. Astrophys.*, *35*, 33, 1997.

Lean, J. L., G. J. Rottman, H. Lee Kyle, T. N. Woods, J. R. Hickey, and L. C. Puga, Detection and parameterization of variations in solar mid- and near-ultraviolet radiation (200–400 nm), *J. Geophys. Res.*, *102*, 29,939-29,956, 1997.

Lean, J., M. Vanhoosier, G. E. Brueckner, D. Prinz, L. Floyd, and K. Edlow, SUSIM/UARS observations of the 120 to 300 nm flux variations during the maximum of the solar cycle—Inferences for the 11-year cycle, *Geophys. Res. Lett.*, *19*, 2203, 1992.

Lee III, R.B., Long-term solar irradiance variability: 1984–1989 observations, in *Proceedings of the Conference on the Impact of Solar Variability on Climate*, ed. by K. H. Schatten and A. Arking, NASA CP #3086, NASA/GSFC, Greenbelt, p.301-308, 1990.

Lemaire, P., C. Emerich, W. Curdt, U. Schuehle, and K. Wilhelm, Solar Hi Lyman alpha full disk profile obtained with the SUMER/SOHO spectrometer, *Astron. Astrophys.*, *334*, 1095, 1998.

Livingston, W. and L. Wallace, An atlas of the solar spectrum in the infrared from 1850 to 9000 cm^{-1} (1.1 to 5.4 mm), NSO Technical Report; 91-01 National Optical Astronomy Observatories, Tucson, 1991.

Maltby, P., E. H. Avrett, M. Carlsson, O. Kjeldseth-Moe, R. L. Kurucz, and R. Loeser, A new sunspot umbral model and its variation with the solar cycle, *Astrophys. J.*, *306*, 284, 1986.

Mewe, R., J. S. Kaastra, and D. A. Liedahl, Update of MEKA: MEKAL, *Legacy*, *6*, 16, 1995.

Meyer, J.-P., Solar-stellar outer atmospheres and energetic particles, and galactic cosmic rays, *Astrophys. J. (Supp.)*, *57*, 173-204, 1985.

Morrill, J. S., K. P. Dere, and C. M. Korendyke, The sources of solar ultraviolet variability between 2765 and 2885 Å: Mg I, Mg II, Si I, and continuum, *Astrophys. J.*, *557*, 854, 2001.

Neckel, H. and D. Labs, The solar irradiance between 3300 and 12500 Å, *Solar Phys.*, *90*, 205, 1984.

NIST atomic spectra database, http://physics.nist.gov/cgi-bin/AtData/main-asd

Nusinov, A. A., Dependence of the intensity of lines of short-wave solar emission on the activity level, *Geomag. Aeron.*, *24*, 439, 1984.

Oster, L., K. H. Schatten, and S. Sofia, Solar irradiance variations due to active regions, *Astrophys. J.*, *256*, 768, 1982.

Pap, J. M., Variations in solar Lyman alpha irradiance on short time scales, *Astron. Astrophys.*, *264*, 249, 1992.

Pap, J. M., Long-term solar-irradiance variability, in *Sounding Solar and Stellar Interiors*, ed. by J. Provost and F.-X. Schmeider, Kluwer Academic Publishers, Dordrecht, p.235-250, 1997.

Pap, J. M., M. Turmon, L. Floyd, C. Fröhlich, and C. Wehrli, Total solar and spectral irradiance record, *Adv Space Res.*, in press, 2002.

Pap, J. M. and C. Frölich, Total solar irradiance variations, *J. Atmos. Solar Terr. Phys.*, *61*, 15-24, 1999.

Pap, J. M., H. S. Hudson, G. J. Rottman, R. C. Willson, R. F. Donnelly, and J. London, Modeling solar Lyman alpha irradiance, in *Climate Impact of Solar Variability*, NASA Conf. Proc. #3086, ed. by K. H. Schatten and A. Arking, NASA/GSFC, Greenbelt, p.189-196, 1990.

Pap, J. M., J. London, and G. J. Rottman, Variability of solar Lyman alpha and total solar irradiance, *Astron. Astrophys.*, *245*, 648-653, 1991.

Pap, J. M. and C. Wehrli, Working group 1 research activities report on SOLERS22 1991 workshop, in *Proceedings of the Workshop on Solar Electromagnetic Radiation Study for Solar Cycle 22*, ed. by R. F. Donnelly, NOAA, Boulder, p.90-105, 1992.

Parker, D. G., R. K. Ulrich, and J. M. Pap, Modeling solar UV variations using Mount Wilson Observatory indices, *Solar Phys.*, *177*, 229-241, 1998.

Planck, M., On the theory of thermal radiation, *Ann. Physik*, *4*, 553-563, 1901.

Preminger, D. G., S. R. Walton, and G. A. Chapman, Solar feature identification using contrasts and contiguity, *Solar Phys.*, *202*, 53-62, 2001.

Rast, M. P., P. A. Fox, H. Lin, O. R. White, R. Meisner, and B. Lites, Bright rings around sunspots, *Nature*, *401*, 678, 1999.

Raymond, J. C. and J. G. Doyle, Emissivities of strong ultraviolet lines, *Astrophys. J.*, *245*, 1141, 1981.

Richards, P. G., J. A. Fennelly, and D. G. Torr, UVAC: A solar EUV flux model for aeronomic calculations, *J. Geophys. Res.*, *99*, 8981-8992, 1994.

Rottman, G. J., T. N. Woods, and T. N. Sparn, Solar-Stellar Irradiance Comparison Experiment 1. I—Instrument design and operation, *J. Geophys. Res.*, *98*, 10667, 1993.

Rottman, G. J. and T. N. Woods, Upper Atmosphere Research Satellite (UARS) Solar Stellar Irradiance Comparision Experiment (SOLSTICE), *Proc. SPIE*, *2266*, 317, 1994.

Schatten, K. H., N. Miller, S. Sofia, and L. Oster, Solar irradiance modulation by active regions from 1969 through 1980, *Geophys. Res. Lett.*, *9*, 49, 1982.

Skumanich, A., C. Smythe, and E. N. Frazier, On the statistical description of inhomogeneities in the quiet solar atmosphere. I—Linear regression analysis and absolute calibration of multi-channel observations of the Ca^+ emission network, *Astrophys. J.*, *2000*, 747, 1975.

Sofia, S., K. H. Schatten, and L. Oster, Solar irradiance modulation by active regions during 1980, *Solar Phys.*, *80*, 87, 1982.

Sofia, S. and L. H. Li, Solar variability caused by structural changes of the convection zone, this volume, 2004.

Solanki, S. K., Spectral irradiance variations, in *Recent Insights into the Physics of the Sun and Heliosphere: Highlights from SOHO and other Space Missions, IAU Symposium #2003, Manchester, England*, ed. by P. Brekke, B. Fleck and J. B. Gurman, p. 10, 2001.

Solanki, S. K. and M. Fligge, Solar irradiance since 1874 revisited, *Geophys. Res. Lett.*, *25*, 341, 1998.

Solanki, S. K. and M. Fligge, A reconstruction of total solar irradiance since 1700, *Geophys. Res. Lett.*, *26*, 2465, 1999.

Solanki, S. K. and Y. C. Unruh, A model of the wavelength dependence of solar irradiance variations, *Astron. Astrophys.*, *329*, 747, 1998.

Solanki, S. K., M. Fligge, and Y. C. Unruh, Variations of the solar spectral irradiance, in *Recent Insights into the Physics of the Sun and Heliosphere: Highlights from SOHO and other Space Missions, IAU Symposium #203, Manchester, England*, ed. by P. Brekke, B. Fleck and J. B. Gurman, p. 66, 2001.

Tapping, K. F. and D. P. Charrois, Limits to the accuracy of the 10.7 cm flux, *Solar Phys.*, *150*, 305, 1994.

Tapping, K. F. and K. L. Harvey, Slowly-varying microwave emissions from the solar corona, in *The Sun as a Variable Star: Solar and Stellar Irradiance Variations*, ed. by J. M. Pap, C. Fröhlich, H. S. Hudson and S. Solanki, Cambridge University Press, Cambridge, p. 182-195, 1994.

Thuillier, G., T. N. Woods, L. E. Floyd, R. Cebula, M. Herse, and D. Labs, Solar irradiance reference spectra, this volume, 2004.

Tobiska, W. K., Recent solar extreme ultraviolet irradiance observations and modeling: A review, *J. Geophys. Res.*, *98*, 18879, 1993.

Tobiska, W. K. and F. Eparvier, EUV97: improvements to EUV irradiance modeling, *Solar Physics*, *177*, 147, 1998.

Tobiska, W. K., T. N. Woods, F. Eparvier, R. Viereck, L. E. Floyd, D. Bouwer, G. J. Rottman, and O. R. White, The SOLAR2000 empirical solar irradiance model and forecast tool, *J. Atmos. Terr. Phys.*, *62*, 1233-1250, 2000.

Turmon, M., J. M. Pap, and S. Mukhtar, Automatically finding solar active regions using SOHO/MDI photograms and magnetograms, *Astrophys. J.*, *568*, 396-407, 2002.

Unruh, Y. C. and S. K. Solanki, Diagnostics from spectral irradiance measurements, American Geophysical Union, Spring Meeting 2001, abstract #SP32B–01, 332B01, 2001.

Unruh, Y. C., S. K. Solanki, and M. Fligge, The spectral dependence of facular contrast and solar irradiance variations, *Astron. Astrophys.*, *345*, 635, 1999.

Unruh, Y. C., S. K. Solanki, and M. Fligge, Modelling solar irradiance variations: Comparison with observations, including line-ratio variations, *Space Science Reviews*, *94*, 145, 2000.

Vernazza, J. E., E. H. Avrett, and R. Loeser, Structure of the solar chromosphere. II—The underlying photosphere and temperature-minimum region, *Astrophys. J. (Supp.)*, *30*, 1, 1976.

Vernazza, J. E., E. H. Avrett, and R. Loeser, Structure of the solar chromosphere. III—Models of the EUV brightness components of the quiet-sun, *Astrophys. J. (Supp.)*, *45*, 635, 1981.

Viereck, R., L. Puga, D. McMullin, D. Judge, M. Weber, and W. K. Tobiska, The Mg II index: A proxy for solar EUV, *Geophys. Res. Lett.*, *28*, 1343-1346, 2001.

Vigouroux, A., J. M. Pap, and P. Delache, Estimating longterm solar irradiance variability: A new approach, *Solar Phys.*, *176*, 1-21, 1997.

Vrsnak, B., V. Ruzdjak, and D. Placko, Calcium plage intensity and solar irradiance variations, *Solar Phys.*, *133*, 205, 1991.

Wallace, L., K. Hinkle, and W. C. Livingston, An atlas of the spectrum of the solar photosphere from 13,500 to 28,000 cm^{-1} (3750 to 7405 Å), NSO Technical Report; 98-01, NOAO, Tucson, 1998.

Wallace, L., K. Hinkle, and W. C. Livingston, An atlas of sunspot umbral spectra in the visible, from 15,000 to 25,500 cm^{-1} (3920 to 6664 Å), NSO technical report; 00-01, NOAO, Tucson, 2000.

Wallace, L., W. C. Livingston, and P. Bernath, An atlas of the sunspot spectrum from 470 to 1233 cm^{-1} (8.1 to 21 μm) and the photospheric spectrum from 460 to 630 cm^{-1} (16 to 22 μm), NSO Technical Report; 94-01 NOAO, Tucson, 1994.

Walton, S. R. and D. G. Preminger, Restoration and photometry of full-disk solar images, *Astrophys. J.*, *514*, 959, 1999.

Warren, H. P., J. T. Mariska, J. Lean, W. Marquette, and A. Johannesson, Modeling solar extreme ultraviolet irradiance variability using emission measure distributions, *Geophys. Res. Lett.*, *23*, 2207, 1996.

Warren, H. P., J. T. Mariska, and J. Lean, A new reference spectrum for the EUV irradiance of the quiet Sun 1. Emission measure formulation, *J. Geophys. Res.*, *103*, 12077-12090, 1998a.

Warren, H. P., J. T. Mariska, and J. Lean, A new reference spectrum for the EUV irradiance of the quiet Sun 2. Comparisons with observations and previous models, *J. Geophys. Res.*, *103*, 12091-12102, 1998b.

Warren, H. P., J. T. Mariska, and J. Lean, A new model of solar EUV irradiance variability: 1. Model formulation, *J. Geophys. Res.*, *106*, 15745-15758, 2001.

Warren, H. P., J. T. Mariska, and K. Wilhelm, High-resolution observations of the solar Hydrogen Lyman lines in the quiet sun with the SUMER instrument on SOHO, *Astrophys. J. (Supp.)*, *119*, 105-120, 1998.

White, O. R., The solar spectral irradiances from X-ray to radio wavelengths, in *The Sun as a Variable Star: Solar and Stellar Irradiance Variations, IAU Symposium #143*, ed. by J. M. Pap, C. Fröhlich, H. S. Hudson and S. Solanki, Cambridge University Press, Cambridge, p. 45, 1994.

White, O. R., G. DeToma, G. J. Rottman, T. N. Woods, and B. G. Knapp, Effect of spectral resolution on the Mg II index as a measure of solar variability, *Solar Phys.*, *177*, 89-103, 1998.

White, O. R., J. Fontenla, and P. A. Fox, Extreme solar cycle variability in strong lines between 200 and 400 nm, *Space Science Reviews*, *94*, 67, 2000.

White, O. R. and W. C. Livingston, Solar luminosity variation. III—Calcium K variation from solar minimum to maximum in cycle 21, *Astrophys. J.*, *249*, 798, 1981.

Willson, R. C., Solar Irradiance Variations, in *The many faces of the sun: a summary of the results from NASA's Solar Maximum Mission*, ed. by K. T. Strong, J. L. R. Saba, B. M. Haisch and J. T. Schmelz, Springer, New York, p.19, 1999.

Willson, R. C. and H. S. Hudson, Variations of solar irradiance, *Astrophys. J.*, *254*, L185, 1981.

Willson, R. C. and A. V. Mordvinov, Time-frequency analysis of total solar irradiance variations, *Geophys. Res. Lett.*, *26*, 3613, 1999.

Woods, T. N., W. K. Tobiska, G. J. Rottman, and J. R. Worden, Improved solar Lyman alpha irradiance modeling from 1947 through 1999 based on UARS observations, *J. Geophys. Res.*, *105*, 27195, 2000.

Woods, T. N., G. J. Rottman, S. M. Bailey, S. Solomon, and J. R. Worden, Solar extreme ultraviolet irradiance measurements during solar cycle 22, *Solar Phys.*, *177*, 133-146, 1998.

Worden, J. R., Ph.D. Thesis, University of Colorado, Boulder, 1997.

Worden, J. R., O. R. White, and T. N. Woods, Plage and enhanced network indices derived from Ca II K spectroheliograms, *Solar Phys.*, *177*, 255-264, 1998.

Worden, J. R., T. N. Woods, W. N. Neupert, and J.-P. Delaboudinière, Evolution of chromospheric structures: How chromospheric structures contribute to the solar He II 30.4 nanometer irradiance and variability, *Astrophys. J.*, *511*, 965-975, 1999.

P. Fox, High Altitude Observatory, National Center for Atmospheric Research, P. O. Box 3000, Boulder, CO 80307 USA (e-mail: pfox@ucar.edu)

Solar Irradiance Reference Spectra

Gérard Thuillier[1], Linton Floyd[2], Thomas N. Woods[3], Richard Cebula[4], Ernest Hilsenrath[5], Michel Hersé[1], and Dietrich Labs[6]

The solar spectrum is a key input for the study of the planetary atmospheres. It allows the understanding through theoretical modeling of the atmospheric properties (e.g., composition and variability). Furthermore, a reference model is useful for the preparation of instruments and platforms to be operated in space. New composite solar irradiance spectra are formed from 0.1 to 2400 nm using recent measurements for two distinct time periods during solar cycle 22. These two time periods correspond to the activity levels encountered during the ATmospheric Laboratory for Applications and Science (ATLAS) Space Shuttle missions which were moderately high (ATLAS 1, March 1992) and low (ATLAS 3, November 1994). The two reference times span approximately half of the total solar cycle amplitude in terms of the Mg II and F10.7 indices. The accuracy of the two presented spectra varies from 40% in the X-ray range to a mean of 3% in the UV, visible, and near IR ranges. After integration over all wavelengths, a comparison with the total solar irradiance measured at the same time shows an agreement of the order of 1%.

1. IMPORTANCE OF A REFERENCE SOLAR SPECTRUM

The solar output, composed of particles and electromagnetic radiation (photons), is the main source of energy for planetary atmospheres and, in particular, Earth's climate system. The composition, thermal structure, and dynamics of the Earth's atmosphere are a consequence of the solar energy input on an atmosphere mainly made of oxygen and nitrogen. Reactions such as photodissociation, photoabsorption, and photo-ionization are wavelength dependent. Therefore, knowledge of the absolute value of the solar spectral irradiance and its variability through time is needed to understand Earth's atmospheric properties. Solar irradiance and spectral irradiance variability are described by several authors [*Fröhlich; Rottman et al.; Woods et al.;* this volume] while this article describes the solar spectral irradiance from X-ray ultraviolet (XUV) to the infrared (IR).

The absorption of solar photons in different regions of the Earth's atmosphere as a function of wavelength is summarized in Table 1. Solar photons of wavelengths shorter than 450 nm are involved in many key reactions with the chemical species of the Earth's atmosphere, [e.g. *Meier*, 1991]. Photochemical reactions with ozone and nitrogen dioxide are catalytic reactions which require the knowledge of the solar irradiance with great accuracy, [e.g. *Nicolet*, 1981]. The absorption of solar protons largely determines the atmosphere's thermal structure. For example, the temperature profile of the stratosphere and thermosphere is a consequence of the absorption of solar protons by ozone and

[1]Service d'Aéronomie du CNRS, Verrières-le-Buisson, France
[2]Interferometrics Inc., Chantilly, VA
[3]Laboratory for Atmospheric and Space Physics, University of Colorado, Boulder, CO
[4]Science Systems and Applications, Lanham, MD
[5]NASA Goddard Space Flight Center, Greenbelt, MD
[6]Landessternwarte, Heidelberg, Germany

Solar Variability and its Effects on Climate
Geophysical Monograph 141
Copyright 2004 by the American Geophysical Union
10.1029/141GM13

Table 1. The origin of solar emissions and affected regions of the neutral atmosphere and ionosphere (F, E, D regions) of the Earth.

Spectral range	Solar origination	Earth's atmospheric absorption	Wavelength range (nm)
Visible and IR	photosphere	troposphere and stratosphere	400-10000
UV	upper photosphere and chromosphere	stratosphere, mesosphere, and lower thermosphere	120-400
			10-120
EUV	chromosphere and transition region	thermosphere, ionosphere [E, F regions]	
XUV (X-rays)	corona	ionosphere [E, D regions]	0-10

atomic oxygen [*Haigh*, this volume]. Longward of 450 nm, the solar radiation is absorbed by minor tropospheric constituents (aerosols and clouds) and the Earth's surface (continents and oceans). These absorptions provide the basic energy input that determines the thermal properties of the lower atmosphere.

For climate studies and modeling, until recently, the solar input has been characterized by changes in the received total solar irradiance. However, the observed 0.1% change in total solar irradiance over the 11-year cycle is understood to be too small to produce significant climate change. Nevertheless, larger and/or secular changes cannot be excluded [*Muscheler et al.*, this volume]. In addition, recent studies have suggested that other mechanisms may amplify the effects of solar variability, [e.g. *Soon et al.*, 1996]. For example, variations in UV radiation modulate changes in the troposphere-stratosphere temperature vertical gradient thus affecting the Hadley circulation [*Haigh*; *Hood*, this volume]. Further, the absorption of infrared solar flux by water vapor and carbon dioxide in the Earth's atmosphere plays an important role in determining the Earth's radiation budget. While the immediate effect of UV on the stratosphere and the effect of the visible and IR on the troposphere are rather well understood, the deeper consequences for the evolution of the atmosphere and climate variability are only recently being explored. The existence of catalytic reactions and the possible existence of positive feedback in the climate system require accurate values of the solar spectral irradiance and its variation.

For the understanding of these mechanisms, the solar spectral irradiance is a key input in most atmospheric models. Using the same reference solar irradiance spectrum allows better understanding of the models' properties by comparing their outputs and also by comparing their predictions with observations. For these aeronomic studies, reference solar spectra are useful and sometimes required for many environmental applications, thermal modeling of space instruments and space platforms, and estimations of the material aging in the conditions of space. It is in these ways that the use of a reference solar spectrum services a large community of scientists and engineers.

Therefore, for atmospheric applications, a solar reference spectrum should ideally have the following characteristics:

— absolute spectral irradiance with the best achievable accuracy,
— spectral range from XUV to IR,
— two distinct levels of solar activity (close to minimum and maximum),
— spectral sampling / resolution of 1 nm or better, and
— data reduced to 1 AU.

2. MEASUREMENTS

No single instrument is able to measure from the X-ray to the IR spectral range for which, in addition, the solar spectral irradiance varies by more than a factor of 10^5. Therefore, a composite spectrum must be built from several spectra obtained by different techniques.

2.1 Observations From the Ground

At sea level, no solar photons below 295 nm are observed because of their absorption by ozone. Usually, solar spectral irradiance measurements from the ground have a lower wavelength limit of 330 nm. A careful examination of the ground recorded spectra reveals many absorption features. After eliminating the Fraunhofer lines, several absorptions remain of telluric origin. This may be verified by their seasonal and local time changes as well as by the effect of the altitude of the observations. Despite these difficulties, solar observations in the visible and infrared spectral domains can be successfully made from the ground. Having observed a wavelength domain, free of saturated absorptions bands, the extraterrestrial irradiance is still not directly measured because of Rayleigh and aerosols scattering, and absorptions by species such as ozone, water vapor, and nitrogen and carbon compounds. These absorptions which make accurate solar measurements difficult, are in turn useful for atmospheric studies. Despite these difficulties, measurements from ground (and especially from high altitude observatories) present several advantages:

i) careful measurements can be performed with frequent checks of the instrument,
ii) the instrument calibration can be made as many times as necessary,
iii) few weight, volume, and power limitations exist that are often necessary constraints on satellite instruments.

Nevertheless, due to absorptions by various constituents, atmospheric transmission must be taken into account to derive exospheric spectral irradiance. For that, the Bouguer's method assumes that the extinction can be expressed as

$$I(\lambda) = I_0(\lambda) \, \text{Exp-}N(L)\,\sigma(\lambda) \qquad (1)$$

$I_0(\lambda)$ is the solar irradiance outside the atmosphere at wavelength λ. $N(L)$ is the number of absorbers along the line of sight (L) and $\sigma(\lambda)$ is the absorption cross section. It is assumed that the total number of absorbers in direction (L) can be related to its value at zenith by writing:

$$N(L) = N_0/\cos(\mu) \qquad (2)$$

where μ is the zenith angle.

Consequently, the representation of the logarithm of the measurement is a linear function of $1/\cos(\mu)$ for a given wavelength. A least squares method is generally used to calculate $\log(I_0(\lambda))$ by extrapolating to zero air-mass. $N_0\sigma(\lambda)$ is indeed a sum of terms corresponding to absorptions by aerosols, Rayleigh scattering, and the absorption by several chemical species depending on the spectral domain of observation. This method can be generalized by taking into account the effects of refraction and Earth's curvature [*Link and Neuzil*, 1969]. Data as a function of μ are obtained as a function of the local solar time. For example, in a spectral domain with absorptions by ozone, the diurnal variation [*Brasseur and Solomon*, 1984] of the absorbents has to be taken into account.

2.2 Airborne Observations

From airplanes, the measurement techniques are similar to that used from the ground except that the atmospheric correction is typically much less. Inevitably, the solar irradiance is measured through an airplane window. Accordingly, the transmittance of the window must be taken into account through measurements for different incidence angles. Experimental operations are no less convenient as those from the ground. Calibration devices can be an integral part of the experimental apparatus.

For balloon observations, the equipment should be more compact. The thermal environment has to be carefully planned to avoid excessive temperatures. Further, a greater use of automation is required because obviously, there is no observer to ensure that experimental operations remain nominal. The advantage of balloon observations is that the need for the air-mass correction is considerably reduced. Additionally, at the altitudes where solar measurements are generally performed, there is no weather dependence except for the launch and instrument retrieval (needed for post-flight calibration).

For rocket or spacecraft observations, the instruments face other types of difficulties, ranging from limited resources (volume, mass, power), thermal environment, and mechanical vibrations during launch. Rocket measurements suffer from very limited observation times, but the instrument is usually recovered permitting post-flight calibration. For most space experiments, the instrument is not retrieved with exceptions for payloads aboard the Space Shuttle and the EUropean REtrieval CArrier (EURECA) platform.

Presently, most of the solar irradiance observations are carried out from space. A large effort has been devoted to UV measurements because of the importance of the ozone problem. By comparison, few efforts have been dedicated to visible and very few to the IR domain. It should be noted that for the latter, there is a near total absence of observations between about 1975 and 1990.

2.3 Available Data

2.3.1 Ground-based measurements. The very early measurements are reviewed by *Labs and Neckel* [1968]. *Labs and Neckel* [1962] carried out observations from the Jungfraujoch at 3600 m altitude using a double monochromator. The bandpass was nearly perfectly rectangular. The solar light was collected with a Cassegrain telescope. The whole system was checked through the use of tungsten ribbon lamps having the same optical path as the solar photons except that a neutral density filter was used for the Sun. The tungsten ribbon lamps were, before and after the measurements, calibrated against a blackbody at 2500 K operated at the Heidelberg Observatory. These data have been revised in 1984 through incorporation of high resolution spectral observations from the National Solar Obervatory at Kitt Peak [*Neckel and Labs*, 1984]. Their absolute solar spectral irradiance, given from 330 to 1247 nm, is still extensively used.

Burlov-Vasiljev et al. [1995] carried out measurements from the high altitude observatory of the Ukrainian Academy of Sciences at Terskol Peak (Caucasus, 3100 m).

The equipment consists of a grating spectrometer, a coelostat, and a tungsten ribbon lamp. Special attention is given to effects of polarization because of the diurnal variation of the angle of incidence with respect to the coelostat mirror. Solar spectral irradiance is given with a 5 nm sampling at 1 nm resolution from 332.5 to 667.5 nm which was extended later to 1100 nm.

The starting point of the *Lockwood et al.* [1992] investigation is the absolute calibration of the star Vega. Several Vega irradiance spectra were taken over the wavelength range 320-1000 nm using an instrument calibrated by a known blackbody. These spectra were found to agree to within a few percent. Consequently, by comparing the solar irradiance to the Vega stellar irradiance, the former can be determined. The problem presented by this technique is the achievement of a 10^{10} reduction of the solar flux with a precision of a few percent or better. *Lockwood et al.* [1992] succeeded in this task by setting a tiny precision pinhole (30 μm) to form a starlike source from the Sun which is then analyzed by a stellar spectrometer. The main advantages of this system are an identical optical path in stellar and solar modes, and an attenuation that can be precisely computed using the diffraction theory of a circular pinhole. However, a limitation of the absolute accuracy directly is derived from the knowledge of the pinhole shape and dimension. Solar spectral irradiance were obtained from 330 to 850 nm.

Saiedy and Goody [1959] made observations at 8.63, 11.10 and 12.02 μm using a siderostat, a double monochromator, a chopping system to reduce the thermal background, and a blackbody heated at 1300 K. *Kondratyev et al.* [1965] also performed observations from 3 to 13 μm. *Peyturaux* [1968] made observations from Mount Louis, France at 1600 m altitude from 447 to 863 nm using a siderostat, a prism monochromator, and a blackbody heated at 2600 K. Later, *Koutchmy and Peyturaux* [1968] extended their observations between 3.5 to 35 μm.

Most of the IR observations curiously stopped in the 1970's for unknown reasons.

2.3.2 Airplane measurements. The NASA Convair CV-990 was used for solar observations. It was a research airplane providing extensive facilities and allowing long duration observations and flight stability. *Arvesen et al.* [1969] carried out solar spectral measurements from 300 to 2500 nm using a modified Cary spectrometer, which employed a grating-prism double monochromator. The entrance slit faced a rotating integrating sphere, allowing a direct comparison of the Sun and lamp radiation. Eleven flights were made between 11.6 and 12.5 km altitude. Solar light entering the instrument through a quartz window which was carefully calibrated for its transmissivity (± 0.4%).

On board the same aircraft, total and spectral irradiances were measured using several different instruments [*Thekaekara and Drummond*, 1971; *Thekaekara*, 1974]. A Perkin-Elmer monochromator, a Zeiss monochromator, a filter radiometer as well as a Michelson interferometer were operated. A diffusing mirror and a sapphire window were used to collect the sunlight. The whole system was calibrated using a quartz-iodine standard lamp.

2.3.3 Balloon measurements. *Murcray et al.* [1964] flew a single monochromator at an altitude of 31 km collecting data for wavelengths of 4 to 5 μm. Calibration was performed on the ground by use of a blackbody heated to 2500 K, a temperature that was measured by a pyrometer. This observational method was performed by *Hall and Anderson* [1991] to measure the UV solar spectral irradiance from 200 to 310 nm with 10 pm resolution and a wavelength scale accuracy of 4 pm. The measurements were made in 1983 and were higher by about 5% with respect to the solar spectral irradiances obtained by *Labs et al.* [1987] on SpaceLab I and by *VanHoosier et al.* [1988] on SpaceLab II. Although these balloon measurements were made from 40 km altitude, some corrections due to ozone absorption were required. This likely explains the slight difference between these balloon measurements and similar ones from space. These measured spectra are very useful for wavelength calibration needed for making accurate absorption corrections. However, because of the difficulties in making accurate absorption corrections, most of the current measurements are carried out from space (as described in the next section).

2.3.4 Spacecraft measurements. Solar EUV and UV irradiance measurements are reviewed by *Woods et al.* [this volume] and *Rottman et al.* [this volume], respectively. Several observations of the near UV up to 400 nm have been made on board SpaceLab I [*Labs et al.*, 1987], SpaceLab II [*VanHoosier et al.*, 1988], the Upper Atmosphere Research Satellite (UARS) and the ATmospheric Laboratory for Applications and Science (ATLAS) missions. The UARS measurements include those from the Solar Ultraviolet Spectral Irradiance Monitor (SUSIM) [*Brueckner et al.*, 1993] and the SOLar STEllar Irradiance Comparison Experiment (SOLSTICE) [*Rottman et al.*, 1993]. On board the three ATLAS missions, three spectrometers were operated at the same time, namely, a second SUSIM, similar to the UARS instrument [*VanHoosier et al.*, 1996, *Floyd et al.*, 1998, 2001], the Shuttle Solar Backscatter UltraViolet (SSBUV) [*Cebula et al.*, 1996] and the SOLar SPECtrum instrument (SOLSPEC) [*Thuillier et al.*, 1997, 1998a, 1998b].

In addition, near UV-visible spectral irradiance is also provided by the Solar Backscatter Ultraviolet model 2

(SBUV/2) instruments flying on NOAA spacecraft. Its primary mission is long-term observations of global column ozone and its altitude distribution in the stratosphere [*Cebula et al., 1998*]. The Global Ozone Monitoring Experiment (GOME) is a double spectrometer dedicated to Earth's middle atmosphere physics by observing the backscattered and direct solar light in the near UV and visible [*Weber et al., 1998*]. GOME provides solar spectral irradiance from 240 to 790 nm.

Above 400 nm, the visible and near IR domains have been observed by the SOLSPEC and SOlar SPectrum (SOSP) instruments placed on ATLAS and EURECA platforms, respectively [*Thuillier et al., 1998a, 1998b, 2003*]. SOLSPEC and SOSP are two nearly identical instruments, both composed of three double holographic gratings spectrometers of similar optical design dedicated to the UV, visible, and IR.

Table 2 summarizes the main characteristics of the near UV, visible and IR data.

2.4 Composite Spectra

Composite spectra are made from solar atmospheric modeling and/or through compilations of several independent spectra, and covering large wavelength ranges. However, certain recent composite spectra have also used previously published results derived from relatively old data. This is why we have reported on them in Section 2.2. When two data sets are merged together, a certain smoothing and/or adjustment is made at the junction, and furthermore, a normalization to a given value of the total solar irradiance is applied. This explains why when detailed comparisons are performed, some differences may appear even when the solar spectra are based on the very same measurements or models.

The *Labs and Neckel* [1968] spectrum is constructed for 205 to 330 nm from a compilation of rocket observations made by *Tousey* [1963]. From 328.8 to 656.9 nm, the spectral irradiance is deduced from the *Labs and Neckel* [1962] measurements with center-to-limb variation corrections derived from data obtained by several authors, especially *Goldberg and Pierce* [1959], and *David and Elste* [1962]. From 656.9 to 1250 nm, their own measurements were used with appropriate center-to-limb variations (close to unity). From 1250 to 2500 nm, the measurements made by *Pierce* [1954] are used with a correction factor to fit a model-distribution between 1000 and 1500 nm, taking into account their own measurements as well as those of Pierce. From 2.5 to 100 μm, the *Holweger* [1967] and the Bilderberg models [*Gingerich and De Jager*, 1968] are used after corrections for temperature deduced from various intensity measurements (including IR) by *Saiedy and Goody* [1959], *Saiedy* [1960] and *Murcray et al.* [1964].

Smith and Gotttlieb [1974] proposed a spectrum from 0.2 nm to 2 cm. It is based on several data sets from rocket and satellite measurements below 330 nm. Above 330 nm, it is based on the following spectra: *Labs and Neckel* [1968] from 330 to 1000 nm, *Arvesen et al.* [1969] and *Pierce* [1954] from 1000 to 2400 nm, and above 2400 nm, *Farmer and Todd* [1964], *Koutchmy and Peyturaux* [1968], *Murcray et al.* [1964], and the results of *Saiedy and Goody* [1959], and finally above 13 μm, the data are known in terms of solar temperature, which was converted into irradiance by use of the Planck function.

The World Radiation Center (WRC) spectrum [*Wehrli*, 1985] is a composite spectrum made from several contributions: from 200 to 310 nm [*Brasseur and Simon*, 1981], from 310 to 330 nm [*Arvesen et al.*, 1969], from 330 to 869 nm [*Neckel and Labs*, 1984], and from 869 nm to 20 μm [*Smith and Gottlieb*, 1974].

Kurucz [1995] has also generated a solar irradiance spectral model using the solar atmosphere opacity as a function of wavelength at resolution $\Delta\lambda/\lambda = 500\ 000$, and adding a solar continuum model.

Colina et al. [1996] built a composite solar irradiance spectrum from 120 to 2500 nm using the UARS data in UV up to 410 nm, the spectrum of *Neckel and Labs* [1984] up to 870 nm, *Arvesen et al.* [1969] up to 960 nm, and a solar model from *Kurucz* [1993a].

A zero air-mass solar spectral irradiance standard has been made by the American Society for Testing and Materials [ASTM, 2000]. It is a composite spectrum made as follows:

i)   from 119 to 410 nm, the mean UARS spectrum built using SOLSTICE and SUSIM data at the ATLAS 2 period (April 1993),

ii) from 410 to 825 nm, the spectrum of *Neckel and Labs* [1984],

Table 2. Near-UV, visible and infrared data. $\Delta\lambda$ is for spectral resolution.

Authors	Range (nm)	$\Delta\lambda$(nm)
Arvesen et al. [1969]	300–2495	0.1 to 0.3
Neckel and Labs [1984]	330–1247	2
Burlov-Vasiljev [1995]	332–1062	1
Weber et al. [1998]	240–790	0.25
SOLSPEC	200–870	1
SOSP	850–2500	20

iii) from 825 to 4 μm, the synthetic spectrum computed by *Kurucz* [1993b],

iv) from 4 to 20 μm, the spectrum of *Smith and Gottlieb* [1974],

v) above 20 μm, the values are obtained from the logarithmic irradiances given by *Smith and Gottlieb* [1974].

At each junction a fit is made, in particular between 330 and 410 nm where the UARS mean spectrum has been reduced by 3.2%. The explanation of this reduction is given in Section 4.2.

Thuillier et al. [2003] have generated a solar spectrum from 200 to 2400 nm using only the SOLSPEC and the SOSP data from ATLAS 1 period (March 1992) and the beginning of EURECA mission (August-September 1992). After completing the spectrum above 2400 nm using the *Kurucz* [1995] model, the total solar irradiance (TSI) was calculated and compared with the TSI measured at the time of ATLAS 1 observations (assuming no variation in the IR domain). A difference of 1.4% was found, which is within the estimated uncertainties of the SOLSPEC and SOSP measurements. This spectrum is named hereafter SOLSPEC composite ATLAS 1.

Table 3 summarizes the available composite and model spectra.

2.5 Comparisons of Measurement and Model Spectra

2.5.1 The near UV domain. The availability of coincident measurements from three ATLAS and two UARS instruments provided the opportunity to perform comparisons among them. The results of these comparisons are given by *Woods et al.* [1996], *Cebula et al.* [1996], and *Thuillier et al.* [1997, 1998a, 1998b, 2003]. These studies show that the absolute spectral solar irradiance in the near UV is now known with an upper accuracy limit better than 4%, an amount dominated by systematic uncertainties.

2.5.2 The visible domain. Prior to the work of Labs and Neckel, many spectra in the visible and the near-visible UV

Table 3. Spectral range of models and composite spectra.

Authors	Range (μm)
Labs and Neckel [1968]	0.205–100
Smith and Gottlieb [1974]	0.0002–20000
WRC [1985]	0.1995–20
Kurucz [1995]	0.2–200
Colina et al. [1996]	0.120–2.5
ASTM [2000]	0.1195–1000
SOLSPEC composite ATLAS 1	0.2–2.4
New Composite Spectra ATLAS 1,3	0.0001–2.4

disagreed by about 30% (see for example in *Pierce and Allen* [1977]) especially around 400 nm and below. The reason was that ozone absorption became more important at decreasing wavelengths, providing larger corrections of reduced accuracy.

Comparison between SOLSPEC and the *Burlov-Vassiljev* [1995] spectra shows RMS differences (likely due to the presence of Fraunhofer lines) of the order of 2%. We note that the mean difference is 1% from 350 to 870 nm, including the critical wavelengths below 420 nm. Comparing SOLSPEC with *Neckel and Labs* [1984] spectrum shows a remarkable agreement (below 2%, mean) from 350 to 870 nm. However, below 420 nm the difference increases toward shorter wavelengths. This difference was originally reported by *Peytureaux* [1968] to be about 8%, in agreement with *Shaw* [1982] (from the Mauna Loa observations) reporting 4% at 416 and 460 nm. The SpaceLab II SUSIM observations [*VanHoosier et al.*, 1988] showed greater values (by a few percent) than those given by *Neckel and Labs* [1984], as well as the SSBUV [*Cebula et al.*, 1996] from ATLAS 1 and SOLSTICE data [*Woods et al.*, 1996]. From the ground, *Burlov-Vassiljev et al.*, [1995] found an irradiance deficit decreasing from 330 to about 450 nm having a typical value of 5% at 380 nm. Detailed comparisons between SOLSPEC results with available visible spectra are made by Thuillier *et al.* [1998a, 1998b, 2003].

Figure 1 illustrates the comparisons between the results reported above and the model spectrum of *Kurucz* [1995]. For that, a running mean over 5 nm has been made in order to avoid the large oscillations generated by pairs of spectra having even a very small wavelength scale difference and sampling, and taking the ratio of each spectrum to that of *Kurucz* [1995]. At 450 nm, a deviation of 10% appears no matter what spectrum is used, likely generated by the spectrum of *Kurucz* [1995]. Above 450 nm, all ratios (Figure 1a) stay within -3 to +5% [*Lockwood et al.*, 1992; *Burlov-Vasiljev et al.*, 1995], and within ±3% with respect to the SOLSPEC composite ATLAS 1 and *Neckel and Labs* [1984] (Figure 1b). However, the latter deviates significantly above 850 nm.

2.5.3 IR domain up to 2400 nm. The recent IR solar spectral observations are essentially dedicated to high resolution radiance observations in relative scale [*Grevesse and Sauval*, 1991]. Consequently, most of the absolute spectral solar irradiance observations in this domain are rather old [*Arvesen et al.*, 1969; *Thekaekara*, 1974]. However, *Neckel and Labs* [1984] conducted observations up to 1250 nm and SOSP on board EURECA, providing measurements up to 2400 nm. Figure 2 shows comparisons of these spectra with the model spectrum of *Kurucz* [1995] as for Figure 1, but

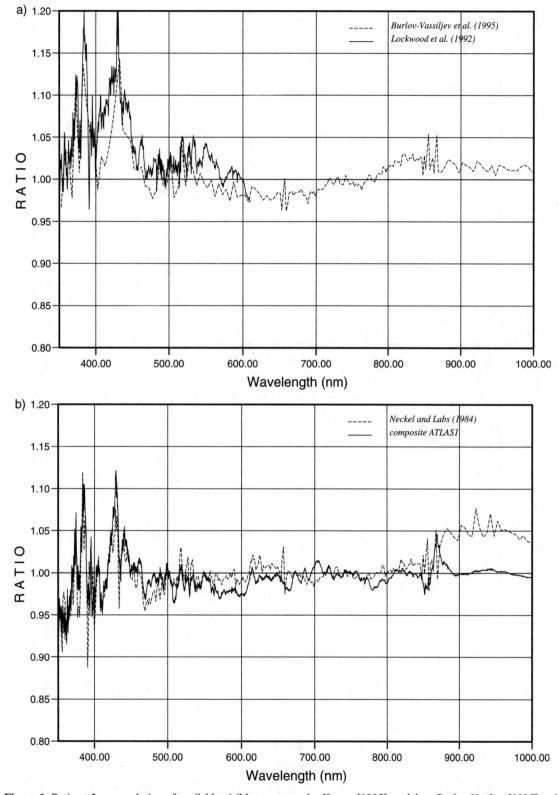

Figure 1. Ratio at 5 nm resolution of available visible spectra to the *Kurucz* [1995] model. a: *Burlov-Vasiljev* [1995] and *Lockwood et al.* [1992]; b: *Neckel and Labs* [1984] and the SOLSPEC composite ATLAS1 spectrum [*Thuillier et al.*, 2003].

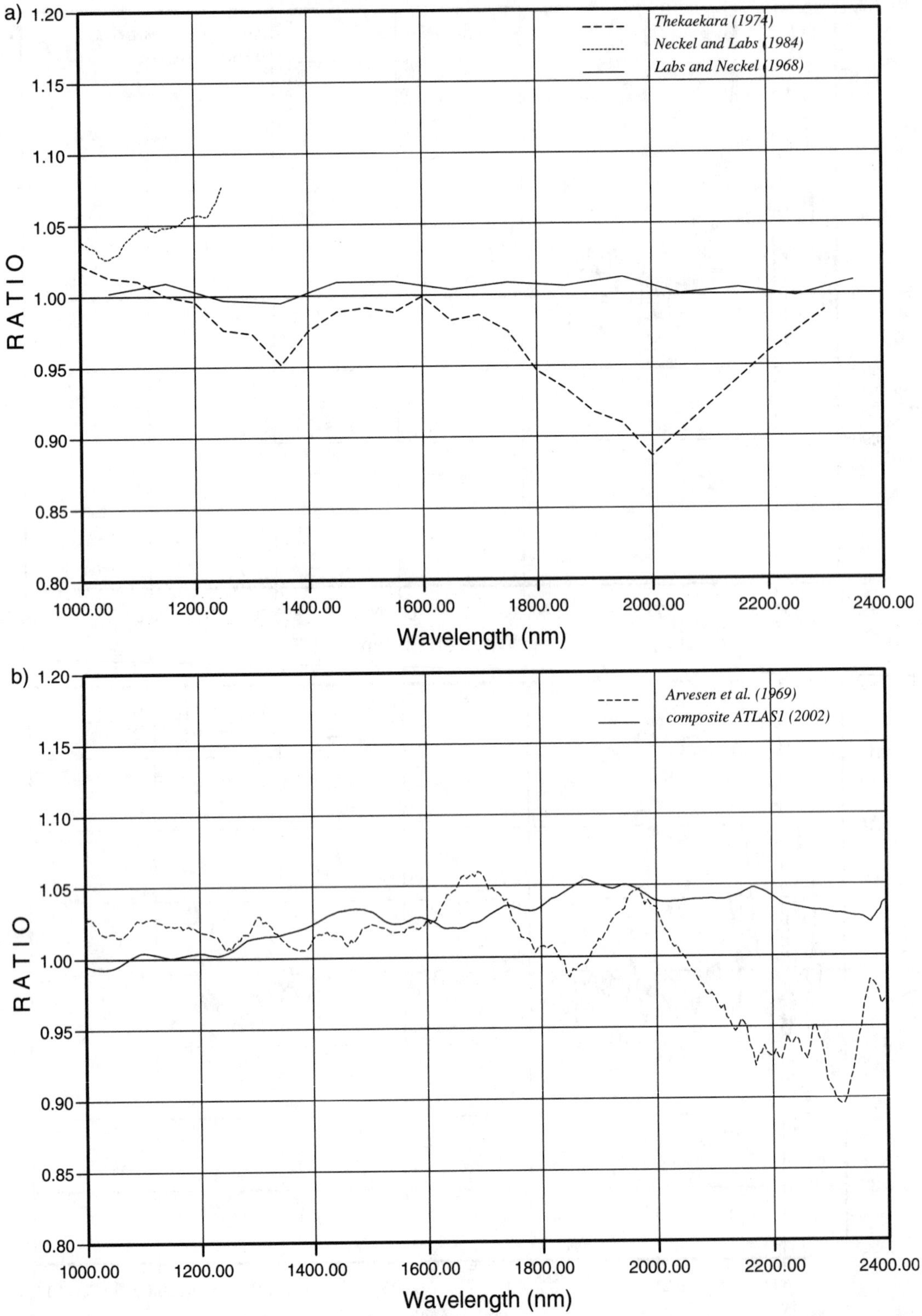

Figure 2. Ratio at 50 nm resolution of available near IR spectra to the *Kurucz* [1995] model. a: *Labs and Neckel* [1968], *Neckel and Labs* [1984] and *Thekaekara* [1974]; b: *Arvesen et al.* [1969] and the SOLSPEC composite ATLAS 1 spectrum [*Thuillier et al.*, 2003].

presented at 50 nm resolution, taking into account the spectral domain.

Figure 2a shows discrepancies between the spectra of *Kurucz* [1995] and *Thekaekara* [1974]. The agreement between the Kurucz and the *Labs and Neckel* [1968] spectra is good, likely because the same base data were used to construct the latter with a level adjustment. We note that the irradiance given by *Neckel and Labs* [1984] is greater than that of all of the other models from 850 to 1250 nm.

Figure 2b shows the behavior of the *Arvesen et al.* [1969] spectra with respect to that of *Kurucz* [1995], suggesting absorption and calibration uncertainties in the former. As reported, the SOLSPEC composite ATLAS 1 spectral irradiance is as much as 5% larger in the 1900 nm region than the model spectrum of *Kurucz* [1995].

2.5.4 Domain above 2400 nm. The few existing datasets are not recent, but nonetheless show a consistent distribution with wavelength (Figure 3a). The *Kurucz* [1995] and *Labs and Neckel* [1968] spectra are displayed together with comparisons showing a close agreement among the data sets. Figure 3b show the ratios of the *Thekaekara* [1974], ASTM [2000], *Labs and Neckel* [1968] spectra to the model spectrum of *Kurucz* [1995] at 50 nm resolution. The particular behavior of the spectrum of *Thekaekara* [1974] below 2400 nm is still observed around 3000 and 4500 nm. Near 4500 nm, some common features are present, which are explained by the data of Figure 3a which shows the irradiance of *Labs and Neckel* [1968] to be greater than the model spectrum of *Kurucz* [1995], and the irradiance of both spectra are above the observations of *Murcray et al.* [1974]. ASTM [2000] also contains this feature likely induced by the spectrum of *Smith and Gottlieb* [1974], which uses that of *Labs and Neckel* [1968].

3. NEW COMPOSITE REFERENCE SOLAR SPECTRA

Among the requirements for a solar spectral model, accuracy is the most difficult to achieve. The UV observations made in the 1970's showed discrepancies up to 20% among them in the near UV and greater at wavelengths below. The origin of these errors was identified as problems resulting from calibration sources and procedures as well as instrument aging. Actions were taken to coordinate the various measurement programs more systematically. First, all instruments were to be intercompared prior to their use in space. Missions would consist of redundant experiments that should each incorporate onboard means for their own calibration. Short-term missions would be designed for instruments' retrieval, thus allowing for their post-flight calibration. Longer-term missions would also be planned, but instrument retrieval would not necessarily be provided.

For most of the published spectra, authors made comparisons with other available spectra and discussed the differences, emphasizing simultaneous measurements. These comparisons revealed the conditions where the best agreement was obtained. In this case, data obtained from different instruments having their own design, their own calibration, if in agreement, ensure that the measurements are made in the absolute scale within their quoted accuracy. For these cases, composite spectra can be constructed. Such spectra should improve the overall accuracy over that of any of its individual component spectra. The idea is that errors, be they random or systematic, are usually reduced when more data are included.

The different types of errors affecting the irradiance spectra have distinct characteristics. Random measurement errors are present, but these are typically much smaller than the systematic uncertainties affecting the instrument responsivity and scattered light effects. This is why combining spectra from different instruments should reduce the systematic uncertainties. The composite spectrum to be built here will necessarily use data from different instruments due to the large wavelength range (XUV to IR) and in each range several spectra will be used, if possible.

3.1 Results of Merging Spectra

A validation effort has been undertaken concerning the UARS and ATLAS spectrometers by comparison of their results obtained in similar conditions. The instruments having different designs and different methods of calibration provide an important opportunity for comparisons from 120 to 400 nm. The validation of the UARS solar irradiances, through error analysis of the pre-flight calibrations, comparison between SUSIM and SOLSTICE, and comparisons to other ATLAS solar measurements, indicate that the UARS solar irradiances have an absolute uncertainty of 2 to 5% (at 1σ) as reported by *Woods et al.* [1996]. The ATLAS 1 and 2 solar irradiances were included in this original validation effort for UARS.

Similarly, *Cebula et al.* [1996] compared the solar spectral irradiance measured by three independent spectrometers placed aboard the Space Shuttle for the ATLAS 1 mission (March, 1992) namely SOLSPEC, SSBUV, and the ATLAS version of SUSIM. They were found consistent within 5%. Furthermore, these three spectra were averaged and compared with the mean SOLSTICE and SUSIM measurements carried out at the same time on board UARS [*Woods et al.*, 1996]. For the 200–350 nm spectral range, the mean devia-

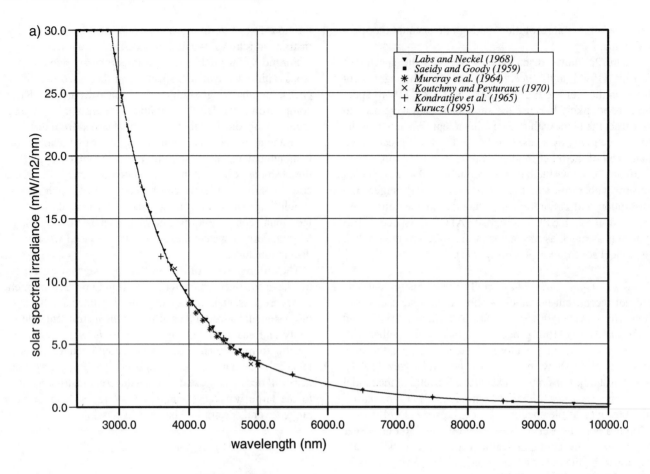

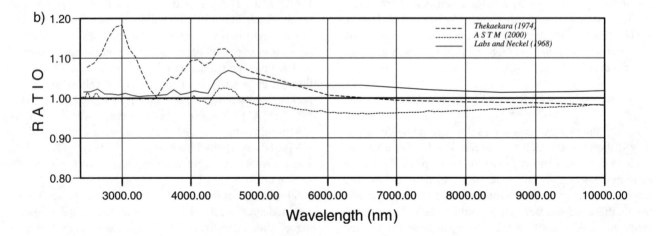

Figure 3. a: A comparison of the *Labs and Neckel* [1968] and *Kurucz* [1995] models with the available data in the 2.4-10 μm range. b: ratio at 50 nm resolution of the *Labs and Neckel* [1968], *Thekaekara* [1974] and ASTM [2000] to the *Kurucz* [1995] model.

tion between these two mean spectra was found to be 0.14% ± 0.20% with a RMS of 1% [*Cebula et al.*, 1996].

The comparison with UARS data is continued here by using the last versions of the SUSIM and SOLSTICE data (versions 20 for SUSIM and 17 for SOLSTICE). The dates chosen for these reference spectra are March 29, 1992, April 15, 1993, and November 11, 1994 for the ATLAS 1, 2, and 3 missions, respectively. For these UARS mean spectra, the SUSIM and SOLSTICE data are smoothed to a spectral resolution of 0.25 nm before combining the different measurements. The differences between the UARS reference spectra and the individual SOLSTICE and SUSIM measurements are typically less than 5% below 150 nm and even smaller at the longer wavelengths. These new UARS reference spectra are very similar to the *Woods et al.* [1996] spectra for the ATLAS 1 and 2 mission periods; however, they are based on improved algorithms. The accuracy of these UARS reference spectra, based on the measured differences in spectral irradiance between the two data sets, is approximately 3.5% and is essentially constant across the 119 to 410 nm range.

These new spectra are compared with the mean spectra calculated from the SOLSPEC, SSBUV and SUSIM measurements observing at the same time during the three ATLAS missions. Figures 4a and 4b show the comparison of the two spectra and their ratio for ATLAS 1, respectively. The ratio of their means at 5 nm resolution is unity while the corresponding RMS is 2.2%.

This ratio is presented in Figure 5 for the ATLAS 3 (for ATLAS 2, the ratio behaves similarly). The ratio departs from unity by one percent while the ratio's RMS remains 2.2% (it is 2.5% for ATLAS 2). The RMS value is driven by the strong Fraunhofer lines (Mg II, Ca II), which have different measured depths since these depend strongly on the resolution and sampling of the spectrometer. These comparisons and analyses confirm that the absolute uncertainty is between 2 to 4%. Furthermore, the mean UARS and ATLAS spectra present less differences than the individual contributing spectra.

3.2 Data Sets Composing the Reference Solar Spectrum

Based on results shown in section 3.1, it appears that merging spectral irradiance spectra is certainly an effective way to build a composite spectrum which improves on individual component spectra. As the spectrum to be built here extends from the XUV to the IR, the level of solar activity must be accounted for in the XUV, EUV, and UV wavelength regions. To take advantage of the UARS and ATLAS missions, we have gathered data corresponding to the

ATLAS 1 and ATLAS 3 missions, allowing the use of five instruments observing simultaneously in the UV and the short wavelength portion of the visible range. The levels of solar activity as indicated by the sunspot number and solar 10.7 cm radio flux (F10.7) are listed in Table 4.

Because of the wide spectral coverage, several data sets are needed and are presented below, broken down into the XUV range shortward of 30 nm, the EUV range between 30 and 120 nm, the FUV range between 120 and 200 nm, the near UV (NUV) between 200 to 400 nm, the visible between 400 and 870 nm, and finally the IR between 870 and 2400 nm.

3.2.1 XUV and EUV (0.5 to 120 nm). Woods and Rottman [2002] have produced an EUV to 200 nm spectrum using UARS SOLSTICE data and the measurements by a rocket observation made in 1994 [*Woods et al.*, 1998a]. Variability was derived from proxy models based primarily on the Atmospheric Explorer-E data. The minimum solar spectral irradiance was provided as well as the 11-year solar cycle variability applicable to solar cycle 22.

Below Ly α, the spectral solar irradiance is calculated for the conditions of the ATLAS missions 1 and 3 (Figure 6) based on Table 4 and taking into account the maximum of cycle 22 estimated when the sunspot number was equal to 200 (August 1990).

3.2.2 Far UV (120 to 200 nm). Three mean spectra from SOLSTICE and SUSIM instruments observing together on board UARS were generated for the three ATLAS periods (see section 3.1). For the construction of the reference spectrum, the mean UARS spectra during the ATLAS 1 and 3 missions are used (Figure 7).

3.2.3 Near UV (200 to 400 nm). In this wavelength region, five data sets are available. These include UARS experiments, SOLSTICE and SUSIM, and SOLSPEC, SSBUV, and SUSIM aboard ATLAS 1 and 3. These data have different spectral resolutions, slit functions and may differ slightly in their wavelength scales. When doing the averaging, this can cause additional errors near wide Fraunhofer lines. Furthermore, adjusting each wavelength scale (within their quoted accuracy) of the five instruments, is not achievable without some ambiguity. Consequently, the mean of the five spectra is calculated by a linear interpolation to the wavelength scale of the spectrum having the smallest sampling interval. The results are shown in Figure 8 for ATLAS 1 and 3. The comparison with the UARS mean spectra (section 3.1) in the 200–400 nm range at 5 nm resolution reveals an agreement better than 0.5% for the mean

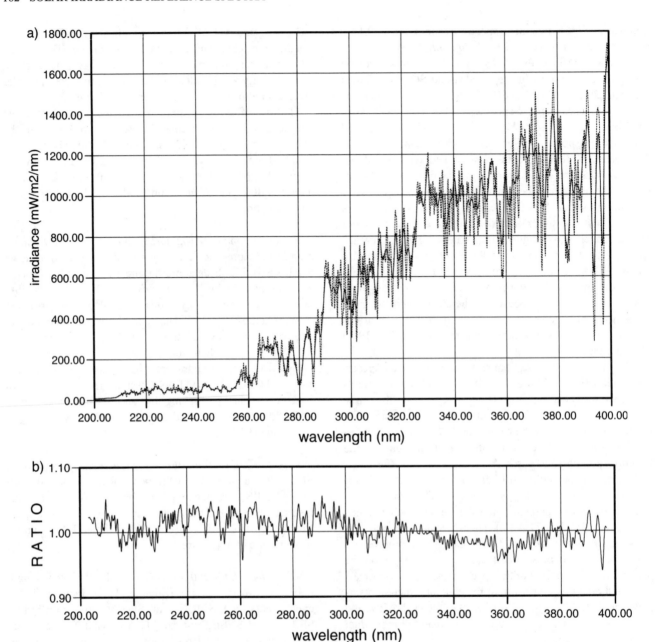

Figure 4. Comparison between the mean UARS spectrum (dashed line) and the mean ATLAS 1 spectrum (solid line). a: these two spectra displayed together. b : ratio of these two spectra at 5 nm resolution.

and a RMS difference of 2%. The 5 nm resolution for smoothing was chosen for consistency with previous works using similar data sets (Figure 8).

3.2.4 Visible (400 to 870 nm). The visible spectrum, measured by the SOLSPEC spectrometer during the three ATLAS missions, shows RMS differences not greater than

1.7%, while the mean difference remains 2%. These three spectra are, therefore, considered to be consistent.

A large portion of solar irradiance variability occurs below 300 nm [*Floyd et al.,* 1998; *DeLand and Cebula,* 1998]. It also exists in the core of certain Fraunhofer lines, e.g., Ca II (393 nm) and He II (1083 nm). For other lines, the equivalent width weakly varies, Mn (539 nm) is quoted

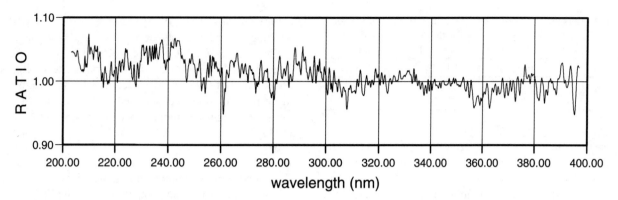

Figure 5. Comparison between the mean UARS spectrum and the mean ATLAS 3 spectrum by their ratio at 5 nm resolution.

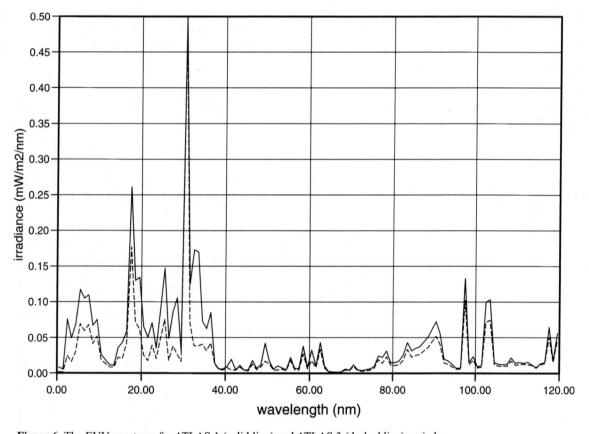

Figure 6. The EUV spectrum for ATLAS 1 (solid line) and ATLAS 3 (dashed line) periods.

to 2%, and much lower for others [*Livingston*, 1992]. Furthermore, for the 11-year cycle, a solar temperature change of about 1.5 K was found by *Gray and Livingston* [1997]. The very small solar activity effect on the visible solar irradiance has been estimated by *Fontenla et al.*

[1999] who calculated a change of 0.1% on the solar continuum. Consequently, at one nanometer resolution, the solar variability in the visible is barely detectable.

In order to reduce noise effects during solar observations and uncertainties occurring in the photometric calibration,

Table 4. Sunspot number (monthly mean, Rz), the daily F10.7, and the 81-day smoothed <F10.7> for the ATLAS 1 and 3 missions on March 29, 1992 and November 11, 1994, respectively.

Mission	Rz	F10.7	<F10.7>
ATLAS 1	121	192	171
ATLAS 3	20	77.5	83.5

we use the mean of the three SOLSPEC visible spectra. The mean SOLSPEC visible spectrum is shown in Figure 9a together with the *Neckel and Labs* [1984] and *Burlov-Vasiljev* [1995] spectra. The ratio to the mean of each of the three SOLSPEC spectra is shown at 5 nm resolution in Figure 9b. Their means are centered on unity with a RMS of 2 and 1.4%, respectively. However, the lower spectral irradiance of the *Neckel and Labs* [1984] spectrum with respect to the two others is shown below 450 nm. The difference of the ratios from unity is mostly due to the strong Fraunhofer lines (Fe I at 427 nm and Ca II at 855 nm). The use of

Burlov-Vasiljev et al. [1995] spectrum was considered, but its sampling at 5 nm was too low for our requirements.

The mean SOLSPEC visible spectrum has a resolution of 1 nm, while below 400 nm the two composite spectra have a resolution of 0.25 nm (see section 3.1). To ensure continuity in resolution, we operated with the following way using the high resolution spectral model from *Kurucz* [1995]:

i)   It has been degraded in resolution with a 0.5 nm running mean (while maintaining the original sampling) to generate a spectrum comparable with the Ca II line profile of the mean of the five UV spectra (see section 3.2.3).

ii) The spectrum of i) was further degraded to the 1 nm SOLSPEC resolution. The comparison of this result with the mean SOLSPEC spectrum is used to derive coefficients for correcting the model spectrum of to the SOLSPEC photometric scale.

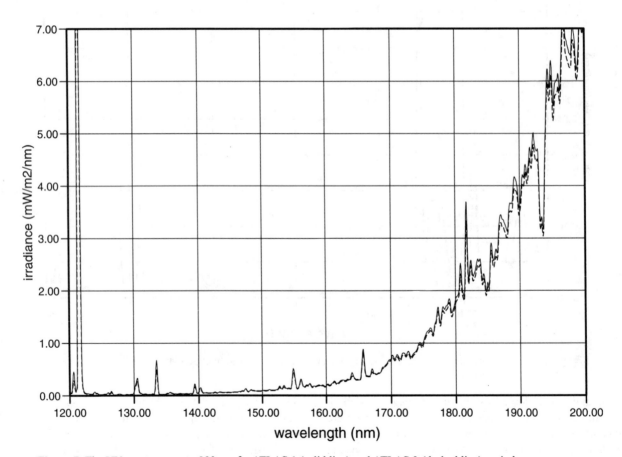

Figure 7. The UV spectrum up to 200 nm for ATLAS 1 (solid line) and ATLAS 3 (dashed line) periods.

As the sampling of the model spectrum of Kurucz is higher than that of SOLSPEC, the set of coefficients is linearly interpolated between each pair of consecutive SOLSPEC measurements. To verify the correctness of the interpolation method, we have afterward verified that the original mean SOLSPEC spectrum was reproduced after integration. To do that, their ratio and RMS were calculated. Their mean was found smaller than 10^{-5} with respect to unity and the RMS was found to be 0.7%.

3.2.5 IR (870 to 2400 nm). The SOSP instrument, the twin instrument of SOLSPEC, was operated on board the EURE-CA platform. The thermal conditions were stable (16°C ± 0.5°C), which allowed IR measurements of better quality than with SOLSPEC during the ATLAS missions where the temperature steadily increased during the solar observations. While scanning from 850 to 2400 nm, different second order filters as well as neutral density filters were employed as a function of wavelength.

The thickness of density filters is reduced as a function of increasing wavelength to compensate for the large decrease of the solar irradiance and instrument responsivity. The filters generally induce a few percent perturbation at the wavelengths where the filter thickness is changed. Taking into account that no Fraunhofer lines are clearly detectable at 20 nm resolution by the SOSP IR spectrometer, the spectrum has been smoothed by a polynomial fit and is shown in Figure 10.

The SOSP spectrum is very close to *Labs and Neckel* [1968], *Colina et al.* [1996], and *Kurucz* [1995] solar models continua, but provides a higher level of irradiance (4% between 1500 and 2000 nm and 3% at 2400 nm [*Thuillier et al.*, 2003]) before normalization.

As no Fraunhofer lines are present in the smoothed IR SOSP spectrum, we have used those of the *Kurucz* [1995] model spectrum and we have applied the same method as in the visible (see section 3.2.4). For that, the Kurucz spectrum has been smoothed to 50 nm resolution in order to generate a spectrum with resolution similar to that of the IR SOSP results shown in Figure 10. The same verification has been also made, resulting in a mean ratio smaller than 10^{-3} with respect to unity, and a RMS equal to 0.2%. The result is illustrated for ATLAS 1 period in a linear scale in Figure 11a, and in a logarithmic scale to display the spectral domain below 250 nm in Figure 11b. Using the procedure to be described in section 3.3.2, it is possible to extend the reference spectrum above 2400 nm by means of the *Kurucz* [1995] model spectrum.

3.3 Normalization to Total Solar Irradiance

The various contributions from 0 to 2400 nm have been assembled into two spectra corresponding to ATLAS 1 and ATLAS 3. They are named RSSV0-ATLAS 1 and RSSV0-ATLAS 3, respectively (RSSV0 stands for Reference Solar Spectrum, version 0).

Spectra such as *Labs and Neckel* [1968] or *Kurucz* [1995] are normalized to values of the total solar irradiance (TSI) as given in Table 5. The normalization rationale is chosen because of the higher accuracy (0.1%) of the TSI radiometric measurements versus the photometric measurements of the spectrometers (2%, at best). Two implementations of the normalization are described in the sections to follow.

3.3.1 Integrated Spectral Irradiance up to 2397.5 nm. We have integrated the RSSV0 ATLAS 1 and ATLAS 3 spectra up to 2397.5 nm as well as the spectra of *Kurucz* [1995], *Colina et al.* [1996], and *Labs and Neckel* [1968], which are used for comparison because they do not exhibit strong identified defects (see Figure 2). These five numbers (named SI2397) are listed in Table 5 which also displays the percentage of the differences between the two RSSV0 spectra and the three spectra listed above. The percentage of adjustment is then between 1.0 and 1.4%. The SI2397 is slightly greater for ATLAS 1 than for ATLAS 3 by an amount of 0.16 Wm^{-2} because of the decreasing solar activity condition. This change is entirely due to the difference in spectral irradiance below 400 nm since the irradiance for longer wavelengths is identical for both RSSV0-ATLAS spectra. The measured TSI difference between the two ATLAS periods is 1 Wm^{-2}, so it may be assumed that the remaining difference (0.84 Wm^{-2}) is due to unmeasured variation above 400 nm. Furthermore, the 0.16 Wm^{-2} difference in the UV is consistent with that expected from the change in solar activity [*Lean et al.*, 1997].

3.3.2 Normalizing with the total solar irradiance. We now calculate the energy above 2397.5 nm (TS>2397) and the solar spectral irradiance at 2397.5 nm (I_{2397}) for RSSV0, *Kurucz* [1995], and *Labs and Neckel* [1968] spectra. (Note that *Colina et al.* [1996] have no data above 2500 nm).

The RSSV0 solar irradiance is 61.33 mWm^{-2}nm^{-2} at 2397.5 nm as shown in Table 6. Its ratio to the *Kurucz* [1995], and *Labs and Neckel* [1968] spectral irradiances at that wavelength is used to estimate the solar irradiance above 2397.5 nm. This allows the calculation of the TSI for the RSSV0. For example, for ATLAS 1 together with the

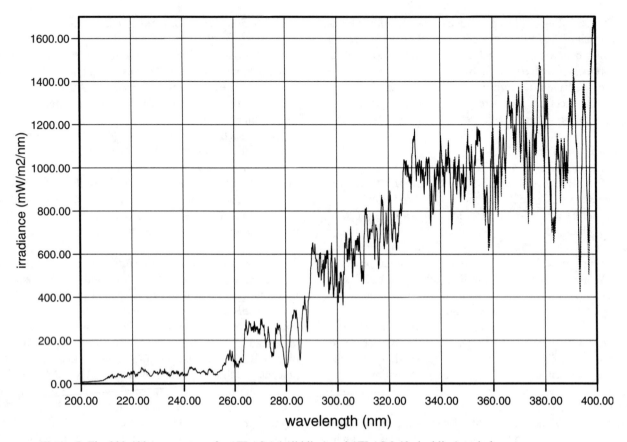

Figure 8. The 200-400 nm spectrum for ATLAS 1 (solid line) and ATLAS 3 (dashed line) periods.

model spectrum of Kurucz, we obtain 1330.28 + (51.46 x 61.33/59.97). Similarly for ATLAS 3, the spectra of *Labs and Neckel* [1968] and *Kurucz* [1995] are used. The results are given in Table 7.

The displayed percentage differences for ATLAS 1 and 3 are very consistent. The 1.11 to 1.16% and 1.17 to 1.22%

Table 5. Integrated solar irradiance in units of Wm^{-2} up to 2397.5 nm (SI2397) is given with the percentage difference with the two RSSV0 reference spectra. Column 4 provides the TSI associated with the spectra of *Kurucz* [1995] and *Labs and Neckel* [1968]. The measured TSI [*Fröhlich and Lean*, 1998] values are listed in column 5 for the two ATLAS periods. K[1995] and LN [1968] indicate *Kurucz* [1995] and *Labs and Neckel* [1968], respectively.

Spectra	SI2397	%	Calculated TSI	Measured TSI
K [1995]	1316.79	1.0	1368.11	
Colina et al. [1996]	1311.58	1.4		
LN [1968]	1314.10	1.2	1366.36	
RSSV0-ATLAS 1	1330.28			1367.7
RSSV0-ATLAS 3	1330.12			1366.7

normalization percentages for ATLAS 1 and ATLAS 3, respectively, shown in Table 7, could be adopted. Furthermore, these results are in agreement with the normalization results in Section 3.3.1. Given true changes in the solar irradiance, the normalization percentage should be greater for ATLAS 3 than for ATLAS 1, because the identical spectrum was used for both above 400 nm.

A mean percentage of adjustment may be chosen based either on the *Labs and Neckel* [1968] or *Kurucz* [1995]

Table 6. Calculated TSI, energy above 2397.5 nm (TS>2397) in Wm^{-2}, and solar spectral irradiance at 2397.5 nm in mWm^{-2}nm^{-1} for the spectra of *Kurucz* [1995] and *Labs and Neckel* [1968], and RSSV0. For the latter, a unique value is given at 2397.5 nm because, by construction, this part is identical for ATLAS 1 and ATLAS 3. K[1995] and LN [1968] stand for *Kurucz* [1995] and *Labs and Neckel* [1968], respectively.

Spectra	TSI	TSI>2397	I$_{2397}$
K [1995]	1368.11	51.46	59.97
LN [1968]	1366.36	52.48	60.44
RSSV0			61.33

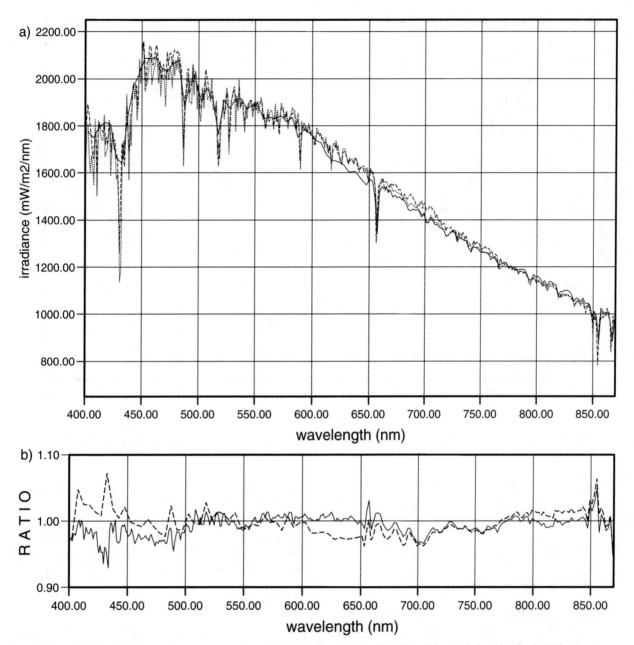

Figure 9. The visible spectrum. a: comparison of the *Burlov-Vasiljev* [1995] (solid), *Neckel and Labs* [1984] (short dashed) spectra and the mean ATLAS 1-2-3 SOLSPEC spectrum long dashed). b: ratio of *Burlov-Vasiljev* [1995] (dashed) and *Neckel and Labs* [1984] spectra (solid) to the mean ATLAS 1-2-3 SOLSPEC spectrum at 5 nm resolution.

spectra. However, for planetary atmospheric studies, the use of a spectrum with Fraunhofer lines may be required. This is why we have chosen to normalize with the model spectrum of *Kurucz* [1995]. The adjustment percentages are 1.11 and 1.17% for the ATLAS 1 and ATLAS 3 periods, respectively. As expected, this adjustment is below the uncertainties of the spectral measurements, which are of the order of 2 to 3%, as compared with the absolute radiometer's accu-

racy quoted to 0.1%. Note that this adjustment is based on the fact that most of the energy of total irradiance measured by a single instrument is in the visible and IR range.

Applying the 1.11 and 1.17% adjustment to the ATLAS 1 and 3 spectra in version 0, we obtain the Reference Solar Spectra in Version 1, which are available on request via e-mail to <gerard.thuillier@aerov.jussieu.fr>. An extension of these two spectra above 2397.5 nm is achievable after tak-

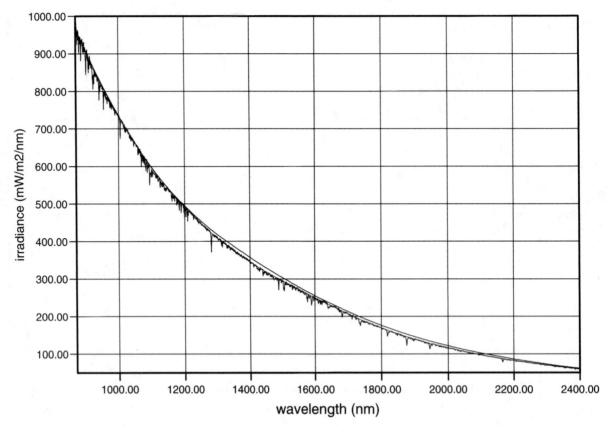

Figure 10. The IR range up to 2400 nm: comparison between spectra from *Kurucz* [1995] (dashed line) and SOSP without Fraunhofer lines (solid line).

ing the ratio of solar spectral irradiance of each RSSV1 to the *Kurucz* [1995] spectrum at 2397.5 nm as we did to normalize the RSSV0 spectra.

4. PROPERTIES OF THE REFERENCE SPECTRA

4.1 Accuracy of the Reference Solar Spectra

The accuracy of the two reference spectra (RSSV1) varies as a function of wavelength. Following the estimate of *Woods and Rottman* [2002], the accuracy is quoted to be about 40% in the XUV and 30% in the EUV. Above the Ly α line at 121.6 nm and below 200 nm, the UARS mean spectra have an accuracy better than 3.5% [*Woods et al.,* 1996]. Between 200 and 400 nm, the availability of five instruments (*Cebula et al.*, 1996; *Woods et al.*, 1996) has made possible several comparisons which show an accuracy between 2 to 4%. A similar study was conducted using *Neckel and Labs* [1984], *Burlov-Vasiljev et al.* [1995], and SOLSPEC spectra, providing an accuracy of about 3%. Above 870 nm, the accuracy is based on error analysis alone [*Thuillier et al.*, 2003] since comparisons with other data

are not practical because of the effect of undercorrected water vapor absorption. The accuracy is found to vary from 2 to 3% between 870 and 2400 nm.

Figures 6 to 10 display the V0 results. Normalization is taken into account in Figures 11a and 11b which illustrate the full spectrum for the ATLAS 1 period.

4.2 Comparison With ASTM [2000]

Figure 11c compares the ATLAS 3 reference spectrum in version 1 (after normalization) with the recent ASTM [2000] spectrum by taking their ratio at 10 nm resolution. Below 400 nm, this ratio is smaller than unity because the UARS spectral irradiance in this wavelength range used in the construction of the ASTM spectrum was reduced by 3.2% to match the *Neckel and Labs* [1984] spectrum. Around 900 and 1200 nm, the ratio of 3% above unity results from the use of the *Smith and Gottlieb* [1974] spectrum. Above 1300 nm, the ratio is a few percent below unity because the SOSP-EURECA data differ by this amount from the solar continuum model spectrum of *Smith and Gottlieb* [1974].

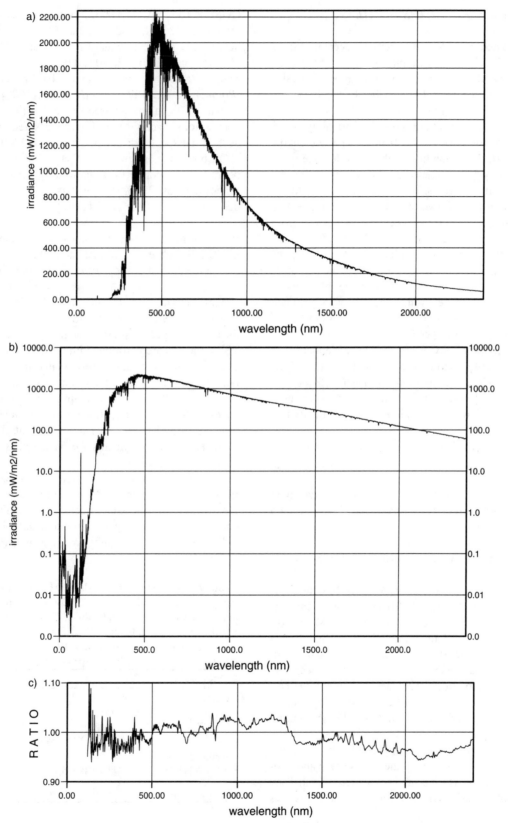

Figure 11. The reference spectrum for ATLAS 1. a: in linear coordinates. b: in logarithmic coordinates showing data at short wavelengths, and the quasi linear logarithmic irradiance above 500 nm. c: its ratio to ASTM [2000].

4.3 Solar Variability From ATLAS 1 to ATLAS 3

As demonstrated in Table 4, solar activity decreased from ATLAS 1 to ATLAS 3 periods. Figure 12 shows the ratio of ATLAS 1 to ATLAS 3 spectra at 1 nm resolution. This ratio decreases as expected from EUV to IR. In the EUV, the variability is obtained from that given by *Woods and Rottman* [2002]. At Ly α (121.6 nm), the variability is 1.5 and reaches a factor 3 for the He II line. For wavelengths longer than Ly α, the variability for the two reference spectra results directly from the measured data sets during the ATLAS 1 and 3 time periods. We also calculated the Mg II indices [*Rottman et al.*, this volume] from our two reference spectra and compared with the other indices derived from SUSIM and SBUV/2 spectrometers (Table 8).

Table 8 shows consistent results. The Mg II indices derived from the composite spectra are within the two other determinations given by the two other single instruments. This also allows us to verify that there is no significant wavelength shift among the five instruments used to build this part of the composite spectrum.

Part of the TSI variability occurs in the wavelength range below 400 nm. Table 9 shows a variation of 0.2 Wm^{-2}. This represents 20% of the TSI variation, consistent with the estimate of *Lean et al.* [1997] which claims a variation of about 30% from the maximum to the minimum of solar activity.

4.4 Sampling and Resolution of the Two Reference Solar Spectra

The sampling and resolution as a function of wavelength is not constant, as listed in Table 10, because they depend on the origin of the data.

5. PRESENT AND FUTURE MISSIONS

The UARS mission started in Septemeber 1991 and it is still operating. The NOAA-16 SBUV/2 instrument was launched in the fall of 2000 and additional SBUV/2 instruments are scheduled for launch throughout the first decade of the 21st century. The SBUV/2 instruments will also continue to provide solar spectral observations in the near UV.

The ENVISAT 1 platform, launched in March 2002 carries two spectrometers, MERIS (Medium Resolution Imaging Spectrometer) and SCIAMACHY (SCanning Imaging Absorption SpectroMeter for Atmospheric CHartographY) [*Bovensmann et al.*, 1999]. By observing from 240 to 2380 nm, they study the land surface and middle atmosphere by measuring the backscattered light. They will also be able to make solar spectral observations as does

GOME. The NASA TIMED satellite was launched on December 7, 2001 and its solar EUV experiment (SEE) began daily measurements of the irradiance between 0 and 200 nm on January 22, 2002 [*Woods et al.*, 1998b]. The SOlar Radiation and Climate Experiment (SORCE) [*Rottman et al.*, 1997; *Woods et al.*, 2000] launched in January 2003 on board a free-flying satellite, is expected to operate for six years. SORCE will measured the TSI as well as the solar spectral irradiance from 1 to 2000 nm.

On board the International Space Station (ISS), a solar pallet to be operated for a duration of three years, is now scheduled for 2006. It consists of three instruments described by *Thuillier et al.* [1999]:

i)   SOVIM measuring the total solar irradiance (TSI),
ii) SACES observing from 17 to 220 nm,
iii) SOLSPEC observing from 180 to 3000 nm.

Atmospheric, climate and solar physics are the basic objectives of SORCE and the ISS solar pallet by measuring the total and spectral solar irradiances and studying how the TSI variations are partitioned into different spectral ranges. These two missions are closely related in terms of spectral ranges, but differ in design and calibration principles. Furthermore, the International Space Station (ISS) solar pallet extends more in the IR range while SORCE observes more toward the EUV. An advantage of the ISS is the retrieval of the instruments for a post-mission laboratory check and calibration.

SORCE will overlap with both the Space Station and ENVISAT 1 missions and, hopefully, also with the two UARS experiments. Consequently, continuity and useful comparisons for both spectral and total solar irradiance data will be achieved. We finally note that the strategy of over-

Table 7. Calculated TSI for ATLAS 1 and 3 reference spectra using the solar irradiance up to 2397.5 nm supplemented by the spectra of *Kurucz* [1995] and *Labs and Neckel* [1968]. Percentages of difference is given for each ATLAS period with respect to these spectra (last line). K[1995] and LN [1968] stand for *Kurucz* [1995] and *Labs and Neckel* [1968], respectively.

TSI Source	ATLAS 1	ATLAS 3
measured TSI	1367.7	1366.7
calculated TSI with K[1995]	1382.91	1382.74
calculated TSI with LN[1968]	1383.53	1383.37
difference in %	1.11 / 1.16	1.17 / 1.22

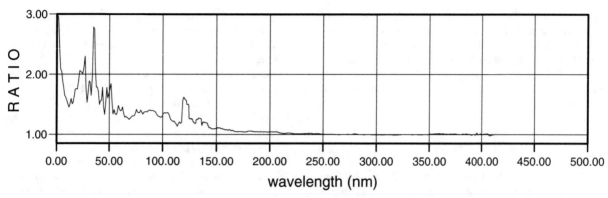

Figure 12. Ratio of the ATLAS 1 to ATLAS 3 spectra at 1 nanometer resolution.

lapping missions should continue, thus ensuring that TSI and absolute spectral irradiance and their specific variability will be accurately measured as required for atmospheric, climate and solar physics. We expect that these new data sets will provide the opportunity to improve the reference solar spectra.

6. CONCLUSION

The existing measurements and reference spectra in the UV, visible and IR solar irradiances have been reviewed. Comparisons of the different data sets assess the accuracy of the recent measurements to be less than 4% at most wave-

Table 8. Mg II indices derived from the two reference spectra, SUSIM (V19r5) and SBUV/2 experiment data.

	RSS	SUSIM	SBUV/2
ATLAS 1	0.2748	0.2708	0.2756
ATLAS 3	0.2600	0.2596	0.2625

lengths. Recent observations made by several instruments

Table 9. Energy per spectral ranges in Wm^{-2}.

	0-200 nm	200-400 nm	> 400 nm
ATLAS 1	0.116	109.1	1258.51
ATLAS 3	0.106	108.91	1257.74

Table 10. Sampling (s) and resolution (r) of the reference spectra. From 400 to 2400 nm, the sampling intervals increase monotonically.

Ranges	s (nm)	r (nm)
XUV-EUV	1	1
Ly α to 400 nm	0.05	0.25
400 to 2400 nm	0.2 to 0.6	0.5

are available particularly in the UV range, and to a lesser extent in the visible. Most of the observations in the IR domain are old with a few exceptions.

With the most recent existing data, we have built two reference solar spectra close to moderately high and low solar activity conditions as encountered during the ATLAS 1 and 3 periods, made by assembling data from different instruments, primarily from the UARS, ATLAS and EURECA missions. They extend from XUV to IR. Their accuracy depends of the spectral range, typically 30% below Ly α, 3.5% up to 200 nm and about 2 to 4% above. The sampling varies with respect to wavelength. The measure of solar variability presented here is based on only two reference spectra; different levels of activity would correspond to different spectra. This caveat has no practical effect for 300-400 nm at continuum wavelengths and universally above 400 nm since spectral irradiance variability in those wavelength regions has not been reliably detected. At shorter wavelengths, the variability generally increases for decreasing wavelengths. In the UV domain, there is a correlation between irradiance variability and activity. Variability is particularly strong in the EUV and XUV domains. The epochs of the two reference spectra described here correspond to about half of the solar cycle amplitude in terms of F10.7 and Mg II indices. After an extension into the IR, the calculated TSI values for the initial reference spectra are about 1% higher than the measured TSI values during the ATLAS missions; the final reference solar spectra are normalized to these measured TSI values. Future missions will surely improve these results by using new data with better accuracy and simultaneous measurements (e.g., SORCE and ISS solar pallet).

Acknowledgments. Each participant has provided the necessary data to build this reference spectrum. Important discussions took place in recent meetings to define the principles of the spectrum construction. Data handling was carried out by Georges Azria from Service d'Aéronomie. R. P. Cebula was supported by NASA contract NAS1-98106. T. N. Woods was supported by NASA grant NAG5-6850 to the University of Colorado. L. Floyd was supported by NASA-Defense Purchase Requests S14798D and S10108X. We thank the editors and reviewers of this monograph for their patience and diligence.

REFERENCES

Arvesen, J. C., R. N, Jr. Griffin, and B. D. Jr. Pearson, Determination of extraterrestrial solar spectral irradiance from a research spacecraft, *Appl. Opt., 8*, pp. 2215-2232, 1969.

ASTM E 490, Standard Solar Constant and Zero Air Mass Solar Spectral Irradiance Tables, in *Space Simulation and Applications of Space Technology*, 2000.

Bovensmann, H,, J. P. Burrows, M. Buchwitz, J. Frerick, S. Noel, V. V. Rozanof, K. V. Chance, and A. P. H. Goede, SCIA-MACHY: mission objectives and measurements modes, *J. Atm. Sci.*, 56, 127-150, 1999.

Brasseur, G., and P. Simon, Stratospheric chemical and thermal response to long-term variability in solar UV irradiance, *J. Geophys. Res., 86*, pp. 7343-7362, 1981.

Brueckner, G. E., K. L. Eldow, L. E. Floyd, J. L. Lean, and M. E. J. VanHoosier, The Solar Ultraviolet Spectral Irradiance Monitor (SUSIM) experiment on board the Upper Atmosphere Research Satellite (UARS), *J. Geophys. Res., 98*, pp. 10695-10711, 1993.

Burlov-Vasiljev, K. A., E. A. Gurtovenko, and Y. B. Matvejev, New absolute measurements of the solar spectrum 310-685 nm, *Solar Phys., 157*, pp. 51-73, 1995.

Cebula, R. P., G. Thuillier, M. R. J. Vanhoosier, E. Hilsenrath, M. Hersé, P. C. Simon, Observations of the solar irradiance in the 200-350 nm interval during the ATLAS 1 mission: A comparison among three sets of measurements-SSBUV, SOLSPEC, and SUSIM, *Geophys. Res. Lett., 23*, pp. 2289-2292, 1996.

Cebula, R. P., M. T. DeLand, and E. Hilsenrath, NOAA-11 SBUV/2 solar spectral irradiance measurements 1989-1994: I. Observations and long-term calibration, *J. Geophys. Res., 103*, pp. 16235-16249, 1998.

Colina, L., R. C. Bohlin, and F. Castelli, The 0.12-2.5 μm absolute flux distribution of the sun for comparison with solar analog stars, *Astrophys. J., 112*, pp. 307-315, 1996.

David, K. H. and G. Elste, Der Einfluss von Streulicht auf die Photometrie der Sonnenoberflache, *Z. Astrophys.*, 54, 12, 1962.

DeLand, M. T., and R. P. Cebula, Solar Backscatter Ultraviolet, model 2 (SBUV/2) instrument solar spectral irradiance measurements in 1989-1994, 2, Results, validation, and comparisons, *J. Geophys. Res., 103*, pp. 16251-16273, 1998.

Farmer, C. B., and S. J., Todd, Absolute solar spectra 3.5-5.5 microns: 1 experimental spectra for altitude range 0-15 km, *Appl. Opt., 3* , pp. 453-465, 1964.

Floyd, L. E., P. A. Reiser, P. C. Crane, L. C. Herring, D. K. Prinz, and G. E. Brueckner, Solar Cycle 22 UV Spectral Irradiance Variability: Current Measurements by SUSIM UARS, *Solar Phys., 177*, pp. 79-87, 1998.

Floyd, L. E., D. K. Prinz, P. C. Crane, L. C. Herring, Solar UV Irradiance Variation during cycles 22 and 23, *Adv. Space Res.*, 29, pp. 1957-1962, 2002.

Fontenla, J., O. R. White, P. A. Fox, E. H. Avrett, and R. L. Kurucz, Calculation of solar irradiances. I. Synthesis of the solar spectrum, *Astrophys. J., 518*, pp. 480-499, 1999.

Fröhlich, C., Solar Irradiance Variability, *this volume*.

Fröhlich, C., and J. Lean, Total Solar Irradiance Variations, *in New Eyes to see inside the Sun and Stars*, edited by F.L.Deubner *et al.*, pp. 89-102, Proceedings IAU Symposium 185, Kyoto, August 1997, Kluwer Academic Publ., Dordrecht, Netherlands, 1998.

Gingerich, O., and C. de Jager, The Bilderberg model of the photosphere and low chromosphere, *Solar Phys., 3*, pp. 5-25, 1968.

Goldberg A. K., and Pierce: *Hanbuch der Physik*. Ed. S. Flügge, *52*, Berlin-Gottingen-Heidelberg, Springer, 1959.

Gray, D. F. and W. C. Livingston, Monitoring the solar temperature: spectroscopic temperature variations of the sun, *Astrophys. J,, 474*, pp. 802-809, 1997.

Grevesse, N., and A. J. Sauval, The infrared solar radiation, in *Proceedings of the international colloquim held in Montpellier*, Ed. C. Jaschek and Y. Andrillat, Cambridge University Press, pp. 215-233, 1991.

Haigh, J.D., Fundamentals of the Earth's Atmosphere and Climate, *this volume*.

Hall, L. A., and G. P. Anderson, High-resolution spectrum between 2000 and 3100 A, *J. Geophys. Res.*, 96, 12927-12931, 1991.

Heath, D., and B. M. Schlesinger, The Mg II 280 nm doublet as a monitor of changes in solar ultraviolet irradiance, *J. Geophys. Res., 91*, pp. 8672-8682, 1986.

Holweger, H., Ein empirisches Modell der Sonnenatmosphaere im lokalen thermodynamischen Gleichgewicht, *Z. Astrophys.*, 65, pp. 365-417, 1967.

Hood, L., Effect of Solar UV Variability on the Stratosphere, *this volume*.

Kondratyev, K. Y., S. D. Andreev, I. Y. Badinov, V. S. Grishechkin, and L. V. Popova, Atmospheric optics investigation on Mt Elbrus, *Appl. Opt.*, 4, pp. 1069-1076, 1965.

Koutchmy, S., and R. Peyturaux, On the intensity of the solar continuum at λ = 20.15 μm. Absolute measurements of the brightness temperature at the centre of the disk, *C.R. Acad. Sci.*, Paris, *267*, series B, pp. 905-908, 1968.

Kurucz, R., Smithonian Astrophys. Obs., CD rom # 13, 1993a.

Kurucz, R., Smithonian Astrophys. Obs., CD rom # 19, 1993b.

Kurucz, R., Smithonian Astrophys. Obs., CD rom # 23, 1995.

Labs, D. and H. Neckel, Die absolute Strahlungsintensitat der Sonnenmitte im Spektralbereich 4010<λ<6569A, *Z. Astrophys.* 55, pp. 269-289, 1962.

Labs, D., and H. Neckel, The radiation of the solar photosphere from 2000 Å to 100 μm, *Z. Astrophys.* 69, pp. 1-73, 1968.

Labs, D., H. Neckel, P. C. Simon, and G. Thuillier, Ultraviolet solar irradiance measurement from 200 to 358 nm during spacelab 1 mission, *Sol. Phys., 107*, pp. 203-219, 1987.

Lean, J., G. J. Rottman, H. L. Kyle, T. N. Woods, J. R. Hichey, and L. C. Puga, Detection and parametrization of variation in solar mid-and near-ultraviolet radiation (200-400 nm), *J. Geophys. Res., 102*, pp. 29939-29956, 1997.

Link, F., and Neuzil, L., in Tables of light trajectories in the terrestrial atmosphere, Ed. Hermann, Paris, 1969.

Livingston, W. C., Observations of solar irradiance variations at visible wavelengths, in *Proceedings of the Workshop on the Solar Electromagnetic Radiation Study for Solar Cycle 22*, edited by R. F. Donnelly, pp. 11-19, 1992.

Lockwood, G. W., H. Tüg., and N. M. White, *Astrophys. J., 390*, pp. 668-678, 1992.

Meier, R. R., Ultraviolet Spectroscopy and Remote Sensing of the Upper Atmosphere, *Space Science Reviews, 58*, pp. 1-185, 1991.

Murcray, F. H, D. G., Murcray, and W. J., Williams, The spectral radiance of the Sun from 4 μm to 5 μm, *Appl. Opt., 3*, pp. 1373-1377, 1964.

Muscheler, R., J. Beer, and P. W. Kubik, Long-Term Solar Variability and Climate Change Based on Radionuclide Data from Ice Cores, *this volume.*

Neckel, H., and D. Labs, The solar radiation between 3300 and 12500 Å, *Solar Phys., 90*, pp. 205-258, 1984.

Nicolet, M., The solar spectral irradiance and its action in the atmospheric photodissociation processes, Planet. Sp. Sci., 9, 951-974, 1981.

Peytureaux, R., La mesure absolue de l'énergie émise par le centre du disque solaire de 4477 à 8638 Å, *Ann. Astrophys. 31*, pp. 227-235, 1968.

Pierce, A. K., Relative solar energy distribution in the spectral region 10,000 - 25,000 Å, *Astrophys. J., 119*, pp. 312-327, 1954.

Pierce, A. K. and R. G. Allen, in *The solar output and its variation*, edited by O. R. White, Colorado Associated University Press, Boulder, pp. 169-192, 1977.

Rottman, G. J., T. N. Woods, and T. P. Sparn, Solar Stellar Irradiance Comparison Experiment: instrument design and operation, *J. Geophys. Res., 98*, pp. 10667-10677, 1993.

Rottman, G. J., G. Mount, G. Lawrence, T. N. Woods, J. Harder, and S. Tournois, Solar spectral Irradiance measurements: visible to near infrared, *Metrologia, 35*, pp. 707-712, 1997.

Rottman, G., L. Floyd, and R. Viereck, Measurement of solar ultraviolet irradiance, *this volume.*

Saiedy, F., and R. M., Goody, Mon., The solar emission intensity at 11 μm, *Monthly Notices Roy. Astron. Soc., 119*, p. 213-222, 1959.

Saiedy, F., Solar intensity and limb darkening between 8.6 and 13 μm, *Monthly Notices Roy. Astron. Soc., 121*, pp. 483-495, 1960.

Shaw, G. E., Solar spectral irradiance and atmospheric transmission at Mauna Loa observatory, *Appl. Opt., 21*, pp. 2006-2011, 1982.

Smith, E. V. P., and D. M. Gottlieb, Solar Flux and its Variations, *Space Sci. Rev., 16*, pp. 771-802, 1974.

Soon, W. H., E. S. Posmentier, and S. L. Baliunas, Inference of solar irradiance variability from terrestrial temperature changes, 1880-1995: an astronomical application of the sun-climate connection, Astrophys. J., *472*, pp. 891-902, 1996.

Thekaekara, M. P., and A. J. Drummond, *Nat. Phys. Sci., 229*, pp. 6-9, 1971.

Thekaekara, M. P., Extraterrestrial solar spectrum, 3000-6100 Å at 1-Å resolution, *Appl. Opt., 13*, pp. 518-522, 1974.

Thuillier, G., M. Hersé, P. C. Simon, D. Labs, H. Mandel,. and D. Gillotay, Observation of the UV solar spectral irradiance between 200 and 360 nm during the ATLAS I mission by the SOLSPEC spectrometer, *Solar Phys., 171*, pp. 283-302, 1997.

Thuillier, G., M. Hersé, P. C. Simon, D. Labs, H. Mandel, and D. Gillotay, Observation of the visible solar spectral irradiance between 350 and 850 nm during the ATLAS I mission by the SOLSPEC spectrometer, *Solar Phys., 177*, pp. 41-61, 1998a.

Thuillier, G., M. Hersé, P. C. Simon, D. Labs, H. Mandel, and D. Gillotay, Observation of the solar spectral irradiance from 200 to 870 nm during the ATLAS 1 and 2 Missions by the SOLSPEC Spectrometer, *Metrologia, 35*, pp. 689-675, 1998b.

Thuillier G., C. Fröhlich, and G. Schmidtke, Spectral and total solar irradiance measurements on board the international Space Station, *in Utilisation of the International Space Station,* pp. 605-611, Second european Symposium, ESA-SP433, 1999.

Thuillier, G., M. Hersé, D. Labs, T. Foujols, W. Peetermans, D. Gillotay, P. C. Simon, and H. Mandel, The solar spectral irradiance from 200 to 2400 nm as measured by the SOLSPEC spectrometer from the ATLAS 1-2-3 and EURECA missions, *Solar Phys., 214.*, pp. 1-22, 2003.

Tousey, R., The extreme ultraviolet spectrum of the sun, *Sp. Sci. Rev., 2*, pp. 3-69, 1963.

VanHoosier, M. E., J-D. F. Bartoe, G. E. Brueckner, and D. K. Prinz, Absolute solar spectral irradiance 120nm-400nm (Results from the Solar Ultraviolet Spectral Irradiance Monitor-SUSIM-Experiment on board Spacelab 2), *Astrophys. Lett., 27*, pp. 163-168, 1988.

VanHoosier, M. E., Solar ultraviolet spectral irradiance data with increased wavelength and irradiance accuracy, *SPIE Proceedings, 2831*, pp. 57-64, 1996.

Weber, M., J. P. Burrows, and R. P. Cebula, GOME solar UV/VIS irradiance measurements between 1995 and 1997 - First results on proxy solar activity studies, *Solar Phys., 177*, pp. 63-77, 1998.

Wehrli, C., Extraterrestrial Solar Spectrum, *PMOD publication 615*, 1985.

Woods, T. N., D. K. Prinz, G. J. Rottman, J. London, P. C. Crane, R. P. Cebula, E. Hilsenrath, G. E. Brueckner, M. D. Andrews, O. R. White, M. E. VanHoosier, L. E. Floyd, L. C. Herring, B. G. Knapp, C. K. Pankratz, and P. A. Reiser, Validation of the UARS solar ultraviolet irradiances: Comparison with the ATLAS 1 and 2 measurements, *J. Geophys. Res., 101*, pp. 9541-9569, 1996.

Woods, T. N., G. J. Rottman, S. M. Bailey, S. C. Solomon, and J. Worden, Solar extreme ultraviolet irradiance measurements during solar cycle 22, *Solar Phys., 177*, pp. 133-146, 1998a.

Woods, T. N. , F. G. Eparvier, S. M. Bailey, S. C. Solomon, G. J. Rottman, G. M. Lawrence, R. G. Roble, O. R. White, J. Lean, and W. K. Tobiska, TIMED SOLAR EUV Experiment, SPIE Proceedings, 3442, pp. 180-191, 1998b.

Woods, T. N., G. J. Rottman, J. Harder, G. Lawrence, B. McClintock, G. Kopp, and C. Pankratz, Overview of the EOS SORCE mission, *SPIE Proceedings, 4135,* pp.192-203, 2000.

Woods, T. N., and G. J. Rottman, Solar ultraviolet variability over time periods of aeronomic interest, in *Comparative Aeronomy in the Solar System,* edited by M. Mendillo, A. Nagy, and J. Hunter Waite, Jr., Geophys. Monograph Series, Wash. DC, 221-234, 2002.

Woods, T. N., L. W. Acton, S. Bailey, F. Eparvier, H. Garcia, D. Judge, J. Lean, D. McMullin, G. Schmidtke, S. C. Solomon, W. K. Tobiska, and H. P. Warren, Solar extreme ultraviolet and X-ray irradiance variations, *this volume.*

Gérard Thuillier, Michel Hersé, Service d'Aéronomie du CNRS, Bp 3, F 91371 Verrières-le-Buisson, France

Linton Floyd, Interferometrics Inc., 14120 Parke Long Court, Suite 103, Chantilly, VA 20151, USA

Thomas Woods, Laboratory for Atmospheric and Space Physics, University of Colorado, 1234 Innovation Drive, Boulder, CO 80303-7814, USA

Richard Cebula, Science Systems and Applications, Inc. 10210 Greenbelt Road, Suite 400, Lanham, MD 20706, USA

Ernest Hilsenrath, Mail Code 916, NASA Goddard Space Flight Center, Greenbelt, MD 20771, USA

Dietrich Labs, Landessternwarte, Königstuhl, D 69117 Heidelberg, Germany

Solar Energetic Particle Variations

David Lario

The Johns Hopkins University, Applied Physics Laboratory, Laurel, Maryland, USA

George M. Simnett

School of Physics and Astronomy, University of Birmingham, Birmingham, United Kingdom

The population of energetic particles in the heliosphere changes from solar maximum to solar minimum. The ultimate driver for those variations is the Sun. Solar variability is reflected in the dynamics of the large-scale structure of the heliosphere, the solar output of energetic particles and most definitively in the intensity, energy and composition of the population of energetic particles observed by spacecraft and ground-based detectors. We describe the classification, origin, intensity and properties of the solar energetic particle events, in terms of both energetic ions and electrons. We review the long-term databases created to compare the occurrence frequency and intensity distribution of events among the solar cycles. Models for the production of the largest gradual solar energetic particle events are discussed.

1. ENERGETIC PARTICLE POPULATIONS IN THE HELIOSPHERE

The energetic charged particle population of the heliosphere drastically changes from solar maximum to solar minimum. Spacecraft and ground-based observations show that the intensity of energetic particles in the solar system largely varies with the solar activity cycle [*Lanzerotti*, 1977; *Lin*, 1977]. The sources of these particles are diverse, depending on the phase of the solar cycle and the energy of the particles. Proton intensities above about 200 MeV are mainly dominated by galactic cosmic rays which originate in interstellar space [*Jokipii*, 1983; *Burger*, 2000 and references therein]. Below ~100 MeV the proton intensity averaged over a solar cycle is mainly dominated by events of solar origin. Energetic ions associated with the interaction

between fast and slow solar wind streams (i.e., corotating interaction regions, CIRs) can also be observed at energies as high as several MeV/nucleon at all heliolatitudes [*Lanzerotti et al.*, 1995; *Sanderson et al.*, 1995]. During quiet times it is also possible to observe energetic ions accelerated at the heliospheric termination shock at energies as high as 100 MeV/nucleon (i.e., anomalous cosmic rays, ACRs) [*Leske et al.*, 2000 and references therein].

The energetic electron population in the heliosphere also changes over the solar cycle. At energies above ~1 GeV the electron intensity at 1 AU is dominated by galactic cosmic-ray electrons which continuously penetrate the heliosphere. From 3 MeV to 1 GeV galactic cosmic-ray electrons are still observed but their fluxes are attenuated and modulated by solar activity. At solar quiet times, most of the electrons observed in the range from a few hundred keV to a few MeV are of Jovian origin. At lower energies, ~50 keV, corotating interaction regions accelerate electrons [*Simnett et al.*, 1995]. During the active phases of the solar cycle the emission of solar electrons exhibits a considerable increase. Those electrons are observed at 1 AU in the form

Solar Variability and its Effects on Climate
Geophysical Monograph 141

of transient events from a few keV to an upper limit of ~100 MeV [*Simnett*, 1974; *Lin*, 1985].

1.1. Variations Within a Solar Cycle

The contribution of each one of these sources (i.e., galactic, solar and interplanetary) to the heliospheric energetic particle population changes throughout the solar cycle and is ultimately determined by the variations in the large-scale structure of the heliosphere. The out-of-ecliptic orbit of the Ulysses spacecraft has revealed important differences between the 3-D structure of the heliosphere observed at solar minimum and at solar maximum [see individual papers in *Marsden*, 2001]. During solar minimum the heliosphere is dominated by the interaction between fast and slow solar wind which usually occurs in the vicinity (<30°) of the Sun's equatorial plane. During solar maximum, the more complicated and time evolving structure of the solar corona, as well as the more frequent occurrence of coronal mass ejections (CMEs) lead to a more complex and dynamic heliosphere [*McComas et al.*, 2001]. Energetic particle observations during solar minimum and solar maximum are a faithful reflection of the changing Sun [*Hakura*, 1974; *Roelof et al.*, 1997; *Simnett*, 2001; *Lario et al.*, 2001a]. Figure 1 shows the 38–53 keV electron fluxes as observed by the HI-SCALE experiment [*Lanzerotti et al.*, 1992] on board the Ulysses spacecraft for a 3-year period during its first (top panel) and second (bottom panel) high latitude excursions. The first Ulysses orbit occurred during the decaying phase of the solar cycle 22 (years 1993–1995). The remarkable feature during

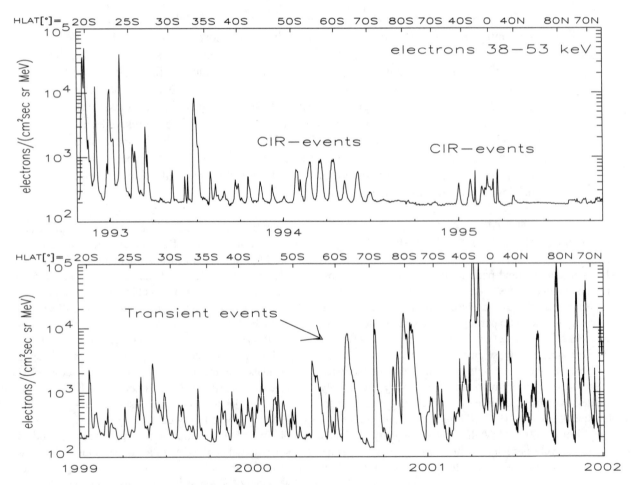

Figure 1. Daily averages of the 38–53 ke V electron intensity observed by Ulysses during three-year intervals of the solar minimum orbit (upper panel) and solar maximum orbit (lower panel). Upper axis shows the heliographic latitude of Ulysses. During the 1st orbit the 26-day periodicity is visible up to practically the highest latitude reached by Ulysses. During the 2nd orbit large transient events of solar origin were observed, especially during the solar maximum years of 2000 and 2001.

the descent to southern high heliolatitudes was the regular appearance, at the solar rotation period (26 days sidereal), of electron increases right up to S80° [*Roelof et al.*, 1997]. These have been interpreted by *Simnett and Roelof* [1995] and *Roelof et al.* [1996] as electrons, accelerated in the middle heliosphere at CIRs, which are able to propagate to high heliographic latitudes. The intensity of the CIR-events, however, decreased as Ulysses moved to high latitudes. When Ulysses returned to low heliospheric latitudes (<40°) early in 1995, the intensity of the recurrent electron events associated with CIRs recovered. The rest of the period shown in the top panel of Figure 1 (>40°N) was characterized by a flat electron intensity profile without significant particle flux increases [see details in *Roelof et al.*, 1997].

The second Ulysses orbit occurred during the rising phase (year 1999) and maximum phase (years 2000–2001) of the solar cycle 23. The electron intensity profiles (bottom panel of Figure 1) showed a completely different pattern, with numerous transient events of solar origin. The electron fluxes showed big increases above S50° as a result of major solar events identified by *Simnett* [2001]. We note that the elevated electron fluxes were observed even at latitudes as

high as S80°. The complex pattern of the energetic particle profiles during the second Ulysses orbit, also observed in the low-energy ions [*Lario et al.*, 2001a], is the result of the increasing level of solar activity and the dynamic evolution of the solar corona and heliospheric structure. Additional analysis of the low-energy ion intensities, composition and anisotropies over different phases of the solar cycle also show that at solar minimum the low-energy (≤2 MeV/nucleon) particle population is mainly dominated by CIRs, while at solar maximum, the transient events of solar origin increase their contribution [*Shields et al.*, 1985; *Richardson et al.*, 1993; *Lario et al.*, 2001b].

1.2. Variations Over the Solar Cycles

To study energetic particle variations over several solar cycles, it is important to have long-time series of direct and reliable measurements of the energetic particle intensities at standardized energy levels. The long-duration mission of the Interplanetary Monitoring Platform 8 (IMP-8) offers us the opportunity to assemble a uniform and almost continuous database for nearly three solar cycles. Figure 2 shows an

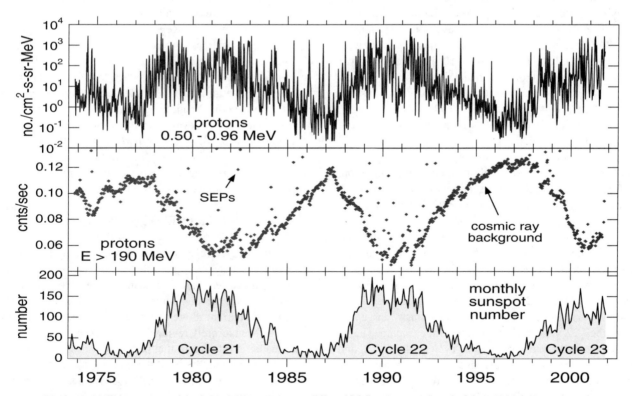

Figure 2. 10-day averages of the 0.50–0.96 MeV proton differential flux (top panel) and of the >190 MeV proton count rate (mid-panel) as measured by the CPME experiment on board IMP-8 [*Sarris et al.*, 1976] from 26 October 1973 to 24 October 2001. Bottom panel: Monthly sunspot number for the period October 1973– November 2001.

overview of the energetic particle observations from the Charged Particle Measurement Experiment (CPME) on board IMP-8 [*Sarris et al.*, 1976] from 26 October 1973 to 24 October 2001. This figure provides a complete perspective of the effects that solar cycle variations produce on several energetic particle populations. The top panel shows the 0.5–0.96 MeV proton intensity; the middle panel the count rates of the protons at energies above 190 MeV; and the lowest panel the monthly sunspot number.

The low-energy (0.5–0.96 MeV) proton channel shows an oscillating trace modulated by the sunspot number. Superimposed on this global trend there is an abundance of relatively short-lived particle flux increases (from hours to several days) that are sporadic transient events of solar origin or associated with recurrent CIRs. Noteworthy is the fact that the intensity minima are sustained at a higher level during the active phase of the solar cycles and only return to instrumental level when the solar activity is minimum. This behavior is the result of multiple and frequent particle injections from the Sun. Occasionally, during periods of sustained high solar activity, the inner heliosphere may act as a reservoir of low-energy particles [*Roelof et al.*, 1992]. During these periods, the heliosphere may be filled uniformly over large spherical volumes of radius ≤5 AU with energetic electrons, protons and heavy ions [*Maclennan et al.*, 2001]. That leads to the simultaneous observation by several spacecraft of long-lived (> 10 days) high particle intensities with essentially null radial, longitudinal or latitudinal gradients, especially during the decaying phase of large particle events [*Roelof et al.*, 1992; *McKibben et al.*, 2001]. A likely origin of these reservoirs is the formation of enhanced magnetic field regions created by the superposition of plasma structures driven by CMEs [*Roelof et al.*, 1992]. Recent studies have dealt with the origin, formation, location and effects of these reservoirs over different phases of the solar cycle [*McKibben et al.*, 2001; *Roelof et al.*, 2002].

The mid-panel of Figure 2 shows the count rate of galactic cosmic rays. This count rate varies inversely with the sunspot number. Occasionally the Sun is also a sporadic source of high-energy particles (> 190 MeV). These particles appear in Figure 2 as isolated points above the galactic cosmic ray background. The cosmic ray intensity curve also shows the ~22 year cycle with alternative maxima, being flat-topped in the minima of solar cycles 20 and 22, and peaked at the minimum of solar cycle 21. This behavior has been reproduced by models of cosmic ray modulation based on the observed reversal of the Sun's magnetic field polarity every ~11 years and curvature and gradient drifts of the particles in the large-scale magnetic field of the heliosphere

[*Jokipii et al.*, 1977; *Jokipii and Thomas*, 1981]. During solar minimum, when the large-scale heliospheric field is relatively ordered, gradient and curvature particle drifts dominate the cosmic ray modulation. During solar maximum, the increase of the interplanetary magnetic field (IMF) strength, the more complex structure of the heliosphere (with a highly tilted heliospheric current sheet), the higher frequency of the CMEs and shocks propagating outward from the Sun and also the increase in the strength of the shocks have been indicated as possible causes of the cosmic ray modulation [*Cliver and Ling*, 2001 and references therein].

Cosmic-ray data from neutron monitor measurements exist since 1954. Differences in the cosmic-ray intensity from 1954 to 2000 can be found in *Lockwood et al.* [2001]. Several attempts have been made to extend back in time the cosmic ray intensity at Earth, by combining data from different ion chambers [*Ahluwalia*, 1997], or by using the existing anticorrelation between the strength of the heliospheric magnetic field and the cosmic ray intensity [*Lockwood*, 2001]. *Lockwood* used the geomagnetic index 'aa' as a proxy to estimate that the IMF has increased since ~1800. As a result, he claimed that the cosmic ray fluxes above 3 GeV were 15% higher, on average, around 1900 than they are now. However, historical records of other geomagnetic indices do not show such an increase [*Ponyavin*, 2001], and it is not clear how the local measure of a geomagnetic index can be related to infer the variations of the IMF strength in the whole heliosphere [*Ponyavin*, 2001]. On the other hand, the cosmic ray intensity seems to be modulated by the level of solar activity, perhaps through the number of CMEs [*Cliver and Ling*, 2001]. Thus, an increase in the number of CMEs would be reflected in an increase of the geomagnetic indices. Further investigation in this field is warranted.

2. SOLAR ENERGETIC PARTICLES

The most powerful signatures of solar activity in terms of energetic particles are the solar energetic particle (SEP) events. The current paradigm distinguishes two basic types of SEP events: the impulsive and the gradual events. The origins and current usage of the terms impulsive and gradual have been recently reviewed by *Cliver and Cane* [2002]. The reader is referred to that paper for a description of the usefulness and limitations of these terms. Under the current paradigm, it is believed that the impulsive events have their origin during rapid flares (lasting from a few minutes to an hour) and are particularly rich in electrons, ^{3}He and heavy ions. The gradual events are proton-rich and are well associated with CMEs. Impulsive events are observed in a narrow cone of longitudes corresponding to observers magnetically well-connected to the site of the progenitor solar flare. Conversely, gradual events are observed in a wide spread range of longitudes independently of the associated solar flare location (if a flare can be identified at all). Impulsive events show high Fe charge states (+20), consistent with a localized flare source at high (10 MK) temperature. Alternatively the source may be dominated by non-thermal ionization. Gradual events usually show low Fe charge states (+11–14) that are indicative of a cooler (1–2 MK) source material. Impulsive events are about 100 times more frequent than gradual events at maximum activity. However, impulsive events have typical durations of the order of hours and are less intense than gradual events which can last several days. The detailed characteristics and properties of the two classes of events are described elsewhere, see for example *Reames* [1999] for a review.

Figure 3 is an example of two recent SEP events observed by the ACE and IMP-8 spacecraft. The gradual event (right panel) was associated with a CME observed by SOHO/LASCO coronagraph at 1511 UT on day 95 of 2000; at N16W66 a C9.7/2F solar flare was observed with X-ray onset at 1512 UT and maximum at 1541 UT. Low-energy protons reached a small peak near the time of a shock passage at 1604 UT on day 97. Elemental abundances (relative to carbon) for this specific SEP event are characteristic of a gradual event [*von Rosenvinge et al.*, 2001]. The left panel shows an impulsive event associated with an impulsive M1.1 X-ray flare at N20W54 with onset at 1016 UT and maximum at 1027 UT. This event was of shorter duration (less than one day), rich in Fe and ³He and showed a high electron to proton ratio [*Kahler et al.*, 2001].

In the current two-class paradigm, the flare process accounts for acceleration in the impulsive events whereas particle acceleration in interplanetary CME-driven shocks dominates the gradual events. To rule out the possibility that both processes contribute to a given energetic particle event, it is essential to find pure cases of gradual events not associated with solar flares. Those events are usually associated

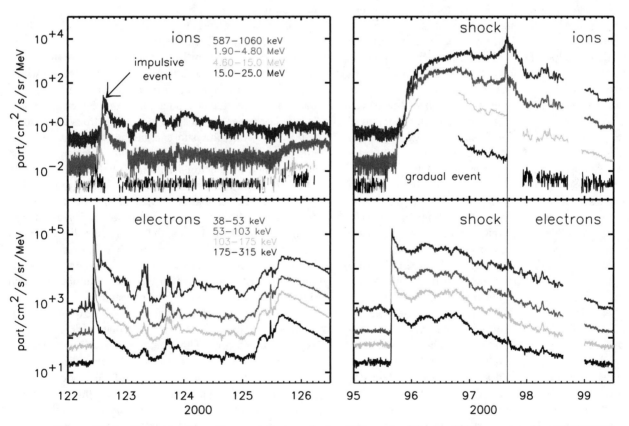

Figure 3. Intensity-time profiles of ions and electrons for two SEP events of the year 2000 as measured by ACE/EPAM [*Gold et al.*, 1998]. The two lower traces (high energy channels) of the two top panels are proton observations from IMP-8/CPME [*Sarris et al.*, 1976].

with filament eruptions [*Domingo et al.*, 1979, 1981; *Sanahuja et al.*, 1983, 1986, 1991; *Kahler et al.*, 1986], and in one case with a huge X-ray arcade [*Kahler et al.*, 1998]. However, these events are usually observed at low (≲50 MeV) proton energies. In the same way as large gradual SEP events are usually associated with solar events involving both flares and CMEs, it is interesting to note that the impulsive SEP event in Figure 3 was reported as a case of an impulsive event with an associated CME [*Kahler et al.*, 2001]. On the other hand, several recent observations of gradual events have also shown compositions and charge states consistent with impulsive events [*Cohen et al.*, 1999], challenging the simple classification of SEP events into gradual or impulsive. Whether there really are only two classes of events or whether the two classes represent extreme cases of a more continuous change from impulsive to gradual is still under debate [*Cliver et al.*, 2002; *Cliver and Cane*, 2002].

One of the concerns about the two-class paradigm is the ability of the shocks to accelerate coronal and solar wind particles rapidly to GeV energies [*Cliver et al.*, 2002]. The very largest events accelerate protons to >20 GeV. Those events are detected by ground-based neutron monitors and are known as ground-level enhancements (GLEs). In terms of their short duration (a few hours), these events may be classified as impulsive. However, such events are often accompanied by CMEs and may well develop the signatures of a gradual event at MeV energies as the CME-driven shock propagates through the interplanetary medium. Because of the long time taken for lower energy (~ 1 MeV) protons to travel 1 AU, it is difficult to separate unambiguously acceleration issues from transport effects.

Electrons have two advantages over protons in terms of their diagnostic capability. Firstly, gyrosynchrotron radiation in the microwave spectral range is emitted from relativistic (> 1 MeV) electrons in the low corona/chromosphere [*Bastian et al.*, 1998], which provides an accurate time marker for the relativistic electron acceleration. Secondly, when electrons travel on open magnetic field lines into the interplanetary medium, type III plasma radiation may be emitted. This provides a second time marker indicating when the electrons are released, while the drift rate, coupled with a reasonable coronal density model, gives the energy of the beam. *Dulk et al.* [1998] and *Haggerty et al.* [2002] deduced from the drift rate of decametric type III bursts that this energy is only a few keV. Therefore, the microwave and the type III radiation are being emitted by electrons in vastly different energy ranges. In addition, whereas the low-energy (2–19 keV) impulsive electron events observed in interplanetary space are well associated with interplanetary type III bursts [*Lin*, 1985; *Benz*

et al., 2001], at higher energies (>25 keV) impulsive electron events are delayed in relation to the interplanetary type III bursts and thus appear to be a different population of electrons [*Haggerty et al.*, 2002], and/or accelerated at a later stage [*Krucker et al.*, 1999; *Simnett et al.*, 2002; see discussion in section 2.2].

2.1. Solar Energetic Electron Events

Emission of solar electrons has been monitored for over three decades. At low energies (starting at a few keV) the emission is impulsive and the characteristics have been reviewed by *Lin* [1985]. The spectrum of the impulsive events is a power law (typically γ ~4) extending down to 2 keV, which shows that the origin of these events is in the high corona. They tend not to be associated with Hα flares. If the events extend to 100 keV or so, then they are more likely to be flare associated, and the spectrum tend to be represented by a double power law, with a spectral flattening below 100–200 keV. *Lin* [1985] has interpreted this as due to ionization losses of a population originating in the low corona.

In the relativistic region, above a few MeV, the spectrum of flare associated events tends to have a spectrum with γ ~3.0–3.5 [see review by *Simnett*, 1974]. For non flare associated events the spectrum is noticeably steeper, which is possibly a consequence of propagation from a backside flare, and may include some trapping effects [*Simnett*, 1974]. The upper energy limit to flare electrons seems to be ~100 MeV. At relativistic energies the events tend to be impulsive. However, the gradual ion events are often accompanied by electrons up to tens or hundreds of keV. *Moses et al.* [1989] studied electron events from 0.1–100 MeV. They found that the events could be categorized according to the type of associated soft X-ray event. For short duration X-ray events, the electron rigidity spectrum was a double power law, steepening below around 2 MV/*c*. Electron events associated with long duration X-ray events tended to have spectra which were a single power law in rigidity over the whole energy range studied.

Occasionally, observations at 1 AU show extremely long-lived relativistic events, such as that of Figure 4 in September/October 1979. The electron intensity had an e-folding decay constant of ~76 hours. The event in March/April 1969 [*Simnett*, 1974] had a decay constant of 125 hours, and the upper energy reached ~100 MeV. Thus it is clear that during these events the inner heliosphere was acting like a particle reservoir [*Roelof et al.*, 1992].

Solar energetic electron events observed at two distant heliocentric locations may well have quite different intensity-time profiles. Plate 1 shows the intensity-time profile of 53-103 keV electrons at both ACE and Ulysses for a 25 day period in 1998. The major event is that following a fast CME (1550 km s⁻¹ at the leading edge) associated with a flare over the west limb at S23W90 on 20 April 1998. This was the first major particle event in 1998, so the heliosphere was not experiencing significant remnants of recent activity. Ulysses was at 5.4 AU and a heliolatitude of S7°, and for a nominal solar wind speed of 400 km s⁻¹ Ulysses was magnetically connected to longitude W35 with respect to the Earth-Sun line. Plate 1 shows that the inner heliosphere gradually filled with electrons such that between 26–30 April the electron intensity was significantly higher at Ulysses than at ACE. The lower trace in Plate 1 shows the approximate ratio of outward to inward flowing electrons at Ulysses; it is clear that during the long decay, the ratio was constant and small. Detailed analysis of the pitch angle distribution shows that the intensity was isotropic. Note that the flare related activity seen at ACE from 30 April–5 May did not appear to reach Ulysses. There were 21 other CMEs observed between 20–30 April with a variety of velocities and widths. It is likely that the superposition of these interplanetary disturbances effectively impeded the escape of the solar particles seen at ACE into the heliosphere at 5 AU. Plate 1 also shows that for periods of intense solar activity

there is a higher frequency of events observed at ACE than at Ulysses. The events at ACE have higher peak amplitudes and shorter time-scales compared with those at Ulysses. This latter point should be interpreted with caution since global solar active periods may be reflected at Ulysses as a single event but in fact are the result of multiple solar particle injections. Similar observations for other periods of time are described in detail in *Simnett* [2001] and for protons in *Lario et al.* [2000].

2.2. Electron Acceleration Associated With CMEs

Recent detailed studies by *Haggerty and Roelof* [2002] and *Simnett et al.* [2002] made use of detailed timing from the electromagnetic signatures of electrons at the Sun, the evolution through the corona of the associated CMEs, and the arrival of ~40–300 keV electrons at ACE. They recognized that particle intensities from transient solar events were not necessarily emitted over a large area, and that often the onset of an electron event with a poor magnetic connection from ACE to the acceleration site might appear virtually isotropic, indicating that the heliosphere around 1 AU was receiving particles that must have been scattered back to the Sun from beyond 1 AU. When trying to understand acceleration and propagation processes, it is crucial to take into account the electron anisotropy. Fast electrons are the ideal particle to use for this, as they are too fast for the Compton-Getting effect to be important, and they are "tied" to the magnetic field. Therefore *Haggerty and Roelof* [2002] and *Simnett et al.* [2002] carefully selected electron events where ACE was magnetically connected to the source. This was recognized by the appearance at ACE of strong, magnetic field aligned beams. Figure 5 shows an example of such a beam seen on 28 June 2000. The intensity-time distribution in the 42–65 keV energy band from the LEFS60 telescope of ACE/EPAM [*Gold et al.*, 1998] is shown for all eight sectors. Inset is the pitch angle distribution for the period 1925–1930 UT. As the full opening angle of the LEFS60 telescope is 53°, then this event is consistent with a very strong electron beam traveling along the magnetic field.

From the study of 79 electron beam events associated with electromagnetic emission and the study of 52 beam events with associated CMEs, *Haggerty and Roelof* [2002] and *Simnett et al.* [2002] concluded the following:

(1) The electron injection time was typically delayed by ~20 minutes from the CME launch time, and ≥10 minutes after the onset of the electromagnetic radio and X-ray signatures of the flare (when present).

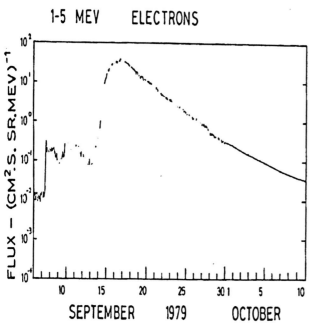

Figure 4. The long duration relativistic electron event that started on 13 September 1979 and lasted over one solar rotation [from *Simnett*, 1986].

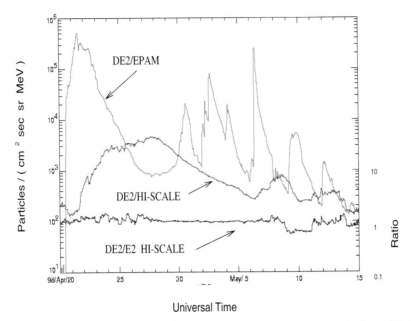

Plate 1. The intensity-time history of the 53–103 KeV electrons (DE2) at ACE/EPAM and Ulysses/HI-SCALE from 20 April – 15 May 1998. The lowest trace, DE2/E2 (right hand scale) shows the approximate ratio of outward to inward flowing electrons at Ulysses [*Simnett et al.*, 2002]

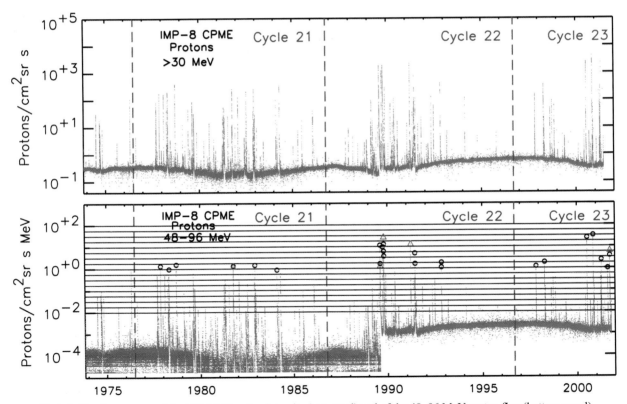

Plate 2. Hourly averages of the >30 MeV proton intensity (top panel) and of the 48–96 MeV proton flux (bottom panel) from 26 October 1973 to 24 October 2001 as measured by IMP-8/CPME [*Sarris et al.*, 1976]. The change of the background level in August 1989 was due to the failure of the anticoincidence scintillator of the instrument that had eliminated the cosmic rays penetrating from off-axis directions. The intensities at the peak of the SEP events are not affected by the change of background.

(2) The radial distance of most CMEs at the electron release time was between 1.5 and 3.5 $R_\odot$.

(3) Half the CMEs had projected speeds >600 km s^{-1} and there was some anticorrelation between the electron delay time and the CME speed.

(4) All electron events were accompanied by decametric type III emission. This emission is consistent with the presence of electron beams having exciter energies of ~few keV.

The consequence of (1) is that the relativistic (>1 MeV) electrons that must be present to produce the chromospheric electromagnetic emission do *not* escape promptly. The consequence of (2) and (3) is that most of the near-relativistic (30–300 keV) electrons observed by ACE/EPAM in interplanetary space are accelerated later by the shock driven by the coronal transient and are released at a radial dis-

tance around 2–3 $R_\odot$. The implication of (1) through (4) is that the escaping type III-associated electrons constitute yet a third population that must have a very steep energy spectrum, because usually no prompt near-relativistic or relativistic electrons are measured at 1 AU. The complete relationship among these three populations has not yet been established. For example, for each one of these 52 electron beam events, there was always an associated CME and type III emission; however, not all westward emitted CMEs are associated with beam-like energetic electron events observed near Earth. By being able to concentrate on magnetically well connected events, these studies have determined the timing, location and necessary conditions for electron acceleration in association with CMEs.

3. OCCURRENCE OF SOLAR ENERGETIC PARTICLE EVENTS

It is well-known that the occurrence rate of SEP events is higher during the active periods of the solar cycle. In order to determine the number of events occurring within each solar cycle, it is necessary to establish standard criteria to identify each single event. These criteria usually depend on the availability, quality and energy threshold of the analyzed data. There have been a number of attempts to assemble available solar proton data into catalogs [*Malitson*, 1963; *Bailey*, 1964; *King*, 1974; *van Hollebeke et al.*, 1975; *Švestka and Simon*, 1975; *Akinyan et al.*, 1983; *Goswami et al.*, 1988; *Shea and Smart*, 1990; *Feynman et al.*, 1990; *Bazilevskaya et al.*, 1990; *Sladkova et al.*, 1998; *Miroshnichenko et al.*, 2001; *Watari et al.*, 2001; *Gerontidou et al.*, 2002 and references therein]. The chronology of these lists of events reflects the evolution of the detection techniques [*Smart and Shea*, 1989]. In order to investigate the solar cycle effects in solar particle events it is necessary to have long-term, uniform and homogeneous lists of SEP events. One of these lists is the list of ground-level enhancements (GLEs) detected by neutron monitors. The sensitivity of these monitors has been approximately the same since 1953 [*Smart and Shea*, 1989]. The GLE list is continuously updated [*Duldig and Watts*, 2001; *Shea et al.*, 2001]. The annual frequency of GLEs is plotted together with the monthly sunspot number in Figure 6. The reader is referred to the above papers for a study of the GLE frequency over several solar cycles.

Shea and Smart [2001] have assembled a list of solar proton events that covers more than four solar cycles from May 1954 (start of cycle 19) to May 2001. A significant proton event is defined by *Shea and Smart* [2001] as one having 10 particles (cm^2 sec ster)$^{-1}$ above 10 MeV, in agreement with

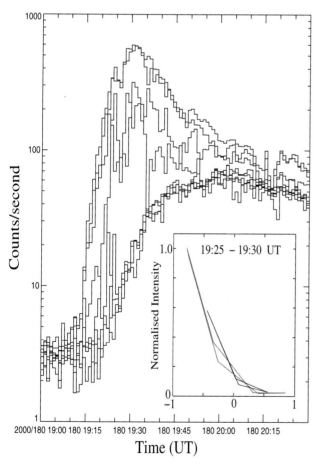

Figure 5. The electron beam event seen by EPAM in the 42–65 keV energy band from the LEFS60 telescope on 28 June 2000. Inset is the pitch angle distribution for the period 1925–1930 UT.

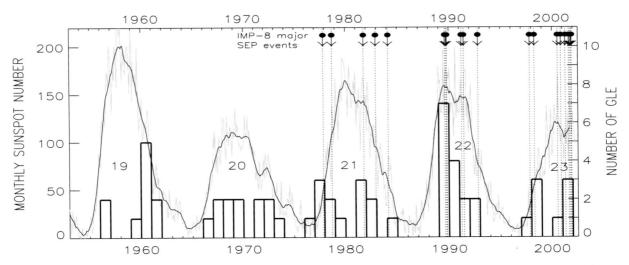

Figure 6. Monthly smoothed sunspot number (thin line), monthly sunspot number (gray line) and annual frequency of GLEs (histogram) for the period 1953–2001. Arrows mark the occurrence of the major proton events observed by IMP-8 (see section 4 and Table 1.1 for a description of such events).

the criterion established by the NOAA Space Environment Center. Unlike the NOAA/SEC list [http://umbra.nascom.nasa.gov/SEP/seps.html], *Shea and Smart* [2001] identify each unique solar proton injection into the interplanetary medium and detected at Earth as a discrete event. Thus, each event in an episode of solar proton events that may be associated with the same active solar region as it traverses the solar disk is counted as a separate event. They found that while the distribution of events in time differs from cycle to cycle, the total number of events during each cycle (as per their definition) is remarkably constant (with an average number of 75 events per solar cycle). More solar proton events generally occur during the maximum of solar activity (within around three years of the maximum of the solar sunspot cycle) than during the remaining portion of the solar cycle. Solar cycle 21 was an exception to this trend since events were observed throughout the cycle with a significant number of events in the year two after solar minimum. However, the intensity of the events and the total fluence of particles during solar cycle 21 were less than in the other studied solar cycles. It is important to note that in the study by *Shea and Smart* [2001] significant solar proton events (as per their definition) can occur at almost any time of the solar cycle. On the other hand, approximately 16% of the events for each solar cycle (cycles 19–22) contain relativistic solar protons as recorded by ground-based neutron monitors. This percentage is also relatively constant from cycle to cycle.

There have been several attempts to extend back in time the database of solar particle events. These techniques are based on the search for signatures that cosmic rays and SEPs leave on various natural terrestrial environments. When penetrating into the upper atmosphere, SEPs and cosmic rays trigger two main physical processes: ionization and nuclear interactions with the main atmospheric constituents, primarily N and O. The first process results in the production of nitrate ions. The second process creates isotopes such as the cosmogenic isotopes ^{10}Be and ^{14}C. The nitrate ions NO_3^-, produced by dissociation and ionization of the atmospheric oxygen and nitrogen, precipitate in the snow layers of the polar ice cores [*Dreschhoff and Zeller*, 1990]. *McCracken et al.* [2001a] showed that thin nitrate-rich layers found in both Arctic and Antarctic ice cores are the consequence of the occurrence of large fluence solar proton events at Earth. Using this high correlation between nitrate-rich layers and SEP events, the authors were able to date, within ± 2 months, the occurrence of 125 solar proton events for the interval 1561–1950. They estimated that the fluence of these events was higher than 1.0×10^9 cm^{-2} for >30 MeV protons. In addition, *McCracken et al.* [2001b] found that the frequency of these large SEP events has varied by a factor >10 in the interval 1561–1994. They found a well-defined "Gleissberg" (80–90 years) periodicity in the data, with six well-defined minima, two in close association with the Maunder (1650–1699) and Dalton (1810–1830) minima in sunspot numbers, and four other minima near 1560–80, 1750, 1910 and in the vicinity of 1980. The present "satellite era" of observations coincides with a recurrence of this persistent series of minima in solar proton event frequency. On the other hand, concentration of the cosmogenic isotope ^{10}Be in polar ice has also revealed great minima of activity [*McHargue and Damon*, 1991; *Beer et al.*, 1998]. However, the ^{10}Be atmospheric residence time (1–2 years; cf. *Beer et al.* [1990]) is not sufficiently short to date specific SEP events.

The radiocarbon atom ^{14}C is oxidized in the atmosphere to form $^{14}CO_2$. This form of carbon dioxide is chemically indistinguishable from the ordinary carbon dioxide $^{12}CO_2$ found in large amounts in the air. Through photosynthesis, live trees assimilate both types of carbon dioxide and deposit them into their outer rings. The ratio of the amounts of the two kinds of carbon, $^{14}C/^{12}C$, provides an inventory of atmospheric radiocarbon and, therefore, of solar activity. Since the residence time of ^{14}C in the atmosphere is a little bit longer than the 11-year solar cycle and the annual production of ^{14}C is only about 1% of the total atmospheric content [*Peristykh and Damon*, 1999], it is not possible to date individual SEP events. However, it is possible to study the differences between solar cycles and to confirm the periodicities of 11 (Schwabe cycle), 88 (Gleissberg cycle) and 208 (Suess "wiggles") years [*Lang*, 2000].

Armstrong et al. [1983] assembled a collection of data from several satellite observations and constructed a nearly time-continuous record of the flux of protons above the thresholds of 10, 30, and 50 MeV. They used data from different experiments on board the last 8 satellites of the Interplanetary Monitoring Platforms (IMP 1 to 8) and on the Orbiting Geophysical Observatory number 1 (OGO-1). Figures 1 through 4 of their paper contain the >30 MeV proton fluxes observed by these spacecraft from 1963 to mid-1982. The longevity of the IMP-8 allows us to extend these figures until 24 October 2001 point when IMP-8 operations were discontinued. The top panel of Plate 2 shows the hourly averaged proton flux above 30 MeV, as measured by the CPME experiment on board IMP-8, from the end of 1973 to mid 2001. The background level is determined by galactic cosmic rays. In August 1989, the anticoincidence scintillator of the instrument which had eliminated the cosmic rays penetrating from off-axis directions failed. The result was an increase of the background level which, nevertheless, did not affect the observations of SEP events with intensities well above this level. We have established the divisions between solar cycles in Plate 2 following *Harvey and White* [1999]. SEP events during solar cycle 21 were less intense and occurred throughout the solar cycle with the exception of the solar minimum years. We also note the small number of SEP events in 1980 which has been known as the *Gnevyshev* gap [*Bazilevskaya et al.*, 1999]. Solar cycle 22 was marked by the absence of significant events in its rising phase (years 1987–88). A

peak in the number of events occurred in 1989 and 1991. The rising phase of solar cycle 23 showed two events in November 1997 and April 1998 and three series of events in May, August and September 1998 [*Lario et al.*, 2000]. The year 1999 was marked by the complete absence of significant SEP events. However, the still ongoing solar cycle 23 has shown its splendor of activity in years 2000 and 2001 when the largest SEP events of the whole IMP-8 mission were observed.

4. MAJOR SOLAR PROTON EVENTS

4.1. Proton Peak Intensities

In order to compare individual events in the different solar cycles we need a uniform and homogeneous database. In addition, if we want to study the most intense SEP events, it is necessary to use instrumentation that does not saturate at high particle intensities. Fortunately, the count rate logic used in the 48.0–96.0 MeV proton channel of the CPME experiment on board IMP-8 [*Sarris et al.*, 1976] provides us with the adequate data for this study. We have scanned these data for the period 26 October 1973 to 24 October 2001. The bottom panel of Plate 2 shows the 48.0–96.0 MeV proton flux for this period. To study the most intense events we have selected those events with 48.0–96.0 MeV proton peak intensities greater than 1 proton (cm^2 sr s MeV)$^{-1}$. According to the proton peak flux distribution of SEP events for specific energy ranges [*van Hollebeke et al.*, 1975; *Cliver et al.*, 1991], our selection criterion reduces the study to a few events with really high intensities.

Table 1.1 summarizes the periods that meet the condition described above. We have separated those periods in individual solar proton events. To make this distinction we have associated each proton event with a specific solar event using published studies of such events as referred in the last column of Table 1.1. Observations of coronal mass ejections (CMEs) for these events were restricted to the operational periods of the SMM and LASCO coronagraphs as described in the respective references. We note that when it was possible to compute the plane-of-sky (POS) speeds of these CMEs, they were always higher than 1000 km s^{-1}. We have added in Table 1.1, the last three large SEP events of 2001. Data from IMP-8 were not recorded at that time, but inspection of GOES data leads us to think that those events would have met our selection criterion. The majority of events in Table 1.1 (17 out of 26) were GLE events. This higher percentage with respect other analyses [e.g., *Shea and Smart*, 2001] is a result of our restrictive criterion for selection to events more intense at higher energies.

Table 1. Major solar proton events observed by IMP-8.

SEP Event	Solar Flare			CME		Shock at 1 AU	Peak Flux 48–96 MeV	IMP-8 Loc.	Ref.
	X-ray Maximum	Class/Brtns X-ray/Opt	AR Location	Speed POS [km s^{-1}]	Observation Time [UT]				
Nov'77[GLE]	326/1006	X1/2B	N23W40	–	–	329/1213	1.34	S	[1]
Sep'78[GLE]	266/1023	X1/3B	N35W50	–	–	268/0718*	1.56	SW	[2]
Oct'81[GLE]	285/0627	X3/2B	S17E30	–	–	286/2240*	1.32[O]	T	[3]
Dec'82[GLE]	341/2354	X2/1B	S19W86	–	–	344/0720*	1.44	SW	[4]
Feb'84[GLE]	047/0858	–	~S12W130	1260	047/0844	NO	>0.87	T	[5]
Aug'89	224/1427	X2/2B	S16W38	>1200	224/<1445	226/0612*	11.5	SW	[6,7]
Aug'89[GLE]	228/0107	X20/2N	S15W85	1377	228/~0050	229/1540*	1.63[O]	SW	[7,8]
Sep'89[GLE]	272/1133	X9/	~S24W105	1828	272/1122	NO	9.54	T	[8,9]
Oct'89[GLE]	292/1258	X13/4B	S27E10	–	–	293/~1650	6.04[O]	SW	[10]
Oct'89[GLE]	295/1755	X2/1N	S27W32	–	295/<1917	297/0215*	12.8	SW	[8,11]
Oct'89[GLE]	297/1831	X5/2N	S29W57	1453	297/1756	299/1427*	3.60	T	[8,11]
Mar'91	081/2247	X9/3B	S26E28	–	–	083/0342*	>14.2[Δ]	T	[12]
Jun'91[GLE]	162/0229	X12/3B	N31W17	–	–	164/0016*	4.92	S	
Jun'91[GLE]	166/0821	X12/3B	N33W69	–	–	168/1018*	>1.63	SW	[13]
Oct'92	304/1816	X1/2B	S22W61	–	–	306/2146	>1.11	SW	
Nov'92	307/0308	X9/2B	S23W90	–	–	?	>1.85	S	
Nov'97[GLE]	310/1155	X9/2B	S18W63	1560	310/1210	313/1003	1.21	SW	[14]
Apr'98	110/1021	M1/	~S23W90	1638	110/1007	113/1730	1.86	S	[14]
Jul'00[GLE]	196/1024	X5/3B	N22W07	>1450	196/1054	197/1417	24.5	SW	[15]
Nov'00	313/2304	M7/3F	N10W77	1345	313/2306	315/0605	32.0	S	
Apr'01[GLE]	105/1350	X14/2B	~S20W85	1199	105/<1406	108/0005	2.40	SW	
Aug'01	NO	–	~N16W180	1575	227/2354	NO	>0.98	SW	[16]
Sep'01	267/1038	X2/2B	S16E23	~1900	267/<1030	268/2003	3.68[O]	SW	
Nov'01[GLE]	308/1620	X1/3B	N06W18	~1723	308/<1630	310/0124	NO IMP	–	
Nov'01	326/2308	M9/2N	S15W34	~1500	326/<2330	328/0540	NO IMP	–	
Dec'01[GLE]	360/0540	M7/2B	N08W54	~1435	360/<0530	363/0456	NO IMP	–	

[a]References: [1] *Burlaga et al.* [1980]; [2] *Cane* [1985]; [3] *Richardson et al.* [1991]; [4] *Cane et al.* [1986]; [5] *Kahler et al.* [1990]; [6] *Kahler* [1993]; [7] *Richardson et al.* [1994]; [8] *Kahler* [1994]; [9] *Klein et al.* [1999]; [10] *Lario and Decker* [2002]; [11] *Cane and Richardson* [1995]; [12] *Shea and Smart* [1993]; [13] *Kocharov et al.* [1994]; [14] *Lario et al.* [2000]; [15] *Smith et al.* [2001]; [16] *Lawrence and Thompson* [2001]. The index *GLE* indicates that the SEP event was also observed by neutron monitors. IMP-8 location is indicated as [T], [S] or [SW] for magnetotail, magnetosheath and solar wind, respectively.
*Arrival time of the shock based on the occurrence of a SSC. Peak fluxes are given in protons (cm^{2}sr s MeV)$^{-1}$. [O] Peak flux evaluated outside the ESP event. Δ Peak flux evaluated at the shock passage.

We have also identified the arrival of interplanetary shocks associated with these major SEP events. For those periods with an IMP-8 data gap, we use the occurrence of Sudden Storm Commencements (SSCs) as a proxy for the arrival of interplanetary shocks at Earth (these cases are indicated by an asterisk in the seventh column of Table 1). Note that the shock for the second event in August 1989 was not associated with the main solar event responsible for the SEP event at Earth [*Richardson et al.*, 1994]. On the other hand, the events in August 1989, March 1991, and June 1991 occurred during periods of intense solar activity and, most probably, some additional solar events (not specified in Table 1) contributed to the observed proton flux [*Shea and Smart*, 1993; *Richardson et al.*, 1994].

In Table 1 we have included the X-ray classification, brightness and location of the flares temporally associated with the origin of the events. The majority of events were associated with X-class X-ray flares and/or brilliant (B-class) Hα flares. Among the 26 events in Table 1, seven were associated with flares occurring at heliographic longitudes between E30–W30, seven were associated with flares at heliolongitudes between W30–W60, nine in the range W60–W90, and three events were associated with flares originated, presumably, behind the west limb of the Sun. The events with larger proton fluence were associated with solar events occurring near the central meridian of the Sun [*Shea and Smart*, 1996; *Lario et al.*, 2001c]. For these latter SEP events, a strong interplanetary shock was usually observed at Earth. This observation is consistent with a fast CME-driven shock able to accelerate (and trap) particles continuously as it travels from the Sun to Earth. In addition, for these events the observer is connected to stronger regions of the shock front as it moves away from the Sun (see section 5).

Column 8 of Table 1.1 contains the peak intensity (based on 1 hour averages) of the 48.0–96.0 MeV protons for each one of the 23 events observed by IMP-8. For those events with data gaps, we have supplemented the profiles with those from the GOES energetic particle detectors, to infer a lower limit to the peak flux detected by IMP-8. Note that IMP-8 was in the solar wind for 13 of the 23 events. In the bottom panel of Figure 8, we have identified by open circles the peak flux for each one of these events. For some of these events, the flux profile shows a second peak usually associated with the arrival of the interplanetary shock. We have identified this second peak by a triangle in Figure 8. Note that the event in March 1991 showed only a single peak associated with the shock.

It is significant to note that the two events in 2000 showed the highest 48.0–96.0 MeV proton peak intensities ever observed by the CPME experiment on IMP-8. It has been argued that there is an upper bound on the intensities of particles that arrive early in SEP events [*Reames*, 1990]. This effect, known as the streaming limit, determines the maximum proton intensity observed in the prompt component of the SEP event [*Reames and Ng*, 1998]. Before reaching the observer, energetic particles injected from an intense source near the Sun have to propagate along the IMF lines. These particles generate resonant Alfvén waves that scatter other particles that follow [*Lee*, 1983]. According to the theory, as the intensity of the streaming particles increases, the wave generation also increases until there is enough scattering to restrict the streaming of the particles and, thus, impede their escaping from the injection region. If the source of particles is a traveling CME-driven shock propagating towards the spacecraft, it should be possible to observe an increase of the particle flux at the time of the shock arrival. *Reames and Ng* [1998] examined the largest particle events seen on GOES in solar cycle 22. They determined, observationally, the streaming limit at several energy levels. A value of ~ 1.2 × 10^1 protons (cm^2 sr s MeV)$^{-1}$ was inferred for the streaming limited intensity of the 39–82 MeV proton channel. This limit was exceeded in the solar cycle 22 only during the intense energetic storm particle event (ESP) on October 20, 1989. This specific particle flux increase was due to the arrival of a complex plasma structure formed in front of the associated CME-driven shock and not by local shock acceleration of particles at these high energies [*Lario and*

Decker, 2002]. We have plotted in Figure 7 the 39–82 MeV proton intensities measured by GOES-7 and GOES-8 during the large SEP events of September and October 1989 (included in the analysis by *Reames and Ng* [1998]) together with the two major SEP events of the year 2000 (July and November 2000).

Figure 7 reveals that for the two major SEP events in 2000, the initial value of the streaming limit at this energy was amply exceeded. Inspection of the GOES data during the first part of solar cycle 23 shows that the ESP events associated with the events in November 2001 (Table 1.1) also exceeded this limit (not shown in Figure 7). It is interesting to note that the two events in 2000 exceeded the limit even during the prompt component of the SEP events. Comparison of the corrected GOES intensities (Figure7) with corresponding data from the CPME experiment on the IMP-8 spacecraft shows similar differences (up to a factor of 5) [*Lario et al.*, 2001c]. Several arguments can be used to

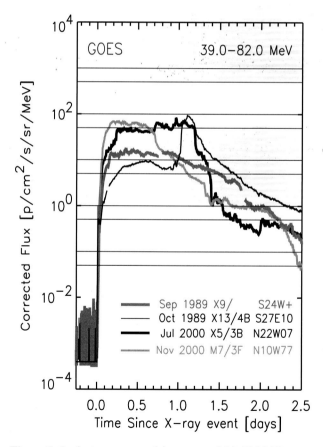

Figure 7. 5-minute averages of the corrected 39–82 MeV proton intensity as measured by GOES-7 and GOES-8 for the large SEP events of September and October 1989 and July and November 2000.

explain the exceeding of this limit. The event of July 2000 (known as the Bastille Day event) occurred just before the arrival of an interplanetary shock and associated CME at Earth [*Smith et al.*, 2001]. The trapping and/or reflection of particles between the first shock and the CME-driven shock associated with the big SEP event may have produced the plateau of the proton intensity that lasted longer than a day [*Reames et al.*, 2001]. *Kallenrode and Cliver* [2001] showed that the existence of two converging interplanetary shocks is a necessary condition to produce long-lasting high particle intensities. That phenomenon was first identified during the big SEP event of August 1972 [*Pomerantz and Duggal*, 1974].

The prompt component of the November 2000 SEP event occurred under relatively different conditions. Two days before the occurrence of the parent solar event on day 313 of 2000 (cf. Table 1.1), a CME crossed the Earth. Two days after this solar event, a high-speed, low-density solar wind stream, originating in an equatorial coronal hole, was observed at Earth. The magnetic field observed at 1 AU during the development of the SEP event was also quite disturbed. In this specific case, it is possible that the trapping and/or reflection of particles between the preceding CME (far beyond 1 AU) and the shock driven by the CME on day 313 produced the plateau of high proton intensity observed during the first 14 hours of the SEP event (Figure 7). It is

also possible that the interplanetary conditions for particle transport were different than the rest of events inhibiting the amplification of resonant waves and thus increasing the value of the streaming-limited intensity. This is a phenomenon that warrants further investigation.

4.2. Temporal Distribution of the Major SEP Events and the High-Intensity Periods

To compare the variations of proton intensities over the solar cycles, we have computed the number of hours that the 48.0–96.0 MeV proton intensity was within a particular intensity range (Figure 8). We have divided the proton intensities in several bins as plotted in the bottom panel of Plate 2. Since the time coverage of IMP-8 is not complete throughout the solar cycles, we have normalized the number of hours spent in each specific bin to the total coverage of each solar cycle. From Figure 8 it is clear that solar cycle 21 was marked by the absence of intense SEP events. During solar cycle 22, IMP-8 spent just 5 hours measuring 48.0–96.0 MeV proton intensities above 1.8×10^1 protons $(\text{cm}^2 \text{ sr s MeV})^{-1}$, mainly due to the ESP event on October 20, 1989 [*Lario and Decker*, 2002]. The first portion of solar cycle 23 (September 1996–October 2001) shows a clear excess at high intensities due to the largest SEP events observed in 2000. Those high intensities were observed dur-

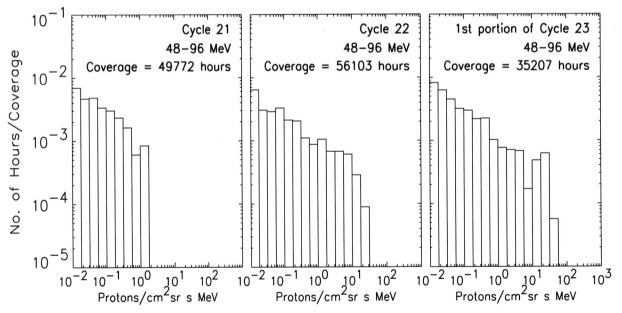

Figure 8. Number of hours that the 48–96 MeV proton channel of the CPME experiment on board IMP-8 spent at a given proton intensity during the solar cycles 21, 22 and the first portion of cycle 23. The temporal divisions of each solar cycle were taken from *Harvey and White* [1999]. The number of hours has been normalized to the time coverage of each time interval.

ing the prompt component of the SEP event and not necessarily associated with the arrival of interplanetary shocks (Figure 7).

We have also indicated in Figure 6 the temporal distribution of the major SEP events described in Table 1. These events tend to occur during the active periods of the solar cycles, being completely absent during the years of sunspot solar minimum. We note that during solar cycle 21 the major SEP events (as per our definition) were not observed near the maximum of the cycle. The sunspot number in solar cycle 22 showed a double-peaked structure which has been observed repeatedly in other solar cycles [*Gnevyshev*, 1977]. This double peak seems also to appear in the current solar cycle 23. At the time of the local sunspot minimum no major events occurred. For the moment solar cycle 23 has been shown to be quite active, from mid 2000 and throughout 2001, and more intense than the two preceding solar cycles.

5. MODELING GRADUAL SEP EVENTS

The largest long-lasting SEP events are associated with fast CMEs [*Kahler*, 2001]; although the converse is not true. Those SEP events last for several days or longer and are found to be associated with CMEs originating from virtually anywhere on the visible solar disk [*Cliver et al.*, 1995; *Reames*, 1999]. The study of these events is important mainly for two reasons: their space weather implications [*Kahler*, 2001], and their dominant contribution to the fluence of energetic particles observed throughout a solar cycle [*Lanzerotti*, 1977; *Shea and Smart*, 1996]. The proposed scenario to account for these events involves the presence of a fast CME able to drive shocks. Modelers assume that the injection of energetic particles into the interplanetary medium starts when a perturbation, originated as a consequence of a solar eruption, generates a shock wave that propagates across the solar corona. If the conditions are appropriate, this shock is able to accelerate particles from the ambient plasma (or to accelerate particles from contiguous or previous solar events), and to inject them at the base of the IMF lines. These energetic particles stream out along these lines en route to Earth and to spacecraft located in the interplanetary medium. The perturbation that generated the shock into the corona may also expand through the heliosphere driving a shock wave across the interplanetary medium. In order to explain observations of SEP events by multiple spacecraft magnetically connected to regions of the Sun distant from the parent solar active region, it is assumed that the shocks may extend up to 300° in longitude near the corona [*Cliver et al.*, 1995]. However, interplanetary shocks observed at 1 AU extend at most 180° [*Cane*, 1988].

As the shock propagates away from the Sun, it crosses many IMF lines and may be responsible for accelerating particles out of the solar wind and/or from remnant particles of previous SEP events [*Desai et al.*, 2001]. These energetic particles propagate along the IMF lines flowing outward from the shock. When these particles arrive at the spacecraft, particle intensity increases are detected which constitute the SEP events. The particle intensity profiles of the SEP events take different forms depending on (1) the heliolongitude of the source region with respect to the observer location, (2) the strength of the shock and its efficiency at accelerating particles, (3) the presence of a seed particle population to be further accelerated, (4) the evolution of the shock (its speed, size, shape and efficiency in particle acceleration), (5) the conditions for the propagation of shock-accelerated particles, and (6) the energy considered [*Heras et al.*, 1988, 1995; *Cane et al.*, 1988; *Lario et al.*, 1998; *Kahler et al.*, 1999].

The details of the proton flux profiles during these gradual SEP events are consistent with the presence of a traveling CME-driven shock which continuously injects energetic particles as it propagates away from the Sun. Figure 9 shows the proton flux profiles as a function of longitude for several recent events observed by the ACE and IMP-8 spacecraft. This figure is derived from Figure 15 of *Cane et al.* [1988]. The concept of "cobpoint" (Connecting with the OBserver point), defined by *Heras et al.* [1995] as the point of the shock front which magnetically connects to the observer, is very useful to describe the different types of SEP flux profiles. Solar events from the western hemisphere have rapid rises to maxima because, initially, the cobpoint is close to the nose of the shock near the Sun. These rapid rises are followed by gradual decreasing intensities because the cobpoint is at the eastern flank of the shock just where and when the shock is weaker. The observation of the shock at 1 AU in these western events depends on the width and strength of the shock. Near central meridian the cobpoint is initially located on the western flank of the shock and progressively moves toward the nose of the shock. Low-energy proton fluxes usually peak at the arrival of the shock, being part of what are known as energetic storm particle (ESP) events. For events originating from eastern longitudes, connection with the shock is established just a few hours before the arrival of the shock and the cobpoint moves from the weak western flank to the central parts of the shock; connection with the shock nose is only established when the shock is beyond the spacecraft and, usually, it is at this time when the peak particle flux is observed. The evolution of the low-energy ion flow anisotropy profiles throughout the SEP events reflects also the cobpoint motion along the shock front [*Domingo et*

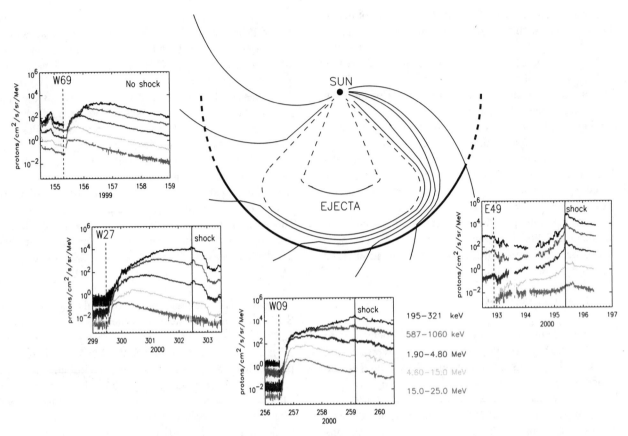

Figure 9. Ion intensity-time profiles for four different SEP events observed by ACE/EPAM [*Gold et al.*, 1998] and IMP-8/CPME [*Sarris et al.*, 1976]. Those profiles are typical of the SEP events generated from different solar longitudes relative to the observer. Dashed vertical lines indicate the occurrence of the parent solar event and solid vertical lines the arrival time of CME-driven shocks.

al., 1989]. For additional examples see *Cane et al.* [1988]; *Heras et al.* [1995]; or *Kahler* [2001].

The simulation of these particle events requires a knowledge of how particles and shocks propagate through the interplanetary medium, and how shocks accelerate and inject particles into interplanetary space. The modeling of particle fluxes and fluences associated with SEP events has to consider (1) the changes in shock characteristics as it travels through the interplanetary medium, (2) the different points of the shock where the observer is connected to, and (3) the conditions under which particles propagate. There have been several attempts to model these events. Each model presents its own simplifying assumptions in order to tackle the series of complex phenomena occurring during the development of SEP events. Two main approximations have been used to describe the particle transport: the cosmic ray diffusion equation [*Jokipii*, 1966] and the focusing-diffusion transport equation [*Roelof*, 1969; *Ruffolo*, 1995]. To describe the shock propagation, approximations range from

considering a simple semicircle centered at the Sun propagating radially at constant velocity, to fully developed magnetohydrodynamic (MHD) models.

Lee and Ryan [1986] adopted an analytical approach to solve the time-dependent cosmic ray diffusion equation for an evolving interplanetary shock which was modeled as a spherically-symmetric blast wave propagating into a stationary surrounding medium. Besides the inapplicability of the diffusion approximation outside the shock region, some strong assumptions were needed to retain a tractable model, in particular, very high blast wave velocities, the neglect of solar wind and a radial mean free path, independent of the particle energy, that increased with r^2, where r denotes the radial heliocentric distance. None of these assumptions is especially well supported observationally in the inner heliosphere.

Heras et al. [1992, 1995] were the first to adopt the focused-diffusion transport equation, including a source term, Q, which represents the injection rate of particles accelerated at the traveling shock. The use of this transport equation is more adequate for these SEP events since it allows us to reproduce the large and long-lasting anisotropies usually observed at low-energies in gradual SEP events [Heras et al., 1994]. The injection of particles is considered to take place at the cobpoint. To track this point with time, the authors used an MHD model that describes the shock propagation from a given inner boundary close to the Sun up to the observer. The IMF is described upstream of the shock by the usual Parker spiral. This model has been refined by including solar wind convection and adiabatic deceleration effects into the particle transport equation and the corotation of the IMF lines [Lario et al., 1998]. It has been successfully applied to reproduce the low-energy (<20 MeV) proton flux and anisotropy profiles of a number of SEP events simultaneously observed by several spacecraft [Heras et al., 1995; Lario et al., 1998].

Kallenrode and Wibberenz [1997] and Kallenrode [2001] adopted the same scheme as the previous works. However, these authors use a semicircle propagating radially from the Sun at constant speed to describe the shock. They also parameterize the injection rate, Q, in terms of a radial and azimuthal variation which represents the temporal and spatial dependences of the shock efficiency in accelerating particles. They allow also for particle propagation in the downstream region of the shock just by changing the magnitude of the focusing length, however they do not modify the actual IMF topology behind the shock which may lead to different results [Lario et al., 1999]. They also allow for a transmission of particles across the shock, but not a change of the particle energy when they are reflected and/or transmitted. On the other hand, Torsti et al. [1996] and Antilla et al. [1998] adopted a similar scheme as the above-mentioned works but assumed, in order to locate the cobpoint, that the distance of the cobpoint to the observer along the IMF line connecting with the observer decreases linearly with time. They also used a complex parametric function to describe, Q, including energetic, temporal and spatial dependences. Differences among the above models have been described in Sanahuja and Lario [1998] and Kallenrode [2001].

Ng et al. [1999a, 1999b, 2001] have used a similar approach to deal with SEP simulations. The authors have developed a numerical model where the particle transport includes proton-generated Alfvén waves. Whereas the above-described models assume that the scattering of particles may be parametrized by a given mean free path (which may depend on the particle energy and time), Ng et al. [1999a] consistently solve the focused-diffusion transport equation for the particles and the equation describing the evolution of differential wave intensity. Assuming that particles are accelerated out of a constant source plasma with a specific composition, Ng et al. [2001] successfully describe the evolution of abundance ratios in some SEP events. No quantitative agreement of the predicted wave spectrum has yet been presented. Several simplifications were made in the model such as the assumption of radial IMF and the use of several phenomenological parameters in the equations. The shock was assumed to travel radially away from the Sun at a constant speed. The injection, Q, was also parametrized to account for temporal, radial and energy dependence. This model allows for a better description of self-generated scattering processes throughout the transport of particles of different species. Nevertheless, the use of radial IMF does not allow the reproduction of the longitudinal dependence shown in Figure 9. On the other hand, observation of shock speeds in different directions [Cane, 1988] and dynamic studies from MHD simulations [Smith and Dryer, 1990] indicate a decrease of the shock speed towards its flanks and a weakening of its front as it expands. Those models describing the shock as a semicircle propagating at constant speed oversimplify the shock geometry and evolution and therefore misplace the cobpoint and neglect the physical conditions at the point where particles are accelerated and injected.

None of the above models treats the fundamental nature of particle acceleration at the evolving interplanetary shocks. The details of how the MHD conditions at the shock front translate into an efficiency in particle acceleration and how it evolves as the shock expands are not completely understood. Lario et al. [1998] proposed a parameterization to relate the evolution of the injection rate of shock-accelerated particles to the dynamic properties of the shock. That relation yields a quantification of the injection rate, its energy spectrum and its evolution; however, it does not address the physical mechanism of particle shock acceleration.

Recently, theoretical efforts have been addressed to incorporate the mechanisms of shock-acceleration of particles into traveling interplanetary shocks [Zank et al., 2000; Lee, 2001; Berezhko et al., 2001]. In particular, Zank et al. [2000] have developed a dynamical time-dependent model of particle acceleration at the propagating shock. This model assumes a spherically symmetric solar wind into which a blast wave propagates. Both the wind and shock are modeled numerically using hydrodynamic equations and assuming a Parker spiral for the IMF. The local characteristics of the shock, such as the shock strength or the Mach number, are dynamically computed. Those parameters are used to determine the distribution of particles injected into the diffusive shock acceleration mechanism. Shock-accelerated particles propagate diffusively in the vicinity of the shock generating resonant Alfvén waves which are included in the model. However, when these particles are far enough upstream of the shock, they are allowed to freely escape without experiencing any scattering, deceleration or convection effects. In that way, *Zank et al.* [2000] neglected the complications associated with a detailed transport model, such as those developed by *Ruffolo* [1995], *Lario et al.* [1998] and *Ng et al.* [1999a]. An interesting point of the *Zank et al.* [2000] model is that, for extremely strong shocks, particle energies of the order of 1 GeV can be achieved when the shock is still close to the Sun. As the shock propagates outward, the maximum accelerated particle energy decreases sharply. Other shock acceleration models [*Berezhko et al.*, 2001] also suggest the possibility that 1 GeV protons can be accelerated when extremely strong shocks are close to the Sun (< 3 solar radii). Comparisons of these models of particle shock-acceleration with specific observations have not yet been reported. For the moment, no theoretical and/or numerical model treats SEP acceleration and transport near its full complexity.

6. CONCLUSIONS

The population of energetic particles in the heliosphere is modulated by the solar activity. Under solar minimum conditions the main sources of the energetic particles observed at 1 AU are: (1) the interstellar medium in the form of galactic cosmic rays observed at energies above ~200 MeV for protons and above ~3 MeV for electrons; (2) the termination shock in the form of anomalous cosmic rays; (3) the corotating interaction regions which accelerate electrons up to around 300 keV and ions up to a few MeV/nucleon; and (4) the Jovian magnetosphere that generates electrons observed at 1 AU during quiet times in the range from a few hundreds keV to a few MeV. During solar maximum the Sun becomes the dominant source of energetic particles. These particles may fill the heliosphere even up to the highest latitudes. The processes under which solar particles are accelerated and released into the interplanetary medium are related to flares and shocks. The contribution of these processes in the production of large solar energetic particle events is still under debate. Measurements of the charge states of heavy ions in SEP events give important information on the source population and acceleration processes. Coronal shocks may accelerate near-relativistic electrons, provided a suitable seed population is present. CMEs play an important role in accelerating the bulk of non-relativistic ions seen around 1 AU and such acceleration takes place as a CME-driven shock propagates both through the corona and then out into the interplanetary medium. Electron and ion acceleration is usually most effective at a few solar radii from the Sun, but interplanetary acceleration plays an important role in large gradual events.

Acknowledgments. We are grateful to A. Aran, R. B., Decker, D. K. Haggerty, G. C. Ho, E. C. Roelof and B. Sanahuja for their comments and help while writing the manuscript. DL was partially supported by NASA grant NAG5–10787. This review article was written in December 2001 and does not contain the most recent developments in the study and modeling of SEP events published following this date as well as the major energetic particle events occurring during the decay phase of the solar cycle 23 (e.g., the large events of October-November 2003).

REFERENCES

Ahluwalia, H.S., Galactic cosmic ray intensity variations at a high latitude sea level site 1937-1994, *J. Geophys. Res., 102*, 24229-24236, 1997.

Akinyan, S. T., et al., Catalog of solar proton events 1970-1979, edited by Y.I. Logachev, Moscow, Nauka, IZMIRAN, 184, 1983.

Antilla, A., L. G. Kocharov, J. Torsti, and R. Vainio, Long-duration high-energy proton events observed by GOES in October 1989, *Ann. Geophysicae, 16*, 921-930, 1998.

Armstrong, T. P., C. Brungardt, and J. E. Meyer, Satellite observations of interplanetary and polar cap solar particle flux from 1973 to the present, in *Weather and Climate Responses to Solar Variations*, edited by B. M. McCormac, *Colorado Associated Univ. Press., Boulder*, 71-79, 1983.

Bailey, D. K., Polar cap absorption, *Planetary and Space Science, 12*, 495-539, 1964.

Bastian, T.S., A.O. Benz and D.E. Gary, Radio emission from solar flares, *Ann. Rev. Astron. Astrophys., 36*, 131-188, 1998.

Bazilevskaya, G. A., et al., Catalog of solar proton events 1980-1986, edited by Y.I. Logachev, Moscow, World Data Center B-2, 160, 1990.

Bazilevskaya, G.A., et al., The Gnevyshev gap effect in solar cosmic rays, *Conf. Pap. Int. Cosmic Ray Conf. 26th, 6*, 240-243, 1999.

Beer, J., T. Steven, N. Weiss, An active Sun throughout the Maunder minimum, *Solar Physics, 181*, 237-249, 1998.

Beer, J., et al., Use of ^{10}Be in polar ice to trace the 11-year cycle of solar activity, *Nature, 347*, 164-166, 1990.

Benz, A.O., et al., The source regions of impulsive solar electron events, *Solar Physics, 203*, 131-144, 2001.

Berezhko, E.G., S. I. Petukhov, and S. N. Taneev, Shock acceleration of energetic particles in solar corona, *Conf. Pap. Int. Cosmic Ray Conf. 27th, 8*, 3215-3218, 2001.

Burger, R.A., Galactic cosmic rays in the heliosphere, in *26th International Cosmic Ray Conference*, edited by B.L. Dingus, et al., *AIP Conf. Proc. 516*, 83-102, 2000.

Burlaga, L., et al., Interplanetary particles and fields, November 22 to December 6, 1977: Helios, Voyager and Imp observations between 0.6 and 1.6 AU, *J. Geophys. Res., 85*, 2227-2242, 1980.

Cane, H.V., The evolution of interplanetary shocks, *J. Geophys. Res., 90*, 191-197, 1985.

Cane, H.V., Two classes of solar energetic particle events associated with impulsive and long-duration soft X-ray flares, *Astrophys. J., 301*, 448-459, 1986.

Cane, H.V., The large-scale structure of flare-associated interplanetary shocks, *J. Geophys. Res., 93*, 1-6, 1988.

Cane, H.V., and Richardson, I.G., Cosmic ray decreases and solar wind disturbances during late October 1989, *J. Geophys. Res., 100*, 1755-1762, 1995.

Cane, H. V., D. V., Reames, and T. T. von Rosenvinge, The role of interplanetary shocks in the longitude distribution of solar energetic particles, *J. Geophys. Res., 93*, 9555-9567, 1988.

Cliver, E. W., and A. G. Ling, Coronal mass ejections, open magnetic flux, and cosmic-ray modulation, *Astrophys. J., 556*, 432-437, 2001.

Cliver, E. W., and H.V. Cane, The last word, *EOS Trans. AGU, 83*, 61-68, 2002.

Cliver, E. W., et al., Size distribution of solar energetic particle events, *Conf. Pap. Int. Cosmic Ray Conf. 22nd, 3*, 25-28, 1991.

Cliver, E. W., et al., Extreme "propagation" of solar energetic particles, *Conf. Pap. Int. Cosmic Ray Conf. 24th, 4*, 257-260, 1995.

Cliver, E. W., B., Klecker, M.-B. Kallenrode, and H. V. Cane, Researchers discuss role of flares and shocks in SEP events, *EOS Trans. AGU, 83*, 132, 2002.

Cohen, C. M. S., et al., New observations of heavy-ion-rich solar particle events from ACE, *Geophys. Res. Lett., 26*, 2697-3000, 1999.

Desai, M. I., et al., Acceleration of ^{3}He Nuclei at Interplanetary Shocks, *Astrophys. J., 553*, L89-L92, 2001.

Domingo, V., R.J. Hynds, and G. Stevens, A solar proton event of possible non-flare origin, *Conf. Pap. Int. Cosmic Ray Conf. 16th, 5*, 192-197, 1979.

Domingo, V., B. Sanahuja, and K.–P. Wenzel, Non-flare injection of protons into interplanetary space, *Conf. Pap. Int. Cosmic Ray Conf. 17th, 3*, 109-112, 1981.

Domingo, V., B. Sanahuja, and A.M. Heras, Energetic particles, interplanetary shocks and solar activity, *Adv. Space Res., 9*, 191-195, 1989.

Dreschhoff, G.A.M., and E.J. Zeller, Evidence of individual solar proton events in antarctic snow, *Solar Physics, 127*, 337-346, 1990.

Duldig, M.L., and D.J., Watts, The new international GLE database, *Conf. Pap. Int. Cosmic Ray Conf. 27th, 8*, 3409-3412, 2001.

Dulk, G.A., et al., Electron beams and radio waves of solar type III bursts, *J. Geophys. Res., 103*, 17223-17234, 1998.

Feynman, J., T.P. Armstrong, L. Dao-Gibner, and S. Silverman, Solar proton events during solar cycles 19, 20, and 21, *Solar Physics, 126*, 385-401, 1990.

Gerontidou, M., et al., Frequency distributions of solar proton events, *J. Atm. Solar-Terr. Phys., 64*, 489-496, 2002.

Gnevyshev, M. N., Essential features of the 11-year solar cycle, *Solar Physics, 51*, 175-183, 1977.

Gold, R.E., et al., Electron, Proton, and Alpha Monitor on the Advanced Composition Explorer spacecraft, *Space Sci. Rev., 86*, 541-562, 1998.

Goswami, J.N., et al., Solar flare protons and alpha particles during the last three solar cycles, *J. Geophys. Res., 93*, 7195-7205, 1988.

Haggerty, D.K, and E.C. Roelof, Impulsive near-relativistic solar electron events: Delayed injection with respect to solar electromagnetic emission, *Astrophys. J., 579*, 841-853, 2002.

Haggerty, D.K, E.C. Roelof and M.L. Kaiser, Relative Timing of Impulsive Solar Electron Injections and Solar Electromagnetic Emissions, *Planetary Radio Emissions V*, Austrian Acad. Sciences, 437-444, 2002.

Hakura, Y., Solar cycle variation in energetic particle emissivity of the Sun, *Solar Physics, 39*, 493-497, 1974.

Harvey, K.L., and O.R. White, What is solar cycle minimum?, *J. Geophys. Res., 104*, 19759-19764, 1999.

Heras, A. M., B. Sanahuja, V. Domingo, and J.A. Joselyn, Low-energy particle events generated by solar disappearing filaments, *Astron. and Astrophys., 197*, 297-305, 1988.

Heras, A. M., et al., The influence of the large-scale interplanetary shock structure on a low-energy particle event, *Astrophys. J., 391*, 359-369, 1992.

Heras, A.M., et al., Observational signatures of the influence of the interplanetary shocks on the associated low-energy particle events, *J. Geophys. Res., 99*, 43-51, 1994.

Heras, A. M., et al., Three low-energy particle events: modeling the influence of the parent interplanetary shock, *Astrophys. J., 445*, 497-508, 1995.

Jokipii, J. R., Cosmic-Ray Propagation. I. Charged Particles in a Random Magnetic Field, *Astrophys. J., 146*, 480-487, 1966.

Jokipii, J. R., Solar modulation of energetic particles and cosmic-ray deposition, in *Weather and Climate Responses to Solar Variations*, edited by B. M. McCormac, *Colorado Associated Univ. Press., Boulder*, 57-70, 1983.

Jokipii, J. R., and B. Thomas, Effects of drift on the transport of cosmic rays IV: Modulation by a wavy interplanetary current sheet, *Astrophys. J., 243*, 1115-1122, 1981.

Jokipii, J. R., E. H. Levy, and W. B. Hubbard, Effects of particle drift on cosmic-ray transport. I - General properties, application to solar modulation, *Astrophys. J., 213*, 861-868, 1977.

Kahler, S. W., Coronal mass ejections and long risetimes of solar energetic particle events, *J. Geophys. Res., 98*, 5607-5615, 1993.

Kahler, S. W., Injection profiles of solar energetic particles as functions of coronal mass ejection heights, *Astrophys. J., 428*, 837-842, 1994.

Kahler, S. W., Origin and properties of solar energetic particles in space, in *Space Weather*, edited by Song et al., *AGU Monography, 125* 109-122, 2001.

Kahler, S.W., et al., Solar filament eruptions and energetic particle events, *Astrophys. J., 302*, 504-510, 1986.

Kahler, S.W., Reames, D.V., and Sheeley, Jr., N.R., Coronal mass ejections and the injection profiles of solar energetic particle events, *Conf. Pap. Int. Cosmic Ray Conf. 21st, 5*, 183-186, 1990.

Kahler, S.W., et al., The solar energetic particle event of April 14, 1994, as a probe of shock formation and particle acceleration, *J. Geophys. Res., 103*, 12069-12076, 1998.

Kahler, S.W., J.T. Burkepile, and D.V. Reames, Coronal/Interplanetary factors contributing to the intensities of E>20 MeV gradual SEP events, *Conf. Pap. Int. Cosmic Ray Conf. 26th, 6*, 248-251, 1999.

Kahler, S. W., D. W. Reames, and N. R. Sheeley, Jr., Coronal mass ejections associated with impulsive solar energetic particle events, *Astrophys. J.*, *562*, 558-565, 2001.

Kallenrode, M.–B., Shock as a black box 2. Effects of adiabatic deceleration and convection included, *J. Geophys. Res.*, *106*, 24989-25003, 2001.

Kallenrode, M.–B., and G. Wibberenz, Propagation of particles injected from shocks: A black box model and its consequences for acceleration theory and data interpretation, *J. Geophys. Res.*, *102*, 22311-22334, 1997.

Kallenrode, M.–B., and E. W. Cliver, Rogue SEP events: Observational aspects, *Conf. Pap. Int. Cosmic Ray Conf. 27th*, *8*, 3314-3317, 2001.

King, J., Solar proton fluences for 1977-1983 space missions, *J. Spacecraft*, *11*, 401-408, 1974.

Klein, K.–L. et al., Flare-associated energetic particles in the corona at 1 AU, *Astron. Astrophys.*, *348*, 271-285, 1999.

Kocharov, L.G., et al., Electromagnetic and corpuscular emission from the solar flare of 1991 June 15: Continuous acceleration of relativistic particles, *Solar Phys.*, *150*, 267-283, 1994.

Krucker, S., D.E. Larson, R.P. Lin, and B.J. Thompson, On the origin of impulsive electron events observed at 1 AU, *Astrophys. J.*, *519*, 864-875, 1999.

Lang, K. R., in *The Sun from Space*, edited by Springer Verlag, pp. 251, 2000.

Lanzerotti, L.J., Measures of energetic particles from the Sun, in *The Solar Output and Its Variation*, O.R. White, ed., University of Colorado Press, Boulder, CO, 383-403, 1977.

Lanzerotti, L.J., et al., Heliosphere instrument for spectra, composition and anisotropy at low energies, *Astron. Astrophys.*, *92*, 349-363, 1992.

Lanzerotti, L.J., et al., Over the southern solar pole: Low-energy interplanetary charged particles, *Science*, *268*, 1010-1013, 1995.

Lario, D., and R.B. Decker, The energetic storm particle event of October 20, 1989, *Geophys. Res. Lett.*, *29*, 10.1029/2001GL014017, 2002.

Lario, D., B. Sanahuja, and A. M. Heras, Energetic particle events: efficiency of interplanetary shocks as 50 keV < E < 100 MeV proton accelerators, *Astrophys. J.*, *509*, 415-434, 1998.

Lario, D., M. Vandas, and B. Sanahuja, Energetic Particle Propagation in the Downstream Region of Transient Interplanetary Shocks, in *Solar Wind Nine*, edited by S.R. Habbal, et al., *AIP Conf. Proc. 471*, 741-744, 1999.

Lario, D., et al., Energetic proton observations at 1 and 5 AU. 2. Rising phase of the solar cycle 23, *J. Geophys. Res.*, *105*, 18251-18274, 2000.

Lario, D., E.C. Roelof, R.J. Forsyth, and J.T. Gosling, 26-day analysis of energetic ion observations at high and low heliolatitudes: Ulysses and ACE, *Space Sci. Rev.*, *97*, 249-252, 2001a.

Lario, D., et al., High-latitude Ulysses observations of the H/He intensity ratio under solar minimum and solar maximum conditions, in *Solar and Galactic Composition*, edited by R. F. Wimmer, *AIP Conf. Proc. 598*, 183-188, 2001b.

Lario, D., R. B. Decker, and T. P. Armstrong, Major solar proton events observed by IMP-8 (from November 1973 to May 2001), *Conf. Pap. Int. Cosmic Ray Conf. 27th*, *8*, 3254-3257, 2001c.

Lawrence, G.R., Thompson, B.J., Proton Storms Associated with Far-Sided Coronal Mass Ejections EOS. Trans. AGU 82 (47), Fall Meet. Suppl., Abstract SH41B-0753, 2001.

Lee, M. A., Coupled hydromagnetic wave excitation and ion acceleration at traveling interplanetary shocks, *J. Geophys. Res.*, *88*, 6109-6119, 1983.

Lee, M.A., The Theory of Particle Acceleration at Coronal/Interplanetary Shocks, *EOS Trans. AGU*, *82(47)*, Fall Meet. Suppl., Abstract SH12C-09, 2001.

Lee, M. A., and J.M. Ryan, Time-dependent coronal shock acceleration of energetic solar flare particles, *Astrophys. J.*, *303*, 829-842, 1986.

Leske, R.A., et al., Observations of anomalous cosmic rays at 1 AU, in *Acceleration and Transport of Energetic Particles Observed in the Heliosphere: ACE 2000 Symposium*, edited by R.A. Mewaldt et al., *AIP Conf. Proc. 528*, 293-300, 2000.

Lin, R.P., Energetic particles from the Sun and solar modulation of galactic cosmic rays, in *The Solar Output and Its Variation*, O.R. White, ed., University of Colorado Press, Boulder, CO, 39-41, 1977.

Lin, R.P., Energetic electrons in the interplanetary medium, *Solar Physics*, *100*, 537-561, 1985.

Lockwood, M., Long-term variations in the magnetic field of the Sun and the heliosphere: Their origin, effects and implications, *J. Geophys. Res.*, *106*, 16021-16038, 2001.

Lockwood, J.A., W.R. Webber, and H. Debrunner, Differences in the maximum intensities and the intensity-time profiles of cosmic rays in alternate solar magnetic field polarities, *J. Geophys. Res.*, *106*, 10635-10644, 2001.

Maclennan, C.G., L.J. Lanzerotti, and S.E. Hawkins III, Populating an inner heliosphere reservoir (<5 AU) with electrons and heavy ions, *Int. Cosmic Ray Conf. 27th*, *8*, 3265-3268, 2001.

Malitson, H. H., Table of solar proton events, in *Solar Proton Manual*, edited by F. B. McDonald, *NASA, Washington, DC, NASA TR-169, Sept*, 109-117, 1963.

Marsden, R.G. (ed.), *The 3-D Heliosphere at Solar Maximum*, edited by Kluwer Academic Publishers, 2001.

McComas, D.J., R. Goldstein, J.T. Gosling, and R.M. Skoug, Ulysses' second orbit: Remarkably different solar wind, *Space Sci. Rev.*, *97*, 99-103, 2001.

McCracken, K. G., et al., Solar cosmic ray events for the period 1561-1994 1. Identification in polar ice, 1561-1950, *J. Geophys. Res.*, *106*, 21585-21598, 2001a.

McCracken, K. G., G. A. M. Dreschhoff, D. F. Smart, and M. A. Shea, Solar cosmic ray events for the period 1561-1994 2. The Gleissberg periodicity, *J. Geophys. Res.*, *106*, 21599-21609, 2001b.

McHargue, L.R., and P.E. Damon, The Global Beryllium 10 Cycle, *Rev. Geophys.*, *29*, 141-158, 1991.

McKibben, R.B., et al., Ulysses COSPIN observations of the energy and charge dependence of the propagation of solar energetic particles to the Sun's south polar regions, *Int. Cosmic Ray Conf. 27th*, *8*, 3281-3284, 2001.

Miroshnichenko, L. I., B. Mendoza, and R. Pérez-Enríquez, Size distributions of the > 10 MeV solar proton events, *Solar Physics*, *202*, 151-171, 2001.

Moses, D., W. Droege, P. Meyer, P. Evenson, Characteristics of energetic solar flare electron spectra, *Astrophys. J.*, *346*, 523-530, 1989.

Ng, C. K., D. V. Reames, and A. J. Tylka, Effects of proton-amplified waves on evolution of solar energetic particle composition in gradual events, *Geophys. Res. Lett.*, *26*, 2145-2148, 1999a.

Ng, C. K., D. V. Reames, and A. J. Tylka, Model for the evolution of the elemental abundances of solar energetic particles, *Conf. Pap. Int. Cosmic Ray Conf. 26th*, *6*, 151-154, 1999b.

Ng, C. K., D. V. Reames, and A. J. Tylka, Evolution of abundances and spectra in the large solar energetic particle events of 1998 Sep 30 and 2000 Apr 4, *Conf. Pap. Int. Cosmic Ray Conf. 27th*, *8*, 3140-3143, 2001.

Peristykh, A. N., and P. E. Damon, Multiple evidence of intense solar proton events during solar cycle 13, *Conf. Pap. Int. Cosmic Ray Conf. 26th*, *6*, 264-267, 1999.

Pomerantz, M. A., and S. P. Duggal, Interplanetary acceleration of solar cosmic rays to relativistic energy, *J. Geophys. Res.*, *79*, 913-919, 1974.

Ponyavin, D.I., Geomagnetic tracing of the inner heliosphere, *Space Sci. Rev.*, *97*, 225-228, 2001.

Reames, D.V., Acceleration of energetic particles by shock waves from large solar flares, *Astrophys. J.*, *358*, L63-L67, 1990.

Reames, D.V., Particle acceleration at the Sun and in the heliosphere, *Space Sci. Rev.*, *90*, 413-491, 1999.

Reames, D. V., and C. K. Ng, Streaming-limited intensities of solar energetic particles, *Astrophys. J.*, *504*, 1002-1005, 1998.

Reames, D.V., Ng, C.K., and Tylka, A.J., Heavy ion abundances and spectra and the large gradual solar energetic particle event of 2000 July 14, *Astrophys. J.*, *548*, L233-236, 2001.

Richardson, I.G., Cane, H.V, and von Rosenvinge, T.T., Prompt arrival of solar energetic particles from far eastern events: The role of large-scale interplanetary magnetic field structure, *J. Geophys. Res.*, *96*, 7853-7860, 1991.

Richardson, I.G., Farrugia, C.J. and Winterhalter, D., Solar activity and coronal mass ejections on the western hemisphere of the Sun in mid-August 1989: Association with interplanetary observations at the ICE and IMP8 spacecraft, *J. Geophys. Res.*, *99*, 2513-2529, 1994.

Richardson, I.G., L.M. Barbier, D.V. Reames, and T.T. von Rosenvinge, Corotating MeV/amu ion enhancements at 1 AU or less from 1978 to 1986, *J. Geophys. Res.*, *98*, 13-32, 1993.

Roelof, E.C., Propagation of solar cosmic rays in the interplanetary magnetic field, in *Lectures in High Energy Astrophysics*, edited by H. Ögelman and J.R. Wayland, *NASA Spec. Publ., SP-199*, 111, 1969.

Roelof, E. C., Near-relativistic electrons: Corotation interaction regions and particle reservoirs, EOS. Trans. AGU 83, Spring Meet. Suppl., Abstract, 2002.

Roelof, E.C., et al., Low energy solar electrons and ions observed at Ulysses February-April 1991: The inner heliosphere as a particle reservoir, *Geophys. Res. Lett.*, *19*, 1243-1246, 1992.

Roelof, E.C., G.M. Simnett, and S.J. Tappin, The regular structure of shock-accelerated ~40-100 keV electrons in the high latitude heliosphere, *Astron. Astrophys.*, *316*, 481-486, 1996.

Roelof, E. C., et al., Reappearance of recurrent low energy particle events at Ulysses/HI-SCALE in the northern heliosphere, *J. Geophys. Res.*, *102*, 11251-11262, 1997.

Ruffolo, D., Effect of adiabatic deceleration on the transport of solar cosmic rays *Astrophys. J.*, *442*, 861-874, 1995.

Sanahuja, B., and D. Lario, Low-energy cosmic rays: Modeling gradual events, in *Conf. Proc. 16th European Cosmic Ray Symp.*, edited by J. Medina, *Alcalá University Editions*, 129-140, 1998.

Sanahuja, B., et al., A large proton event associated with solar filament activity, *Solar Physics*, *84*, 321-337, 1983.

Sanahuja, B., A.M. Heras, V. Domingo, and J.A. Joselyn, Low-energy particle events and solar filament eruptions, *Adv. Space Res.*, *6*, 277-280, 1986.

Sanahuja, B., A.M. Heras, V. Domingo, and J.A. Joselyn, Three solar filament disappearances associated with interplanetary low-energy particle events, *Solar Physics*, *134*, 379-394, 1991.

Sanderson, T.R., et al., The Ulysses south polar pass: Energetic ion observations, *Geophys. Res. Lett.*, *22*, 3357-3360, 1995.

Sarris, E.T., S.M. Krimigis, and T.P. Armstrong, Observations of magnetospheric bursts of high-energy protons and electrons at approximately 35 earth radii with Imp 7, *J. Geophys. Res.*, *81*, 2341-2355, 1976.

Shea, M. A., and D. F. Smart, A summary of major solar proton events, *Solar Physics*, *127*, 297-320, 1990.

Shea, M.A., and Smart, D.F., March 1991 solar-terrestrial phenomena and related technological consequences, *Conf. Pap. Int. Cosmic Ray Conf. 23rd*, *3*, 739-742, 1993.

Shea, M.A., and Smart, D.F., Solar proton fluxes as a function of the observation location with respect to the parent solar activity, *Adv. Space Res.*, *17*, 225-228, 1996.

Shea, M.A., and D. F. Smart, Solar proton and GLE event frequency: 1955-2000, *Conf. Pap. Int. Cosmic Ray Conf. 27th*, *8*, 3401-3404, 2001.

Shea, M.A., et al., Update on the GLE database: solar cycle 19, *Conf. Pap. Int. Cosmic Ray Conf. 27th*, *8*, 3405-3408, 2001.

Shields, J.C., Armstrong, T.P., Eckes, S.P., and Briggs, P.R., Solar and interplanetary ions at 2-4 MeV/nucleon during solar cycle 21–Systematic variations of H/He and He/CNO ratios and intensities, *J. Geophys. Res.*, *90*, 9439-9453, 1985.

Simnett, G.M., Relativistic Solar Electrons, *Space Sci. Rev.*, *16*, 257-316, 1974.

Simnett, G.M., Interplanetary phenomena and solar radio bursts, *Solar Phys.*, *104*, 67-91, 1986.

Simnett, G.M., Energetic particle characteristics in the high-latitude heliosphere near solar maximum, *Space Sci. Rev.*, *97*, 231-242, 2001.

Simnett, G.M., and E.C. Roelof, Reverse shock acceleration of electrons and protons at mid-heliolatitudes from 5.3-3.8 AU, *Space Sci. Rev.*, *72*, 303-308, 1995.

Simnett, G.M., K. Sayle, S.J. Tappin, and E.C. Roelof, Corotating particle enhancements out of the ecliptic plane, *Space Sci. Rev.*, *72*, 327-330, 1995.

Simnett, G.M., E.C. Roelof and D.K. Haggerty, The Acceleration and Release of Relativistic Electrons by CMEs, *Astrophys. J.*, *579*, 854-862, 2002.

Sladkova, A. I., et al., Catalog of solar proton events 1987-1996, edited by Y.I., Logachev, Moscow, Moscow Univ. Press, 1998.

Smart, D.F., and M.A. Shea, Solar proton events during the past three solar cycles, *J. Spacecraft*, *26*, 403-415, 1989.

Smith, Z., and M. Dryer, MHD study of temporal and spatial evolution of simulated shocks in the ecliptic plane within 1 AU, *Solar Physics*, *129*, 387-405, 1990.

Smith, C.W., et al., ACE observations of the Bastille day 2000 interplanetary disturbances, *Solar Physics*, *204*, 227-252, 2001.

Švestka, Z., and P. Simon (Eds.), *Catalog of solar proton events 1955-1969*, D. Reidel Publishing Company, 1975.

Torsti, J., et al., The 1990 May 24 solar cosmic ray event, *Solar Physics*, *166*, 135-158, 1996.

van Hollebeke, M. A. I., L. S. Ma Sung, and F. B. McDonald, The variation of solar proton energy spectra and size distribution with heliolongitude, *Solar Physics*, *41*, 189-223, 1975.

von Rosenvinge, T. T., et al., Time variations in elemental abundances in solar energetic particle events, in *Solar and Galactic Composition*, edited by R. F. Wimmer-Schweingruber, *AIP Conf. Proc. 598*, 343-348, 2001.

Watari, S., et al., The Bastille Day (14 July 2000) event in historical large Sun-Earth connection events, *Solar Physics*, *204*, 423-436, 2001.

Zank, G.P., W. K. M. Rice, and C. C. Wu, Particle acceleration and coronal mass ejection driven shocks: a theoretical model, *J. Geophys. Res.*, *105*, 25079-25095, 2000.

D. Lario, The Johns Hopkins University, Applied Physics Laboratory, 11100 Johns Hopkins Rd., Laurel, MD 20723, USA.

G.M. Simnett, School of Physics and Astronomy, The University of Birmingham, Edgbaston Birmingham B15 2TT, United Kingdom.

Section 3

Solar Variability and Climate

Section 3. Solar Variability and Climate

John P. McCormack

E. O. Hulburt Center for Space Research, Naval Research Laboratory, Washington, D.C.

Gerald R. North

Department of Atmospheric Sciences, Texas A&M University, College Station, Texas

This chapter addresses two basic scientific questions: what are the effects of solar variability on the Earth's climate, and how are these climatic effects manifested through changes in the radiative, photochemical, and dynamical processes in the atmosphere? When considering solar-terrestrial interactions, the term "solar variability" can include phenomena ranging from day-to-day fluctuations in the solar wind to changes in total solar irradiance on time scales of centuries and millennia. Accordingly, scientific investigations into the origins of these interactions should consider a wide range of physical processes extending from microscopic to celestial scales.

The first half of this chapter focuses on the detection and quantification of a solar signal in the Earth's climate through the use of long-term observations, statistical methods, and state-of-the-art climate models. *Damon and Peristykh* present evidence of climate forcings on centennial and millennial time scales from ^{14}C fossil records. The reconstructions of past climate variability from ice core samples reviewed by *Muscheler et al.* offer additional evidence that solar forcing plays an important role in long-term climate variability. Despite the availability of over 100 years' worth of measurements in some locations, the length of modern instrumental records are still too short to unambiguously identify a solar influence on climate. *North et al.* outline a statistical technique for detection of such an influence in available surface temperature measurements. Their results show that the amplitude of the global solar signal is consistent with theoretical expectations based on a simple

Solar Variability and its Effects on Climate
Geophysical Monograph 141
Published in 2004 by the American Geophysical Union
10.1029/141GM15

energy balance model. Signal detection studies carried out with more complex coupled atmosphere/ocean models, as reviewed by *Schlesinger and Andronova*, confirm that solar forcing did contribute to the observed climate variability during the first half of the twentieth century.

Despite the growing body of evidence of the Sun's influence on climate, the exact origins of the observed climate variability related to solar forcing remain unclear. This problem is complicated by the fact that the largest percentage changes in solar irradiance occur at shorter wavelengths (see Figure 5 of *Rottman et al.*, this volume). The energy at these wavelengths represents a small fraction of the Sun's total output and is largely deposited in the upper stratosphere (see Figure 4 of *Haigh*, this volume), and so these changes do not directly impact the troposphere and surface. Likewise, the effects of most high energy particles from the Sun are limited to the upper stratosphere and mesosphere. For these processes to be responsible for the 11-year climate signals, there must be some mechanism linking the solar variability observed in the middle atmosphere with the lower atmosphere and surface. An alternative explanation may be that the effects of solar variability bypass the middle atmosphere altogether and interact directly with the climate system, e.g., through cloud microphysical processes.

The second half of this chapter addresses both these possibilities, presenting a review of current research into the effects of solar variability on the atmosphere produced by ultraviolet (UV) and extreme ultraviolet (EUV) radiation, by energetic solar proton events (SPE's), and by solar modulated fluxes of galactic cosmic rays (GCR's). *Hood* examines the response of stratospheric ozone and temperature to solar UV variations over both the 27-day orbital period and the 11-year cycle and discusses the possible importance of internal stratospheric circulation modes in amplifying the stratospheric response to solar UV forcing. *Jackman and McPeters* describe the impact of several large SPE events on the photochemistry of the upper stratosphere and mesosphere at high latitudes using both observations and model calculations. *Fuller-Rowell et al.* discuss the implications of recently updated observational estimates of soft X-ray fluxes for modeling solar variability in thermospheric temperature. These articles highlight the physical processes linking the different regions of the middle atmosphere (i.e, stratosphere, mesosphere, thermosphere). *Tinslev and Yu* propose two cloud microphysical processes that may be affected by the 11-year cycle in GCR fluxes. Together these four articles outline current understanding of the atmospheric response to solar variability and point to promising areas of future research.

John McCormack, E. O. Hulburt Center for Space Research, Naval Research Laboratory, Code 7641.5, Washington DC 20375, USA (mccormack@uap2.nrl.navy.mil)

Gerald North, Department of Atmospheric Sciences, Texas A&M University, College Station, TX 77843–3150 (g-north@tamu.edu)

Long-Term Solar Variability and Climate Change Based on Radionuclide Data From Ice Cores

Raimund Muscheler[1], Jürg Beer[2], and Peter W. Kubik[3]

The cosmogenic radionuclides ^{14}C, ^{10}Be and ^{36}Cl allow us to trace solar variability several tens of millennia back in time. Here we describe some approaches to reconstructing past solar variability based on radionuclide concentrations measured in ice cores. The similarity in the variability of measurements of solar magnetic activity and solar irradiance during the last 20 years makes cosmogenic radionuclides a promising tool for reconstructing the variability of solar irradiance in the past. The analysis of many well-dated, high-resolution climate records clearly indicates that solar forcing plays an important role in climate change. However, the link between solar activity and climate is not yet well understood, partly because it is likely to involve non-linear processes within the Earth's climate system, but also possibly at the surface of the Sun.

INTRODUCTION

Very early, the observation of the variable number of sunspots led to the hypothesis of a solar influence on climate (e.g [*Herschel*, 1801]). Nevertheless, the expression "solar constant" illustrates nicely that the solar radiation was generally considered to be constant and hence of no importance to climate changes on Earth. Only in recent years has one taken solar variability into account in global circulation models to understand the climate change we are experiencing at the moment. This is due to the fact that satellite-based measurements reveal small changes in solar irradiation and because the model calculations show that the recent warming cannot be explained by the increase in greenhouse gases alone. In addition, paleoclimate records from natural archives such as ice and sediment cores clear-ly show that climate never was stable, with increasing evidence that the variable solar forcing played an important role in past climate change. Especially on longer time scales, an influence on climate may be expected since certain constituents of the climate system (e.g. ice sheets) react slowly to changing conditions. However, going back in time, observational data of the Sun and the climate becomes increasingly sparse and one has to rely on indirect proxies.

The Sun has experienced periods of much larger variability than the last decades of observational satellite data show. To understand changes in solar forcing and the corresponding climatic responses it is crucial to reconstruct the solar activity as well as the climate changes in the past. Studies of extreme solar conditions, such as for example the Maunder Minimum, provide information about the mechanisms and magnitude of the solar impact on climate change. Here, we will discuss the two fundamentally different causes for changes in solar forcing. One is the so-called Milankovitch forcing, which relates to changes in the Earth's orbit around the Sun. The corresponding changes in insolation (incoming solar radiation) can be calculated back in time for some millions of years. In contrast, the history of solar variability is not yet well understood but can be reconstructed over many millennia by means of cosmogenic nuclides (^{10}Be, ^{14}C etc.). In spite of the fact that many changes in climate correlate well with changes in solar activity, the relationship between

[1]Department of Quaternary Geology Lund University, Sweden
[2]Department of Surface Waters, EAWAG, Dübendorf, Switzerland
[3]Paul Scherrer Institute c/o ETH Hönggerberg, Zürich, Switzerland

Solar Variability and its Effects on Climate
Geophysical Monograph 141
10.1029/141GM16

solar variability and solar forcing is still unclear. However, the observed relationship between cosmogenic nuclides and climate change allows us to discuss the mechanisms which have been proposed as the connection between Sun and climate.

1. VARIABILITY OF SOLAR FORCING

Changes in solar activity as well as changes in the orbital parameters of the Earth on its way around the Sun are responsible for variations in solar forcing. While the changes in the orbital parameters cause changes in total and seasonal insolation on time scales of 10^4 to 10^5 years, changes in solar activity can cause more complex and shorter-term changes. The changes due to the variable solar activity are described in Chapter 2. In particular, changes in total irradiance, solar UV variation and cosmic ray intensity have been presumed to affect the climate. In the following, we will discuss the insolation changes due to the Milankovitch cycles in more detail, since it is believed that the Milankovitch forcing caused the strongest climate changes during the last few millions of years, including the glacial–interglacial cycles.

Since the solar system consists of the Sun and several planets, the Earth's orbit around the Sun and its obliquity are not constant through time. There are three cyclic variations operating on different time scales (details of which can be found in [*Berger et al.*, 1984]). First, there is the variation of the eccentricity of the Earth's orbit, i.e. the deviation of the Earth's orbit from a circle. The orbit changes from an elliptical shape to a shape closer to a circle with periods of 400,000 and 100,000 years. The second variation is in its obliquity. The tilt of the Earth's rotational axis with respect to the ecliptic (the plane defined by the Earth's orbit around the Sun) varies between 22° to 25°. Seasonality increases with the tilt. This change from warmer summers and colder winters to cooler summers and warmer winters takes place within a period of approximately 41,000 years. The third cyclic variation is the precession of the equinoxes, which occurs with periods of 19,000 and 23,000 years. The result of this is that the timing between the changing Earth-Sun distance and the seasons varies: at the moment, for example, in the Northern Hemisphere the distance to the Sun is shorter during the winter season than during summer. In 11,000 years from now, the Earth will be closer to the Sun in summer in the Northern Hemisphere, causing stronger seasonal variations in the Northern Hemisphere and smaller variations in the Southern Hemisphere.

The variation in eccentricity causes only a relatively small change in the annual mean global insolation. The changes in eccentricity, obliquity and precession, however, have a strong influence on the latitudinal and seasonal distribution of the radiation. Close to the poles, the annual average insolation can vary by up to 5%. Seasonal variations can even exceed this range. The prominent 100,000-year cycle found in climate data over approximately the last 900,000 years (e.g. [*Elkibbi and Rial*, 2001]) apparent in the glacial and interglacial cycles appears to be related to the eccentricity cycle. It is interesting to compare the changes in mean global insolation with reconstructed climate changes. Figure 1 shows the relative change in annual mean global solar insolation [*Laskar*, 1990; *Paillard et al.*, 1996] due to the changes in eccentricity together with the climate parameter δD measured in the Vostok ice core [*Petit et al.*, 1999]. Obviously the changes in mean global insolation are extremely small. With a mean global solar insolation of 342 W/m², the change of 1‰ corresponds to a change in forcing of approximately 0.35 W/m². Therefore, if only the change in global irradiance was responsible for the glacial–interglacial cycle, this change would indicate a strong sensitivity of more than 10°C/(W/m²), assuming a mean global temperature change of 5°C during the deglaciation [*Aeschbach-Hertig et al.*, 2000; *Thompson et al.*, 1995]. This sensitivity is two orders of magnitude larger than a simple energy balance model would suggest. However, the importance of the eccentricity lies not in the change in the amount of total radiation, but in its coupling with the precession. As matter of fact, the greater the eccentricity of the Earth's orbit, the greater the amplitude of the precessional variation. This means that the precessional amplitudes of the insolation at any given location on the Earth depend very much on the

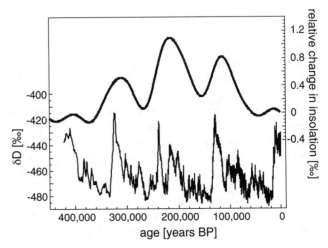

Figure 1. Comparison of the changes in annual mean global insolation [*Laskar*, 1990; *Paillard et al.*, 1996] with a climate record from Antarctica. The lower panel shows δD from the Vostok ice core [*Petit et al.*, 1999] which illustrates the changes associated with ice age – warm period cycles for the last 400,000 years.

eccentricity. Hence, the eccentricity period is reflected in the amplitudes of the precessional effects (amplitude modulation of the precession by the eccentricity). These variations strongly influence the seasonal insolation. The complex interplay between seasonal and latitudinal changes in insolation due to Milankovich forcing is further illustrated by the fact that during certain periods in Earth's history, climate reacted more sensitively to the obliquity cycle, while during others, like today for example, it reacted more sensitively to the eccentricity cycle [*Elkibbi and Rial*, 2001]. Simple climate models with an energy input that is averaged over years, and which also neglect feedback mechanisms, are therefore not appropriate for the study of the climatic effects of orbital forcing. The same is probably true for changes in solar forcing and its relationship to climate changes. For this reason it is necessary to include feedback mechanisms such as the build-up of ice sheets and the subsequent change in albedo, i.e. in the reflectivity of the Earth, in order to explain long-term climate changes in terms of Milankovitch forcing. Apart from ice sheet changes, feedback mechanisms involve, for example, sea level changes, which influence the albedo and the amount of aerosols in the atmosphere, or changes in the flux of fresh water to the oceans, which influence deep-water formation and hence the "conveyor belt", with significant consequences for the global distribution of heat. Temperature changes influence the atmospheric content of the most important "greenhouse gas", viz. water vapour. The Earth's albedo is strongly influenced by cloud cover, and this again depends on atmospheric water vapour. Other processes also exist, and a combination of different mechanisms is probably ultimately responsible for the strong climatic changes that occurred during the glacial and interglacial cycles.

2. RECONSTRUCTION OF PAST SOLAR VARIABILITY

The most detailed information about solar activity changes comes from modern satellite measurements. However, there have been periods when the solar activity was much lower than during the last 20 years for which satellite data are available. Further back in time, the data are much more sparse and incomplete. Observational sunspot numbers provide a reliable indirect proxy of solar activity back to approximately 1600 years AD [*Eddy*, 1976].

2.1 Indirect Proxies - Cosmogenic Radionuclides

Cosmogenic radionuclide records have the potential to extend the record of solar activity much further back in time. These nuclides are produced by the interaction of

galactic cosmic rays with the atoms of the atmosphere. Depending on the intensity of the solar wind and the enclosed magnetic fields, the galactic cosmic rays are more or less deflected from our solar system. High solar activity corresponds to strong solar wind, less cosmic rays entering the atmosphere and hence a reduced production rate of cosmogenic radionuclides. The neutron flux as shown in Figure 2 is a proxy for the incoming cosmic ray flux and varies similarly as the production rates of the cosmogenic radionuclides [*Masarik and Beer*, 1999]. At least for the period of direct solar irradiance measurements, the relatively good correlation between irradiance and neutron flux points to a relationship between the magnetic field emitted from the Sun and solar irradiance.

The best-known cosmogenic radionuclide is radiocarbon (^{14}C). It is produced mainly by the interaction of thermal neutrons with nitrogen in the atmosphere. After production, ^{14}C oxidises to $^{14}CO_2$ and becomes incorporated in the global carbon cycle. Therefore, the atmospheric ^{14}C concentration does not only depend on the ^{14}C production rate but also on processes in the carbon cycle. Due to the $^{14}CO_2$ exchange with the biosphere and the oceans, ^{14}C variations in the atmosphere are dampened and delayed compared to the changes in the ^{14}C production rate. In addition, strong climatic changes can (for example via changes in oceanic mixing) influence the atmospheric ^{14}C concentration [*Muscheler et al.*, 2000; *Siegenthaler et al.*, 1980] making it difficult to evaluate the origin of ^{14}C changes. In contrast to ^{14}C, ^{10}Be and ^{36}Cl can potentially provide a much more

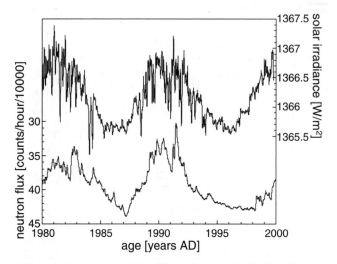

Figure 2. Comparison of irradiance changes [*Fröhlich and Lean*, 1998] with changes in the neutron flux. The neutron data (proxy of atmospheric cosmic rays as measured by the neutron monitor at Climax, Colorado) and the irradiance changes are averaged over 30 days.

direct signal of past variations in the production rate. ^{10}Be, for example, becomes attached to aerosols and is removed from the atmosphere within 1–2 years mainly by wet deposition. Measuring ^{10}Be in natural archives such as for example polar ice cores potentially allows us to reconstruct the history of production rate variations due to variations in solar activity. However, because of the relatively short tropospheric residence of 1–2 weeks, the atmosphere cannot be considered as globally well mixed with respect to ^{10}Be and ^{36}Cl. This might lead to local and/or climatic effects on the ^{10}Be and ^{36}Cl records. In the next section two examples illustrate the advantages and disadvantages of ^{10}Be and ^{14}C as proxies for the past solar activity.

Figure 3a shows ^{10}Be concentrations measured in the Dye3 ice core [*Beer et al.*, 1990; *Beer et al.*, 1994]. The

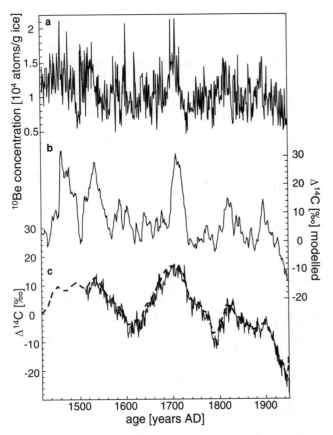

variability can be attributed to variations in solar activity as for example during the 11-year sunspot cycle and potentially to variations in transport and deposition of ^{10}Be. It is possible that, for example, a year with a high precipitation rate leads to a decreased ^{10}Be concentration in ice simply by an increased dilution of ^{10}Be in snow and ice. Such effects can be corrected for by using the ^{10}Be flux instead of the ^{10}Be concentration to reconstruct solar activity. However, since for the last 600 years the accumulation rate is more or less constant at Dye3, ^{10}Be flux and ^{10}Be concentration are essentially interchangeable. Potential changes in the ^{10}Be transport or deposition on the way to Dye3 cannot be assessed with this method.

Figure 3c shows tree ring Δ^{14}C measurements [*Stuiver and Braziunas*, 1993; *Stuiver et al.*, 1998] (Δ^{14}C indicates the relative deviation of the atmospheric ^{14}C/^{12}C ratio with respect to a standard [*Stuiver and Polach*, 1977]). Since ^{14}C has a half-life of 5730 years, only approximately 0.1‰ of ^{14}C is replaced every year by freshly produced ^{14}C. In addition, freshly produced ^{14}C and 'older' ^{14}C from the biosphere and ocean reservoirs become mixed due to the carbon cycle. This is the reason for the strong dampening of short-term ^{14}C production rate variations in the Δ^{14}C record. For example, a box-diffusion model [*Siegenthaler*, 1983] indicates that for the 11-year sunspot cycle the amplitude in the atmospheric ^{14}C concentration is two orders of magnitude less than the amplitude of ^{14}C production rate variations of typically 20%. Δ^{14}C variations during the 11-year cycle, therefore, are expected to be in the order of 2‰. However, in reconstructing long-term solar activity variations, ^{14}C has the advantage that the mean atmospheric residence time of 6–7 years is long enough to guarantee a globally well-mixed atmospheric ^{14}C.

Figure 3b shows Δ^{14}C derived from the ^{10}Be data: The calculation was based on the assumption that the ^{10}Be concentration measured in the Dye3 ice core is proportional to the global ^{10}Be production rate. The ^{10}Be production rate can then be converted to a global ^{14}C production rate using production rate calculations [*Masarik and Beer*, 1999]. In a next step a carbon cycle model [*Siegenthaler*, 1983] is necessary to calculate the ^{10}Be-based Δ^{14}C. The comparison of modelled and measured Δ^{14}C reveals common signals which are caused by variations in the production rates of ^{10}Be and ^{14}C. The increases in production rate around 1500, 1700, 1800 and 1900 yr AD are present in both data sets. However, the agreement is not perfect. Especially during the Spoerer Minimum around 1500 yr AD and the Maunder Minimum around 1700 yr AD significant differences are visible. Two arguments indicate that these differences are caused by climatic effects in the ^{10}Be system and not by changes in the carbon cycle: Short-term Δ^{14}C changes due

Figure 3. Comparison of ^{10}Be data from the Dye3 ice core [*Beer et al.*, 1990] and Δ^{14}C data [*Stuiver and Braziunas*, 1993; *Stuiver et al.*, 1998] from 1420 to 1950 yr AD. Panel a shows the ^{10}Be concentration which was measured in the Dye3 ice core. Assuming the ^{10}Be concentration to be proportional to the global ^{10}Be production rate it is possible to calculate a ^{10}Be-based Δ^{14}C (panel b). Panel c displays Δ^{14}C which was measured in tree rings with annual [*Stuiver and Braziunas*, 1993] and decadal (dashed line [*Stuiver et al.*, 1998]) resolution.

to changes in the carbon cycle are explicable only with very strong changes in ocean ventilation, since $\Delta^{14}C$ is most sensitive to changes in the ^{14}C transfer to and within the oceans [*Siegenthaler et al.*, 1980]. The difference of approximately 10‰ in $\Delta^{14}C$ that evolves during the first 50 years of the Maunder Minimum can be explained by a reduction in ocean ventilation by 30%. However, this scenario seems rather unlikely since such changes are observed during the strong climate deterioration during the Younger Dryas at the end of the last glacial period [*Muscheler et al.*, 2000]. Furthermore, compared to the Dye3 data, the ^{10}Be record from the South Pole indicates a broader production rate maximum during the Maunder Minimum [*Bard et al.*, 1997]. Therefore, we think that only changes in the ^{10}Be transport to Dye3 can explain the main differences between modelled and measured $\Delta^{14}C$ shown in Figure 3. In contrast to Dye3, ^{10}Be measured in the ice cores from Summit in Central Greenland seems to mirror very closely the global ^{10}Be production rate changes. This can be inferred from the close agreement between measured and modelled $\Delta^{14}C$ for the early Holocene period [*Muscheler et al.*, 2000].

The Younger Dryas is an excellent example to further illustrate the differences between the geochemical behaviour of ^{10}Be and ^{14}C. In spite of the fact that the accumulation rate varies up to a factor of two during the deglaciation [*Johnsen et al.*, 1995] and also the ^{10}Be concentration [*Finkel and Nishiizumi*, 1997], it can be shown that the ^{10}Be flux to Summit is independent of the climate proxy $\delta^{18}O$ measured in the Summit ice cores for the period around the Younger Dryas [*Muscheler et al.*, 2000]. This indicates a minor or even no climatic influence on the ^{10}Be flux. During the Younger Dryas, a strong reduction in North Atlantic Deep Water (NADW) formation decreased the ^{14}C transport to the deep sea and caused a corresponding $\Delta^{14}C$ increase. This can be seen in Figure 4, where the measured $\Delta^{14}C$ is plotted together with the modelled $\Delta^{14}C$, which is based on ^{10}Be measurements in the GISP2 ice core [*Finkel and Nishiizumi*, 1997]. The middle panel shows the results of a calculation based on the assumption that the ^{10}Be flux to Summit is proportional to the global ^{10}Be production rate and that no changes in the oceanic circulation occurred. The lower panel shows the ^{10}Be-based $\Delta^{14}C$ which was calculated assuming a decrease in global deep-water formation of 30% during the Younger Dryas [*Muscheler et al.*, 2000]. This comparison clearly shows that during the Younger Dryas $\Delta^{14}C$ exhibits a strong signal related to changes in the carbon cycle making it difficult to use ^{14}C as a direct proxy for solar activity.

The ^{10}Be flux to Summit is an example that certain regions receive a ^{10}Be deposition which appears to be proportional to the global production rate, making it possible to

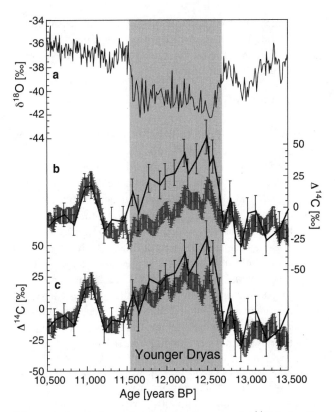

Figure 4. Comparison of modelled and measured $\Delta^{14}C$ during the Younger Dryas. Panel a shows the climate parameter $\delta^{18}O$ which was measured in the GRIP ice core [*Johnsen et al.*, 1997]. The Younger Dryas cold period is indicated by the shaded area. Panel b shows the measured $\Delta^{14}C$ as measured in sediments from the Cariaco basin [*Hughen et al.*, 1998] (black line). The grey line shows $\Delta^{14}C$ as calculated on the basis of the ^{10}Be flux to Summit under the assumption that no changes in the carbon cycle occurred [*Muscheler et al.*, 2000]. Both curves are linearly detrended. Panel c shows the result of a calculation under the assumption that the oceanic circulation was reduced by 30% during the Younger Dryas.

derive past changes in solar activity. Another advantage in comparison to ^{14}C is that the much longer half-lives of ^{10}Be and ^{36}Cl enable us to study changes in solar activity further back in time. ^{10}Be has a half-life of 1.51 million years [*Hofmann et al.*, 1987] (^{36}Cl: 301,000 years [*Goldstein*, 1966]) and is therefore a suitable tracer for the reconstruction of solar activity over very long periods of time. $\Delta^{14}C$ records, on the other hand, have the advantage that they provide global signals. However, short-term variations are strongly attenuated and changes in the carbon cycle can influence the atmospheric ^{14}C content. For a reliable reconstruction of past solar activity as many radionuclide records as possible should be compared to detect and remove possible system effects in the data. Especially if one wants to infer solar induced climate changes from cosmogenic radionuclide data it must be ensured that climatic changes did not influence the radionuclide records and thus feign a solar influence on the climate.

2.2 Separating the Solar Signal

Even if ^{10}Be, ^{36}Cl and ^{14}C records were not influenced by changes in climate, these records do not exclusively show changes in solar activity. The galactic cosmic rays are not only deflected by the solar magnetic field but also by the geomagnetic dipole field. In consequence, changes in the geomagnetic field leave their imprint in cosmogenic radionuclide records [*Lal and Peters*, 1967]. Furthermore, potential changes in the galactic cosmic ray intensity lead to changes in the production rates of the cosmogenic radionuclides. To separate the solar signal different strategies can be applied: Assuming that the radionuclide signal is composed only of a geomagnetic and a solar component it is possible to subtract the geomagnetic signal if the past geomagnetic field is known. The very good agreement between geomagnetic field reconstructions and the ^{10}Be and ^{36}Cl flux to Summit Greenland for the period of the last ice age [*Wagner et al.*, 2001a; *Wagner et al.*, 2000b] justifies this approach for the Summit radionuclide data. The limited time resolution and uncertainties of sedimentary geomagnetic field reconstructions as well as potential changes in the radionuclide transport and deposition limits this approach.

An alternative way to isolate the solar signal from the radionuclide data is to assume that solar activity variations occur on shorter time scales than geomagnetic field variations. The ^{14}C data for the Holocene period points to changes in solar activity on time scales from decades to millennia. The comparison with geomagnetic field reconstructions indicates an influence of geomagnetic field changes on the radionuclide records on time scales longer than 2000-3000 years [*Laj et al.*, 1996; *Wagner et al.*, 2000a; *Wagner et al.*, 2000b]. Assuming that shorter-term changes in the cosmogenic radionuclide data are caused by changes in solar activity, it is possible to derive a history of (short-term) solar activity changes by high-pass filtering the radionuclide data. However, it is not sure that all solar cycles are shorter than typical geomagnetic field changes and therefore it is not clear where to set the cut-off frequencies.

The most accurate way to infer past changes in solar activity is to identify known solar cycles in the radionuclide records and to isolate such variations. This approach is of course restricted to known variations, while non-cyclic changes are not assessable with this method. In the following section, this approach will be discussed in more detail.

One of the most prominent cycles in the Δ^{14}C data for the Holocene is the 207-year (Suess or de Vries) cycle [*Damon and Sonett*, 1991]. For the last 1000 years the Δ^{14}C data show strong changes on this time scale (see Figure 5). The agreement with an artificial 207-year cycle is not perfect but it illustrates, for example, that the Maunder Minimum, which is characterised by an almost complete absence of sunspots, can be regarded as a manifestation of the 207-year cycle. For the Maunder Minimum, we have observations of solar variability [*Eddy*, 1976] which allow us to link the 207-year cycle to changes in solar activity.

Going further back in time to the last ice age the Δ^{14}C data exhibit strong scatter and are not suitable, at least for the time being, to trace back the solar activity. The ^{10}Be data from the GRIP ice core allow us to reconstruct this cycle further back in time. The spectral analysis of the ^{10}Be flux to Summit for the period from 25,000 to 50,000 years BP reveals that the 207-year cycle is also present during this period [*Wagner et al.*, 2001a]. Figure 6a shows this cycle isolated from the ^{10}Be flux by band-pass filtering the data. Only variability with periodicities from 195 to 215 years remains in the filtered data. Panel b in Figure 6 shows that there is an interplay between geomagnetic field intensity and solar activity changes on the ^{10}Be production rate. The production calculations reveal that the smaller the geomagnetic field, the larger the amplitude of cycles in the radionuclide data caused by the Sun [*Wagner et al.*, 2001a]. This connection can be seen nicely in Figure 6 when, during the

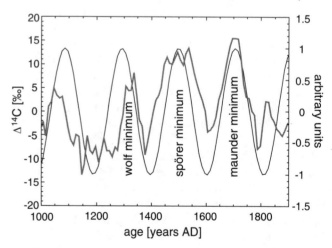

Figure 5. Comparison of Δ^{14}C tree ring data with an artificial 207-year cycle for the last 1000 years. The changes in solar activity recorded by the cosmogenic radionuclides show periodic changes which can be detected by spectral analysis of radionuclide time series [*Beer et al.*, 1990; *Beer et al.*, 1994; *Damon and Sonett*, 1991; *Wagner et al.*, 2001a].

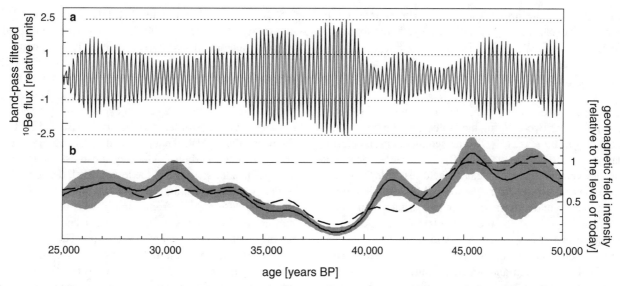

Figure 6. Comparison of the amplitude of the solar 207-year cycle with changes in geomagnetic field intensity. The upper panel shows the band-pass filtered ^{10}Be flux $\{(215\ \mathrm{yr})^{-1} \leq \tau \leq (195\ \mathrm{yr})^{-1}\}$ for the time period from 25,000 to 50,000 years BP. The lower panel shows paleomagnetic field intensities during the last ice age: The first (dotted line) is derived from remanence measurements of several Mediterranean sediment cores [*Tric et al.*, 1992]. The second (solid line) is based on remanence measurements of several North Atlantic sediment cores [*Laj et al.*, 2000]. The shaded area corresponds to the ± 2 σ uncertainties.

Laschamp event around 40,000 years BP, the geomagnetic field almost vanished and the amplitude of the 207-year cycle was roughly twice as large as during other periods. This is in quantitative accordance to production rate calculation [*Masarik and Beer*, 1999] and corroborates the hypothesis of a solar origin of the 207-year cycle in the radionuclide data. The mean amplitude of the 207-year cycle in the ^{10}Be data during the last ice age is comparable to the Δ^{14}C changes during the Holocene, taking into account that the Δ^{14}C signal is attenuated by a factor of roughly 20 for ^{14}C production rate variations associated with the 207-year cycle.

However, there are difficulties in interpreting radionuclide and also climate data with respect to solar cycles: Very accurate time scales are necessary for the detection of cycles. Even small relative errors in the time scales can cause the cycles not to be visible in the data. Variability in the cycle length as found for the 11-year cycle [*Gleissberg*, 1944] further complicates the situation. In addition, the time resolution of the measurements has to be better than half of the cycle length to allow the detection of the cycle. This is why the 207-year cycle cannot be detected in GRIP data older than 50,000 years. The thinning of the annual layers due to the ice flow reduces the time resolution towards the bottom of the ice core and hence the range of detectable periodicities.

3. EVIDENCE FOR SOLAR INFLUENCE ON CLIMATE CHANGES

There is increasing evidence for an influence of solar variability on climate change. However, the importance of this influence is still a matter of debate. The estimates for the solar contribution to the current global warming range from minor [*Forest et al.*, 2001] to "very significant" [*Svensmark and Friis-Christensen*, 1997]. This indicates that there is still much uncertainty about the solar influence on climate. Furthermore, not only the impact of solar forcing but also the mechanisms of the Sun acting on the Earth's climate are subject of current debate. In this section, we will show some examples for a correlation between climate change and solar variability during the Holocene. This discussion is confined to examples of solar activity changes inferred from cosmogenic radionuclide records.

To study the solar influence on climate in the past with cosmogenic radionuclides there is a choice of approaches. One method is to concentrate on special solar and/or climatic events and to study such events with respect to their worldwide impact. Another approach consists of analysing time-series of climate proxies and comparing them with time-series of the cosmogenic radionuclides. This can be done both in the time and the frequency domain (spectral analysis). Even without accurate information on the past solar activity a solar influence on climate can be inferred if known solar cycles (as e.g. the 207-year cycle) are found in climate records.

3.1 Ice Rafted Debris Cycles in the North Atlantic

The first example refers to the North Atlantic "quasi-periodic" 1500-year cycle during the Holocene which was discovered by *Bond et al.* [1997]. The sediments in the North Atlantic region contain tracers (IRD: "ice rafted debris") which are related to amounts and trajectories of drift ice circulating in the surface waters [*Bond et al.*, 1997]. Therefore, these tracers yield information on the surface wind and ocean hydrography in the subpolar North Atlantic. Percentage increases of such tracers in sediment cores can be interpreted as eastward advection of cooler surface water from the Labrador sea and southward advection from the Nordic seas, possibly accompanied by increased northerly winds [*Bond et al.*, 2001]. Figure 7 shows a comparison of radionuclide records with a stacked sediment record from four individual records indicating the amount of drift ice reaching the location of sediment cores MC52VM29-191

(55°28'N, 14°43'W) and MC21GGC-22 (44°,18'N, 46°,16'W) [*Bond et al.*, 2001]. The records are filtered to remove fluctuations with periodicities longer than 1800 years and shorter than 70 years. In the upper panel, the IRD is compared with the ^{14}C production rate which was derived from Δ^{14}C [*Stuiver and Braziunas*, 1998; *Stuiver and Braziunas*, 1993] using a carbon cycle model [*Oeschger et al.*, 1975]. Since (similarly to the Dye3-^{10}Be and Δ^{14}C differences) these Δ^{14}C changes hardly can be assigned to changes in the carbon cycle they are most likely due to changes in solar activity. In the lower panel, the comparison of the ^{10}Be flux to Summit with IRD is shown. The similarity of the ^{10}Be records from Summit with the Δ^{14}C record during the Holocene is the strongest evidence for an external origin of the Δ^{14}C and ^{10}Be variations. The agreement of the radionuclide with the IRD records points to a dominant influence of solar activity changes on North Atlantic climate on centennial to millennial time scales during the Holocene.

3.2 δ^{18}O From Stalagmites From the Oman

Stalagmites from the Oman are another high-resolution climate archive which can be studied with respect to solar forcing of the climate [*Neff et al.*, 2001]. The fractionation of the oxygen isotopes (δ^{18}O) measured in stalagmites from the Northern Oman can be interpreted as a proxy for variations in the tropical circulation and monsoon rainfall [*Neff et al.*, 2001]. The lower panel of Figure 8 shows the running average over 20 years of the δ^{18}O record from the Oman. The upper panel shows Δ^{14}C after applying the same low-pass filter plus removing the long-term trend by subtracting a 1000 years average. The grey lines indicate the corresponding forcing (Δ^{14}C) and climate (δ^{18}O) events as suggested by *Neff et al.* [2001]. It is clearly visible that considerable adjustments to the time scale of the stalagmite record are necessary but these changes are still within the uncertainties of the time scale constrained by U–Th dating [*Neff et al.*, 2001]. The relatively large and not systematic corrections can be explained by the typically non-linear growth of stalagmites [*Hill and Forti*, 1997]. After the correction of the stalagmite time scale there is a good agreement between the δ^{18}O and Δ^{14}C records [*Neff et al.*, 2001]. Nevertheless, this example illustrates the problem that uncertainties in the time scales of climate records are often larger than the duration of events to be studied. In spite of the fact that the time scale of the Δ^{14}C record can be regarded as an absolute time scale with errors smaller than 1 year for the last approximately 11,000 years [*Spurk et al.*, 1998], the uncertainties in sedimentary or stalagmite records are often in the order of 100 years or more. This compares to the duration of typical

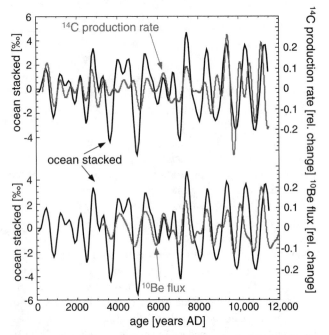

Figure 7. Comparison of changes in drift ice in the North Atlantic with cosmogenic radionuclide records. The upper panel shows the comparison with a reconstructed ^{14}C production rate and the lower panel shows the comparison with a ^{10}Be record from the Summit ice cores [*Finkel and Nishiizumi*, 1997; *Yiou et al.*, 1997]. The records are band-pass filtered {$(1800 \text{ yr})^{-1} \leq \tau \leq (500 \text{yr})^{-1}$}. Differences to Bond et al. are due to slightly different filter methods.

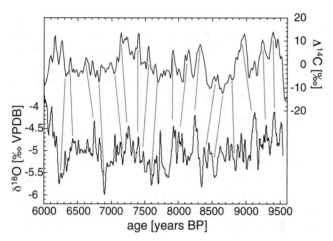

Figure 8. Comparison of Δ14C changes [*Stuiver et al.*, 1998] with changes in δ18O in stalagmites from the Oman which can be interpreted in terms of monsoon variability [*Neff et al.*, 2001]. Both data sets are filtered to remove short-term variation shorter than 20 years and long-term variation longer than 1000 years. The grey lines indicate common variability as suggested by *Neff et al.* [2001].

solar minima such as the Maunder Minimum. Of course, after the procedure of matching forcing and climate events it is impossible to study leads and lags of forcing and climate.

In the following, we will present two climate events which have been reconstructed from very well dated archives and which therefore provide valuable information on changes in solar activity and their influence on climate, especially with respect to the timing of the observed changes.

3.3 Solar Forcing Around 2650 Calendar Years BP

Radiocarbon is an excellent tool to date organic samples. There are, however, uncertainties induced by the method itself. 14C dating is based on the radioactive decay of 14C. To calculate the age from a measured 14C concentration it is essential to know the initial 14C concentration. For plants on land, for example, this initial concentration is set by the atmospheric 14C concentration. As discussed earlier, changes in solar activity, geomagnetic field or in the carbon cycle cause changes in the atmospheric 14C concentration. This makes it difficult to infer a unique date for organic samples, especially from periods with decreasing Δ14C. Measuring a sequence of 14C ages with a known relative chronology can overcome this problem and very precise absolute ages can be inferred by matching the sequence with the calibration curve ("wiggle matching") (*Van Geel et*

al. [1998] and references therein). This method was successfully applied to northwest European raised bogs formed around 2700 years BP, a period when a strong increase in Δ14C is observed. During this time, an abrupt climate change from relatively warm and continental to cooler and wetter conditions in northern Netherlands can be inferred from paleoecological and archaeological evidence [*Van Geel et al.*, 1996]. Van Geel et al. show evidence that this event has a global imprint with similar changes in the temperate and boreal zones as in the northern Netherlands and a change to dryer conditions in the Caribbean and in the tropical Africa [*Van Geel et al.*, 1996]. The relationship to the Δ14C increase and a corresponding 10Be increase in the Summit records points to a solar activity change related to the climate change. The advantage of the wiggle-matching dating is that a very good chronology can be inferred. The relative timing between solar activity and climate changes is surprising: At 2650 years BP the climate change happens during the increase of Δ14C and not during the maximum of Δ14C [*Van Geel et al.*, 1996]. Similar results have been found for the Little Ice Age [*Mauquoy et al.*, 2002].

3.4 Early Holocene Climatic Changes

During the beginning of the Holocene, three prominent peaks are visible both in the 10Be data from the Summit ice cores and in the Δ14C tree ring data [*Finkel and Nishiizumi*, 1997; *Stuiver et al.*, 1998] (Figure 9). The agreement between 10Be and Δ14C points to a common origin of the observed changes [*Muscheler et al.*, 2000], which again are most probably caused by changes in solar activity. A geomagnetic origin cannot be completely ruled out. However, the relatively short duration of the events and the amplitudes which compare to typical solar variations such as for example during the Maunder Minimum are strong indications of a solar origin. At least two of the Δ14C changes seem to be related to changes in climate. *Björck et al.* [1996] discuss the so-called Preboreal oscillation (the climate deterioration around 11,250 yr BP) in detail. This discussion illustrates the important contribution of 10Be and 36Cl data to pinpoint the origins of Δ14C changes. At the time of publication of the paper (1996), the 10Be data had not yet been available. The most obvious scenario at this time was to associate both Δ14C and the climate changes to changes in ocean ventilation [*Björck et al.*, 1996]. The good agreement between 10Be and Δ14C as shown in Figure 9 indicates a different relationship. Changes in solar activity induced the Δ14C and 10Be changes and triggered the climate change. Changes in ocean ventilation cannot be completely ruled out, however, significant changes should be reflected in the

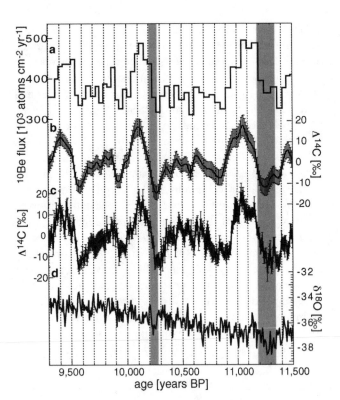

Figure 9. $\Delta^{14}C$ and ^{10}Be variation during the early Holocene. Panel a shows the ^{10}Be flux inferred from ^{10}Be measurements from the GISP2 ice core [*Finkel and Nishiizumi*, 1997]. The data are plotted on the GRIP timescale. Panel b shows the calculated $\Delta^{14}C$ on the basis of the ^{10}Be data [*Muscheler et al.*, 2000]. Panel c shows $\Delta^{14}C$ from tree ring measurements [*Stuiver et al.*, 1998]. Both $\Delta^{14}C$ curves are detrended by removing a 500 years running average. Panel d shows the climate proxy $\delta^{18}O$ which was measured in the GRIP ice core [*Johnsen et al.*, 1997].

$\Delta^{14}C$ tree ring record leading to a difference with the ^{10}Be-based $\Delta^{14}C$ record which does not exist. At 10,300 yr BP, another cold phase can be detected [*Björck et al.*, 2001]. The timing of this event is known very precisely since tephra layers can be used to synchronise lacustrine and ice core records [*Björck et al.*, 2001]. In addition to the northern latitude cooling, a disturbance in the thermohaline circulation can be reconstructed which again was not strong enough to cause a significant difference between ^{10}Be-based and measured $\Delta^{14}C$. Different climate records point to a more global relevance of this event. However, the time scales of these records have to be improved to strengthen this hypothesis [*Björck et al.*, 2001]. The time shift between $\Delta^{14}C$ and climate change excludes that a decreased North Atlantic deep-water formation is responsible for both the $\Delta^{14}C$ increase and the climate deterioration. Again, the agreement between $\Delta^{14}C$ and ^{10}Be shows that common production rate changes are related to this event. The two cold phases

around 11,250 and 10,300 yr BP correspond to the first two "Bond-cycles" during the Holocene shown in Figure 7 [*Björck et al.*, 2001]. $\delta^{18}O$ records from the Summit ice cores are not sensitive enough to reflect small changes in climate during the Holocene well. In addition, due to some discrepancies between the GRIP and GISP2 $\delta^{18}O$ records for the Holocene period, additional climate proxy records are needed to confirm the climate changes indicated by $\delta^{18}O$. Nevertheless, the $\delta^{18}O$ records do show minima around 11,300 yr BP and possibly also around 10,300 yr BP [*Grootes and Stuiver*, 1997; *Johnsen et al.*, 1997] confirming the other climate records. A lack of a distinct $\delta^{18}O$ minimum around 9,500 yr BP (which would correspond to the third "Bond-cycle") is not necessarily an indication that there was no change in climate.

If climate and forcing proxies can be measured in the same core, one has the possibility to study precisely the relative timing of solar activity and climate changes. The Summit ice cores offer us this comfortable situation. From the two cold phases during the early Holocene (Figure 9) and from the climate change observed around 2,700 yr BP it seems that solar activity and climate changes are not synchronous. The maximum cosmic ray intensities lag behind the respective cold phases, pointing to a trigger mechanism for climate changes. This information has to be considered when discussing the potential mechanisms linking Sun and climate.

3.5 A Solar Forced Younger Dryas ?

It has been proposed that the most recent pronounced cold phase of the last ice age (the Younger Dryas) was caused by solar forcing [*Van Geel et al.*, 1999]. This was inferred from the agreement of ^{10}Be concentration in the GISP2 ice core and $\delta^{18}O$ records from Summit. However, the ^{10}Be concentration in the Summit ice cores during the last ice age is strongly influenced by the changing accumulation rate [*Wagner et al.*, 2001b]. During cold periods with less precipitation we observe an increased ^{10}Be concentration. Therefore, it is crucial to remove the climatic component from the ^{10}Be record in order to assess changes in the production rate due to solar and geomagnetic field changes. It has been shown that for the period from 20,000–60,000 yr BP the combined flux of ^{10}Be and ^{36}Cl to Summit reflects the galactic cosmic ray flux modulated by the varying geomagnetic field on time scales longer than 2000–3000 years [*Baumgartner et al.*, 1998; *Wagner et al.*, 2000b]. Furthermore, the ^{10}Be flux to Summit does not correlate with $\delta^{18}O$ over the period from 15,000 to 9,300 years BP, when the strong climate changes are present in the $\delta^{18}O$ data [*Muscheler et al.*, 2000]. Therefore only the radionuclide flux allows us to study the solar influence on climate. The

[10]Be flux shows an increase at the beginning of the Younger Dryas [*Muscheler et al.*, 2000] pointing to a relationship between production rate changes and climate. Furthermore, the Δ^{14}C increase [*Hughen et al.*, 1998; *Hughen et al.*, 2000] during this period cannot be explained by changes in the carbon cycle alone [*Marchal et al.*, 2001; *Muscheler et al.*, 2000]. The origin of the production rate changes at the beginning of the Younger Dryas cannot be unambiguously attributed to changes in solar activity since geomagnetic field reconstruction on shorter time scales reveal considerable differences [*Tauxe*, 1993]. Nevertheless, whether or not the [10]Be and Δ^{14}C increase around 12,700 can be attributed to a decrease in solar activity, this change is synchronous with the strong climatic changes observed during this period.

Many different climate records have been studied with respect to a solar influence. Lake level changes ([*Magny*, 1993; *Verschuren et al.*, 2000]), glacier advances and retreats ([*Denton and Karlén*, 1973; *Hormes et al.*, 1998]) or changes in atmospheric salt and dust transport [*O'Brien et al.*, 1995] can be reconstructed and compared with radionuclide records. Such records yield information on past temperatures, precipitation and atmospheric circulation. *Damon and Peristykh* [this volume] discuss other examples which indicate that solar activity is an important factor influencing climate.

4. PROPOSED MECHANISMS

In most climate studies, uncertainties in time scales or the in matching of forcing and climate changes do not allow us to draw final conclusions about the leads and lags of solar variability and climate. However, some studies do indicate a direct influence of the Sun on climate with no time delay (see e.g. [*Stuiver et al.*, 1997]). Nevertheless, for the climate events during early Holocene and around 2700 yr BP the observed relative timing of radionuclide records and climate changes points to a non-linear relationship between solar activity and climate changes. It indicates that, at least for these events, the underlying mechanisms linking Sun and climate are not straightforward to evaluate. The delay between climate and solar magnetic field could be due either to processes in the Sun [*Beer et al.*, 2000] or to the non-linear response of the climate system to solar forcing. In the following we will discuss some mechanisms proposed for a solar-climate connection in the context of the above examples.

4.1 Irradiance Changes

One possible cause for the influence of the changing Sun on the climate could be changes in the solar irradiance. However, the irradiance changes related to solar activity are relatively small. For example, the increase in total solar irradiance from the Maunder Minimum to today is probably less than 0.5% [*Lean et al.*, 1995]. If the relationship between irradiance changes and changes in cosmic ray flux is constant in time, this corresponds to the range of irradiance changes expected during the Holocene from the radionuclide records. It has been discussed before that the global annual insolation changes due to the Milankovitch forcing alone should not suffice to significantly influence the climate. Feedback mechanisms have to be invoked for both Milankovitch and solar forcing to explain the observed climate changes. The timing between solar activity and climate as discussed in the previous section points to a "solar trigger" which induces both cosmic ray flux and climate changes. Another possibility is that the solar magnetic field changes (which are measured by the cosmogenic radionuclides) are not synchronous with irradiance changes on centennial time scales. However, such a delay is not suggested by the observations for the last 20 years on decadal time scales (see Figure 2). In both cases, a straightforward relationship between irradiance and climate cannot be inferred.

4.2 Cosmic Ray Intensity and Clouds

Clouds play an important role in the radiation budget of the Earth. Especially low clouds reflect a large amount of the incoming solar radiation back to space and thus have a cooling effect on the Earth's climate. It has been suggested that cosmic rays contribute significantly to the production of cloud condensation nuclei [*Dickinson*, 1975; *Svensmark and Friis-Christensen*, 1997]. As a consequence, changes in the galactic cosmic ray flux should play an important role for the Earth's radiation budget and therefore also for climate. This hypothesis was based on satellite observations of the cloud cover during one sunspot cycle and the close relationship to the neutron flux which reflects the incoming cosmic ray flux [*Svensmark and Friis-Christensen*, 1997]. It was stated that this might be the "missing link" of the solar influence on the Earth's climate, which might potentially explain a significant part of the warming we experience at the moment [*Svensmark and Friis-Christensen*, 1997]. However, there are several observations that challenge this hypothesis [*Wagner et al.*, 2001c].

The cosmogenic radionuclide records provide useful information to test this hypothesis: Assigning changes in cloud cover to solar induced changes in the cosmic ray flux might be problematic since irradiance changes could also be responsible for the correlation. Therefore, looking at cosmic ray intensity variations due to variations in the geomagnetic field is a perfect test for the postulated relationship between cosmic rays and clouds. The GRIP ice core provides an ideal archive to test this interpretation: A large change in the galactic cosmic ray flux penetrating into the atmosphere occurred approximately 40,000 years ago during the so-called Laschamp event, when the geomagnetic field intensity was only roughly 10% of its present value [*Meynadier et al.*, 1992; *Tric et al.*, 1992; *Wagner et al.*, 2000b] (see Figure 6). If the link between cosmic ray flux and clouds exists, we would expect significant global cooling to have occurred at this time. Combining and low-pass filtering the ^{10}Be and ^{36}Cl fluxes reduces transport and solar activity effects, which are assumed to occur on time scales shorter than 3000 years [*Damon and Sonett*, 1991]. This record reflects the cosmic ray flux modulated by the geomagnetic field and can be compared with the δ^{18}O and CH$_4$ data from ice cores which can be regarded as proxies for the climate of the North Atlantic region at least [*Blunier et al.*, 1998; *Bond et al.*, 1993; *Dansgaard et al.*, 1993]. It is suggested that they are controlled by the hydrological cycle at low latitudes [*Chappellaz et al.*, 1993] and are therefore proxies for a supra-regional climate. Both records are available as a function of time [*Blunier et al.*, 1998; *Dansgaard et al.*, 1993]. In Figure 10 the combined, low-pass filtered flux of ^{10}Be and ^{36}Cl is compared with the δ^{18}O and CH$_4$ data. A high value of the combined flux of ^{10}Be and ^{36}Cl corresponds to a high cosmic ray flux, which, according to the proposed link, should result in an increase in global cloud cover and a lower mean global temperature. This, however, is clearly not the case. During the Laschamp event (36,000–41,500 year BP), marked by a peak in the radionuclide data, the combined flux of ^{10}Be and ^{36}Cl is not significantly ($p < 0.1$) correlated neither with δ^{18}O ($r^2 = 0.07\%$) nor with CH$_4$ ($r^2 = 0.09\%$). The same applies over the entire time interval shown in Figure 10 ($r^2 = 0.3\%$ and 0.4%, respectively). This suggests that the relationship between geomagnetic field, cosmic rays and climate proposed by [*Svensmark and Friis-Christensen*, 1997] is unlikely to be valid for the period investigated here. This conclusion is supported by the fact that the variation in the galactic cosmic ray flux that occurred during the Laschamp event was substantially greater than the variations associated with the 11-year solar modulation that occurred during the 1980s and 1990s [*Wagner et al.*, 2000b]. The result that the enhanced galactic cosmic ray flux prevailing during the Laschamp event had no significant impact on climate is inconsistent with the direct link between cosmic ray flux and the production of cloud condensation nuclei. However, it is not inconsistent with a scenario in which the link between solar activity and cloud formation acts via cosmic ray induced changes in the global electric circuit [*Tinsley*, 1996, 2000; *Tinsley and Yu*, this volume]. Nevertheless, the relative tim-

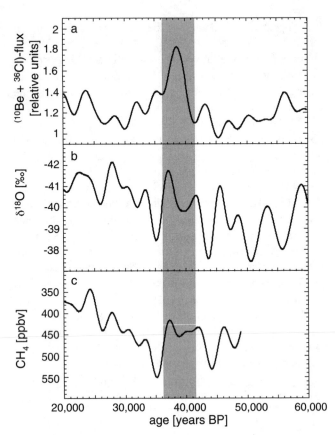

Figure 10. Comparison of combined flux of ^{10}Be and ^{36}Cl [*Wagner et al.*, 2000b] and climate proxies δ^{18}O [*Johnsen et al.*, 1997] and methane [*Blunier et al.*, 1998]. The data was low-pass filtered (cut-off frequency = 1/3000 yr^{-1}). The proportion of variance shared between the combined flux of ^{10}Be and ^{36}Cl and the δ^{18}O and the CH$_4$ data respectively during the Laschamp (which corresponds to the maximum in the radionuclide flux) event is essentially zero (shaded area). The proportion of shared variance over the period 20,000–60,000 yr B.P. is also essentially zero. The δ^{18}O and CH$_4$ data are well correlated ($r^2 = 0.61$).

ing between observed climate changes and cosmic ray variation as shown in Figure 9 questions both mechanisms: If the cosmic ray intensity is directly related to changes in cloud cover why don't we observe a synchronous climate change? Therefore, invoking this mechanism to explain climate changes also needs internal mechanisms to explain all of the observed climate changes.

4.3 Changes in UV Radiation

During a solar 11-year cycle the relative change in the UV output from the Sun is much larger than the relative change of the total solar irradiance [*Haigh*, 1994]. Therefore, solar induced climate changes are expected to be more pronounced if they are mediated via changes in UV radiation than via changes in total solar irradiance. Changes in the UV output from the Sun cause changes in the amount of heat deposited in the ozone layer of the atmosphere. This might cause changes in the global atmospheric circulation with significant influence on climate [*Haigh*, 1994, 1996, and this volume]. This mechanism implies a cooling in the high northern latitude atmosphere during periods of low solar activity. *Bond et al.* [2001] suggest that this might lead to increased drift ice and cooling of the surface ocean and atmosphere. A simultaneous change in the thermohaline circulation might act as an important amplifying mechanism [*Bond et al.*, 2001]. However, the relative timing again poses the question why we do not observe synchronous changes in solar activity and climate.

5. CONCLUSIONS

We have shown that there is growing evidence that changes in solar activity can influence the climate. Especially for the Holocene period there is an increasing amount of high-resolution data that yields information about climate, for instance in the North Atlantic or in the monsoon belt. Combining such information may provide a global perspective for climatic changes caused by the Sun. In our opinion, high-resolution data with very precise timescales provide the key in order to understand the mechanisms linking the Earth's climate to solar forcing. As well illustrated in the case of the Milankovitch forcing feedback processes are fundamental and must be taken into account. The non-synchronicity of some climate deteriorations with the corresponding solar activity minima is a sign of the complexity of the climate system and its forcings.

Acknowledgments. We would like to thank Claire Wedema, David Livingstone and Paul Damon for helpful comments on this manuscript. IRD data were provided by Gerard Bond. The neutron data were provided by the University of Chicago (National Science Foundation Grant ATM-9613963). Solar irradiance data were obtained from PMOD/WRC, Davos, Switzerland, including unpublished data from the VIRGO Team (Version 15). This work was supported by the Swiss National Science Foundation.

REFERENCES

Aeschbach-Hertig, W., F. Peeters, U. Beyerle, and R. Kipfer, Paleotemperature reconstruction from noble gases in groundwater taking into account equilibration with entrapped air, *Nature, 404*, 1040-1044, 2000.

Bard, E., G. M. Raisbeck, F. Yiou, and J. Jouzel, Solar modulation of cosmogenic nuclide production over the last millenium: com-parison between ^{14}C and ^{10}Be records, *Earth. Planet. Sci. Let., 150*, 453-462, 1997.

Baumgartner, S., J. Beer, J. Masarik, G. Wagner, L. Meynadier, and H.-A. Synal, Geomagnetic Modulation of the ^{36}Cl Flux in the GRIP Ice Core, Greenland, *Science, 279* (5355), 1330-1332, 1998.

Beer, J., A. Blinov, G. Bonani, R. C. Finkel, H. J. Hofmann, B. Lehmann, H. Oeschger, A. Sigg, J. Schwander, T. Staffelbach, B. Stauffer, M. Suter, and W. Wölfli, Use of ^{10}Be in polar ice to trace the 11-year cycle of solar activity, *Nature, 347* (6289), 164-166, 1990.

Beer, J., S. Baumgartner, B. Dittrich-Hannen, J. Hauenstein, P.W. Kubik, C. Lukasczyk, W. Mende, R. Stellmacher and M. Suter, Solar Variability Traced by Cosmogenic Isotopes, in *The Sun as a Variable Star*, edited by J.M. Pap, C. Fröhlich, H.S. Hudson and S.K. Solanki, 291-300, Cambridge University Press, 1994.

Beer, J., W. Mende and R. Stellmacher, The role of the Sun in climate forcing, *Quat. Sci. Rev., 19,* 403-415, 2000.

Berger, A., J. Imbrie, J. Hays, G. Kukla, and B. Saltzman, Milankovitch and Climate, in *NATO ASI Series*, vol. 126, pp. 895, D. Reidel Publishing Company, Dordrecht, 1984.

Björck, S., B. Kromer, S. Johnsen, O. Bennike, D. Hammarlund, G. Lemdahl, G. Possnert, T. L. Rasmussen, B. Wohlfarth, C. U. Hammer, and M. Spark, Synchronized terrestrial-atmospheric deglacial records around the North Atlantic, *Science, 274*, 1155-1160, 1996.

Björck, S., R. Muscheler, B. Kromer, C. S. Andresen, J. Heinemeier, S. J. Johnsen, D. Conley, N. Koc, M. Spurk, and S. Veski, High-resolution analyses of an early Holocene climate event may imply decreased solar forcing as an important climate trigger, *Geology, 29*(no. 12), 1107-1110, 2001.

Blunier, T., J. Chapellaz, J. Schwander, A. Dällenbach, B. Stauffer, T. F. Stocker, D. Raynaud, J. Jouzel, H. B. Clausen, C. U. Hammer, and S. J. Johnsen, Asynchrony of Antarctic and Greenland climate change during the glacial period, *Nature, 394*, 739-743, 1998.

Bond, G., W. S. Broecker, S. Johnsen, J. McManus, L. Labeyrie, J. Jouzel, and G. Bonani, Correlations between climate records from North Atlantic sediments and Greenland ice, *Nature, 365*, 143-147, 1993.

Bond, G., W. Showers, M. Chesby, R. Lotti, P. Almasi, P. deMenocal, P. Priore, H. Cullen, I. Hajdas, and G. Bonani, A pervasisve millenial-scale cycle in north atlantic and glacial climates, *Science, 278*, 1257-1266, 1997.

Bond, G., B. Kromer, J. Beer, R. Muscheler, M. N. Evans, W. Showers, S. Hoffmann, R. Lotti-Bond, I. Hajdas, and G. Bonani, Persistant Solar Influence on North Atlantic Climate During the Holocene, *Science, 294*, 2130-2136, 2001.

Chappellaz, J., T. Blunier, D. Raynaud, J. M. Barnola, J. Schwander, and B. Stauffer, Synchronous changes in atmospheric CH_4 and Greenland climate between 40 and 8 kyr BP, *Nature, 366*, 443-445, 1993.

Damon, P. E., and C. P. Sonett, Solar and terrestrial components of the atmospheric C-14 variation spectrum, in *The Sun in time*, edited by C. P. Sonett, M. S. Giampapa and M. S. Matthews, pp. 360-388, The University of Arizona, Tucson, 1991.

Damon, P. E., and A. N. Peristykh, Solar and Climatic Implications of the centennial and Millennial Periodicities in atmospheric C14 Variations, *this volume.*

Dansgaard, W., S. J. Johnsen, H. B. Clausen, D. Dahl-Jensen, N. S. Gundestrup, C. U. Hammer, C. S. Hvidberg, J. P. Steffensen, A. E. Sveinbjörnsdottir, J. Jouzel, and G. Bond, Evidence for general instability of past climate from a 250-kyr ice core record, *Nature, 364,* 218-220, 1993.

Denton, G. H., and W. Karlén, Holocene climatic variations - their pattern and possible cause, *Quat. Res., 3*(2), 155-205, 1973.

Dickinson, R. E., Solar variability and the lower atmosphere, *Bull. Am. Met. Soc., 56*(12), 1240-1248, 1975.

Eddy, J. A., The Maunder Minimum, *Science, 192*(4245), 1189-1201, 1976.

Elkibbi, M., and J. A. Rial, An outsider's review of the astronomical theory of the climate: is the eccentricity-driven insolation the main driver of the ice ages?, *Earth-Science Reviews, 56,* 161–177, 2001.

Finkel, R. C., and K. Nishiizumi, Beryllium 10 concentrations in the Greenland ice sheet project 2 ice core from 3-40 ka, *J. Geophys. Res., 102*(C12), 26699-26706, 1997.

Forest, C. E., P. H. Stone, A. P. Sokolov, M. R. Allen, and M. D. Webster, Quantifying Uncertainties in Climate System Properties with the Use of Recent Climate Observations, *Science, 295,* 113-117, 2001.

Fröhlich, C., and J. Lean, The Sun's total irradiance: Cycles, trends and related climate change uncertainties since 1976, *J. Geophys. Res., 25*(NO. 23), 4377-4380, 1998.

Gleissberg, W., A table of secular variations of the solar cycle, *Terr. Magn. Atm. Electr. 49,* 243-244, 1944.

Goldstein, G., Partial half-life for b-decay of ^{36}Cl, *J. Inorg. Nucl. Chem., 28,* 937-939, 1966.

Grootes, P. M., and M. Stuiver, Oxygen 18/16 variability in Greenland snow and ice with 10^{-3} to 10^{5}-year time resolution, *J. Geophys. Res., 102*(C12), 26,455-26,470, 1997.

Haigh, J., The impact of solar variability on climate, *Science, 292,* 981-984, 1996.

Haigh, J. D., The role of stratospheric ozone in modulating the solar radiative forcing of climate, *Nature, 370,* 544-546, 1994.

Haigh, J. D., Fundamentals of the Earth's Atmosphere and Climate, *this volume.*

Herschel, W., Observations Tending to Investigate the Nature of the Sun, in Order to Find the Causes or Symptoms of its Variable Emission of Light and Heat., *Philos. Trans. R. Soc. London, 91,* 265-318, 1801.

Hill, C., and P. Forti, Cave Minerals of the World. Second Edition, pp. 463, National Speleological Society, Huntsville., 1997.

Hofmann, H. J., J. Beer, G. Bonani, H. R. Von Gunten, S. Raman, M. Suter, R. L. Walker, W. Wölfli, and D. Zimmermann, ^{10}Be: Half-Life and AMS-Standards, *Nucl. Instr. Meth. Phys. Res., B29,* 32-36, 1987.

Hormes, A., C. Schlüchter, and T. F. Stocker, Minimal extension phases of Unteraargletscher (Swiss Alps) during the Holocene based on ^{14}C analysis of wood, *Radiocarbon, 40*(2), 807-817, 1998.

Hughen, K., J. T. Overpeck, S. Lehmann, M. Kashgarian, J. Southon, L. C. Peterson, R. Alley, and D. M. Sigman, Deglacial changes in ocean circulation from an extended radiocarbon calibration, *Nature, 391,* 65-68, 1998.

Hughen, K. A., J. R. Southon, S. J. Lehmann, and J. T. Overpeck, Synchronous Radiocarbon and Climate Shifts During the Last Deglaciation, *Science, 290,* 1951-1954, 2000.

Johnsen, S. J., D. Dahl-Jensen, W. Dansgaard, and N. Gundestrup, Greenland palaeotemperatures derived from GRIP bore hole temperature and ice core isotope profiles, *Tellus, 47B,* 624-629, 1995.

Johnsen, S. J., H. B. Clausen, W. Dansgaard, N. S. Gundestrup, C. U. Hammer, U. Andersen, K. K. Andersen, C. S. Hvidberg, D. Dahl-Jensen, J. P. Steffensen, H. Shoji, A. E. Sveinbjörnsdottir, J. White, J. Jouzel, and D. Fisher, The $\delta^{18}O$ record along the Greenland Ice Core Project deep ice core and the problem of possible Eemian climatic instability, *J. Geophys. Res., 102,* 26,397-26,410, 1997.

Laj, C., A. Mazaud, and J.-C. Duplessy, Geomagnetic intensitiy and ^{14}C abundance in the atmosphere and ocean during the past 50 kyr, *Geophys. Res. Let., 23*(16), 2045-2048, 1996.

Laj, C., C. Kissel, A. Mazaud, J. E. T. Channell, and J. Beer, North Atlantic paleointensity stack since 75 kal (NAPIS-75) and the duration of the Laschamp event, *Phil. Trans. R. Soc. Lond., 358*(A), 1009-1025, 2000.

Lal, D., and B. Peters, Cosmic Ray Produced Radioactivity on the Earth, in *Handbuch für Physik,* vol. 46/2, edited by S. Flügge, pp. 551-612, Springer, Berlin, 1967.

Laskar, J., The chaotic motion of the solar system: A numerical estimate of the chaotic zones, *Icarus, 88,* 266-291, 1990.

Lean, J., J. Beer, and R. Bradley, Reconstruction of solar irradiance since 1610: implications for climate change, *Geophys. Res. Lett., 22*(23), 3195-3198, 1995.

Magny, M., Solar influences on Holocene climatic changes illustrated by correlations between past lake-level fluctuations and the atmospheric ^{14}C record, *Quat. Res., 40,* 1-9, 1993.

Marchal, O., T. F. Stocker, and R. Muscheler, Atmospheric radiocarbon during the Younger Dryas: Production, ventilation, or both?, *Earth Plan. Sci. Lett., 185,* 383-395, 2001.

Masarik, J., and J. Beer, Simulation of particle fluxes and cosmogenic nuclide production in the Earth's atmosphere, *J. Geophys. Res., 104*(D10), 12,099-12,111, 1999.

Mauquoy, D., B. van Geel, M. Blaauw and J. van der Plicht, Evidence from northwest European bogs shows 'Little Ice Age' climatic changes driven by variations in solar activity, *The Holocene 12,1,* 1-6, 2002

Meynadier, v., J. P. Valet, R. Weeks, N. J. Shackleton, and V. Lee Hagee, Relative geomagnetic intensity of the field during the last 140 ka, *Earth Planet. Sci. Lett., 114,* 39-57, 1992.

Muscheler, R., J. Beer, G. Wagner, and R. C. Finkel, Changes in deep-water formation during the Younger Dryas cold period inferred from a comparison of ^{10}Be and ^{14}C records, *Nature, 408,* 567-570, 2000.

Neff, U., S. J. Burns, A. Mangini, M. Mudelsee, D. Fleitmann, and A. Matter, Strong coherence between solar variability and the monsoon in Oman between 9 and 6 kyr ago, *Nature, 411,* 290-293, 2001.

O'Brien, S. R., P. A. Mayewski, L. D. Meeker, D. A. Meese, M. S. Twickler, and S. I. Whitlow, Complexity of holocne climate as reconstructed from a Greenland ice core, *Science, 270,* 1962-1964, 1995.

Oeschger, H., U. Siegenthaler, U. Schotterer, and A. Gugelmann, A box diffusion model to study the carbon dioxide exchange in nature, *Tellus, 27*(2), 168-192, 1975.

Paillard, D., L. Labeyrie, and P. Yiou, Macintosh program performs time series analysis, *Eos Trans. AGU, 77,* 379, 1996.

Petit, J. R., J. Jouzel, D. Raynaud, N. I. Barkov, J.-M. Barnola, I. Basile, M. Bender, J. Chapppellaz, M. Davis, G. Delaygue, M. Delmotte, V. M. Kotlyakov, M. Legrand, V. Y. Lipenkov, C. Lorius, L. Pépin, C. Ritz, E. Saltzmann, and M. Stievenard, Climate and atmospheric history of the past 420,000 years from the Vostok ice core, Antarctica, *Nature, 399,* 429-436, 1999.

Siegenthaler, U., M. Heimann, and H. Oeschger, ^{14}C variations caused by changes in the global carbon cycle, *Radiocarbon, 22*(2), 177-191, 1980.

Siegenthaler, U., Uptake of excess CO_2 by an outcrop-diffusion model ocean, *J. Geophys. Res., 88*(C6), 3599-3608, 1983.

Spurk, M., M. Friedrich, J. Hofmann, S. Remmele, B. Frenzel, H. H. Leuschner, and B. Kromer, Revisions and Extension of the Hohenheim Oak and Pine Chronologies: New Evidence about the Timing of the Younger Dryas / Preboreal Transition, *Radiocarbon, 40*(3), 1107-1116, 1998.

Stuiver, M., and H. A. Polach, Discussion reporting of ^{14}C data, *Radiocarbon, 19*(3), 355-363, 1977.

Stuiver, M., and T. F. Braziunas, Sun, Ocean, Climate and Atmospheric [14]CO_2, an evaluation of causal and spectral relationships, *The Holocene, 3,* 289-305, 1993.

Stuiver, M., T. F. Braziunas, and P. M. Grootes, Is there evidence for solar forcing of climate in the GISP2 oxygen record?, *Quat. Res., 48,* 259-266, 1997.

Stuiver, M., and F. Braziunas, Anthropogenic and solar components of hemispheric ^{14}C, *Geophys. Res. Let., 25*(3), 329-332, 1998.

Stuiver, M., P. J. Reimer, E. Bard, J. W. Beck, G. S. Burr, K. A. Hughen, B. Kromer, G. McCormac, J. Van der Plicht, and M. Spurk, INTCAL98 radiocarbon age calibration, 24,000-0 cal BP, *Radiocarbon, 40*(3), 1041-1083, 1998.

Svensmark, H., and E. Friis-Christensen, Variation of cosmic ray flux and global cloud coverage - a missing link in solar-climate relationships, *J. Atm. Terr. Phys., 59*(11), 1225-1232, 1997.

Tauxe, L., Sedimentary records of relative paleointensity of the geomagnetic field: theory and practice, *Rev. Geophys., 31,* 319-354, 1993.

Thompson, L. G., E. Mosley-Thompson, M. E. Davis, P.-N. Lin, K. A. Henderson, J. Cole-Dai, J. F. Bolzan, and K.-B. Liu, Late glacial stage and Holocene tropical ice core records from Huascaran, Peru, *Science, 269,* 46-50, 1995.

Tinsley, B. A., Correlations of atmospheric dynamics with solar wind induced changes of air-earth current densitiy into cloud tops, *J. Geophys. Res., 101*(D23), 29701-29714, 1996.

Tinsley, B. A., Influence of solar wind on the global electric circuit, and inferred effects on cloud microphysics, temperature, and dynamics in the troposphere, *Space Sci. Rev., 94,* 231-258, 2000.

Tinsley, B. A., and F. Yu, Atmospheric Ionization and Clouds as Links Between Solar Activity and Climate, *this volume.*

Tric, E., J. P. Valet, P. Tucholka, M. Paterne, L. LaBeyrie, F. Guichard, L. Tauxe, and M. Fontugne, Paleointensity of the geomagnetic field during the last 80,000 years, *J. Geophys. Res., 97,* 9337-9351, 1992.

Van Geel, B., J. Buurman, and H. T. Waterbolk, Archaeological and palaeoecological indications of an abrupt climate change in Netherlands, and evidence for climatological teleconnections around 2650 BP, *J. Quat. Sci., 11*(6), 451-460, 1996.

Van Geel, B., J. Van der Plicht, M. R. Kilian, E. R. Klaver, J. H. M. Kouwenberg, H. Renssen, I. Reyneau-Farrera, and H. T. Waterbolk, The sharp rise of Δ^{14}C ca. 800 cal BC: possible causes, related climatic teleconnections and the impact on human environments, *Radiocarbon, 40*(No. 1), 535-550, 1998.

Van Geel, B., O. M. Raspopov, H. Renssen, J. Van der Plicht, V. A. Dergachev, and H. A. J. Meijer, The role of solar forcing upon climate change, *Quat. Sci. Rev., 18,* 331-338, 1999.

Verschuren, D., K. R. Laird, and B. F. Cumming, Rainfall and drought in equatorial east Africa during the past 1,100 years, *Nature, 403,* 410-414, 2000.

Wagner, G., J. Beer, C. Laj, C. Kissel, R. Muscheler, and H.-A. Synal, Chlorine-36 evidence for the Mono Lake event in the Summit GRIP ice core, *Earth Planet. Sci. Let., 181,* 1-6, 2000a.

Wagner, G., J. Masarik, J. Beer, S. Baumgartner, D. Imboden, P. W. Kubik, H.-A. Synal, and S. Suter, Reconstruction of the geomagnetic field between 20 and 60 kyr BP from cosmogenic radionuclides in the GRIP ice core, *Nucl. Instr. Meth. Phys. Res., B172,* 597-604, 2000b.

Wagner, G., J. Beer, P. W. Kubik, C. Laj, J. Masarik, W. Mende, R. Muscheler, G. M. Raisbeck, and F. Yiou, Presence of the solar de Vries cycle (205 years) during the last ice age, *Geophys. Res. Lett., 28*(No. 2), 303-306, 2001a.

Wagner, G., C. Laj, J. Beer, C. Kissel, R. Muscheler, J. Masarik, and H.-A. Synal, Reconstruction of the paleoaccumulation rate of central Greenland during the last 75 kyr using the cosmogenic radionuclides [36]Cl and [10]Be and geomagnetic field intensity data, *Earth Planet. Sci. Let., 193,* 515-521, 2001b.

Wagner, G., D. M. Livingstone, J. Masarik, R. Muscheler, and J. Beer, Some results relevant to the discussion of a possible link between cosmic rays and the Earth's climate, *J. Geophys. Res., 106,* 3381-3387, 2001c.

Yiou, F., G. M. Raisbeck, S. Baumgartner, J. Beer, C. Hammer, S. Johnsen, J. Jouzel, P. W. Kubik, J. Lestringuez, M. Stiévenard, M. Suter, and P. Yiou, Beryllium 10 in the Greenland Ice Core Project ice core at Summit, Greenland, *J. Geophys.Res., 102*(C12), 26783-26794, 1997.

J. Beer, Department of Surface Waters, Swiss Federal Institute of Environmental Science and Technology (EAWAG), Überlandstr. 133, CH-8600 Dübendorf, Switzerland. (beer@eawag.ch)

P.W. Kubik, Paul Scherrer Institute, c/o ETH Hönggerberg, HPK-H33, CH 8093 Zürich, Switzerland. (kubik@particle.phys. ethz.ch)

R. Muscheler, GeoBiosphere Science Center, Dept. of Quaternary Geology, Lund University, Solvegatan, 12, 22362 Lund, Sweden. (raimund.muscheler@geol.lu.se)

Solar and Climatic Implications of the Centennial and Millennial Periodicities in Atmospheric $\Delta^{14}C$ Variations

Paul E. Damon and Alexei N. Peristykh

Department of Geosciences, University of Arizona, Tucson

Temporal changes in atmospheric ^{14}C require high precision calibration of the radiocarbon time scale. The calibration process also provides $\Delta^{14}C$ which is the age and fractionated corrected deviation of past atmospheric ^{14}C relative to an international standard. The dominant cause of atmospheric ^{14}C fluctuation is changes in the geomagnetic dipole moment. Detrending yields residual $\Delta^{14}C$ which can be related to solar activity and climate change. The biogeochemical cycle of ^{14}C acts as a lowpass filter that greatly attenuates the high frequency components of solar activity. This attenuation is compensated by higher production rate and higher measurement precision relative to other cosmogenic isotopes. The spectrum of $\Delta^{14}C$ contains many periods, overtones and combination tones. We are concerned here with the millennial and centennial periods. The Suess cycle (208 yr) and Gleissberg (88 yr) are discrete solar cycles that can be related to forcing of climate change whereas the millennial cycles do not appear to have an independent existence but are required by Fourier analysis to reconstruct the variance of the century scale cycles. We find the most effective way of demonstrating the relationship between residual $\Delta^{14}C$ and climate is to match the pattern of residual $\Delta^{14}C$ with the pattern of climate change.

1. INTRODUCTION

1.1. Manifestation of the Solar Activity by Cosmogenic ^{14}C

Radiocarbon is generated in the atmosphere by the reaction $^{14}N(n,p)^{14}C$. Fast neutrons are produced in spallation reactions on atmospheric gases during nuclear cascade processes initiated dominantly by galactic cosmic rays (GCR) and to a lesser extent solar cosmic rays (SCR) [*Lal and Peters*, 1967]. About 65% of the fast neutrons are thermalized, producing ^{14}C [*Lingenfelter and Ramaty*, 1970]. The intensity of GCR entering the Earth's atmosphere depends on the state of solar activity via the magnetic field embedded in the solar wind. As solar activity increases, fewer GCR enter the atmosphere to produce ^{14}C. Hence, there is an inverse relationship between ^{14}C production and solar activity.

Radiocarbon rapidly equilibrates with atmospheric CO_2 and enters the biogeochemical cycle of carbon including the life cycle of biosphere (flora and fauna) (different models are shown in Figure 1). After the organism dies or a part ceases to participate in biogeochemical cycle of carbon like tree or coral rings, $^{14}CO_2$ is no longer replaced and decays with its characteristic half-life of 5730 yrs.

As a result of the circulation of carbon through its biogeochemical cycle the radiocarbon variations in all components of the exchange system of the carbon cycle $\vec{n}_{14C}(t)$ are the convolution of the input signal of generation rate $\vec{Q}_{14C}(t)$ and the time response of the exchange system denoted by linear operator $\tilde{H}_{sys}$:

Solar Variability and its Effects on Climate
Geophysical Monograph 141

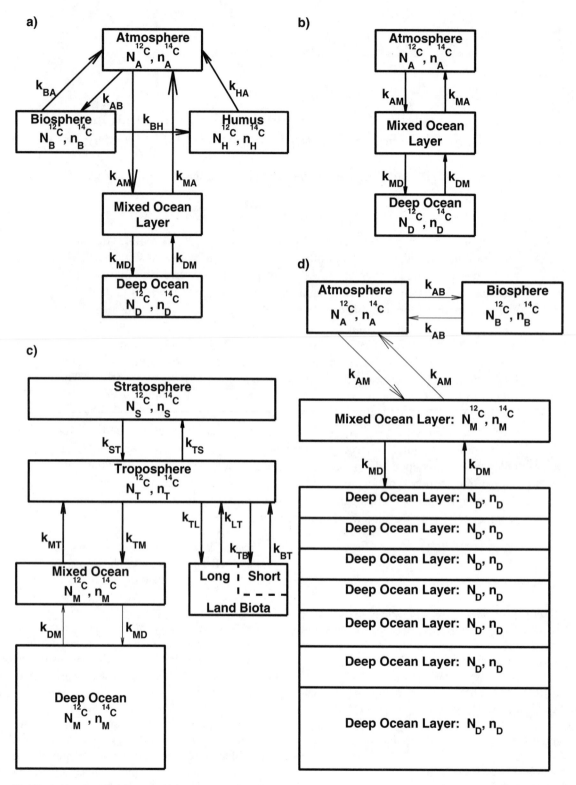

Figure 1. Simple 'box' models of the biogeochemical cycle of radiocarbon: a) 5-box model by *Craig* [1957]; b) 3-box model by *Houtermans et al.* [1973]; c) 6-box model by *Ekdahl and Keeling* [1973]; d) box-diffusion model by *Oeschger et al.* [1975].

$$\vec{n}_{14_C}(t) = \overline{H}_{sys} * \vec{Q}_{14_C} = \int_{-\infty}^{t} \hat{H}_{sys}(t-t')\vec{Q}_{14_C}(t')dt' \qquad (1)$$

where the matrix representation [*Peristykh*, 1988] of the operator of system response is used and that describes the state of the system in linear (vector) space $\mathbf{R}^n$.

In studies of quasi-periodical processes in systems, it is essential to obtain the frequency response functions of the exchange system: amplitude frequency response (or gain) $g(\omega)$ and phase frequency response (or phase lag) $\varphi(\omega)$ of the exchange system vs. frequency (or period of variations). Those are the amplitude and the phase of the complex Fourier transform $\tilde{F}_\omega\{\cdot\}$ of the system time response $\tilde{H}_{sys}(t)$:

$$g(\omega) = \text{abs } \tilde{F}_\omega\{\tilde{H}_{sys}(\cdot)\}$$
$$\varphi(\omega) = \text{Arg } \tilde{F}_\omega\{\tilde{H}_{sys}(\cdot)\} \qquad (2)$$

Examples of these transforms are given in Figure 2. The amplitude for each model in Figure 2a reaches a plateau approaching the DC gain which is the gain at zero frequency $g(0)$. The ratio of the gain at any frequency to DC gain, $g(f)/g(0)$, yields the attenuation. The DC gain as can be obtained from radiocarbon data, g_{obs}, is proportional to the steady state number of ^{14}C atoms in the atmosphere and inversely proportional to the ^{14}C production rate. Thus, if an average production rate of 2.2 at/(cm$^2 \cdot$ sec) [*Lingenfelter and Ramaty*, 1970] is used g_{obs} will be ca. 110 [*Lazear et al.*, 1980], whereas if the low value of 1.8 at/(cm$^2 \cdot$ sec) [*O'Brien*, 1979] is used g_{obs} will be 150. If the model parameters are not balanced, the model will seek a new steady state. This can be rectified by changing the production rate or the reservoir parameters. *Damon and Sternberg* [1989] point out that ^{14}C biogeochemical models neglect the relatively large amount of ^{14}C sequestered in lagoons, marshes, bays, deltas, freshwater lakes and wetlands and consequently add a sedimentary sink to all the models [*Damon et al.*, 1983].

It is quite obvious from Figure 2 that the biogeochemical cycle of ^{14}C acts as a low pass filter with a 10 yr cycle being attenuated by a factor of ca. 100 with a time lag of ca. 2 yr, a 100 yr cycle by a factor of ca. 20 with a time lag of 14 yr and a 1000 yr cycle by a factor of ca. 14 with a time lag of 190 yr, etc. (Figures 2b-c). This attenuation of ^{14}C is compensated relative to other cosmogenic isotopes by its much higher production rate and measurement precision. Consequently, the Schwabe (ca. 11 yr) and Hale (ca. 22 yr) cycles can be observed clearly in $\Delta^{14}C$ measurements from single year tree-ring records [*Stuiver and Brazinas*, 1998; *Damon and Peristykh*, 2000].

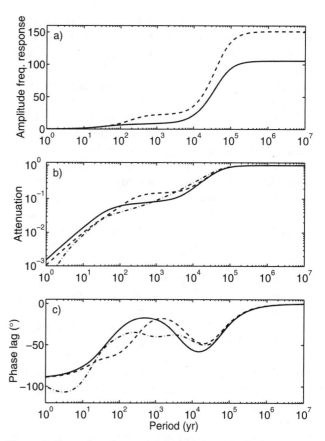

Figure 2. Frequency response characteristics: amplitude frequency response (a), attenuation (b), phase lag (c), of the carbon cycle models: 5-box model by *Craig* [1957] with parameters from *Dergachev and Ostryakov* [1980] (solid line), 3-box model by *Houtermans et al.* [1973] (dashed line), box-diffusion model by *Siegenthaler et al.* [1980] with stratosphere and troposphere separated (broken line).

1.2. Radiocarbon Dating

Willard Libby received the Nobel Prize in Chemistry on December 10, 1960 for developing radiocarbon dating. Prior to that, the second edition of his book entitled Radiocarbon Dating had been published in 1955 [*Libby*, 1955]. Two of the basic assumptions underlying the method were that the specific activity of atmospheric ^{14}C is geographically and temporally constant; and the half-life of ^{14}C was 5570±30 yrs.

However, in 1958 *de Vries* had demonstrated the first evidence for temporal fluctuation of pre-anthropogenic ^{14}C by natural consequences. This was quickly confirmed by four laboratories by 1962 and by 1963 three laboratories [*Ralph and Stuckenrath*, 1960; *Suess*, 1961; *Damon et al.*, 1963] had confirmed that the de Vries effect fluctuations are superimposed on a higher amplitude, longer period trend. By the time of the 5th International Radiocarbon Conference that was held in 1962, three different laboratories [*Mann et al.*, 1961; *Watt et al.*, 1961; *Olsson et al.*, 1962] had made precise and accurate measurements of the ^{14}C half-life with the mean of the three values being 5,730±40 yr. By that time many ^{14}C dates had been published and because correcting the half-life alone would not yield correct calendar ages, it was decided to not change the half-life for dating purposes but to report dates in Libby years BP (before 1950). Since many years of calibration would follow it would be necessary to also publish the best available calibrated date subject to change as calibration work proceeded. Dendrochronologically dated tree-rings that can be accurate with much work to the calendar year provided the best means of calibration.

Lastly, Libby's assumption that the carbon isotopes are well mixed on a global basis and, as a consequence, the specific activity of ^{14}C would be independent of geographic location is strictly not correct, although sufficiently correct for the early phase of ^{14}C dating. With the attainment of high precision measurements, geographic differences in contemporaneous measurements became apparent. For example, there is a hemispheric latitude effect causing dates in the Southern Hemisphere to be older than in the Northern Hemisphere [*Braziunas et al.*, 1995]. The latitude effect is negligible from the Arctic circle to the equator but increases from the equator to about 65°S when it has been calculated to be −6‰ (50 years older than above the equator). There are other geographic effects on a smaller scale (see *Stuiver and van der Plicht* [1998, p.1045–1046] for a brief review of "Atmospheric and Regional Effect").

Thus, although high precision ^{14}C measurements reveal second order deviations from the first order assumptions, we should not lose sight of the revolutionary impact of ^{14}C dating on such fields as archaeology and Quaternary geology during the first two decades of its application. We are simply able to ask increasingly refined questions and extend our work to solar physics and astrophysics [*Damon and Peristykh*, 2000].

1.3. Basic Equations

The reporting of ^{14}C data is thoroughly discussed in a paper [*Stuiver and Polach*, 1977]. There are two commonly accepted radiocarbon standards originally produced by the National Bureau of Standards (NBS) which is now the National Institute of Science and Technology (NIST). These are NIST Oxalic acid 1 and 2 (NIST1 and NIST2). We will only refer to NIST1 that we primarily use in our laboratory.

NIST1 is normalized to $\delta^{13}C = -19‰$ with respect to the PDB standard of the stable isotopes of carbon:

$$A_{on} = 0.95 A_{ox} \cdot [1 - 2 \cdot (19 + \delta^{13}C) / 1000] \qquad (3)$$

A_{on} changes exponentially between 1950 and the time of counting y but so does the normalized sample activity A_{sn} and so no correction to time y is required. Isotopic fractionation in all samples is taken into account by normalizing to −25‰ PDB:

$$A_{sn} = A_s \cdot [1 - 2 \cdot (25 + \delta^{13}C) / 1000] \qquad (4)$$

The conventional age BP with the Libby half-life is given by

$$t = -8033 \cdot \ln(A_{sn} / A_{on}) \qquad (5)$$

The calibrated calendar age is then obtained relative to a graph or preferably a computer program.

In geophysical studies, we are interested in the activity of ^{14}C in the past relative to the normalized standard A_{on}. In this calculation, the more precise half-life $T_{1/2} = 5,730$ yr is used and a quantity $\Delta^{14}C$ is obtained as follows:

$$\Delta^{14}C = [A_{sn} / A_{on} \cdot e^{\lambda \cdot \tau_s} - 1] \cdot 1000, ^\circ/_{\circ\circ} \qquad (6)$$

where τ_s is the age of the sample prior to AD 1950. In Equation (3) the normalization factor 0.95 makes $\Delta^{14}C = 0$ in the mid-19th century prior to the large increase of CO_2 as a result of the combustion of fossil fuels. At the University of Arizona ^{14}C is measured by state-of-the-art scintillation counting (Quantulus) in a underground laboratory with a precision equal to or less than 0.2% or by accelerator mass spectrometry (AMS) with a precision of 0.3% (when using 4 targets).

2. PRESENT STATUS OF CALIBRATION AND $\Delta^{14}C$ TIME SERIES

The *Calibration Issue* of the journal *Radiocarbon* [*Stuiver and van der Plicht*, 1998] contains the most recent published information on the calibration of the radiocarbon time scale. As previously discussed, Equation (5), used to calculate conventional radiocarbon dates, is not exact but has been maintained to be consistent with past publications. In practice, to obtain a date in calendar years, we must relate the date to a calendar year time scale. Dendrochronology has provided an ideal calendar year time scale and the cellulose from tree rings is an ideal substance for radiocarbon dating. The INTCAL98 calibration curve is shown in

Figure 3. The dendrochronologic data extends back to about 9900 BC. Before that time, ^{14}C is calibrated by Cariaco varves and U/Th dating on coral samples. A reservoir correction must be made to convert from the marine coral to the atmospheric environment. This is accomplished by deducting 500 ^{14}C years from the marine ages. As shown in Figure 3, a radiocarbon age of 6000 years converts into a calibrated calendar year of *ca.* 4900 BC or *ca.* 6850 cal BP. At the beginning of the Holocene following the end of the Younger Dryas climatic event a radiocarbon age of 10,120 years BP converts into a calibrated calendar year of *ca.* 9,720 BC or 11,670 cal BP. These conversions are not trivial and become greater prior to the Holocene. Such corrections are important for archaeologic, paleoclimatic and many other studies. The inset of Figure 3 shows that what are mere ripples viewed from the longer time scale are conspicuous fluctuations when viewed on a shorter time scale (inset). It was the necessity of calibrating the radiocarbon time scale that provided the data base for important geophysical and solar physical research but this becomes obvious when converting to Δ^{14}C (Eq.(6)).

By the time of the Twelfth Nobel Symposium entitled *"Radiocarbon Variations and Absolute Chronology"* held in 1969 [*Olsson*, 1970], the three main causes of atmospheric ^{14}C fluctuation had been identified. These were modulation by changes in the geomagnetic dipole moment, heliomagnetic modulation by changes in the magnetic field imbedded in the solar wind and changes in the biogeochemical cycle of carbon. Referring to Figure 4 there has been general agreement that the trend line is primarily the result of

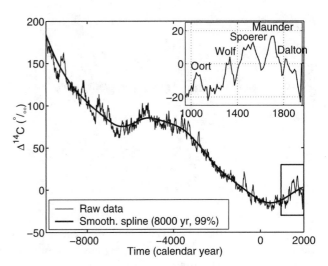

Figure 4. INTCAL98 Δ^{14}C calibration curve.

changes in the geomagnetic dipole moment whereas the fine structure ('wiggles') around the smooth trend line are the result of solar activity. This can be seen in the inset where the fine structure after removal of the trend line is shown. Five solar minima (Δ^{14}C maxima), named after astronomers, are clearly apparent. Surprisingly, it was possible to model these solar minima successfully very early in the history of radiocarbon geophysics [*Grey and Damon*, 1970] using a compendium of sunspot numbers by *Schove* [1955]. In his trenchant paper on the Maunder Minimum, *Eddy* [1976] placed these solar minima in a historic astronomic perspective bringing to our attention the solar phenomena that had been observed during these events. In his summary and conclusions he observed that "the long-term envelope of sunspot activity carries the indelible signature of slow changes in solar radiation which surely affect our climate". Eddy's paper on the Maunder Minimum was followed a year later by a paper entitled *"Climate and the changing Sun"* in which he pointed out the climatic significance of the existence of such solar maxima and minima as demonstrated by the seven millennia of proxy radiocarbon data available at that time [*Eddy*, 1977]. Eddy's paper was interesting and brought the significance of the ^{14}C data as a proxy indicator of solar activity and climate to a wider audience but the relationship was already well known, if not to a wider audience. For example, *Damon* [1968] in commenting on a paper by *Suess* [1968] stated "The relatively high atmospheric C^{14} concentration during the Little Ice Age and the relatively low C^{14} concentration during the medieval warm period is also in qualitative agreement with the expected inverse relationship between sunspot activity and $\bar{Q}$" (production rate of ^{14}C).

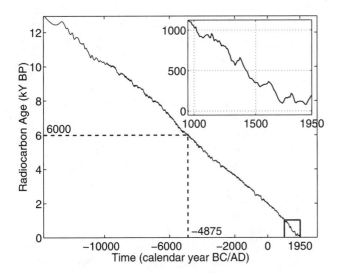

Figure 3. INTCAL98 age calibration curve. The dendrodated time goes back to *ca.* 10,000 yrs BC. Older ages are derived from ^{230}Th and ^{14}C dating of corals.

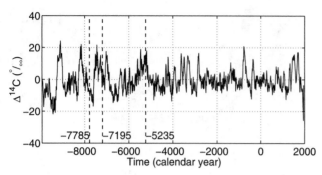

Figure 5. INTCAL98 $\Delta^{14}C$ calibration curve with long-term trend removed.

Figure 5 shows the residual $\Delta^{14}C$ with the trend removed. Back to 5235 BC, the dendrochronology has been satisfactorily replicated and cross dated. Single year, decadal and bidecadal data have been merged into a decadal sequence. The correction in the German oak chronology of 95 years at 7785 BC is still under consideration and some decadal intervals are missing before 7195 BC. There are still hopes of pushing the tree-ring chronology back beyond 9905 BC.

Even before the cogent paper by *Broecker and Denton* [1989] entitled "The role of ocean-atmosphere reorganizations in glacial cycles", it was pointed out by *Damon* [1970] "that the changes in climate and sea-level during the last eight millennia are relatively small when compared to the extreme conditions during the last major glaciation (wurm). It is quite possible that the climatic effect on $\Delta^{14}C$ may be much greater in magnitude than during more recent times". *Stuiver et al.* [1998] combine marine and varve data from *ca.* 11,640 BP to *ca.* 15,550 BP to derive a residual $\Delta^{14}C$ record for that period of time to compare with $\delta^{18}O$. In their resultant Figure 14 (p.1061) during the Bølling warm epoch qualitatively an increase of ca. 5‰ in $\delta^{18}O$ is accompanied by a decrease of over 30‰ in $\Delta^{14}C$ which is the expected inverse relationship between temperature and ^{14}C. An increase of 5‰ in $\delta^{18}O$ corresponds to about a 7.5°C increase in temperature [*Johnsen et al.*, 1992]. *Stuiver et al.* [1998] ascribe the decrease in $\Delta^{14}C$ to increased deep water formation. This was followed by decreased deep water formation at the onset of the Younger Dryas with the resultant increase in $\Delta^{14}C$ and decrease in $\delta^{18}O$ accompanying this marked North Atlantic cold epoch. Interestingly, an oscillating increase in $\Delta^{14}C$ and deep water formation is not accompanied by warming until the sudden end of the Younger Dryas. Also, see paper by Muscheler, Beer and Kubik in this Monograph for the possible role of solar forcing in initiating the Younger Dryas. In contrast to the exteme instability during the ice age, the relatively small fluctuations of $\delta^{18}O$ during the Holocene suggest that the North Atlantic hydros-

aline conveyor has been surprisingly strong and steady. *Stuiver et al.* [1995] pointed out, as an exception, a decrease of 2‰ in $\delta^{18}O$ near 8230 yr BP that was not accompanied by an increase in $\Delta^{14}C$. The authors suggest a scenario where "Summer time surface water cooling by a large number of floating icebergs near Southeast Greenland together with lower sea water $\delta^{18}O$ values (increased freshwater dilution), could easily cause a 2‰ drop lasting 200 years". *Bond et al.* [1997] demonstrate the existence of eight such ice rafting events in the sedimentary record. Only one of these at *ca.* 8100 yr BP is close to the 2‰ decrease in the GISP2 ice core. No such dramatic decrease occurs elsewhere in the Holocene record. The Holocene's freedom from large-scale changes in the thermohaline conveyer system makes identification of the solar signal much less equivocal as will be seen in the following section.

3. SPECTRUM OF ATMOSPHERIC $\Delta^{14}C$ VARIATIONS

Spectral analysis of variations can give one information on presence of an important class of natural processes as cyclicity in a phenomenon under study. At the same time we must note that this is not an infallible tool if applied and especially interpreted unwisely. Presence of statistically significant peaks in the spectrum can be an indication of the presence of cyclic components in the signal under study.

Previous work analysing earlier sets of data made use of the Maximum Entropy Method (MEM) and Bayesian spectral analysis in addition to the conventional Discrete Fourier Transform (DFT) used in this paper [*Sonett and Finney*, 1990; *Damon and Sonett*, 1991]. This analysis has the advantage of the availability of a newer and improved dataset (INTCAL98). Here we essentially confirm the previous work but with the addition of several new features.

In this work the Fourier-spectrum analysis of $\Delta^{14}C$ time variations was performed by Discrete Fourier Transform (DFT) of time variations. Before that the raw data series (Figure 4) were detrended via subtraction of the long-term variation component (Figure 5) obtained from the former by performing lowpass filtering by a smoothing spline method. The amplitude Fourier spectrum is depicted in Figure 6. The peaks one can readily see in the spectrum can be divided into four spectral bands:

1. *Millennial time scale.* These long periods from ~2300 yr through ~700 yr, may not have an independent existence but are required to obtain the variance of the shorter term variations. For example, if the 2300 yr period were an independent solar cycle, it would require a large time lag, *ca.* 250 yr (see Figure 2c)

rather than simply following the variance of the century scale cycles and concurrent climate events as will be observed later. Longer periodicities, if they exist, can not be detected by the time series analysis of this record due to its insufficient length. The peak at ~700 yr appears to have a first overtone at ~350 yr. The peak of ~1433 yr and adjacent 'bulges' are too insignificant compared to the noise level to be counted as real. Other significant millennial peaks will be discussed later.

2. *Centennial time scale.* The 208-yr periodicity with its mysterious companion peak ~225 yr, its first overtone (104 yr), the 88-yr Gleissberg cycle and its first overtone (44 yr) are all present at significant levels.

   We would like to note that the Gleissberg cycle is of great interest because it is not only manifested by its quasi-sinusoidal modulation of the 11-yr sunspot cycle

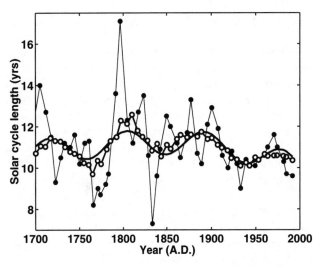

Figure 7. Solar cycle length variations: raw (filled circles) and smoothed (open circles) by Gleissberg's 5-point symmetric trapezoidal 'secular smoothing' lowpass filter. Mono-harmonic regression curve (thick line) is obtained via harmonic analysis over the basic period 88.8 yrs.

envelope (amplitude modulation, AM) but it also modulates the Schwabe cycle length (frequency modulation, FM). To show the effect of FM on solar cycle length *Gleissberg* [1944] used a 5-point trapezoidal symmetric lowpass digital filter applied to sunspot number record (Figure 7). *Friis-Christensen and Lassen* [1991] applied this procedure to solar cycle length (SCL) in an attempt to find a correlation with the Northern Hemisphere temperature variations. However, it is obvious that the irradiance component that has been associated with it by *Friis-Christensen and Lassen* [1991] can not alone explain the 20th century warming [*Damon and Peristykh*, 1999].

3. *Combination tones.* One can recognize in side-bands of 88-yr peak ($f_1^G = 1/88 \ yr^{-1} \approx 0.0114 \ yr^{-1}$) manifestation of its amplitude modulation (AM) by a cyclic component with frequency of Suess cycle $f_1^S = 1/208 \ yr^{-1} \approx 0.00481 \ yr^{-1}$: $f_1^{G \times S} = f_1^G \pm f_1^S$ so the correspondent periods of the peaks are $T_{1-}^{G \times S} = (f_1^G - f_1^S)^{-1} \approx 152 \ yr$ and $T_{1+}^{G \times S} = (f_1^G + f_1^S)^{-1} \approx 61.8 \ yr$. Moreover, one can notice more closely located and less pronounced sidebands of 88-yr peak as manifestation of its AM by a cyclic component with frequency of Hallstattzeit cycle $f_1^H = 1/2200 \ yr^{-1} \approx 0.00045 \ yr^{-1}$: $f_1^{G \times H} = f_1^G \pm f_1^H$ [*Peristykh and Damon*, 2001] so we have periods of the peaks $T_{1-}^{G \times H} = (f_1^G - f_1^H)^{-1} \approx 91.7 \ yr$ and $T_{1+}^{G \times H} = (f_1^G + f_1^H)^{-1} \approx 84.6 \ yr.$

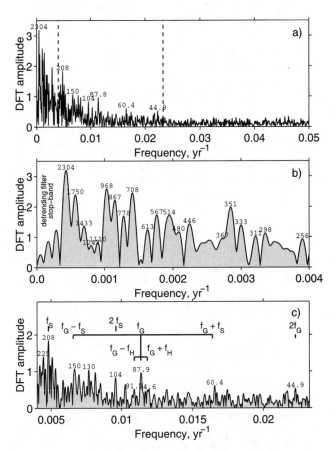

Figure 6. Fourier-spectra of time variations of the decadal data on $\Delta^{14}C$ in the Earth's atmosphere (INTCAL98): a) whole spectrum up to the Nyquist limit; b) long-term spectral band; c) centennial solar cycles band [*Peristykh and Damon*, 2001].

4. RADIOCARBON AND CLIMATE

As briefly discussed above, the literature concerning solar forcing of global climate change goes back to more than thirty years ago. However, there has been renewed interest during the last decade. A recent paper by *Bard et al.* [2000] not only contains many references to the recent literature but reconstructs the total solar irradiance making the explicit assumption that there is a linear relationship between solar irradiance and the magnetic field imbedded in the solar wind. Since the production of cosmogenic isotopes is also heliomagnetically modulated, it then follows that the ^{14}C and [10]Be records provide the fluctuation record of total solar irradiance (TSI) but not the magnitude. They appeal to assumptions made by other authors to obtain a range of scaling factors for the TSI [*Zhang et al.*, 1994; *Lean et al.*, 1995; *Reid*, 1997; *Cliver et al.*, 1998; *Solanki and Fligge*, 1998]. One aspect of solar forcing is that it must be global in extent and they leave this in doubt. However, the mountain glacial record of the Andes and the Southern Alps of New Zealand demonstrates that the Little Ice Age was not limited to the Northern Hemisphere [*Thompson et al.*, 1986; *Broecker and Denton*, 1989]. Climate is complex involving the immense thermohaline currents of the oceans, global circulation of air masses, cyclical phenomena such as El Niño-La Niña quasi-decadal circulation of the oceans, aerosols injected into the stratosphere by volcanos, etc. Even during a climatic event like the Little Ice Age only regionally sensitive areas will be observably cooler. When the signal that we are looking for is global forcing of climate, these other important factors can be noise added to the incomplete record of climate on the same scale especially as we go back in time where there is no historic record and proxy data must be relied upon. Also Fourier analysis reveals various fundamental frequencies, beat frequencies and harmonics (see Figure 6) that together may form a complex pattern.

Thus, it may be necessary to match climatic patterns with patterns of solar activity (Δ^{14}C) as has been done by *Magny* [1993, 1995]. Figure 8 is largely adapted from his work on lake level fluctuations in Jura and Subalpine ranges. The cooling associated with low solar activity results in transgression of lake levels above *ca.* 5‰ Δ^{14}C. Above *ca.* 10‰ cooling is sufficient to produce not only rising levels but at higher altitudes major glacial advances and lowering of timber line occur. The last Little Ice Age occurred in the 14th through 18th century and is so very well known as to be the subject of a book by *Grove* [1988]. Figure 8 shows that cooling was greatest during the Maunder and Spörer Minima split by a temporary warm period and information on alpine glaciation discussed as follows. An earlier little

ice age occurred in the first millennium BC. *Lamb* [1965] refers to it as the early iron age cold epoch. The significance of the *ca.* 2300 yr spacing of these two events will be discussed later.

This epoch had been preceded by gradual cooling following the post-glacial Climatic Optimum that according to thermal studies from the GRIP borehole lasted for about 4000 yr (6000–2000 BC) and reached a temperature at the summit of the Greenland ice sheet that was 2.5°K above the temperature during the beginning of the 20th century [*Dahl-Jensen et al.*, 1998]. According to a thorough study by many authors this event and subsequent slow cooling are the result of orbital parameters COHMAP [*Kutzbach and COHMAP members*, 1988]. Consequently, climate changes resulting from changes in solar irradiance must be superimposed on these long-term events. According to *Lamb* [1965] this "decline from the Climatic Optimum was at first gradual but became abrupt and accompanied catastrophe to some of the human civilizations of the time about 500 BC". The Alpine glaciers advanced during Lamb's early iron age cold epoch. *Schmidt and Gruhle* [1988] refer to this climate epoch in the last millennium BC as the Hallstattzeit. Throughout Central Europe it was cool and wet with effective precipitation, i.e. excess of precipitation over evaporation. As a consequence, lake dwellings were flooded and abandoned, oak trees died out and the wood preserved under expanding bogs, and the marshes were no longer inhabitable. These events from sites in the Friesland province of the Netherlands has been thoroughly documented and the beginning of the water table rise was also precisely dated at

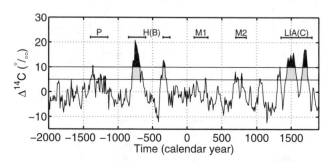

Figure 8. The last 4,000 years of detrended INTCAL98 data compared with high lake level events (with duration marked by segments) in the Jura and Subalpine ranges from [*Magny*, 1993] and alpine glacial advances. P—Pluvus high lake stand, H(B)—the Hallstattzeit little ice age and Bourget high lake level, M1, M2—Petit Maclu high lake stands, LIA(C)—Little Ice Age and Petit Clairvaux high lake level. In this area a Δ^{14}C of ca. 5‰ is sufficient to define a high lake level event. Above *ca.* 10‰ the climate is severe enough to be called a little ice age. See text for discussion and other references.

2730±25 BP (780 BC) [*van Geel et al.*, 1998]. An important conclusion of Schmidt and Gruhle's is "from the correlation already discussed between dendrochronology, climate and ^{14}C fluctuations it ought to be possible to obtain climatic information out of the ^{14}C fluctuations". The correlations in Figure 8 support this conclusion. From this, it may be valid to predict that, except for the warm intervals between the cold events during the Hallstattzeit and Little Ice Age, climate change was comparitively mild from about 200 BC to 1250 AD. *Lamb* [1977] mentions the discovering of Iceland by the Vikings in about A.D. 860 followed soon after by settlement of both Iceland and Greenland. He points out that "the Viking colonies in west and southwest Greenland were able to bury their dead deep in soil that has since been permanently frozen" (p.438). It should be kept in mind that the North Atlantic is a climatically sensitive region. The magnitude of the change in irradiance is not so great as to be unequivocally observed in less sensitive regions and given the complexity of climate the response may not only be less but different as pointed out in [*van Geel et al.*, 1998]. The cold and wet response to the decreased irradiance associated with high Δ^{14}C anomalies in Northern Europe has resulted in decreased monsoonal precipitation in other regions. This has produced serious impact on agricultural communities dependent on the monsoonal rains. The response of decreased irradiance in those areas has been cool and dry rather than cold and wet [*van Geel et al.*, 1998; *Petersen*, 1994; *Davis*, 1994]. In many regions, the response may be much less than forcing by other climatic phenomena.

Dansgaard et al. [1971] in analyzing δ^{18}O data from the Camp Century ice core by the MEM method obtained the significant periods shown in Table 1. Also shown are closely equivalent periods obtained using data from an earlier "*Calibration Issue of Radiocarbon*" [ed. *Stuiver and Kra*, 1986] and the updated data used in this paper. The method of obtaining the age depth relationships was the generally accepted glacier flow theory. The Gleissberg cycle has a generally accepted 88 year period as obtained from the Δ^{14}C data and the historical record of auroral data [*Feynman*, 1988]. If we assume that the δ^{18}O period of 78 yr is the Gleissberg period shortened by under estimating the age vs. depth and multiply each δ^{18}O period by the the ratio 88/78 = 1.13, all periods closely match. After lengthening the Camp Century ice core chronology by the factor 1.13, the resultant δ^{18}O sequence is compared with Δ^{14}C record in Figure 9. *Stuiver et al.* [1997] found a similar match between δ^{18}O and Δ^{14}C for all of the solar minima from the Maunder Minimum to the Oort Minimum.

Black et al. [1999] compare the Δ^{14}C record with the record of the abundance of *G. bulloides* in the varved sediments from Carioca Basin and find a correlation that is, of course, not perfect but surprisingly good. This correlation they point out "suggests that small changes in solar output may influence Atlantic variability on centennial time scales, a possibility supported by climate model experiments that indicate that solar variations can influence Hadley circulation and hence trade wind variability [*Rind and Overpeck*, 1993]".

An important paper by *Verschuren et al.* [2000] shows a strong correlation of effective rainfall and draught in equatorial East Africa with Δ^{14}C during the past millennium. The site of their investigation is the partially submerged Crescent Island Crater basin in Lake Naivasha, Kenya. The authors selected this site because it is unique for equatorial East Africa in combining the climatic sensitivity of a shallow Rift Valley lake with the wind sheltered sedimentation regime of a deep crater lake. During the 11th and 12th Century Medieval Solar Maximum the lowest lake levels occurred. Lake levels were highest during the Maunder and Spörer Minima. Culturely these correlate with the 1st and 2nd ages of prosperity. Their was a lesser rise of lake level during the Wolf Minimum. In between these solar minima were times of lowered lake level and draught. Thus, again, the relationship between low solar activity, high Δ^{14}C and more effective rainfall is repeated.

Above, we have shown reasonably successful examples of the matching of the pattern of Δ^{14}C with the pattern of climate events during the last millennium. Another approach is to look for solar periods in the climate record. Ocassionally, dominant Δ^{14}C are found in climate record.

Table 1. Significant periodicities in Δ^{14}C (tree rings) and δ^{18}O (Camp Century ice core)[a].

Method	Significant Periods, yr	Reference
δ^{18}O (MEM)	78, 181, 2000	*Dansgaard et al.* [1971]
δ^{18}O (rescaled)	88, 205, 2260	*Damon and Sonett* [1991]
Δ^{14}C(MEM, AR=120)	88, 208, 2240	*Damon and Sonett* [1991]
Δ^{14}C (DFT)	88, 207, 2272	*Damon and Sonett* [1991]
Δ^{14}C (DFT)	88, 208, 2310	this paper

[a]δ^{18}O rescaled for the period of the Gleissberg cycle (88 yr).

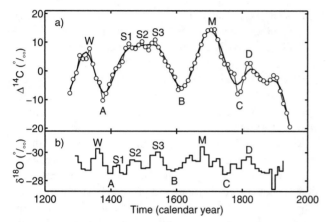

Figure 9. Detrended variations of INTCAL98 $\Delta^{14}C$ (tree rings) compared with $\delta^{18}O$ from Camp Century ice core. D–Dalton Minimum; M–Maunder Minimum; S1, S2, S3–3 peaks of the Spörer minimum; W–Wolf minimum of solar activity. (After *Damon and Jirikowic* [1992])

For example, the ca. 2300 year period has been reported in the $\delta^{18}O$ record of upper water foraminifera in at least three marine cores [*Pestiaux et al.*, 1988]. Also, evidence for a recurrent series of droughts have been observed in lake sediment cores from the Yucatan Peninsula in Mexico [*Hodell et al.*, 2001]. Time series analysis of these draughts shows a dominant periodicity of 208 yr.

Lingenfelter and Romaty [1970] estimated the change in radiocarbon production rate during the 19th solar cycle when it went from 2.42 ± 0.48 at/(cm² · sec) at minimum ($S = 4.4$) to 1.93 ± 0.39 at/(cm² · sec) at maximum solar activity ($S = 190.2$). From this one can approximate a linear relationship between production rate Q_{14C} and Wolf sunspot number S as follows:

$$Q_{14C} = 2.434 - 0.00264 \cdot S, \text{ at }/(\text{cm}^2 \cdot \text{sec}) \qquad (7)$$

Damon [1977] pointed out that this relationstxhip could explain only *ca.* half of the rise in $\Delta^{14}C$ during the Maunder Minimum. *Stuiver and Quay* [1980] attempted to resolve this problem by assuming that solar modulation of Q_{14C} continued below $S = 0$. This also required the *Aa* index to approach zero with the magnetic field embedded in the solar wind having "no measurable effect" on Earth's magnetic field. However, as previously pointed out, there is general agreement that an irradiance change accompanies the changes in the solar wind that results in changes in Q_{14C} as observed during the Maunder Minimum. Cooling as a result of decrease in irradiance can also effect $\Delta^{14}C$ through the carbon cycle. The cooling of the atmosphere and mixed layer of the ocean would result in a decreased thermal gradient between the mixed layer and the deeper ocean. As a

consequence, there would be a decreased transfer rate of CO_2 to the deeper ocean (Soret effect) enhancing the increase in $\Delta^{14}C$ resulting from increased production of ^{14}C.

In climate change positive feedbacks must be taken into consideration. These feedbacks are a function of atmospheric temperature. For example, *Hansen et al.* [1984] estimate that the positive feedbacks during the ice age at 18 kY would be about twice that at present. Referring to the detrended data (Figure 5), one might be tempted to ascribe the greater amplitude of the peaks prior to 7200 BC compared to the Maunder and Spörer Minima to greater Q_{14C}. However, this was during the transition from ice age to the climatic optimum when prevailing temperatures were still less than during the Little Ice Age. Consequently, positive feedbacks were greater and all or part of the difference in amplitude could be explained by the difference in climate rather than a difference in solar activity. A slow increase in the Earth's dipole moment must also be taken into consideration. It should be pointed out that these peaks are not acompanied by significant negative $\delta^{18}O$ anomalies so the increased amplitudes can not be explained by a decrease in thermohaline circulation [*Stuiver et al.*, 1995].

In the paper by *Bond et al.* [1997], previously mentioned, despite the relatively stable Holocene climate, they found evidence for ice rafting events in sedimentary cores as far south as Britain. According to these authors "those drift ice cycles comprise part of an enigmatic, at best quasi-periodic '1500-year' cycle". However, there are a number of periods in the $\Delta^{14}C$ spectrum that are on or close to millennial scale (Figure 6b). These periodicities have been combined by harmonic analysis in Figure 10. We have included a period of 1433 yr, even though it has a very low amplitude, because it is closest to "1500-year". One should not make the mistake of ascribing an independent existence to these millennial periodicities. They are simply the periods required by Fourier analysis to reconstruct the varying amplitudes of the dominant century-scale fluctuations of $\Delta^{14}C$ in Figure 10.

In a subsequent paper, *Bond et al.* [2001] compare the nuclides ^{10}Be and ^{14}C with the ice rafting events and state: "The correlation coefficients between the nuclide production records and both the individual and the stacked records range from $r = 0.44$ to 0.56". There are similarities between the stacked records and our Figure 10, that suggests the millennial variance of the century scale solar cycles may account for the ice rafting events. Time intervals of high amplitude century-scale solar cycles would correspond to cool periods.

5. CONCLUDING REMARKS

In order to reconstruct climate at any time in the past, we need the time dimension. Radiocarbon dating combined

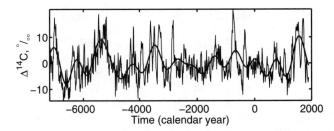

Figure 10. Detrended variations of INTCAL98 $\Delta^{14}C$ approximated by polyharmonic model (harmonic analysis) with set of basic periods {2304, 1750, 1433, 968, 867, 778, 708} yr. Note that the polyharmonic curve follows the amplitude variations of the century scale peaks without phase shift.

with dendrochronology has played an important role in providing this time record. The fact that the $^{14}C/^{12}C$ ratio in atmospheric CO_2 is not constant requires calibration of the ^{14}C time scale to convert Libby years into calendar years. The conversion also provides data to calculate $\Delta^{14}C$ that contains information concerning the changing production rate of ^{14}C as a result of the modulation of primarily galactic cosmic by solar activity and also accompanying changes in solar irradiance that affect the Earth's climate. The variable irradiance component amplifies the $\Delta^{14}C$ changes in production rate.

As can be seen from Figure 6, Fourier analysis requires many periods to reproduce the $\Delta^{14}C$ record. Some of these like the Schwabe (11-yr), Hale (22-yr), Gleissberg (88-yr) and the Suess (208-yr) cycle have an independent existence and others particularly the long periods may only be required to reproduce the variance of the shorter periods as in Figure 10.

In order to demonstrate the relationship of $\Delta^{14}C$ to climate we have relied on the pattern of $\Delta^{14}C$ during the last four millennia and its correlation with climate as shown in Figure 8. We have found in the literature similar matches of climate and $\Delta^{14}C$ from the equator to northern Greenland. Given the complexity of both climate and solar activity finding solar periods in a climate record can be less convincing particularly if the periods are only approximately the same.

As suggested by *Schmidt and Gruhle* [1988] $\Delta^{14}C$ may serve as a means of obtaining information about climate. For example, a large $\Delta^{14}C$ peak at *ca.* 800 BC, according to our argument, represents a significant decrease in solar irradiance. However, the response from region to region may be very different. For example, in one region it may be accompanied by more effective rainfall such as in northern Europe. In another region it may possibly result in draught resulting from decreased water vapor in the clouds that produce seasonal storms. The disparate responses leads to questions and the questions lead to further research.

Acknowledgments. We are grateful to Professors Julia Cole, Oven Davis, Malcolm Hughes, Austin Long and Jonathan Overpeck for their help in our literature search and to Raimund Muscheler for a helpful review. This work was supported by National Science Foundation Grant ATM-9819228.

REFERENCES

Bard, E., G. Raisbeck, F. Yiou, and J. Jouzel, Solar irradiance during the last 1200 years based on cosmogenic nuclides, *Tellus B, 52*, 985-992, 2000.

Black, D. E., L. C. Peterson, J. T. Overpeck, A. Kaplan, M. N. Evans, and M. Kashgarian, Eight centuries of North Atlantic Ocean atmosphere variability, *Science, 286*, 1709-1713, 1999.

Bond, G., W. Showers, M. Cheseby, R. Lotti, P. Almasi, P. deMenocal, P. Priore, H. Cullen, I. Hajdas, and G. Bonani, A pervasive millennial-scale cycle in North Atlantic Holocene and glacial climates, *Science, 278*, 1257-1266, 1997.

Bond, G., B. Kromer, J. Beer, R. Muscheler, M. Evans, W. Showers, S. Hoffmann, R. Lotti-Bond, I. Hajdas, and G. Bonani, Persistent solar influence on North Atlantic climate during the Holocene, *Science, 294*, 2130-2136, 2001.

Braziunas, T. F., I. Y. Fung, and M. Stuiver, The preindustrial atmospheric $^{14}CO_2$ latitudinal gradient as related to exchanges among atmospheric, oceanic, and terrestrial reservoirs, *Global Biogeochem. Cycles, 9*, 565-584, 1995.

Broecker, W. S., and G. H. Denton, The role of ocean-atmosphere reorganizations in glacial cycles, *Geochim. Cosmochim. Acta, 53*, 2465-2501, 1989.

Cliver, E. W., V. Boriakoff, and J. Feynman, Solar variability and climate change: Geomagnetic aa index and global surface temperature, *Geophys. Res. Lett., 25*, 1035-1038, 1998.

Craig, H., The natural distribution of radiocarbon and the exchange time of carbon dioxide between atmosphere and sea, *Tellus, 9*, 1-17, 1957.

Dahl-Jensen, D., K. Mosegaard, N. Gundestrup, G. D. Clow, S. J. Johnsen, A. W. Hansen, and N. Balling, Past temperatures directly from the Greenland Ice Sheet, *Science, 282*, 268-271, 1998.

Damon, P. E., Radiocarbon and climate (A comment on a paper by H.Suess), *Meteorological Monographs, 8*, 151-154, 1968.

Damon, P. E., Climatic versus magnetic perturbation of the atmospheric C14 reservoir, in *Radiocarbon Variations and Absolute Chronology: Proc. Twelfth Nobel Symp.*, edited by I.U.Olsson, pp. 571-593, Almqvist & Wiksell, Stockholm, 1970.

Damon, P. E., Solar induced variations of energetic particles at one AU, in *The Solar Output and its Variation*, edited by O.R. White, pp. 429-448, Colorado Associated Univ. Press, Boulder, Co., 1977.

Damon, P. E., and J. L. Jirikowic, Solar forcing of global climate change ?, in *Radiocarbon After Four Decades: An Interdisciplinary Perspective*, edited by R.E. Taylor, A.Long, and R.S.Kra, pp. 117-129, Springer, New York, 1992.

Damon, P. E., and A. N. Peristykh, Solar cycle length and 20th century Northern Hemisphere warming: Revisited, *Geophys. Res. Lett., 26*, 2469-2472, 1999.

Damon, P. E., and A. N. Peristykh, Radiocarbon calibration and application to geophysics, solar physics, and astrophysics, *Radiocarbon, 42*, 137-150, 2000.

Damon, P. E., and C. P. Sonett, Solar and terrestrial components of the atmospheric ^{14}C variation spectrum, in *The Sun in Time*, edited by C.P.Sonett, M.S.Giampapa, and M.S.Matthews, pp. 360-388, Univ. of Arizona Press, Tucson, 1991.

Damon, P. E., and R. S. Sternberg, Global production and decay of radiocarbon, *Radiocarbon, 31*, 697-703, 1989.

Damon, P. E., and A. Long, and J. J. Sigalove, Arizona radiocarbon dates IV, *Radiocarbon, 5*, 283-301, 1963.

Damon, P. E., R. S. Sternberg, and C. J. Radnell, Modeling of atmospheric radiocarbon fluctuations for the past three centuries, *Radiocarbon, 25*, 249-258, 1983.

Dansgaard, W., S. J. Johnsen, H. B. Clausen, and C. C. Langway Jr., Climatic record revealed by the Camp Century ice core, in *The Late Cenozoic Glacial Ages*, edited by K. K. Turekian, pp. 37-56, Yale University Press, New Haven; London, 1971.

Davis, O. K., The correlation of summer precipitation in the Southwestern U.S.A. with isotopic records of solar activity during the Medieval Warm Period, *Climatic Change, 26*, 271-287, 1994.

de Vries, H., Variation in concentration of radiocarbon with time and location on Earth, *Proc. Koninkl. Ned. Akad. Wetenschap., ser.B, 61*, 94-102, 1958.

Dergachev, V. A., and V. M. Ostryakov, The rate of radiocarbon production during the last ~8 millennia using available experimental data on ^{14}C concentration in the known age samples, *Preprint No.658*, A.F.Ioffe Institute of Physics & Technology, USSR Academy of Sciences, Leningrad, 1980.

Eddy, J. A., The Maunder Minimum, *Science, 192*, 1189-1202, 1976.

Eddy, J. A., Climate and the changing Sun, *Climatic Change, 1*, 173-190, 1977.

Ekdahl, C. A., and C. D. Keeling, Atmospheric carbon dioxide and radiocarbon in the natural carbon cycle: I. Quantitative deductions from records at Mauna Loa Observatory and at the South Pole, in *Carbon and the biosphere (Proc. 24th Brookhaven Symp. Biology, Upton, NY, 1972)*, edited by G. Woodwell and E. Pecan, pp. 51-85, U.S. Atomic Energy Commission, Springfield, Va., 1973.

Feynman, J., Solar, geomagnetic and auroral variations observed in historical data, in *Secular Solar and Geomagnetic Variations in the Last 10,000 Years*, edited by F.R.Stephenson and A.W. Wolfendale, pp. 329-340, Kluwer Academic Publishers, Dordrecht, 1988.

Friis-Christensen, E., and K. Lassen, Length of the solar cycle: An indicator of solar activity closely associated with climate, *Science, 254*, 698-700, 1991.

Gleissberg, W., A table of secular variations of the solar cycle, *Terr. Magn. & Atm. Electr., 49*, 243-244, 1944.

Grey, D. C., and P. E. Damon, Sunspots and radiocarbon dating in the Middle Ages, in *Scientific Methods in Medieval Archaeology*, edited by R.Berger, pp. 167-182, Univ. of California Press, Berkeley, CA, 1970.

Grove, J. M., *The Little Ice Age*, Methuen, London; New York, 1988.

Hansen, J. E., A. Lacis, D. Rind, G. Russell, P. Stone, I. Fung, R. Ruedy, and J. Lerner, Climate sensitivity: Analysis of feedback mechanisms, in *Climate processes and climate sensitivity*, edited by J. E. Hansen and T. Takahashi, no. 29 in Geophysical monograph, pp. 130-163, Lamont-Doherty Geophysical Observatory, Palisades, N.Y., American Geophysical Union, Washington, D.C., 1984.

Hodell, D. A., M. Brenner, J. H. Curtis, and T. Guilderson, Solar forcing of drought frequency in the Maya lowlands, *Science, 292*, 1367-1370, 2001.

Houtermans, J. C., H. E. Suess, and H. Oeschger, Reservoir models and production rate variations of natural radiocarbon, *J. Geophys. Res., 78*, 1897-1908, 1973.

Johnsen, S. J., H. B. Clausen, W. Dansgaard, K. Fuhrer, N. Gundestrup, C. U. Hammer, P. Iversen, J. Jouzel, B. Stauffer, and J. P. Steffensen, Irregular glacial interstadials recorded in a new Greenland ice core, *Nature, 359*, 311-313, 1992.

Kutzbach, J. E., and COHMAP members, Climatic changes of the last 18,000 years: observations and model simulations, *Science, 241*, 1043-1052, 1988.

Lal, D., and B. Peters, Cosmic ray produced radioactivity on the Earth, in *Handbuch der Physik (Encyclopaedia of Physics)*, edited by K.Sitte, vol. 46/2, pp. 551-612, Springer-Verlag, New York, 1967.

Lamb, H. H., The Early Medieval Warm Epoch and its sequel, *Palaeogeogr., Palaeoclimatol., Palaeoecol., 1*, 13-37, 1965.

Lamb, H. H., *Climate: Present, Past and Future*, vol. 2. Climatic history and the future, Methuen; Barnes & Noble, London; New York, 1977.

Lazear, G., P. E. Damon, and R. S. Sternberg, The concept of DC gain in modeling secular variations in atmospheric ^{14}C, *Radiocarbon, 22*, 318-327, 1980.

Lean, J., J. Beer, and R. Bradley, Reconstruction of solar irradiance since 1610: implications for climate change, *Geophys. Res. Lett., 22*, 3195-3198, 1995.

Libby, W. F., *Radiocarbon dating*, 2 ed., University of Chicago Press, 1955.

Lingenfelter, R. E., and R. Ramaty, Astrophysical and geophysical variation in C14 production, in *Radiocarbon Variations and Absolute Chronology: Proc. Twelfth Nobel Symp.*, edited by I.U.Olsson, pp. 513-535, Almqvist & Wiksell, Stockholm, 1970.

Magny, M., Solar influences on Holocene climatic changes illustrated by correlations between past lake-level fluctuations and the atmospheric ^{14}C record, *Quat. Res., 40*, 1-9, 1993.

Magny, M., *Une Histoire du Climat*, Collection des Hesperides, Editions Errance, Paris, 1995.

Mann, W. B., W. F. Marlow, and E. E. Hughes, The half-life of carbon-14, *Int. J. Applied Radiation Isotopes, 11*, 57-67, 1961.

Muscheler, R., Beer, J. and Kubik, W., Long-Term Solar Variability and Climate Change Based on Radionuclide Data From Ice Cores, *this volume*

O'Brien, K., Secular variations in the production of cosmogenic isotopes in the Earth's atmosphere, *J. Geophys. Res. (Space Phys.), 84*, 423-431, 1979.

Oeschger, H., U. Siegenthaler, U. Schotterer, and A. Gugelmann, A box diffusion model to study the carbon dioxide exchange in nature, *Tellus, 27*, 168-192, 1975.

Olsson, I. U. (Ed.), *Radiocarbon Variations and Absolute Chronology: Proc. Twelfth Nobel Symp.*, Almqvist & Wiksell, Stockholm, 1970.

Olsson, I. U., I. Karlén, A. H. Turnbull, and N. J. D. Prosser, A determination of the half-life of ^{14}C with a proportional counter, *Arkiv für Fysik, 22*, 237-255, 1962.

Peristykh, A. N., The program for carbon cycle simulation with allowance for vertical turbulent diffusion of CO_2 in the atmosphere, *Preprint No.1243*, A.F.Ioffe Institute of Physics & Technology, USSR Academy of Sciences, Leningrad, 1988.

Peristykh, A. N., and P. E. Damon, Persistence of the Gleissberg (88-yr) solar cycle over the last 10,000 years: Evidence from cosmogenic isotopes, In: *Conf. Int. Solar Cycle Studies 2001 "Solar variability, Climate and Space Weather", Longmont, June 13–16, 2001*, Fox, P. (ed.)., ICSU/SCOSTEP, p.65-66, 2001.

Pestiaux, P., I. van der Mersch, A. Berger, and J. C. Duplessy, Paleoclimatic variability at frequencies ranging from 1 cycle per 10000 years to 1 cycle per 1000 years: evidence for nonlinear behaviour of the climate system, *Climatic Change, 12*, 9-37, 1988.

Petersen, K. L., A warm and wet little climatic optimum and a cold and dry Little Ice Age in the Southern Rocky Mountains, U.S.A., *Climatic Change, 26*, 243-269, 1994.

Ralph, E. K., and R. Stuckenrath, Carbon-14 measurements of known age samples, *Nature, 188*, 185-187, 1960.

Reid, G. C., Solar forcing of global climate change since the mid-17th century, *Climatic Change, 37*, 391-405, 1997.

Rind, D., and J. Overpeck, Hypothesized causes of decade-to-century-scale climate variability: Climate model results, *Quart. Sci. Rev., 12*, 357-374, 1993.

Schmidt, B., and W. Gruhle, Klima, Radiokohlen-stoffgehalt und Dendrochronologie (Climate, radio-carbon contents and dendrochronology), *Naturwissenschaftliche Rundschau, 41*, 177-182, 1988.

Schove, D. J., The sunspot cycle, 649-BC to AD-2000, *J. Geophys. Res., 60*, 127-146, 1955.

Siegenthaler, U., M. Heimann, and H. Oeschger, ^{14}C variations caused by changes in the global carbon cycle, *Radiocarbon, 22*, 177-191, 1980.

Solanki, S. K., and M. Fligge, Solar irradiance since 1874 revisited, *Geophys. Res. Lett., 25*, 341-344, 1998.

Sonett, C. P., and S. A. Finney, The spectrum of radiocarbon, in *The Earth's Climate and Variability of the Sun Over Recent Millennia: Geophysical, Astronomical and Archaeological Aspects*, edited by J.-C.Pecker and S.K.Runcorn, pp. 15-27, Roy. Soc., London, 1990.

Stuiver, M., and T. F. Braziunas, Anthropogenic and solar components of hemispheric ^{14}C, *Geophys. Res. Lett., 25*, 329-332, 1998.

Stuiver, M., and R. Kra, Calibration Issue, *Radiocarbon, 28*, 805-1030, 1986.

Stuiver, M., and H. A. Polach, Reporting of ^{14}C data (discussion), *Radiocarbon, 19*, 355-363, 1977.

Stuiver, M., and P. D. Quay, Changes in atmospheric carbon-14 attributed to a variable Sun, *Science, 207*, 11-19, 1980.

Stuiver, M., and J. van der Plicht, INTCAL98: Calibration Issue, *Radiocarbon, 40*, 1041-1164, 1998.

Stuiver, M., P. M. Grootes, and T. F. Braziunas, The GISP2 δ^{18}O climate record of the past 16,500 years and the role of the Sun, ocean, and volcanoes, *Quat. Res., 44*, 341-354, 1995.

Stuiver, M., T. F. Braziunas, P. M. Grootes, and G. A. Zielinski, Is there evidence for solar forcing of climate in the GISP2 oxygen isotope record?, *Quat. Res., 48*, 259-266, 1997.

Stuiver, M., P. J. Reimer, E. Bard, J. W. Beck, G. S. Burr, K. A. Hughen, B. Kromer, G. McCormac, J. van der Plicht, and M. Spurk, INTCAL98 radiocarbon age calibration, 24,000-0 cal BP, *Radiocarbon, 40*, 1041-1083, 1998.

Suess, H. E., Secular changes in the concentration of atmospheric radiocarbon, in *Problems Related to Interplanetary Matter (Proc. Informal Conf., Highland Park, Illinois, June 20–22, 1960)*, no. 33 in Nucl. Sci. Ser., pp. 90-95, NAS-NRC, Washington, D.C., 1961.

Suess, H. E., Climatic changes, solar activity, and the cosmic-ray production rate of natural radiocarbon, *Meteorological Monographs, 8*, 146-150, 1968.

Thompson, L. G., E. Mosley-Thompson, W. Dansgaard, and P. M. Grootes, The Little Ice Age as recorded in the stratigraphy of the tropical Quelccaya ice cap, *Science, 234*, 361-364, 1986.

van Geel, B., J. van der Plicht, M. R. Kilian, E. R. Klaver, J. H. M. Kouwenberg, H. Renssen, I. Reynaud-Farrera, and H. T. Waterbolk, The sharp rise of Δ^{14}C *ca.* 800 cal BC: Possible causes, related climatic teleconnections and the impact on human environments, *Radiocarbon, 40*, 535-550, 1998.

Verschuren, D., K. R. Laird, and B. F. Cumming, Rainfall and drought in equatorial east Africa during the past 1,100 years, *Nature, 403*, 410-414, 2000.

Watt, D. E., D. Ramsden, and H. W. Wilson, The half-life of carbon-14, *Int. J. Appl. Radiation Isotopes, 11*, 68-74, 1961.

Zhang, Q., W. H. Soon, S. L. Baliunas, G. W. Lockwood, B. A. Skiff, and R. R. Radick, A method of determining possible brightness variations of the Sun in past centuries from observations of solar-type stars, *Astrophys. J., 427*, L111-L114, 1994.

P.E.Damon, Department of Geosciences, Gould-Simpson Bldg., University of Arizona, Tucson, AZ 85721, U.S.A. (e-mail: damon@geo.arizona.edu)

A.N.Peristykh, Department of Geosciences, Gould-Simpson Bldg., University of Arizona, Tucson, AZ 85721, U.S.A. (e-mail: peristy@geo.arizona.edu)

Detecting the 11-year Solar Cycle in the Surface Temperature Field

G. R. North

Department of Atmospheric Sciences Texas A&M University, College Station, Texas

Q. Wu

Center for Ocean, Land, Atmosphere Studies, Calverton, Maryland

M. J. Stevens

Climate and Global Dynamics Division, National Center for Atmospheric Research, Boulder, Colorado

This chapter reviews work conducted by the authors over the last few years attempting to detect the climatic response of the solar 11-year cycle signal in the Earth's temperature field. After explaining the multivariate regression technique used we present our results, namely that there is a faint response to the solar cycle and its amplitude is roughly what we would expect (few hundredths of a degree C) based upon simple energy balance model estimates. The detection is at about the 25% significance level.

1. INTRODUCTION

Speculation that near surface climate varies with an 11 year period corresponding to the 11 year cycle of sunspots has intrigued weather and climate forecasters for many years. (See the very lucid discussion of the history of this problem by *Burroughs* [1992]) These early speculations and attempts at correlation have not met with much support in the scientific community. Significance of the correlation between climate and solar cycle ranges from virtual certainty in the upper atmosphere to nearly null at the surface. There is also abundant speculation on long-term effects of presumed variation of the solar irradiance, which might be implicated in climate change on the centennial time scale. We focus in this chapter on the near-surface thermal response of the system to the 11 year cycle. One problem is

that usually the time series of observations have been too short to reveal the (presumed) very weak 11 year signal in the near-surface weather variables. History seems to show that as soon as a relationship neared community acceptance, it was dispelled by disagreeable data collected in the next cycle.

The signals near the surface are so weak that it takes many years of data to detect a significant response at the 11 year period. There are data (e.g., National Center for Environmental Prediction Reanalysis) for only about 50 years for variables above the ground—hence, less than four cycles, giving very little to work with. On the other hand, there are about 100 years of surface temperature data at quite a few stations well distributed around the world. Moreover, the surface temperature field is among the easier weather variables to model; such modeling efforts can in principle provide some hint as to how the signal's space-time signature should look. Basically, if we know what signature to look for and we trust it, we can attempt to detect the signal using optimal techniques, estimate its strength, and test for its statistical significance. These two factors, promisingly long record lengths,

Solar Variability and its Effects on Climate
Geophysical Monograph 141
10.1029/141GM18

and hopes for reliable model-based signatures, lead us to select the surface temperature field for a detection variable. In addition to these favorable circumstances, the surface temperature is relevant to human concerns and the successful detection of it in the data stream constitutes a test of sorts of our current climate models.

Estimates of the response of the global average temperature to a sinusoidal variation of the "solar constant" (referred to conventionally now as the solar irradiance) with an amplitude of roughly 0.1% yield only a response amplitude of a few hundredths of a degree C. Figure 1 shows the response amplitude (upper panel) in degrees C and the phase lag in months (lower panel) as calculated from an Energy Balance Climate Model (EBCM) [*Stevens and North*, 1996; *Stevens*, 1997]. Hence, as far as the surface temperature is concerned, there is very little practical con-

Amplitude of Response (˚C)

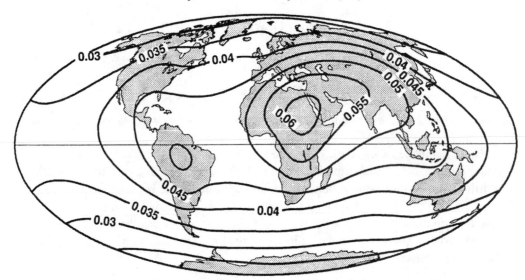

Phase Lag of Response Behind Forcing (months)

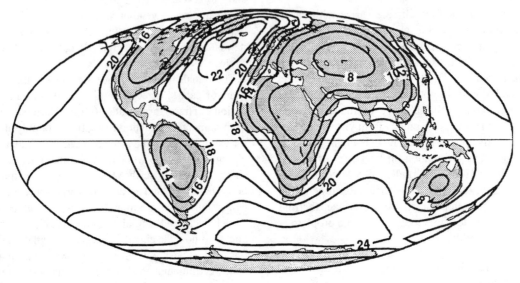

Figure 1. Upper Panel: Map of the surface thermal response (degrees C) to a sinusoidal heating corresponding to a 0.1% amplitude of solar irradiance. Lower Panel: Map of the phase lag behind the forcing in months.

sequence of such a benign "climate change". On the other hand, detection of this signal and establishing its amplitude would have practical consequences for the community of climate modelers. This is because we have essentially no examples of known natural forcings on the system at this decadal frequency. Testing model response at this frequency could provide important empirical information about climate feedback mechanisms, which could be different at these much lower (than annual) frequencies.

Several chapters in this volume report measurements of the total solar irradiance over the last few 11 year cycles and this need not be repeated here. Not all of the solar radiation survives to the planetary surface; some is reflected back to space by clouds, very hard UV radiation is removed in the middle atmosphere by ionization and molecular dissociation, softer UV is removed in the stratosphere by dissociation of ozone, and near infrared solar radiation is significantly absorbed by water vapor in the upper troposphere. The details of this vertical distribution of heating and chemical transformation are still being debated in the literature, but this is a problem that in principle will be solved soon (see *Haigh*, this volume). Notwithstanding this ambiguity in our present understanding of the forcing being applied by the sun's 11 year cycle, we in this study will take the full amount of the variation of the solar constant to be impinging on the tropopause from above. To the extent that some of the radiation is removed or augmented by downward radiation due to changes in stratospheric composition can be interpreted after our results are provided by modifying the input amplitude. Our forcing of the climate system is the modulation of the total solar irradiance as indicated by satellite measurements over the last two decades and extrapolated back a century through its relationship to sunspot number. While this is the most direct interpretation of our input, it would be indistinguishable from other forcings that might be correlated with the sunspot numbers (e.g., high energy particles and their effects on climate through cloud modification, etc.; see *Tinsley and Yu*, this volume).

In this study climate signals are the responses of the surface temperature field to the external forcings. We know of at least four such forcings that are likely to be important in examining a century-long record. These include the greenhouse gas (increase) signal (hereafter denoted G); the volcanic dust veil shading of sunlight (V); anthropogenic aerosols (A), and finally, the smallest of the four, changes in the solar cycle especially at the 11 year period (S). We model the four responses in the surface temperature (signals) using an energy balance climate model developed by *Stevens* [1997] and discussed in *Stevens and North* [1996].

Fortunately, our study is preceded by twenty-five years of intensive research and development of climate models of varying levels of complexity. To date we have about a dozen modeling groups who have developed coupled atmosphere/ocean general circulation models (AOGCMs) that can be used in our application. Most of these have made control runs (no external forcing) for hundreds of years – enough to establish the variance statistics that are needed in serious signal processing studies focused on the band around 11 years. These groups have generously made these model-derived statistics available for investigators at large to use in other studies through the Intergovernmental Panel on Climate Change (IPCC).

In this paper we employ a rather crude (but probably commensurate with other uncertainties in the problem) estimate of the signal waveform in space-time and our knowledge of the "background noise" to estimate the actual amplitude of the solar signal in the real surface temperature data stream. We start by describing our statistical approach and our philosophy of how the data are to be handled. Next comes a description of our model-derived forcings for all the influences (all have to be detected simultaneously because of the possibility of interference leading to mistaken identities). At last we present our results and some discussion of their significance. Finally, we look to the future, since at this writing there is still insufficient performance in the detection scheme to guarantee detection of the solar response signal.

Most of the work reported in this chapter is based on the papers by *Stevens and North* [1996] (hereafter SN96), *North and Stevens* [1998] (hereafter NS98); and *North and Wu* [2001] (hereafter NW01). In some cases we present slight extensions of the work reported in those papers, but the details of method and approach are reported therein. Note, however, that the method has evolved slightly over the sequence of these papers and the latest (NW01) represents our most reliable product. We remark here that the amplitude of G estimated in NW01 is only about 75% of the expected amplitude of that signal. We have found that had we used the most recent value of the heating due to the greenhouse gas signal (3.7 W m^{-2} instead of 4.3 W m^{-2}) we would have obtained a value nearer 100% in accordance with the findings reported in the IPCC 2001 Report. This correction only applies to G, not to A, V, or S.

2. REGRESSION TECHNIQUE

The main assumption in our approach is that the signals are small and can be thought of as linearly superimposed on a background of natural variability. We also assume that the presence of one signal does not affect the other signals. We compute the signals in space-time by a technique to be described later – these space-time shapes are taken as given. The goal of the process is to estimate the strengths of these waveforms in the actual data stream. We can express this as

$$T(\mathbf{r}_i, t_j) = \sum_s \alpha_s S_s(\mathbf{r}_i, t_j) + N(\mathbf{r}_i, t_j) \qquad (1)$$

where $T(\mathbf{r}_i, t_j)$ is the surface temperature evaluated at the space-time point indicated, the $S_s(\mathbf{r}_i, t_j)$ are the signal waveforms for s = G, V, A, and S; $N(\mathbf{r}_i, t_j)$ is the natural variability background, and the coefficients α_s are to be estimated in the regression process. In the regression process cross-correlation matrices in space and time are needed. Since the data stream is short (only 100 years) we elect to compute model covariance matrices from long control runs of AOGCMs as explained below. In our analysis we use 36 spatial locations chosen by a prescription described in the next section. We then have 100 entries (annual averages) of a 36 component vector in our space. A 1000 yr control run would provide 10 realizations of the time series on this interval. We can also make use of the stationarity of the time series to augment the statistical sample. Covariance matrices of this size (3600 by 3600) are hard to manage and we seek a means of reducing the dimension. We do this by diagonalizing the covariance matrix and retaining only a limited dimension of the space depending on how much of the signal in question (S here) is projected on that subspace. The eigenvectors of the covariance matrix are the space-time Empirical Orthogonal Functions (EOFs) and the eigenvalues are the variances associated with the corresponding space-time EOF pattern. In the results section we will show the sensitivity of the amplitude estimates to the truncation level of the EOF expansion.

Figure 2 shows a schematic diagram of how our process works. We use variability estimates from control runs of AOGCMs and signals from the Stevens Energy Balance Climate Model (EBCM) as shown in Figure 1. These ingredients are used to design the regression process. The regression process may be thought of as a filter through which the actual data stream is passed. The output is the signal amplitudes for G, A, V and S. In our procedure we use the variability from many different AOGCMs and go through the entire process one at a time. In this way we can examine the final results for each of the variability models in parallel.

A final cautionary note is that we must estimate the signal amplitudes of G, A, V and S simultaneously. In some cases pairs of these amplitude estimates are correlated because they are nearly parallel or anti-parallel in the subspace. This

is especially clear in the case of G and A, since both exhibit a slow monotonic change over the last century. Also V has significant power at the 10 year band because of a (-n accidental) sequence of eruptions in the latter half of the century that are roughly spaced at 10 years.

2.1. Station Locations

In our data selection procedure we chose sites where we had at least 100 years of reliable surface temperature data in a 10° by 10° box. In addition we required that there be at least four such records in each of the enclosed 5° by 5° boxes. We then averaged these data together to form a time series of annual averages comprising 100 times 36 entries. Figure 3 shows a map of the 36 locations that met our criteria. Note in addition that the sites are rather well distributed over the planet with representation over both land and ocean. The number of stations in polar regions is nil because we do not believe models are reliable in those regions and data are also less likely to be reliable. In our results section we will show estimates of the amplitudes using only subsets (tropical only, etc.) of the 36 stations to demonstrate the robustness of the procedure.

It is part of our analysis philosophy to evaluate the data stream and all model quantities at these very same locations and as annual averages. This eliminates an intermediate step taken by some investigators in detection work to "smooth" the data by expressing it as a continuous field and then comparing to correspondingly smoothed modeled output. In our opinion there is no need for such a spatial smoothing step.

2.2. Climate Forcing Inputs

Next we need to prescribe the signal waveforms that are used in the regression process. Two issues arise: 1) Can we make the linear superposition assumption? The answer appears to be affirmative based upon AOGCM experiments by *Ramaswamy and Chen* [1997], and *North et al.* [1992]. 2) How detailed must the model be to get acceptable patterns? Our choice is outlined next. We used the EBCM developed by *Stevens* [1997] and described in SN96 and NS98 to compute all the signal waveforms. We were satisfied that these waveforms were essentially identical to those derived from several-member ensemble averages of the currently fashionable AOGCMs. Some comparisons of EBCM signal results with those from AOGCMs are shown in NW01. An advantage of the EBCM is that signal waveforms are purely deterministic with no natural variability corruption. If a signal is derived from running an AOGCM with forcing, there will obviously be large amounts of

Construction of the Optimal Filter

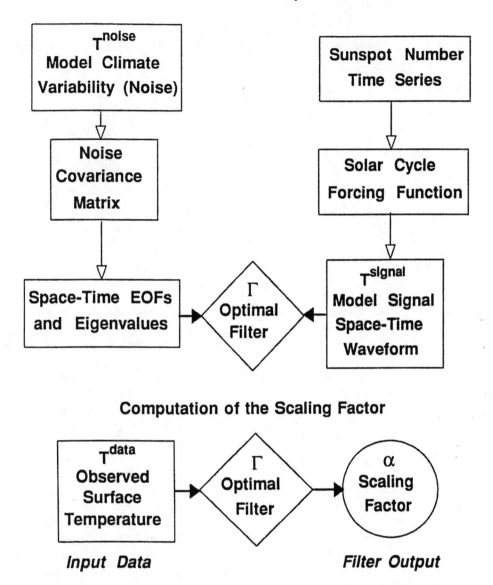

Figure 2. Schematic diagram of the procedure used in estimating the amplitudes of the four signals.

"noise" superimposed. In the case of solar forcing the response is so small it would take tens of ensemble members to get a clean waveform to use.

The greenhouse signal input data can be taken from the carbon dioxide (as well as that of the other GH gases) emissions data and translated into a heating rate by the usual procedures [*IPCC*, 1996]. In fact, we used a forcing function generated from an AGCM by *Kiehl and Briegleb* [1993]. This pattern of forcing has a latitude and a slight longitude dependence mainly due to the distribution of water vapor content in the present atmosphere. We did not change the geographical distribution of the forcing but only allowed the time dependence to modulate it through the Twentieth Century.

2.3. Natural Variability

We have derived our natural variability statistics from a wide variety of AOGCMS as well as a noise-forced EBCM developed by *Stevens* [1997]. In these cases we were able to

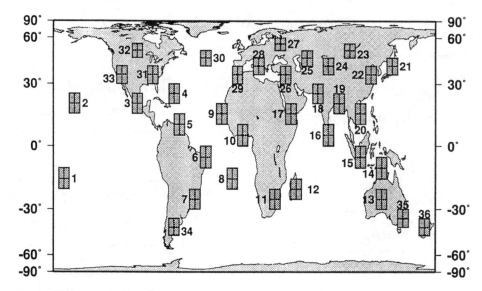

Figure 3. Map showing the 36 sites used in our study. Each site has four associated 5° by 5° boxes with a 100 year time series of surface temperatures in each. The four are averaged together to form the single time series for that box.

use 1000 year control runs and with the EBCM 10,000 years, this latter is important because it allows us to check the sampling error due to finite record length. In each of these cases we computed estimates of the space-time covariance matrices on the 100 year interval (note that this differs slightly from using Fourier analysis on an infinite interval).

The AOGCMs used in this study were as follows: An R-15 AGCM coupled to the mixed layer ocean model from the Geophysical Fluid Dynamics Laboratory (GFDLml); An R-15 AGCM coupled to a fully three dimensional ocean model (GFDLc); Our own EBCM forced by white noise (SN96); The Max Planck Institute ECHAM3/LSG AGCM coupled to a simplified ocean model; The HADCM2 Hadley Centre AGCM coupled to a fully three dimensional ocean model.

3. RESULTS

We proceeded to obtain the space-time EOFs corresponding to the natural variability evidenced by the AOGCMs. In principle, these might have been a set of 3600 eigenvectors. We can expect no more than a third of these to have any physical significance and arguments in NW01 suggest that only about the first 400 are needed to obtain stable answers.

Figure 4 shows estimates of the signal amplitudes with error bars representing 90% confidence intervals (CI). The top panel (a) indicates the estimates for pairs of sample designs (100 years with 20 tropical stations on the left and the 36 global stations as depicted in Figure 3 on the right). Each pair of estimates and associated CI corresponds to a

different noise model employed in computing the space-time covariance matrix. In these estimates we retained different numbers of EOFs according to the stability of the G estimate. Typically the retained number was about 400.

The results in Fig. 4 indicate that using the 20 tropical stations alone is not very different from retaining all 36. This indicates two things: 1) The signal is strongest in the tropics, and 2) The noise is greater outside the tropics, nearly compensating for the greater number of stations. Most of the performance probably comes from a kind of "global average" mode with little added from the geographical expression of the signal and noise characteristics (actually this needs to be explored for S).

As indicated in the figure caption, Figure 5 shows how the detection performance depends on the EOF truncation level of the scheme. We see that after a few hundred modes are retained, the G signal estimate (panel 2e of Figure 5) settles down and a rather stable estimate of the amplitude is obtained. At this point the theoretical signal to noise ratio is around 6.0. On the other hand, the estimate for the solar signal is much less stable because of its very faint amplitude as seen in panel (1e). Still, it is not unreasonable to claim a value of between 1.0 and 2.0.

3.1. Comparing SN96, NS98, NW01

This chapter is mainly a review of the three papers SN96, NS98 and NW01. We compare these papers in this section. SN96 establishes the method of choosing the sites used and it describes the philosophy of using control runs for com-

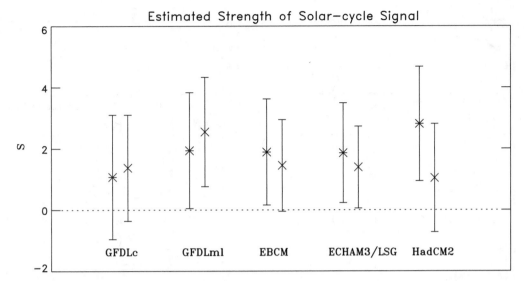

Figure 4. The estimated signal amplification factors for the solar signal using several different climate models for natural variability. For each climate model there are two configurations of sites. The error bars enclose the 90% confidence interval. The left-most point in the pairs of points represents data from 20 tropical stations, while the right-most represents data taken from 36 globally distributed stations.

puting the covariance matrices. By focusing on the solar cycle, SN96 used a narrow frequency band for the signal processing (from periods of 7 years to 20 years). The procedure used Fourier Components (sinusoids) for the temporal part of the signal and noise decomposition along with the same spatial expression of the EOFs in space for each frequency. SN96 used only the single signal S in the analysis. In NS98 it was recognized that other signals can interfere with S and so an attempt was made to include the other three signals by an intuitive procedure that "blocked" out the other three signals at least partially. In NW01 it was recognized that the optimal method of estimation was simply multiple regression, a method which simultaneously removes all signals but the one being sought, resulting in the best linear estimate of all signals simultaneously. NW01 also removed the sinusoid approximation in time and estimated the actual space-time EOFs on the 100 yr interval. In all three papers there appears to be a solar signal, but with slightly different amplitude and significance level in each. We believe the result in NW01 is the most reliable, but of course it is statistically rather unstable, with a confidence level of about 75%.

3.2. Ways of Improving Detection Performance

We believe there are possibilities of improving the performance of our method. One approach is to include the seasons. This has the advantage of taking into account the

seasonality of the forcing—the sun is more overhead in summer and so the signal should be slightly stronger. Seasonality also should improve the optimal estimation since the variance of natural variability is lower in summer. Both of these effects enhance performance. Unfortunately, including the seasons increases the dimensionality of the problem and techniques need to be developed to deal with this problem. We are currently looking into this issue. Note that a small increase in performance could push our significance level into the 95% range.

4. DISCUSSION AND CONCLUSIONS

We have taken a very conservative approach in this paper to detect the 11 year solar cycle in the surface temperature data. We have used as a signal the modeled response of the climate system to the sunspot cycle forcing as extrapolated into the past via the satellite data from the last two decades. This choice of data stream for the analysis has three advantages: 1) It is easy to model, providing a good signature to look for. 2) We have a reliable record of about 100 years duration. 3) We have a fairly robust means of estimating the forcing over that 100 years. We are then able to use existing AOGCM runs to estimate the natural variability covariance matrices and therefore can apply a sensible means of reducing the space-time dimension of the problem.

We applied our technique to the problem and find that we do detect the solar signal to about a 25% level of sig-

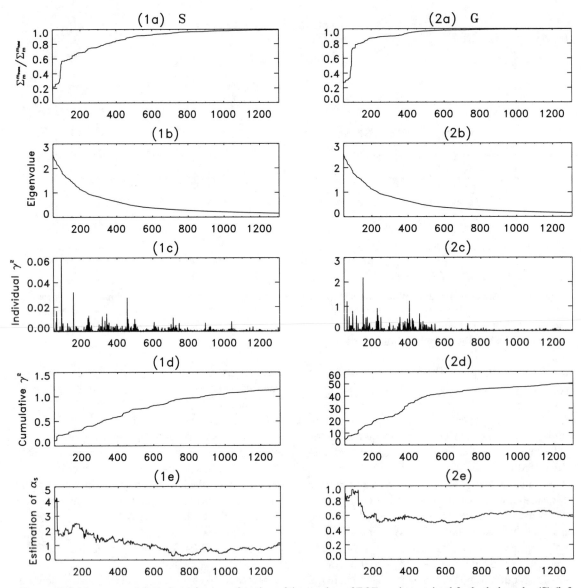

Figure 5. Performance of detection scheme as function of the number of EOF modes retained for both the solar (S) (left column of panels) and greenhouse (G) (right column of panels) signals. The (a) panels indicate the fraction of total signal squared captured by retaining n modes (abscissa). The (b) panels indicate the eigen value associated with mode n. The (c) panel indicates the value of theoretical signal-to-noise squared contributed by each mode. The (d) panels indicate the accumulated SNR squared by truncating at mode n. The (e) panels indicate the estimate a of the amplitude for the signal S or G.

nificance (probability of the null hypothesis being true). Its amplitude is roughly twice our input guess. Because of the low level of confidence we do not speculate on any reason for the factor of two. It could very well differ as we improve performance of our technique. On the other hand, we do suspect that it will not go away based on our study so far.

Acknowledgments. We wish to thank the National Oceano-graphic and Atmospheric Administration for its support through a research grant on climate signal detection.

REFERENCES

Burroughs, W.J., *Weather Cycles: Real or Imaginary?*, Cambridge University Press, Cambridge, UK, 207pp., 1972.

Intergovernmental Panel on Climate Change, *Climate Change 1995: The Science of Climate Change*, Contribution of Working Group I to the Second Assessment Report of the Intergovernmental Panel on Climate Change edited by J. T. Houghton, L. G. Meira Filho, B. A. Callander, N. Harris, A. Kattenberg, and K. Maskell, Cambridge University Press, Cambridge, UK and New York, NY USA, 572pp., 1996.

Intergovernmental Panel on Climate Change, *Climate Change 2001: The Scientific Basis*, Contribution of Working Group I to the Third Assessment Report of the Intergovernmental Panel on Climate Change, edited by J. T. Houghton, Y. Ding, D. J. Griggs, M. Noguer, P. J. van der Linden, X. Dai, K. Maskell, and C. A. Johnson, Cambridge University Press, Cambridge, UK and New York, NY USA, 994pp., 2001.

Kiehl, J. T., and B. P. Briegleb, The relative roles of sulfate aerosols and greenhouse gases in climate forcing, *Science, 260*, 311-314, 1993.

North, G. R., K.-J. Yip, L.-Y Leung and R. M. Chervin, Forced and Free Variations of the Surface Temperature Field in a General Circulation Model, *J. Climate, 5*, 227-239, 1992.

North, G. R. and M. J. Stevens, Detecting climate signals in the surface temperature field, *J. Climate, 11*, 563-577, 1998.

North, G. R. and Q. Wu, Detecting climate signals using space-time EOFs. *J. Climate, 14*, 1839-1863, 2001.

Ramaswamy, V. and C.-T. Chen, Linear additivity of climate response for combined albedo and greenhouse perturbations, *Geophys. Res. Lett., 24*, 567-570, 1997.

Stevens, M. J. and G. R. North, Detection of the climate response to the solar cycle, *J. Atmos. Sci., 53*, 2,594-2,608, 1996.

Stevens, M. J., Optimal estimation of the surface temperature response to natural and anthropogenic climate forcings over the past century, Ph.D. Dissertation, Texas A&M University, 157pp., 1997.

Gerald R. North, Department of Atmospheric Sciences, Texas A&M University, College Station, TX 77843–3150, USA.

Qigang Wu, IGES / COLA, 4041 Powder Mill Road, Suite 302, Calverton, MD 20705–3106, USA.

Mark J. Stevens, National Center for Atmospheric Research, P.O. Box 3000, Boulder, CO 80303, USA.

Has the Sun Changed Climate?
Modeling the Effect of Solar Variability on Climate

Michael E. Schlesinger and Natalia G. Andronova

Climate Research Group, Department of Atmospheric Sciences
University of Illinois at Urbana-Champaign, Urbana, Illinois

Satellite observations since 1978 have shown the sun's total irradiance changed by no more than 0.1% over solar cycles 21–23. However, observations of sunspot numbers, the concentrations of cosmogenic isotopes and sunlike stars suggest that the sun's irradiance may have increased by 0.3% from the 17th to the 20th century. A simple climate model used to determine the climate sensitivity, ΔT_{2x}—the change in global near-surface temperature resulting from doubling the pre-industrial CO_2 concentration—by replicating the observed changes of hemispheric-mean temperatures since 1856 has shown that ΔT_{2x} is nearly halved if the sun's irradiance changed before 1978 as constructed from "surrogate" data. Thus, in terms of future anthropogenic climate change impacts and policy it is important to determine whether the sun changed climate. To do so, fingerprint detection studies have been performed with coupled atmosphere/ocean general circulation models (A/O GCMs) to detect the geographical pattern of solar-induced temperature change. Successive studies performed at the Max-Planck Institute for Meteorology and the Hadley Centre for Climate Research and Prediction have increasingly concluded that the sun did change climate during the first half of the twentieth century. However, these detection studies did not use spectrally dependent changes in solar irradiation. Other GCM studies using the change in solar ultraviolet observed during solar cycle 22 and constructed from 1610 to the present have shown significant changes in near-surface temperature resulting from changes in stratospheric ozone, temperature and wind. Thus it is now time to perform fingerprint detection studies with A/O GCMs that include spectrally dependent solar irradiance variations and comprehensive interactive photochemistry.

1. INTRODUCTION

Because the climate of the earth owes its existence in large part to the sun, it is natural to wonder whether the sun has played a role in changing climate, particularly during the period of instrumental temperature observations in the 19th and 20th centuries. Some scientists have even concluded that most, if not all of the changes in near-surface temperature observed during this time period have been due to changes in the sun's irradiance [e.g., *Friis-Christensen and Lassen*, 1991; *Marshall Institute*, 1992; *Seitz*, 1994].

It is important for two reasons to determine how much of the instrumentally observed near-surface temperature change has been due to the sun. First, if the sun played a role in the historical temperature changes, it may play a role in

Solar Variability and its Effects on Climate
Geophysical Monograph 141
Copyright 2004 by the American Geophysical Union
10.1029/141GM19

future temperature changes and thereby yield temperature changes different from those expected on the basis of anthropogenic effects alone. Such a disparity between the expected and actual future temperature changes could be misinterpreted to mean that anthropogenic influences are small or even nonexistent, and thus lead to possibly inappropriate climate-change-policy responses.

Second, the temperature sensitivity, ΔT_{2x}—the change in global-mean equilibrium near-surface temperature in response to a radiative forcing at the tropopause equivalent to that for doubling the pre-industrial CO_2 concentration [*Schlesinger and Andronova*, 2001]—estimated using the instrumental near-surface temperature record, depends strongly on whether or not the sun's irradiance has changed. In the recent maximum-likelihood estimation of ΔT_{2x} using a simple climate model to reproduce the observed temperatures such that the root-mean-square deviation is minimized [*Andronova and Schlesinger, 2000*], it was found that ΔT_{2x} = 5°C for radiative forcing by greenhouse gases (GHGs) and anthropogenic sulfate aerosols, which exceeds the maximum of the range given by the IPCC, 1.5°C–4.5°C [*Houghton et al., 1990; Kattenberg et al., 1996; Houghton et al., 2001*]. But when solar radiative forcing is included, ΔT_{2x} is reduced by about 50%, to 2.7°C for the solar-radiative-forcing model constructed by Lean and colleagues [*Lean et al., 1995; Fröhlich and Lean, 1998*] and to 2.4°C for the model constructed by *Hoyt and Schatten* [1993], that is, to near the IPCC "best-estimate" value of ΔT_{2x} = 2.5°C. Clearly it is important in terms of climate impacts and policy to determine whether or not the sun's irradiance has varied during the past two centuries. This conclusion is supported by value-of-information studies which show that a large benefit accrues to reducing the uncertainty in ΔT_{2x} [*e.g., Peck and Teisberg*, 1993]. To do this it is necessary to determine if the sun changed climate.

In the following we examine the evidence of a solar effect from empirical/statistical models (Section 2), simple climate models (Section 3), and general circulation models (Section 4). Lastly, we discuss a possible way forward (Section 5).

2. EVIDENCE OF A SOLAR EFFECT BASED ON EMPIRICAL/STATISTICAL MODELS

Empirical/Statistical models have been used to relate changes in climatic quantities with indicators of solar changes that are taken to be "proxies" or "surrogates" for solar-irradiance changes.

The empirical model approach was taken in the study by *Friis-Christensen and Lassen* [1991] wherein the variable length of the "11-year" Schwabe cycle of sunspot numbers,

which varies in duration from about 9.7 to 11.8 years, was correlated with the global-mean temperature as shown in Figure 1. Largely based on this study, *Hoyt and Schatten* [1993] included solar-cycle length as a predictor in their model of solar irradiance changes from 1700 to 1992 (Section 3). *Mitchell* [1979] showed a relationship between three families of drought area indices for the western United States and the 22-year Hale cycle of sunspot magnetic polarity. *Eddy* [1976] conjectured that the virtual absence of sunspots during the Maunder Minimum, from 1645 to 1715, might have resulted in "The Little Ice Age." This concept was the basis, in part, for the construction of solar-irradiance changes from 1610 to the present by *Lean et al.* [1995] (Section 3). *Eddy* [1979] also concluded: " … changes in the level of solar activity and in climate may have a common cause: slow changes in the solar constant, of about 1% amplitude." *Beer et al.* [1988; 1994; 2000] have compared the cosmogenic isotope [10]Be, whose concentration in ice cores is negatively correlated with solar magnetic activity, with a northern hemisphere temperature record and found reasonable agreement. *Labitzke* [1982; 1987] and *Labitzke and van Loon* [1988; 1989; 1992; 1993; 1997] have found associations between the 11-year Schwabe cycle and stratospheric quantities.

Statistical model studies have been performed by *Schönwiese et al.* [1994a,b] and *Tol and Vellinga* [1998].

In the latter study the observed global-mean near-surface temperature record from 1860 to 1996 was represented by the statistical model

$$T_t = (c + \rho T_{t-1}) + \mu_s \alpha \, SUN + \mu_g \beta \, GHG + u_t \, ,$$

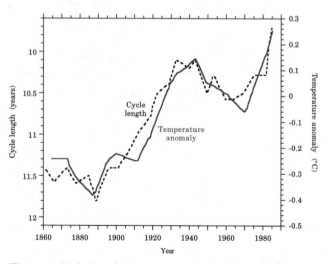

Figure 1. Variation of the sunspot cycle length (left-hand scale) and the Northern Hemisphere land near-surface temperature anomalies (right-hand scale). Adapted from *Friis-Christensen and Lassen* [1991].

where T_t is the temperature for year t, the term in parentheses represents an autoregressive process, u_t represents white noise with zero mean, GHG is the logarithm of the CO_2 concentration, and SUN is one of seven "proxy records" of solar irradiance—sunspot number, Almon-transformation filtered sunspot number, length of the solar cycle based on sunspot minima, solar cycle length based on sunspot maxima, the average of the previous two, the average solar length with a six-year lag, and a Almon-transformation filtered average solar length—and the μ's are either zero or unity. The values for c, ρ, α and β were determine by maximum likelihood estimation.

The pure autoregressive model ($\mu_s = \mu_g = 0$) was found to explain 70% of the variance of the observed temperature changes. Inclusion of SUN ($\mu_s = 1$, $\mu_g = 0$) increased the explained variance by 1 to 7%, and for all proxies but one (untransformed sunspot number) the null hypothesis of no solar influence was rejected with 95% confidence. Inclusion of GHG ($\mu_s = 0$, $\mu_g = 1$) increased the explained variance by

6 to 8%. The significance of GHG was found to be robust to the specification of solar activity ($\mu_s = 1$, $\mu_g = 1$). The converse, however, was not true. Testing solar influence against the model that includes GHG showed that only two of the seven "proxies" significantly influenced temperature, namely, the lagged and transformed records of the average solar cycle length.

A list of some other Empirical/Statistical model studies is presented in Table 1, in chronological order from 1976, together with their corresponding titles.

3. EVIDENCE OF A SOLAR EFFECT BASED ON SIMPLE CLIMATE MODELS

Simplified models of the climate system have been used to examine the effect of putative variations in solar irradiance on the earth's climate, particularly in terms of the global-mean near-surface temperature and ice cover. A list of some of the simple climate model (SCM) studies is present-

Table 1. Some studies using Empirical/Statistical Models; in chronological order from 1976.

Study	Title
Eddy [1976]	The Maunder Minimum
Mass and Schneider [1977]	Statistical evidence on the influence of sunspots and volcanic dust on long-term temperature records
Schuurmans [1978]	Influence of solar activity on solar temperatures
Eddy [1979]	Climate and the changing sun
Hoyt [1979]	Variations in sunspot structure and climate
Mitchell [1979]	Evidence of a 22-year rhythm of drought in the western United States related to the Hale solar cycle since the 17th century
Nastrom and Belmont [1980]	Evidence for a solar cycle signal in tropospheric winds
Hoyt and Siquig [1982]	Possible influences of volcanic dust veils or changes in solar luminosity on long-term local temperature records
Labitzke [1987]	Sunspots, the QBO, and the stratospheric temperature in the north polar region
Labitzke and Van Loon [1988]	Associations between the 11-year solar cycle, the QBO and the atmosphere. Part I: The troposphere and stratosphere in the northern hemisphere in winter
Beer et al. [1988]	Information on past solar activity and geomagnetism from 10Be in the Camp Century ice core
Labitzke and Van Loon [1989]	Associations between the 11-year solar cycle, the QBO and the atmosphere. Part III: Aspects of associations
Friis-Christensen and Lassen [1991]	Length of the solar cycle: An indicator of solar activity closely associated with climate
Labitzke and Van Loon [1992]	Associations between the 11-year solar cycle and the atmosphere. Part V: Summer
Labitzke and Van Loon [1993]	Some recent studies of probable connections between solar and atmospheric variability
Kim and Huang [1993]	Newly found evidence of Sun-climate relationships
Beer et al. [1994]	Solar variability traced by cosmogenic isotopes
Schönwiese et al. [1994]	Analysis and prediction of global climate temperature change based on multiforced observational statistics
Schönwiese [1994]	Solar signals in global climatic change
Crowley and Kim [1996]	Comparison of proxy records of climate change and solar forcing
Lean and Rind [1996]	The sun and climate
White et al. [1998]	Global upper ocean heat storage response to radiative forcing from changing solar irradiance and increasing greenhouse gas/aerosol concentrations
Tol and Vellinga [1998]	Climate change, the enhanced greenhouse effect and the influence of the sun
Beer et al. [2000]	The role of the sun in climate forcing

ed in Table 2, in chronological order from 1969, together with their corresponding titles. Below we illustrate the use of simple climate models to determine whether the sun has influenced climate by presenting the results of *Andronova and Schlesinger* [2000; 2001].

In these recent studies, a simple climate model was used to reproduce the average and difference in northern and southern hemispheric near-surface temperatures for a variety of radiative forcing factors, including greenhouse gases (GHGs), anthropogenic sulfate aerosol (ASA), volcanoes and the sun to estimate: (1) the contribution of these factors to the observed near-surface temperature change from 1856 through 1997, and the values of (2) ΔT_{2x} and (3) the ASA radiative forcing in a reference year, ΔF_{SO4} (1990). Below the simple climate model is described in Section 3.1, its radiative forcing in Section 3.2, the estimation procedure in Section 3.3, and the results in Section 3.4.

3.1. Simple Climate Model

The original, global version of the SCM was developed by Schlesinger, based on the model's original formulation by *Hoffert et al.* [1980], and was used by Schlesinger and colleagues to simulate the global-mean temperature evolution for the different GHG scenarios of the IPCC 1990 report [*Bretherton et al.*, 1990], and for greenhouse-policy studies [*Schlesinger and Jiang*, 1991; *Hammitt et al.*, 1992; *Schlesinger*, 1993; *Lempert et al.*, 1994, 1996; *Lempert and Schlesinger*, 2000; *Lempert et al.*, 2000; *Lempert and Schlesinger*, 2001]. The hemispheric version of the model was developed to study the influence on the climate system of ASA [*Schlesinger et al.*, 1992] and putative solar-irradiance variations [*Schlesinger and Ramankutty*, 1992], and has been used to discover a 65–70 year oscillation in observed surface temperatures for the North Atlantic Ocean and its bordering continental regions [*Schlesinger and Ramankutty*, 1994a, b; *Schlesinger and Ramankutty*, 1995]. A hemispheric version of the model that explicitly calculates the individual temperature changes over land and ocean in each hemisphere [*Ramankutty*, 1994] has been used to investigate the influence on climate of volcanoes [*Ramankutty*, 1994] and the sun [*Schlesinger and Ramankutty*, 1992], to estimate the causes of climate change from 1856 to 1997 [*Andronova and Schlesinger*, 2000], and to objectively estimate the probability distribution for ΔT_{2x} [*Andronova and Schlesinger*, 2001].

The model determines the changes in the temperatures of the atmosphere and ocean, the latter as a function of depth from the surface to the ocean floor. In the model (Figure 2),

Table 2. Some studies using Simple Climate Models; in chronological order from 1969.

Study	Title
Budyko [1969]	The effect of solar radiation variations on the climate of the earth
Sellers [1969]	A global climate model based on the energy balance of the earth-atmosphere system
Schneider and Mass [1975]	Volcanic dust, sunspots, and temperature trends
Robock [1978]	Internally and externally caused climate changes
Nicolis [1979]	The effect of solar output, infrared cooling and latitudinal heat transport on the evolution of the earth's climate
Peng et al. [1982]	Climate studies with a multi-layer energy balance model. Part I: Model description and sensitivity to the solar constant
Gilliland [1982]	Solar, volcanic, and CO_2 forcing of recent climatic changes
North et al. [1983]	Climate response to a time varying solar constant
Gilliland and Schneider [1984]	Volcanic, CO_2 and solar forcing of northern and southern hemisphere surface air temperatures
Sellers [1985]	The effect of a solar perturbation on a global climate model
Reid [1987]	Influence of solar variability on global sea surface temperature
Wigley [1988]	The climate of the past 10,000 years and the role of the sun
Hoffert [1988]	The effects of solar variability on climate
Kelly and Wigley [1990]	The influence of solar forcing on global mean temperature since 1861
Wigley and Raper [1990]	Climate change due to solar irradiance changes
Reid [1991]	Solar total irradiance variations and the global sea surface temperature record
Kelly and Wigley [1992]	Solar cycle length, greenhouse forcing and global climate
Schlesinger and Ramankutty [1992]	Implications for global warming of intercycle solar-irradiance variations
Crowley and Kim 1996]	Comparison of proxy records of climate change and solar forcing
Mendoza et al. [1997]	The impact of solar irradiance on the Maunder Minimum climate
Crowley [2000]	Cause of climate change over the past 1000 years
Andronova and Schlesinger [2000]	Causes of Global Temperature Changes During the 19th and 20th Centuries
Andronova and Schlesinger [2001]	Objective estimation of the probability density function for climate sensitivity

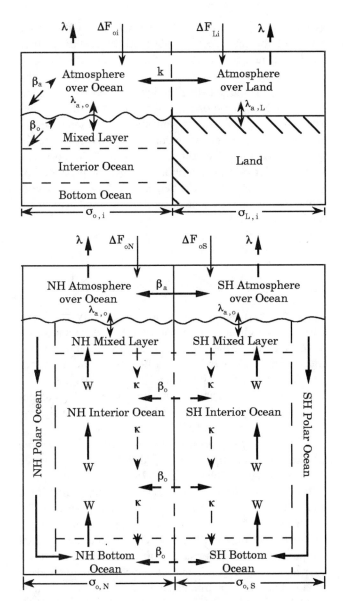

Figure 2. Schematic diagram of the simple climate/ocean model. The top panel shows the general structure of the model, and the bottom panel shows a vertical cross-section through the oceanic part. The symbols by the arrows indicate the following physical processes: ΔF_{Li} and ΔF_{oi}, the tropopause radiative forcing in hemisphere i over land and ocean, respectively; $\lambda = \Delta T_{2x} / \Delta F_{2x}$, the radiative-plus-feedback temperature response of the climate system, expressed as the temperature sensitivity to the radiative forcing of a CO_2 doubling, ΔF_{2x}; k, the atmospheric land-ocean heat exchange; β_a, the atmospheric interhemispheric heat exchange; $\lambda_{a,o}$, the air-sea heat exchange; $\lambda_{a,L}$, the air-land heat exchange; β_o, the oceanic interhemispheric heat exchange; W, the vertical heat transport by upwelling; κ, the vertical heat transport by diffusion. The quantities $\sigma_{L,i}$ and $\sigma_{o,i}$ denote the fractions of hemisphere i covered by land and ocean, respectively. Source: *Schlesinger et al.* [1997].

the ocean is subdivided vertically into 40 layers, with the uppermost being the mixed layer. Also, the ocean is subdivided horizontally into a polar region where bottom water is formed, and a nonpolar region where there is upwelling. In the nonpolar region, heat is transported upwards toward the surface by the water upwelling there and downwards by physical processes whose effects are treated as an equivalent diffusion. Heat is also removed from the mixed layer in the nonpolar region by a transport to the polar region and downwelling toward the bottom, this heat being ultimately transported upward from the ocean floor in the nonpolar region. The atmosphere in each hemisphere is subdivided into the atmosphere over the ocean and the atmosphere over land, with heat exchange between them.

Ten principal quantities must be prescribed in the model: ΔT_{2x}, the radiative forcing due to a doubling of the CO_2 concentration, the upwelling velocity; the diffusivity, the depth of the oceanic mixed layer, the warming of the polar ocean relative to the warming of the nonpolar ocean, and the heat-transfer coefficients which were calibrated against the observed annual cycle of near-surface temperature [*Schlesinger et al.*, 1997].

3.2. Radiative Forcing

The radiative forcing for hemisphere i used by the SCM is

$$\Delta F_i(t) = \Delta F_{GHG}(t) +$$
$$2\beta_i [\mu_{ASA}\Delta F_{ASA}(t) + \mu_{TR03}\Delta F_{TR03}(t)] +$$
$$\mu_{Sun}\Delta F_{Sun}(t) + \mu_{Vol}\Delta F_{Vol,i}(t), i = N, S$$

where $\Delta F_{GHG}(t)$ (Figure 3A) is the radiative forcing by all greenhouse gases other than tropospheric ozone (CO_2, methane, nitrous oxide and the chlorofluorocarbons) based on the IPCC/97 report [*Harvey et al.*, 1997], but updated to *Myhre et al.* [1998]. $\beta_N = \alpha$ (= 0.8) and $\beta_s = 1 - \alpha$ are the fractions of the anthropogenic sulfate aerosol (ASA) and tropospheric ozone (TRO3) radiative forcing that occur in the Northern and Southern Hemispheres, respectively. By choosing the values of the four μ_i's to be either zero or unity, 16 radiative-forcing models (RFMs) were considered. In the following "G", "T", "A", "S" and "V" represent respectively the radiative forcing by GHGs, tropospheric ozone, anthropogenic sulfate aerosol, the sun and volcanos.

3.2.1. Non-solar forcing. $\Delta F_{ASA}(t)$ is the sum of the direct and indirect radiative forcing by anthropogenic and natural sulfate aerosols,

$$\Delta F_{ASA}(t) = \Delta F_{ASA}^{dir}(1990)\left[\frac{E(t)}{E(1990)}\right] +$$
$$\Delta F_{ASA}^{ind}(1990)\left[\frac{\log(1 + E(t) / E_{nat})}{\log(1 + E(1990) / E_{nat})}\right]$$

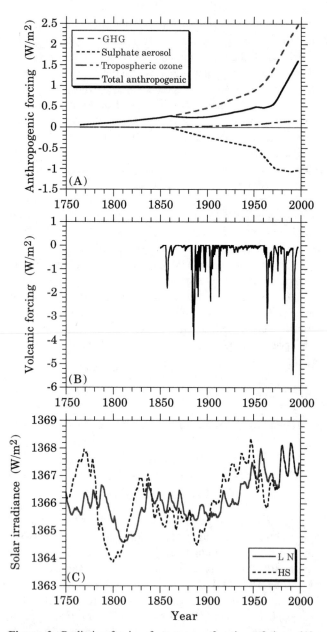

Figure 3. Radiative-forcing factors as a function of time. (A) anthropogenic forcing, (B) volcanic forcing, and (C) solar irradiance forcing for the models by *Lean and colleagues* [*Lean et al.,* 1995; *Fröhlich and Lean,* 1998] (LN) and by *Hoyt and Schatten* [1993] (HS). Source: *Andronova and Schlesinger* [2000].

where E(t) is the global anthropogenic emission rate of sulfur in the form of SO_2, E(1990) = 72.6 TgS/yr, E_{nat} = 22.0 TgS/yr, and ΔF_{ASA}^{dir} (1990) and ΔF_{ASA}^{ind} (1990) are the unknown direct and indirect sulfate forcings in 1990, respectively. A value of ΔF_{ASA}^{ind} (1990)/ ΔF_{ASA}^{dir} (1990) = 0.8/0.3 was chosen following *Harvey et al.* [1997]. Tests with the model showed that the estimated total sulfate radia-

tive forcing in 1990 (Figure 3A) is insensitive within 2% to values of the ratio ΔF_{ASA}^{ind} (1990)/ ΔF_{ASA}^{dir} (1990), at least within the range of 0 to 10. The ASA radiative forcing begins in 1857.

ΔF_{TR03} (t) is the radiative forcing by tropospheric ozone. Because tropospheric ozone is the indirect result of industrial activity, as is ASA, the hemispheric partitioning of the tropospheric-ozone radiative forcing was taken to be the same as that for ASA. Annual values of the tropospheric ozone forcing are interpolation of the data presented by *Stevenson et al.* [1998] and begins in 1860. Stratospheric-ozone forcing due to ozone depletion is ignored here as it is small [*Forster,* 1999].

$\Delta F_{Vol,i}$(t) is the radiative forcing by volcanoes in hemisphere i,

$$\Delta F_{Vol,i} (t) = \Delta F_{Vol,i}^{LW}(t) + \Delta F_{Vol,i}^{SW}(t) , i = N, S$$

where $\Delta F_{Vol,i}^{LW}$(t) and $\Delta F_{Vol,i}^{SW}$(t) are respectively the hemispheric-mean changes in the adjusted longwave and shortwave radiation at the tropopause due to volcanic aerosol, obtained by *Andronova et al.* [1999]. The volcanic hemispheric radiative forcing (Figure 3B) begins in 1850.

The anthropogenic and volcanic radiative forcing components are displayed in Figures 3A and 3B, respectively.

3.2.2. Solar forcing. Satellite observations of the sun have shown that its irradiance varied by about 0.1% over the two full sunspot cycles since 1978 [*Fröhlich and Lean,* 1998], when the observations began [*Willson,* 1994]. If the solar irradiance varied by only this amount prior to 1978, then the sun's effect on climate has been negligible owing to the smallness of the irradiance change and the thermal damping of the ocean for such short-period variations. Over the past seven '11-year' sunspot cycles we know observationally that the solar irradiance did not vary by more than about 1% [*Fröhlich,* 1977], and did vary by about 0.1%, at least over the last two full sunspot cycles.

Longer records of characteristics of the sun other than its irradiance, for example, the number of sunspots and the length of the '11-year' sunspot cycle, do show longer-length variations before 1978. It has been argued that these slowly varying solar characteristics are surrogates for long-period variations of solar irradiance. This argument has been bolstered by observations of sun-like stars. When these stars are divided into magnetically active and magnetically inactive groups, based on observed fluxes from Ca-II H and K lines, the magnetically inactive group shows reduced luminosity of up to 0.4% relative to the magnetically active group [*Baliunas and Jastrow,* 1990]. Thus a reduction in solar irradiance larger than 0.1% during the Maunder

Minimum period (1645–1715), when there were few if any sunspots and the sun was likely magnetically inactive, appears not to be impossible. This has lead to the construction of irradiance models based on supposed 'surrogates', for example, the model of *Hoyt and Schatten* [1993] based on the fraction of penumbral spots, solar-cycle length, equatorial rotation rate, decay rate of the solar cycle, and mean level of solar activity, and the model of *Lean et al.* [1995] based on sunspot areas and locations, He 1083 nm emission, group sunspot numbers, and Ca emissions from the sun and sunlike stars. But, it is not possible to validate these models now against observations of solar irradiance, and it will not be possible to do so until well into the future when additional satellite observations will have been obtained.

The solar-irradiance forcing is

$$\Delta F_{Sun,i}(t) = \left(\frac{1-\alpha_p}{4}\right)[S_i(t) - S_i(t_{o,i})],$$
$$t_{o,i} \leq t \leq 1998$$

where i = LN and HS for the models of solar-irradiance variation proposed respectively by *Lean et al.* [1995] and *Hoyt and Schatten* [1993], $S_{LN}(1610) = 1365.7$ Wm^{-2}, $S_{HS}(1700) = 1363.5$ Wm^{-2}, $\alpha_p = 0.3$ is the planetary albedo, and the factor 4 accounts for the ratio of the area of emission of terrestrial (longwave) radiation to the area for the absorption of solar radiation. (A third construction of total solar irradiance since 1700 has been proposed by *Solanki and Fligge* [1999]. Also, see Sofia and Li [2001].) For the period 1978–1998, when satellite observations of the solar irradiance exist, both solar models were updated to the data of *Fröhlich and Lean* [1998]. The solar radiative forcings are displayed in Figure 3C.

3.3. Estimation Method

For each radiative-forcing model (RFM) the changes in global-mean ((NH+SH)/2) near-surface temperature, $\Delta F_{GL}^{sim}(t)$, and in the interhemispheric (NH–SH) near-surface temperature difference, $\Delta F_{HD}^{sim}(t)$, were calculated from 1765 through 1997 by the simple climate model for many prescribed values of ΔT_{2x} and the direct (clear-sky) ASA radiative forcing in reference year 1990, ΔF_{ASA}^{dir} (1990). The simulated departures in global-mean near-surface temperature, $\delta T_{GL}^{sim}(t) = \Delta T_{GL}^{sim}(t) + C_{GL}$, and in the interhemispheric near-surface temperature difference, $\delta T_{HD}^{sim}(t) = \Delta T_{HD}^{sim}(t) + C_{HD}$, were compared with the corresponding observed temperature departures from the 1961–1990 means [*Jones et al.*, 1999], $\delta T_{GL}^{obs}(t)$ and $\delta T_{HD}^{obs}(t)$, from 1856 through 1997, with the constants, C_{GL} and C_{HD}, determined to minimize the

individual root-mean-square (RMS) differences between $\delta T_{GL}^{sim}(t)$ and $\delta T_{GL}^{obs}(t)$, and $\delta T_{HD}^{sim}(t)$ and $\delta T_{HD}^{obs}(t)$ [*Schlesinger et al.*, 1992; *Schlesinger and Ramankutty*, 1992; *Andronova and Schlesinger*, 2001]. Maximum-likelihood values of ΔT_{2x} and ΔF_{ASA}^{dir} (1990)—hence the total (all-sky) ASA radiative forcing ΔF_{ASA} (1990) = 3.67 ΔF_{ASA}^{dir} (1990) [*Harvey et al.*, 1997]—were determined by simultaneously minimizing the RMS difference between $\delta T_{GL}^{sim}(t)$ and $\delta T_{GL}^{obs}(t)$ and between $\delta T_{HD}^{sim}(t)$ and $\delta T_{HD}^{obs}(t)$. This was done separately for monthly and annual observed temperatures, with negligible differences in the results.

3.4. Results

Figure 4 illustrates the simulated temperatures for the RFM with greenhouse gases, tropospheric ozone, sulfate aerosol and the sun (GTAS) in comparison with the observations.

Figure 5 shows the individual contributions due to the anthropogenic, solar and volcanic radiative forcing for each radiative-forcing model, together with the corresponding residual contribution. For the longest time period, 1856–1990 (Figure 5A), the temperature change simulated for the anthropogenic forcing is 54–84% of the observed temperature change, with the minimum contribution being for the anthro+volc+HS radiative-forcing model and the maximum contribution for the anthro+volc radiative-forcing model. The contribution by the volcanic radiative forcing, for the radiative-forcing models that include it, is negative and thus is of opposite sign to the observed temperature change. The contribution by the solar radiative forcing, for the radiative-forcing models that include it, is positive and 22–27% of the observed temperature change. The contribution by the residual, that is the part of the observed temperature change different from the total simulated temperature change, is positive and ranges from 16% for the anthro forcing model to 37% for the anthro+volc model. According to these results the observed warming over 1856–1990 was predominantly due to anthropogenic radiative forcing plus an unexplained residual warming, with the sun contributing a warming less than half that contributed by the anthropogenic factor – if indeed the solar irradiance varied as constructed, and with volcanoes contributing a small cooling.

In the light of this finding it is of interest to determine whether the warmings observed during 1904–1944 and 1976–1990 were similarly predominantly due to anthropogenic radiative forcing. Figure 5B indicates that this is not the case, with the 1904–1944 warming being predominantly due to the residual factor and volcanic forcing, the latter as a result of the decrease in volcanic eruptions during

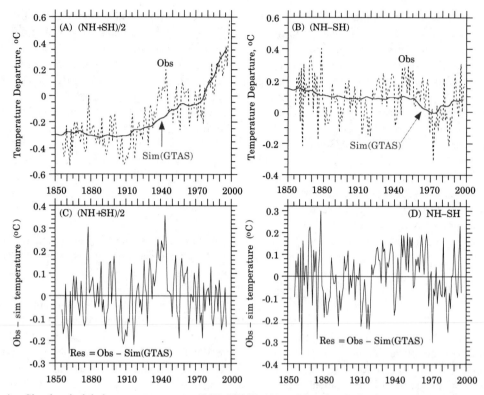

Figure 4. Simulated global-mean temperature ((NH+SH)/2) (A) and interhemispheric temperature difference (NH-SH) (B) for the GTAS radiative forcing model in comparison with the observations. (C) and (D) are the corresponding observed-minus-simulated temperature residuals. Source: *Andronova and Schlesinger* [2001].

1904–1944 relative to the late 19th century (Figure 3B). The role of the residual factor is even more dominant during 1944–1976 (Figure 5C) when the anthropogenically induced warming was in opposition to the observed cooling. During both 1904–1944 and 1944–1976, the sun, if it varied as constructed, played only a minor role in the observed temperature change. The warming observed during 1976–1990 (Figure 5D) is about equally due to anthropogenic radiative forcing and the residual factor, with volcanoes contributing a cooling, and the sun at most a small warming.

In addition to the uncertainty in the estimate of ΔT_{2x} due to the uncertainty in the RFM, there is uncertainty due to the natural variability in the instrumental temperature record. This is illustrated in Figure 4 by the residual temperature change,

$$\delta T_{res}^i(t) = \delta T_{obs}^i(t) - \delta T_{sim}^i(t), i = N, S,$$

where $\delta T_{obs}^i(t)$ and $\delta T_{sim}^i(t)$ are the observed and simulated temperature departures at time t for hemisphere i, the latter for any RFM. Figures 4C and 4D present 0.5 $[\delta T_{res}^N(t) + \delta T_{res}^S(t)]$ and $[\delta T_{res}^N(t) - \delta T_{res}^S(t)]$ for the GTAS RFM.

To measure the "variability-induced" uncertainty, included in $\delta T_{res}(t)$, the bootstrap re-sampling method for correlated data developed by *Solow* [1985] was used to generate 5000 realizations of "natural variability," $\delta T_{boot}^i(t)$, each realization of n = 142 years duration. An ensemble of 5000 surrogate observational temperature records was constructed for each hemisphere by $\delta T_{surobs}^i(t) = \delta T_{boot}^i(t) + \delta T_{sim}^i(t)$, i = N,S, one ensemble for each of the 16 RFMs, and $\delta T_{boot}^i(t)$ is a bootstrap sample of $\delta T_{res}^i(t)$. For each ensemble member the same procedure was used to estimate ΔT_{2x} and ΔF_{ASA} (1990) as for the single real observational record.

Figure 6 displays the resulting ΔT_{2x} probability density function (pdf) and its corresponding cumulative density function (cdf) for the RFMs that contain anthropogenic (GTA) and volcanic (V) forcing, 20,000 realizations each with and without solar forcing (S_{LN}). It is seen that the pdf and the cdf without the sun are shifted toward larger ΔT_{2x} values than the pdf and cdf with the sun. With solar irradiance forcing as constructed by *Lean and colleagues*, the median values of ΔT_{2x} is 3.2°C, the 90% confidence interval is 1.9°C to 5.7°C, and there is only a 16.4% likelihood that ΔT_{2x} lies outside the IPCC range of 1.5°C to 4.5°C, with 1.4% likelihood that $\Delta T_{2x} < 1.5$°C and 15% likelihood that

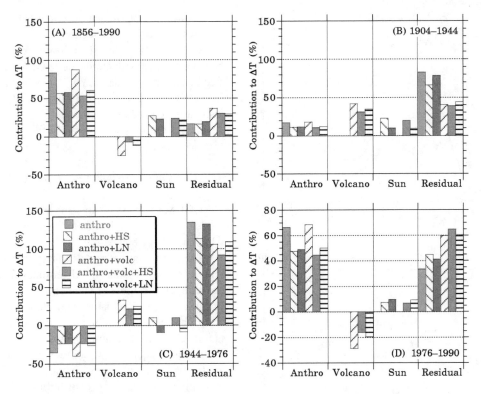

Figure 5. Percentage contribution to the observed temperature change over 1856–1990 (A), 1904–1940 (B), 1944–1976 (C) and 1976–1990 (D), for each radiative-forcing model, by the temperature change simulated individually for the anthropogenic (Anthro), volcanic (Volcano), and solar (Sun) radiative-forcing factors, together with the percentage contribution not explained by the simulated total temperature change (Residual). Source: *Andronova and Schlesinger* [2001].

$\Delta T_{2x} > 4.5°C$. Without solar irradiance forcing, the median value of ΔT_{2x} is 5.8°C, the 90% confidence interval is 2.8°C to 17.3°C, and there is a 70% likelihood that ΔT_{2x} lies outside the IPCC range, all for $\Delta T_{2x} > 4.5°C$. This is a disquieting result that indelibly shows the importance of determining the magnitude and timing of solar irradiance prior to the era of satellite observations thereof.

4. EVIDENCE OF A SOLAR EFFECT BASED ON GENERAL CIRCULATION MODELS

We have seen in Section 3 that it is important for two reasons to determine whether or not the sun's irradiance before the beginning of satellite measurements thereof in 1978 varied by more than the 0.1% measured thereafter during solar cycles 21–23. First, if the sun played a role in the historical temperature changes, it may play a role in future temperature changes and thereby yield temperature changes different from those expected on the basis of anthropogenic effects alone. Second, if the sun played a role in the historical temperature changes, the climate sensitivity required to

reproduce those changes is approximately halved compared to the value required if the sun's irradiance varied only by as much as that measured during the satellite era. But, how can we determine whether the solar irradiance varied before 1978 by more than it varied thereafter?

One way would be to continue to make satellite measurements of solar irradiance well into the future. However, even if the sun's irradiance varied significantly before 1978, it may not do so again for a long time and, conversely, even if it began to do so soon, it would not mean that it did so prior to 1978. Thus this "wait-and-see" approach is not very satisfactory.

A second, less-direct but perhaps more-immediate, way of estimating whether the sun has changed climate during the past two centuries is to search for a solar signal in the observed temperature record, with the solar signal being defined by the most comprehensive type of climate model, the coupled atmosphere-ocean general circulation model (AOGCM). This so-called "fingerprint-detection" method was developed by *Barnett* [1986] and *Barnett and Schlesinger* [1987] to search for the anthropogenic CO_2 sig-

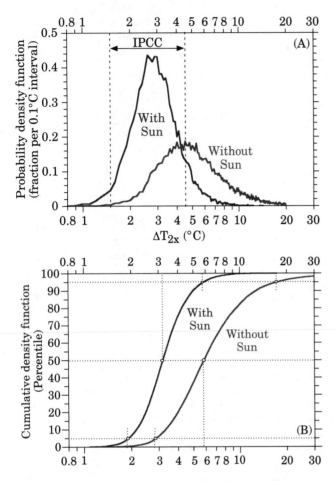

Figure 6. Probability density function (pdf, A) and cumulative density function (cdf, B) pair for ΔT_{2x}. Each curve includes 20,000 realizations for anthropogenic and volcanic radiative forcing, both without and with the solar forcing of *Lean and colleagues* [*Lean et al.*, 1995; *Fröhlich and Lean*, 1998]. Based on the calculations of *Andronova and Schlesinger* [2001].

Accordingly, below we describe: (1) the general characteristics of AOGCMs (Section 4.1), (2) the simulations of the climatic effects of solar irradiance changes that have been performed with atmospheric GCMs coupled to simplified ocean models (Section 4.2), (3) the method of "optimal detection" (Section 4.3), and (4) the results of the three studies that have used this method to detect and attribute a solar signal in the observed near-surface temperature record (Section 4.4).

4.1. General Characteristics of AOGCMs

AOGCMs consist of an atmospheric general circulation model and an oceanic general circulation model, both of which are described below.

General circulation models of the atmosphere are the only type of climate model that determines, in addition to temperature, an ensemble of other climatic quantities, including precipitation, soil moisture and cloud cover. Furthermore, atmospheric GCMs are the only type of climate model that is formulated to determine the geographical distributions of these climatic quantities. Consequently, atmospheric GCMs are the only type of climate model available to estimate, for example, the regional distribution of greenhouse-gas-induced climatic change. To use an atmospheric GCM for this purpose, however, it is necessary to couple it to a model of the ocean so that the geographical distributions of sea-surface temperature and sea-ice extent can be determined.

Atmospheric GCMs are based on the fundamental laws of nature, namely, Conservation of Energy, Conservation of Mass and Newton's Second Law of Motion. These laws respectively govern the time rates of change of temperature, surface pressure and wind velocity; these quantities are therefore called prognostic variables. The large-scale motion pattern of the atmosphere, or atmospheric general circulation, is determined in large part by the release of latent heat when water vapor condenses into liquid (or sublimates into ice). Accordingly, water vapor is a prognostic variable of an atmospheric GCM, governed by a water-vapor conservation law. Because the atmosphere is heated by the underlying surface, the temperature of the ground, the amount of water in the soil, and the mass of snow on the ground are also prognostic variables, governed respectively by energy, water, and snow conservation laws. Atmospheric GCMs also have diagnostic variables such as the vertical velocity component and geopotential height, that is, quantities whose magnitudes rather than time rates of change are determined by the governing laws.

While the governing laws of an atmospheric GCM can be described verbally, such as for Newton's Second Law of

nal in the observational near-surface temperature record. The method was expanded by *Santer et al.* [1993; 1995; 1996] and used to search for the anthropogenic signal in both near-surface temperature and the vertical-latitudinal distribution of zonal-mean temperature due to greenhouse gases and sulfate aerosol. An "optimal detection" method was developed by *Hasselmann* [1979; 1993; 1997] and *Allen and Tett* [1999] which has been used to search for a solar signal by *Hegerl et al.* [1997] using simulations performed by the AOGCM of the Max-Planck Institute for Meteorology and by *Tett et al.* [1999] and *Stott et al.* [2001] using the AOGCM of the Hadley Centre for Climate Research and Prediction.

Motion – force equals mass times acceleration, such a verbal description cannot be used to make a prediction. For this predictive purpose it is necessary to describe the governing laws mathematically. The resulting mathematical expressions of the governing laws are partial differential equations that describe the behavior of the atmosphere continuously in both space and time. These governing equations are so complex, however, that there are no analytical methods available to obtain their solution, as there are, for example, to determine the square root of a number. Accordingly, it is necessary to obtain their solution by numerical methods in which the behavior of the atmosphere is determined discretely in both space and time.

The numerical methods discretize the atmosphere vertically into layers and horizontally into either grid cells, as in finite difference models, or a number of prescribed mathematical functions as in spectral models. The values of the prognostic and diagnostic variables are determined for each vertical layer and for each grid cell or mathematical function. The evolution in time of the prognostic and diagnostic variables is also determined at discrete times by advancing the solution in discrete time steps from some prescribed initial conditions.

Contemporary atmospheric GCMs have up to 100 vertical layers between the earth's surface and their top level, which may be as low as the tropopause (≈ 10 km) or as high as the thermosphere (≈ 100 km), and horizontal resolutions (the distance between grid points) of a few hundred kilometers. These horizontal resolutions are chosen not on the basis of physical grounds—clearly they are severely limiting for climate-impact analyses—but rather by the limitation of present-day computers. For example, if a climate-change simulation required two weeks on a contemporary computer with a horizontal resolution of a few hundred kilometers, increasing the resolution tenfold to a few tens of kilometers would require almost 40 years on the same machine! This thousandfold increase in required computer time for a tenfold increase in horizontal resolution occurs because, in addition to the resulting hundredfold increase in the number of grid cells or mathematical functions, there is an additional multiplicative factor of ten due to the tenfold decrease in time step that is required to maintain computational stability.

The limitation on the horizontal resolution of atmospheric GCMs imposed by present-day computers has a major consequence on which physical processes can be explicitly resolved. In particular, the horizontal scale of the physical processes occurring in the Earth's climate system range from a millionth of a meter for the condensation of water vapor onto condensation nuclei (aerosol particles and dust)

to 40 million meters for a wave that spans the entire circumference of the Earth at the equator. Thus the physical processes range over 14 powers of ten in horizontal size. With present computers, atmospheric GCMs resolve physical processes with a minimum horizontal scale of about 1 million meters (1000 km), thus they cover only 2 of the 14 powers of ten. And, as described above, to go from 2 to 3 of the powers of ten requires a computer 1000 times faster than the fastest computer presently available. To go from 2 to the full 14 powers of ten would require a computer 10^{36} times faster! While a computer 1000 times faster than present-day computers may be available in the near future, a computer fast enough to permit the resolution of the complete range of physical processes in the climate system will, in all likelihood, never exist. Since we cannot therefore hope to resolve all the processes in the climate system, can we ignore those we cannot resolve? Unfortunately, the answer is no. The reason for this is, as noted above, the general circulation of the atmosphere is determined in large part by the release of latent heat which occurs when water vapor condenses into liquid (or sublimates into ice), and this occurs on the smallest unresolved scales. Consequently, the state of affairs is that we cannot resolve the small-scale physical processes and we cannot ignore them. This is the major difficulty in modeling the behavior of the Earth's climate system. The approach taken is to determine the effects of the unresolved, or subgrid-scale, physical processes on the scales resolved by the GCM using information available only on the resolved scales. This approach is called a parameterization.

To simulate time-dependent nonequilibrium climate change it is necessary to couple an atmospheric GCM with an oceanic GCM which is the oceanic counterpart to the atmospheric GCM. In an oceanic GCM the prognostic variables are the temperature, horizontal currents and salinity, and the diagnostic variables include density, pressure and the vertical velocity. There are subgrid-scale processes that must be parameterized in ocean GCMs, including the turbulent transfers of heat, momentum and salt in both the vertical and horizontal directions. The solution of the governing equations is obtained numerically in a manner similar to that used for atmospheric GCMs. The ocean is subdivided vertically into layers and horizontally into grid cells, and the predicted quantities are determined as a function of time by numerical integration.

In oceanic GCMs the thickness of sea ice is predicted based on a thermodynamic energy budget that includes sea water freezing, the accumulation of snowfall, and the melting and sublimation of ice. The transport of sea ice by the upper ocean currents has been included in some oceanic GCMs.

4.2. Simulations with Atmospheric GCMs Coupled to Simplified Ocean Models

The first simulation of the change in the equilibrium climate of the Earth due to a change in solar "constant" was performed by *Wetherald and Manabe* [1975] using an atmospheric GCM coupled to a swamp ocean model. This type of simplified ocean model has zero heat capacity but is perpetually wet. It was chosen to allow the climate model to reach its equilibrium climate using a minimum amount of computer time. Four simulations were made: a control and three experiments in which the solar constant was increased by 2%, decreased by 2% and decreased by 4%. It was found that the temperature of the troposphere increases as the solar constant increases, with the largest increase occurring in the surface layer of higher latitudes due to the positive snow-cover/temperature feedback and the suppression of vertical mixing there. The hydrological cycle was found to be extremely sensitive to increasing the solar constant, with a 27% intensification for a 6% increase in irradiance (given by the +2% simulation minus the -4% simulation). Also the latitude of maximum snowfall retreated poleward with increased solar constant and the snowfall and snow accumulation amounts decreased markedly.

A more-recent study by *Rind et al.* [1999] used a mixed-layer ocean (MLO) model coupled to an atmospheric GCM to simulate the climate response to time-dependent solar forcing since 1600. A MLO model, unlike an oceanic GCM, calculates only the temperature of the uppermost ocean—the mixed layer, wherein the temperature is uniform with depth—and the thickness of sea ice, both as a function of geographical position and time throughout the year. The MLO model that was used included the transport of heat to the deeper ocean as a parameterized vertical heat diffusion through the bottom of the mixed layer, but did not include oceanic currents or salinity. Consequently *Rind et al.* wrote: "Thus these experiments are an exploration, rather than a specification of the role of solar irradiance variations in the climate variations of the past half millennium …" It was found that the solar forcing increase of 0.25% since 1600 constructed by *Lean and colleagues* [*Lean et al.*, 1995; *Fröhlich and Lean*, 1998] (Section 3) accounted for a 0.45°C temperature increase since 1600 and an increase of about 0.2°C over the past 100 years. However, solar forcing by itself was found not to be sufficient to produce the observed rapid warming during the last several decades.

A list of some of the studies using general circulation models and models of complexity intermediate between simple climate and general circulation models is presented in Table 3, in chronological order from 1975, together with their corresponding titles.

4.3. "Optimal Detection" Method

"Optimal detection" is least-squares regression in which the amplitude in observational data of model-simulated spatio-temporal patterns of climate change is estimated. The method is "optimal" because it gives more (less) weight to the less (more) variable components of the patterns. "Optimal detection" generally consists of a detection test and an attribution test.

4.3.1. Detection test. In the detection test a vector of detection variables $\mathbf{d} = d_k$ (k = 1, …, p) is defined by

$$d_k = \sum_{i=1}^{n} f_k^i O_i, \quad k = 1,...,p, \quad (1)$$

where $\mathbf{O} = O_i$ (i = 1, …, n) represents the observed climate state vector in an n-dimensional space—for example, near-surface air temperature at gridpoints, $\mathbf{f_k} = f_k^i$ (i = 1, …, n) is a vector of optimal fingerprints associated with different signal patterns $\mathbf{s_k} = s_k^i$ (i = 1, …, n) simulated by an AOGCM for radiative forcing k – such as GHG, ASA, and the sun – with

$$\mathbf{f_k} = C^{-1} \mathbf{s_k} \quad (k = 1,...,p), \quad (2)$$

where C^{-1} is the inverse of the covariance matrix of random climatic noise given either by the AOGCM or observations. The multiplication of the $\mathbf{s_k}$ by C^{-1} enhances features of the signal pattern for which the natural variability is low and suppresses features for which the variability is high. This approach can be understood as regressing the observations on the AOGCM-simulated signal pattern using a scalar product that takes into account inversely the noise of the climatic quantities. In practice it is useful to transform the signal patterns and optimal fingerprints such that they are orthonormal via

$$\mathbf{s}_l^T C^{-1} \mathbf{s}_m = \delta_{lm} \quad , \quad \mathbf{f}_l^T C^{-1} \mathbf{f}_m = \delta_{lm} \quad , \quad (3)$$

where δ_{lm} is the Kronecker delta.

Detection is said to be achieved at some chosen confidence level P (e.g., 95%) if the null hypothesis that $\mathbf{d}$ can be explained by natural variability alone is rejected at that level, that is, if the probability that the AOGCM-simulated signal is sampled from natural variability is less than $1 \pm P$ (e.g., 5%).

4.3.2. Attribution test. The rejection of the null hypothesis in the detection step implies only that the magnitude of the detection variable is too large to be explained by natural variability. It does not address the question whether the observations are in fact consistent with the model-simulated signal. This is the problem of attribution.

Table 3. Some studies using General Circulation Models and models of complexity intermediate between Simple Climate and General Circulation Models (shown by *); in chronological order from 1975.

Study	Title
Wetherald and Manabe [1975]	The effects of changing the solar constant on the climate of a general circulation model
Gordon and Davies [1977]	The sensitivity of response of climatic characteristics in a two-level general circulation model to small changes in solar radiation
Otto-Bliesner and Houghton [1986]	Sensitivity of the seasonal climate of a general circulation model to ocean surface conditions and solar forcing
Hansen et al. [1990]	Comparison of solar and other influences on long-term climate
Nesme-Ribes et al. [1993]	Solar dynamics and its impact on solar irradiance and the terrestrial climate
Rind and Overpeck [1993]	Hypothesized causes of decade-to-century climate variability: Climate model results
Brasseur [1993]*	The response of the middle atmosphere to long-term and short-term solar variability: A two-dimensional model
Marshall et al. [1994]	A comparison of GCM sensitivity to changes in CO_2 and solar luminosity
Syktus et al. [1994]	Sensitivity of a coupled atmosphere-dynamic upper ocean GCM to variations of CO_2, solar constant, and orbital forcing
Haigh [1994]*	The role of stratospheric ozone in modulating the solar forcing of climate
Felzer et al. [1995]	A systematic study of GCM sensitivity to latitudinal changes in solar radiation
Haigh [1996]	The impact of solar variability on climate
Cubasch et al. [1997]	Simulation of the influence of solar radiation variations on the global climate with an ocean-atmosphere general circulation model
Hegerl et al. [1997]	Multi-fingerprint detection and attribution analysis of greenhouse gas, greenhouse gas-plus-aerosol and solar forced climate change
Lean [1996]	The impact of solar irradiance on the Maunder Minimum climate
Tett et al. [1999]	Causes of Earth's near-surface temperature change in the twentieth century
Bertrand and Ypersele [1999]*	Potential role of solar variability as an agent for climate change
Drijfhout et al. [1999]	Solar-induced versus internal variability in a coupled climate model
Balachandran et al. [1999]	Effects of solar cycle variability on the lower stratosphere and troposphere
Haigh [1999b]	Modelling the impact of solar variability on climate
Haigh [1999a]	A GCM study of climate change in response to the 11-year solar cycle
Bertrand et al. [1999]*	Volcanic and solar impacts on climate since 1700
Rind et al. [1999]	Simulated time-dependent climate response to solar radiative forcing since 1600
Shindell et al. [1999]	Solar cycle variability, ozone, and climate
Soon et al. [2000]*	Climate hypersensitivity to solar forcing?
Stott et al. [2001]	Attribution of twentieth century temperature change to natural and anthropogenic causes

In the attribution test the consistency of a detected climate-change signal with the signal simulated by an AOGCM is performed by representing the observed climate-change signal in terms of the s_k (k = 1, ..., p),

$$O = \sum_{k=1}^{p} \alpha_k s_k + R \quad , \qquad (4)$$

where R is a residual and the amplitude vector α_k (k = 1, ..., p) is obtained from the condition that the mean square residual is minimized

$$R^T C^{-1} R = \min \quad , \qquad (5)$$

where superscript T denotes the transpose. If the signal patterns are statistically orthonormal relative to the noise

covariance matrix (Eq. 3), the detection variable components are statistically uncorrelated and

$$\alpha_k = d_k (k = 1,...,p) \qquad (6)$$

A signal or combination of signals explains the observations if the amplitude of the residual R is consistent, at some prescribed statistical level (e.g., 5%), with internal climate variability estimated from the control simulation or observations.

4.4. Simulation, detection and attribution of a solar signal in the observed temperature record

"Fingerprint-detection" studies to simulate, detect and attribute a solar signal in the observed temperature record have been performed at the Max-Planck Institute for Meteorology by *Hegerl et al.* [1997] and at the Hadley Centre for Climate Prediction and Research by *Tett et al.* [1999] and *Stott et al.* [2001]. These studies are described below.

4.4.1. Max-Planck Institute for Meteorology. The study performed by *Hegerl et al.* [1997] used three simulations performed with an AOGCM of the Max-Planck Institute in Hamburg. The atmospheric model was the ECHAM3 spectral GCM with 19 vertical layers and a horizontal resolution of approximately 5.6° latitude by 5.6° longitude. The oceanic model was the Large Scale Quasi-geostrophic (LSG) model with 11 vertical layers and a horizontal resolution of 5.6° latitude by 5.6° longitude. The AOGCM was flux corrected.

Three anthropogenic climate-change simulations were performed from 1880 to 2049, two (A and B) with radiative forcing by both greenhouse gases and the direct (clear air) effect of sulfate aerosols – treated as a change in surface albedo, and one (C) with only greenhouse-gas forcing. Two simulations (SOL1 and SOL2) were performed by *Cubasch et al.* [1997] from 1700 to 1992 using the radiative forcing of *Hoyt and Schatten* [1993]. Simulations A and B differed by a small perturbation in the aerosol field. Simulations SOL1 and SOL2 started from different state A of the control simulation.

The natural variability used in the detection and attribution tests was obtained from the control simulations of four different AOGCMs and from the observed monthly mean near-surface temperatures from 1854 to 1995 on a 5° × 5° grid [*Jones and Briffa*, 1992; *Jones*, 1994a,b] after subtracting an estimated GHG signal based on the first EOF of simulation C.

The detection test was performed using a single detection variable for which the signal was the first empirical orthogonal function (EOF) for the change in near-surface temperature averaged for simulations A and B. The attribution test was performed using two detection variables. Detection variable 1 used the signal given by the first EOF for the change in near-surface temperature for simulation C. Detection variable 2 used the orthonormalized pattern for the difference between the first EOF for the change in near-surface temperature averaged for simulations A and B and the first EOF for temperature for simulation C. Thus detection variable 2 was equivalent to the change in near-surface temperature caused by the sulfate aerosol. The solar-induced temperature change pattern was not used to define a detection variable because it was poorly defined.

The detection and attribution tests were applied to the 30-year trends of the observed annual near-surface temperature

as well as to the average of simulations A and B, C and the average of SOL1 and SOL2.

The detection test using the single detection variable showed that the null hypothesis that the observed temperature trends are natural can be rejected with 97.5% confidence. It was also indicated that the warming during the first half of the 20th century was partly forced and partly internal climate variability, with the forcing due to the small increase in GHG concentration and possibly to an increase in solar insolation.

The attribution test using the two detection variables yielded inconclusive results, with the difference between the observations and the simulated solar-only forced 30-year trends not significant at the 10% level. Application of the attribution test to 50-year trends of northern hemisphere summer temperature showed that the latest trend patterns was inconsistent at the 95% confidence level with pure GHG forcing and at the 90% confidence level with pure solar forcing.

Based on these tests it was concluded that "the observed climate change is consistent with a combined greenhouse gas and aerosol forcing, but inconsistent with greenhouse gas or solar forcing alone."

4.4.2. Hadley Centre for Climate Prediction and Research. The studies performed by *Tett et al.* [1999] and *Stott et al.* [2001] used 16 simulations performed with the HadCM2 AOGCM of the Hadley Centre. The atmospheric GCM was a gridpoint model with 19 vertical layers and a horizontal resolution of 2.5° latitude × 3.75° longitude. The oceanic GCM had 20 vertical layers and the same horizontal resolution as the atmospheric GCM. The AOGCM was flux corrected.

Two anthropogenic climate change experiments were performed from 1860 to 1996, one (GS) with radiative forcing by both greenhouse gases and the assumed joint direct (clear air) and indirect (cloudy air) effects of sulfate aerosols – treated as a change in surface albedo, and one (G) with only greenhouse-gas forcing. A pure sulfate signal ($S \equiv GS - G$) was generated and combined with G in the attribution test (Eq. 4) to obtain a signal (G+S) in which the GHG and sulfate components could be different from those in the GS experiment. Two non-anthropogenic climate change experiments were performed, one (Vol) with volcanic stratospheric aerosols from 1850 to 1996 and one (Sol) with the *Hoyt and Schatten* [1993] radiative forcing from 1700–1991, with an extension to 1996 using satellite observations. Each of the four experiments consisted of an ensemble of four simulations, each started from different initial conditions.

The natural variability used in the detection and attribution tests was obtained from a 1700-year control simulation by HadCM2. The observed near-surface temperatures on a 5°×5° grid were from *Parker et al.* [1994].

It was found that no more than two signal amplitudes α_k could be estimated simultaneously. Accordingly, the attribution test was performed using both a single detection variable – G, GS, Sol, Vol – and two detection variables: G+S, G+Sol, G+Vol, GS+Sol, GS+Vol and Sol+Vol. The test was performed on 10-year-mean temperatures over the five 50-year periods: 1906–1956, 1916–66, 1926–76, 1936–86, and 1946–96. For a signal or signal combination to explain the temperature changes over a 50-year period, it was required that the amplitude of the estimated residual **R** in Eq. (4) be consistent, at the 5% level, with internal climate variability estimated from the control simulation. This was tested using an F-test with 21 degrees of freedom. For a signal combination to explain the change over the century, it was required that **R** be consistent in all five 50-year periods. A signal was taken to be detected when the null hypothesis that the signal amplitude is zero was rejected, the consistency test was passed, and the signal amplitude was positive.

Table 4 presents the results of the consistency test for the five 50-year periods. Probability values P shown in bold are inconsistent with the observations at the 5% level ($P \leq 0.05$). A signal name is shown in superscript next to the P-value when its amplitude α_k has an entirely positive range, which is equivalent to detecting that the signal is consistent with the observations.

Table 4 shows that internal variability alone cannot explain the early century or late century warmings (Figure 4) and no combination of internal variability and external natural forcings can explain the warming since 1945. Although G and GS are not inconsistent with the observations for all five of the 50-year periods, they are detected

only for the first and last 50-year periods. Sol is detected only for the 1906–56 period. Vol is not detected for any time period. Thus none of the four individual signals alone provides an adequate explanation of twentieth century temperature change.

Figure 7 shows the ellipses containing 90% of the estimated joint distribution of signal amplitudes for G+S and for GS+Sol for 1906–56 and 1945–1996. The G+S signals are detected for 1946–96 but not for 1906–56 because the ellipse encompasses the unit amplitude point (1, 1) for the former period but not the latter period. Conversely, the GS+Sol signals are detected for 1906–56 but not for 1946–96. These results were also obtained for the four ensemble simulations that were made as a sensitivity test using the solar radiative forcing constructed by *Lean and colleagues* [*Lean et al.*, 1995; *Fröhlich and Lean*, 1998] (Section 3).

It was found, however, that during 1906–56 the amplitude of the sulfate aerosol signal relative to the GHG signal in G+S was significantly less than that prescribed in GS, but a solar signal was not detected for G+S. Accordingly, *Tett et al.* [1999] wrote: "thus our detection of a solar signal should be treated with some caution," and concluded that "solar forcing may have contributed to the temperature changes early in the {20th} century, but anthropogenic causes combined with natural variability would also present a possible explanation."

Stott et al. [2001] extended the analysis of *Tett et al.* [1999] to examine the four seasons and found that up to four signal amplitudes could be estimated simultaneously, whereas *Tett et al.* [1999] were able to estimate only two signal amplitudes using annual-mean results. *Stott et al.* [2001] found on the basis of the seasonal analysis (Table 5)

Table 4. Consistency F-test for the climate-change signals for the studies by *Tett et al.* [1999] using decadal-mean, annual-mean data.

Period	1906–1956	1916–66	1926–76	1936–86	1946–1996
Internal var[a]	**0.01**	**0.04**	0.69	0.76	**0.01**
G	0.18[G]	0.11	0.75	0.80	0.10[G]
GS	0.06[GS]	0.23	0.76	0.85	0.30[GS]
Sol[a]	0.11[Sol]	0.06	0.75	0.87	**0.03**[Sol]
Vol[a]	**0.03**	0.08	0.74	0.80	**0.04**[b]
GHG+S	0.21[GHG]	0.17[GHG]	0.68	0.83	0.31[GHG,S]
G+Sol	0.23[G]	0.17[G,Sol]	0.68	0.81	0.07[G]
G+Vol	0.24[G]	0.09	0.67	0.73	0.08[G]
GS+Sol	0.13[Sol]	0.19[GS]	0.68	0.82	0.25[GS]
GS+Vol	0.05[GS]	0.22[GS]	0.68	0.80	0.25[GS]
Sol+Vol[a]	0.08[Sol]	0.09[b]	0.67	0.82	**0.04**[b]

[a]Signal combination is an inadequate explanation of twentieth century temperature change as the F-test fail at least once.
[b]Volcanic amplitude is significantly negative and thus unphysical.

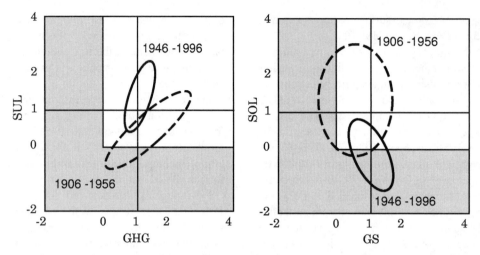

Figure 7. Ellipses containing an estimated 90% of the joint distribution of signal amplitudes for the GHG+SUL (= S) signals (left) and the GS+SOL signals (right). Each ellipse is centered on its best estimate and its size reflects the uncertainty in the amplitudes of the two signals. Adapted from *Stott et al.* [2001].

that the two-signal combinations found by *Tett et al.* [1999], GHG+S ($\alpha_{GHG} \equiv \alpha_G + \alpha_{GS}$ is the best estimate of the pure greenhouse effect) and GS+Sol, had to be rejected because there was a detectable missing signal: Sol in GHG+S+Sol and G in G+GS+Sol. Accordingly, it was concluded that the only consistent explanation of twentieth century temperature change using seasonal data is GHG+S+Sol.

Decadal-mean global-mean temperature changes constructed for the GHG+S+Sol signal combination show a steady warming through the 1906–56 period by GHGs, with a warming peak produced in 1936–46 by the solar influence. In the latter part of the century, cooling by sulfate aerosols balances GHG warming between 1946 and 1976, with warming by GHGs dominating thereafter.

Constructed 50-year trends for the GHG+S+Sol signal combination show the observed temperature changes over the twentieth century result from a combination of steadily increasing GHG forcing, partly balance by steadily increas-

Table 5. Consistency F-test for the climate-change signals for the study by *Stott et al.* [2001] using decadal-mean, seasonal-mean data.

Period	1906–1956	1916–66	1926–76	1936–86	1946–1996
Internal var[a]	**0.01**	**0.02**	0.12	0.05	**0.00**
G[a]	0.26[G]	0.18[G]	0.19	0.10	**0.04**[G]
GS[a]	**0.02**	0.27[GS]	0.25	0.19	0.13[GS]
Sol[a]	0.11[Sol]	0.06	0.21	0.10	**0.01**[Sol]
Vol[a]	**0.03**	0.16[a]	0.23	0.09	**0.01**[b]
GHG+S	0.21[GHG]	0.25[GHG,S]	0.22	0.23[GHG,S]	0.13[GHG,S]
G+Sol[a]	0.47[G,Sol]	0.17[G]	0.16	0.08	**0.03**[G]
G+Vol	0.27[Ga]	0.22[a]	0.18	0.08	0.10[G,Vol]
GS+Sol	0.10[Sol]	0.22[GS]	0.24	0.15	0.12[GS]
GS+Vol[a]	**0.02**	0.40[GSa]	0.21	0.15	0.15[GS]
Sol+Vol[a]	0.08[Sol]	0.13[b]	0.18	0.08	**0.01**[Sol]
GHG+S+Sol	0.41[GHG,Sol]	0.19[GHG,S]	0.22	0.20[GHG,S]	0.11[GHG,S]
GHG+S+Vol	0.21[GHGa]	0.33[GHG,Sa]	0.21	0.17[GHG,S]	0.11[GHG]
GS+Sol+Vol	0.07[Sol]	0.37[GSa]	0.18	0.12	0.21[G,Vol]
G+Sol+Vol	0.81[G,Sol]	0.52[G,Sola]	0.14	0.06	0.28[G,Vol]

[a]Signal combination is an inadequate explanation of twentieth century temperature change as the F-test fail at least once.
[b]Volcanic amplitude is significantly negative and thus unphysical.

ing sulfate cooling, with a relatively small but statistically significant solar contribution to the large early century warming.

Analyzing the seasons separately showed a greater likelihood of a positive solar signal in combination with GS in March-April-May and June-July-August than annual. Moreover, including seasonal information and optimizing, with the highest weight given to MAM and JJA in 1906–56, detection of a solar signal was no longer dependent, as it was in *Tett et al.* [1999], on fixing the relative amplitudes of the GHG and S signals close to those prescribed in the GCM.

Following *Tett et al.* [1999], *Stott et al.* [2001] examined the sensitivity of their solar findings to the uncertainty in solar forcing by performing an ensemble of four simulations for the radiative forcing constructed by *Lean and colleagues* [*Lean et al.*, 1995; *Fröhlich and Lean*, 1998] (LBB) and comparing its results with those obtained for the *Hoyt and Schatten* [1993] solar forcing (Sol).

In terms of annual-mean data, both Sol and LBB are detected when added to GS. However, Sol is detected only in the 1906–56 period while LBB is detected in both 1906–56 and 1916–66. The detection of both Sol and LBB in the 1906–56 period depends on the relative amplitudes of the GHG and sulfate aerosols being fixed close to those in GS; however, this is not so for LBB in 1916–66. In contrast to Sol, there are no detections of anthropogenic signals in combinations with LBB in the early century periods.

In terms of the seasonal data, LBB is not detected in any 50-year period in the three-way regression with GHG+S, and additionally GS+LBB fails the F-test for consistency. The decadal-mean global-mean temperatures constructed for both Sol and LBB show a mid-century peak which occurs in 1946–56 for LBB and in 1936–46 for Sol and the observations. This leads to an inconsistency between the best-fit reconstruction for GS+LBB in 1936–46 that is significant according to the F-test on residuals using seasonal data. Accordingly, it was concluded that detection of a solar influence on early century temperature change is sensitive to which construction of total solar irradiance is used.

5. A WAY FORWARD

In the three detection and attribution studies described above, the total solar irradiance was changed without changing the spectral distribution of irradiance. Yet the satellite observations of the sun show that the change in irradiance over the 11-year sunspot cycle is spectrally non-uniform, with changes for ultraviolet wavelengths (200–400 nm) that are about 3–4 times those for the total irradiance [*Pap*, 2001]. There is increasing evidence from meteorolog-

ical observations and GCM simulations that these variations in UV radiation have an effect on the earth's atmosphere.

The observational studies by *Labitzke* [1982; 1987] and *Labitzke and van Loon* [1988; 1989; 1992; 1993; 1997] have shown strong correlations between the height of isobaric surfaces in the stratosphere and the 11-year sunspot cycle during the easterly phase of the Quasi-Biennial Oscillation (QBO) of the zonal wind in the stratosphere, with higher (lower) heights during solar maximum (minimum).

The simulation studies of *Haigh* [1994; 1996; 1999a, b] with an intermediate complexity climate model or a GCM coupled to an interactive photochemistry model have indicated that the change in UV flux causes a change in the amount of stratospheric ozone. This, in turn, changes the radiative forcing of the underlying troposphere which leads to changes in the subtropical jets, the tropical Hadley circulation and the midlatitude storm tracks.

The simulation study of *Balachandran et al.* [1999] examined the effect of a ±5% change in the solar UV radiation, which was taken to represent the difference between solar maximum and solar minimum, in the Goddard Institute for Space Studies (GISS) GCM without ozone photochemistry or ozone transport. The simulated atmospheric differences included increased geopotential heights in the subtropics at solar maximum relative to solar minimum, and a strong dipole pattern of height differences when the results were partitioned according to the phase of the quasi-biennial oscillation (QBO), which was imposed on the GCM. The solar UV changes directly affected the stratosphere by changing the vertical gradients of temperature and zonal wind. These, in turn, changed the conditions for propagation of tropospheric planetary waves into the stratosphere which thereby influenced the circulation of the troposphere.

A follow-on study by *Shindell et al.* [1999] using the GISS GCM with rudimentary photochemistry and non-interactive ozone transport, found an enhancement of both the response of the stratosphere and troposphere to the 11-year sunspot cycle of solar ultraviolet radiation, solar maximum minus solar minimum, due to the photochemically induced ozone changes. For the near-surface air, zonal-mean warming of up to 0.5°C was found poleward of 55°N latitude.

Rozanov et al. [2001, 2002] have performed a simulation of the effect of the 11-year sunspot cycle of solar ultraviolet radiation (Figure 8) with their coupled stratosphere-troposphere general circulation model with interactive ozone photochemistry and interactive ozone transport. This model calculates the concentrations of 43 chemical species via 199 gas-phase reactions and 8 heterogeneous reactions on type I and type II polar stratospheric clouds which the model also

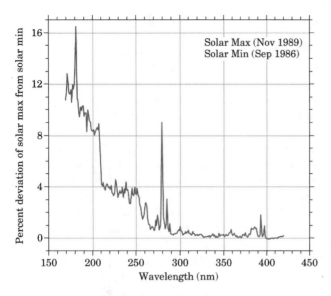

Figure 8. The increase in solar ultraviolet radiation from September 1986 to November 1989. Source: *Rozanov et al.* [2004].

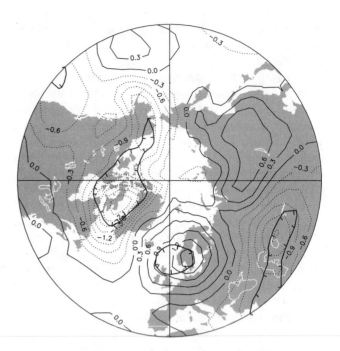

Figure 9. The change in near-surface temperature for December-January-February simulated by the University of Illinois at Urbana-Champaign stratosphere-troposphere general circulation model with interactive photochemistry and interactive ozone transport in response to the change in solar ultraviolet radiation shown in Figure 8. Source: *Rozanov et al.* [2004].

calculates. A control ensemble of five 15-year simulations was made with the average spectral solar irradiance and an experiment ensemble of five 15-year simulations was made with the observed difference in solar irradiance between solar maximum (November 1989) and solar minimum (September 1986) added to the average spectral solar irradiance. The simulations in the ensembles were started from slightly different initial conditions. In all simulations the monthly climatological sea-surface temperature and sea-ice distributions were prescribed and interpolated to each day of the year. The difference in temperature between the experiment and control simulations thus consists of the solar signal and natural variability. Figure 9 shows that there are statistically significant near-surface temperature changes from solar minimum to solar maximum of up to 2.4°C in northern hemisphere winter in middle and high latitudes.

For a much longer timescale than the 11-year sunspot cycle, *Shindell et al.* [2001] have used the GISS GCM with simplified photochemistry and no interactive ozone transport coupled to a mixed-layer ocean to simulate the equilibrium climates of the Maunder Minimum (1680), when both the total and UV solar irradiances constructed by *Lean et al.* [1995] were at a minimum, and the time period 100 years later when solar irradiance was relatively high over several decades (Figure 3). The simulated global-mean near-surface temperature during the Maunder Minimum was about 0.34°C colder than during the late 18th century, a quantitative result that depends on the GCM's climate sensitivity,

the real-world's value of which is highly uncertain as shown in Section 3. Of perhaps greater significance, therefore, was the qualitative change in the simulated regional climate patterns, with a shift during decreased irradiance toward the low-index state of the Arctic Oscillation/North Atlantic Oscillation [e.g., *Hurrell et al.*, 2001], with relatively higher pressures in the Arctic and lower pressures in the middle latitudes of the Northern Hemisphere. This weakened the polar vortex and reduced onshore transport of warm air from the oceans to the continents, with an attendant cooling of 1°C to 2°C over North America and Eurasia.

These recent studies show the importance of varying not only the total solar irradiance, but also the solar UV component. A way forward then to answer the question, Has the Sun changed climate?, is to repeat the detection and attribution studies described in the preceding section using coupled GCM/photochemistry models to simulate the climate changes over the past 150 years caused by the changes in the spectral distribution of solar radiation over that period of time.

Acknowledgments. This research was supported by the National Science Foundation under grant ATM-0084270.

REFERENCES

Allen, M. R., and S. F. B. Tett, Checking for model consistency in optimal fingerprinting, *Climate Dynamics*, *15*, 419–434, 1999.

Andronova, N. G., E. V. Rozanov, F. Yang, M. E. Schlesinger, and G. L. Stenchikov, Radiative forcing by volcanic aerosols from 1850 through 1994, *J. Geophys. Res.*, *104* (D14), 16,807–16,826, 1999.

Andronova, N. G., and M. E. Schlesinger, Causes of Global Temperature Changes During the 19th and 20th Centuries, *Geophys. Res. Lett.*, *27* (14), 2137–2140, 2000.

Andronova, N. G., and M. E. Schlesinger, Objective estimation of the probability density function for climate sensitivity, *J. Geophys. Res.*, *106* (D19), 22,605–22,612, 2001.

Balachandran, N. K., D. Rind, P. Lonergan, and D. T. Shindell, Effects of solar cycle variability on the lower stratosphere and troposphere, *J. Geophys. Res.*, *104* (D22), 27,321–27,339, 1999.

Baliunas, S., and R. Jastrow, Evidence for long-term brightness changes of solar-type stars, *Nature*, *348*, 520–523, 1990.

Barnett, T., Detection of changes in the global tropospheric temperature field induced by greenhouse gases, *J. Geophys. Res.*, *91*, 6659–6667, 1986.

Barnett, T. P., and M. E. Schlesinger, Detecting Changes in Global Climate Induced by Greenhouse Gases, *J. Geophys. Res.*, *92*, 14,772–14,780, 1987.

Beer, J., S. Baumgartner, B. Dittrich-Hannen, J. Hauenstein, P. Kubik, C. Lukasczyk, W. Mende, R. Stellmacher, and M. Suter, Solar variability traced by cosmogenic isotopes, in *The Sun as a Variable Star: Solar and Stellar Irradiance Variations*, edited by J.M. Pap, C. Fröhlich, H.S. Hudson, and S. K. Solanki, pp. 291–300, Cambridge University Press, 1994.

Beer, J., W. Mende, and R. Stellmacher, The role of the sun in climate forcing, *Quat. Sci. Rev.*, *19* (1–5), 403–415, 2000.

Beer, J., U. Siegenthaler, G. Bonani, R. C. Finkel, H. Oeschger, M. Suter, and W. Wölfi, Information on past solar activity and geomagnetism from 10Be in the Camp Century ice core, *Nature*, *331* (675–679), 1988.

Bertrand, C., and J.-P. v. Ypersele, Potential role of solar variability as an agent for climate change, *Climatic Change*, *43*, 387–411, 1999.

Bertrand, C., J.-P. v. Ypersele, and A. Berger, Volcanic and solar impacts on climate since 1700, *Climate Dynamics*, *15*, 355–367, 1999.

Brasseur, G., The response of the middle atmosphere to long-term and short-term solar variability: A two-dimensional model, *J. Geophys. Res.*, *98* (D12), 23,079–23,090, 1993.

Bretherton, F. P., K. Bryan, and J. D. Woods, Time-dependent greenhouse-gas-induced climate change, in *Climatic Change: The IPCC Scientific Assessment*, edited by J. T. Houghton, G.J. Jenkins, and J.J. Ephraums, pp. 173–193, Cambridge University Press, Cambridge, England, 1990.

Budyko, M. I., The effect of solar radiation variations on the climate of the earth, *Tellus*, *21*, 611–619, 1969.

Crowley, T. J., Causes of climate change over the past 1000 years, *Science*, *289*, 270–277, 2000.

Crowley, T. J., and K.-Y. Kim, Comparison of proxy records of climate change and solar forcing, *Geophys. Res. Lett.*, *23* (4), 359–362, 1996.

Cubasch, U., R. Voss, G. C. Hegerl, J. Waszkewitz, and T. J. Crowley, Simulation of the influence of solar radiation variations on the global climate with an ocean-atmosphere general circulation model, *Climate Dynamics*, *13*, 757–767, 1997.

Drijfhout, S. S., R. J. Haarsma, J. D. Opsteegh, and F. M. Selten, Solar-induced versus internal variability in a coupled climate model, *Geophys. Res. Lett.*, *26* (2), 205–208, 1999.

Eddy, J. A., The Maunder Minimum, *Science*, *192*, 1189–1202, 1976.

Eddy, J. A., Climate and the changing sun, *Climatic Change*, *1*, 173–190, 1979.

Felzer, B., R. J. Oglesby, H. Shao, T. W. III, D. E. Hyman, W. L. Prell, and J. Kutzbach, A systematic study of GCM sensitivity to latitudinal changes in solar radiation, *J. Climate*, *8*, 877–887, 1995.

Forster, P. T. D., Radiative forcing due to stratospheric ozone changes 1979–1997, using updated trend estimates, *J. Geophys. Res.*, *104* (D20), 24,395–24,399, 1999.

Friis-Christensen, F., and L. Lassen, Length of the solar cycle: An indicator of solar activity closely associated with climate, *Science*, *254*, 698–700, 1991.

Fröhlich, C., Contemporary measures of the solar constant, in *The Solar Output and its Variation*, edited by O.R. White, pp. 93–109, Colorado Associated University Press, Boulder, 1977.

Fröhlich, C., and J. Lean, The Sun's total irradiance: Cycles, trends and related climate change uncertainties since 1976, *Geophys. Res. Lett.*, *25* (23), 4377–4380, 1998.

Gilliland, R. L., Solar, volcanic, and CO_2 forcing of recent climatic changes, *Climatic Change*, *4*, 111–131, 1982.

Gilliland, R. L., and S. H. Schneider, Volcanic, CO_2 and solar forcing of northern and southern hemisphere surface air temperatures, *Nature*, *310*, 38–41, 1984.

Gordon, H. B., and D. R. Davies, The sensitivity of response of climatic characteristics in a two-level general circulation model to small changes in solar radiation, *Tellus*, *29*, 484–501, 1977.

Haigh, J. D., The role of stratospheric ozone in modulating the solar forcing of climate, *Nature*, *370*, 544–546, 1994.

Haigh, J. D., The impact of solar variability on climate, *Science*, *272*, 981–984, 1996.

Haigh, J. D., A GCM study of climate change in response to the 11-year solar cycle, *Quart. J. Roy. Meteorol. Soc.*, *125*, 871–892, 1999a.

Haigh, J. D., Modelling the impact of solar variability on climate, *J. Atmospheric and Solar-Terrestrial Phys*, *61*, 63–72, 1999b.

Hammitt, J. K., R. J. Lempert, and M. E. Schlesinger, A sequential-decision strategy for abating climate change, *Nature*, *357*, 315–318, 1992.

Hansen, J. E., A. A. Lacis, and R. A. Ruedy, Comparison of solar and other influences on long-term climate, in *Climate Impact of Solar Variability*, edited by K.H. Schatten, pp. 135–145, NASA, NASA Goddard Space Flight Center, 1990.

Harvey, L. D. D., J. Gregory, M. Hoffert, A. Jain, M. Lal, R. Leemans, S. B. C. Raper, T. M. L. Wigley, and J. de Wolde, An introduction to simple climate models used in the IPCC Second Assessment Report, pp. 50, Intergovernmental Panel on Climate Change, Bracknell, U.K., 1997.

Hasselmann, K., On the signal-to-noise problem in atmospheric response studies., in *Meteorology over the tropical oceans*, edited by B. D. Shaw, pp. 251–259, Royal Meteorological Society, Bracknell, Berkshire, England, 1979.

Hasselmann, K., Optimal fingerprints for the detection of time dependent climate change, *J. Climate, 6*, 1957–1971, 1993.

Hasselmann, K., Multi-pattern fingerprint method for detection and attribution of climate change, *Climate Dynamics, 13*, 601–611, 1997.

Hegerl, G. C., K. Hasselmann, U. Cubasch, J. Mitchell, E. Roeckner, R. Voss, and J. Waszkewitz, Multi-fingerprint detection and attribution analysis of greenhouse gas, greenhouse gas-plus-aerosol and solar forced climate change, *Climate Dynamics, 13*, 613–634, 1997.

Hoffert, M. I., The effects of solar variability on climate, *Climatic Change, 13*, 267–285, 1988.

Hoffert, M. I., A. J. Callegari, and C.-T. Hsieh, The role of deep sea heat storage in the secular response to climatic forcing, *J. Geophys. Res., 85*, 6667–6679, 1980.

Houghton, J. T., Y. Ding, D. J. Griggs, M. Noguer, P. J. v. d. Linden, and D. Xiaosu, *Climate Change 2001: The Scientific Basis*, 944 pp., Cambridge University Press, Cambridge UK, 2001.

Houghton, J. T., G. J. Jenkins, and J. J. Ephraums, *Climate Change: The IPCC Scientific Assessment*, 364 pp., Cambridge University Press, Cambridge, 1990.

Hoyt, D. V., Variations in sunspot structure and climate, *Climatic Change, 2*, 79–92, 1979.

Hoyt, D. V., and K. H. Schatten, A discussion of plausible solar irradiance variations, 1700–1992, *J. Geophys. Res., 98*, 18,895–18,906, 1993.

Hoyt, D. V., and R. A. Siquig, Possible influences of volcanic dust veils or changes in solar luminosity on long-term local temperature records, *J. Atmos. Sci., 39*, 680–685, 1982.

Hurrell, J. W., Y. Kushnir, and M. Visbeck, The North Atlantic Oscillation, *Science, 291*, 603–604, 2001.

Jones, P., M. New, D. E. Parker, S. Martin, and I. G. Rigor, Surface air temperature and its changes over the past 150 years, *Rev. Geophys., 37* (2), 173–199, 1999.

Jones, P. D., Recent warming in global temperature series, *Geophys. Res. Lett., 21*, 1149–1152, 1994a.

Jones, P. D., and K. R. Briffa, Global surface air temperature variations during the twentieth century: Part 1, spatial, temporal and seasonal details, *The Holocene, 2*, 165–179, 1992.

Jones, P. P., Hemispheric surface air temperature variability—a reanalysis and update to 1993, *J. Climate, 7*, 1794–1802, 1994b.

Kattenberg, A., F. Giorgi, H. Grassl, G. A. Meehl, J. B. F. Mitchell, R. J. Stouffer, T. Tokioka, A. J. Weaver, and T. M. L. Wigley, Climate Models—Projections of Future Climate, in *Climate Change 1995: The Science of Climate Change*, edited by J. T. Houghton, L.G. Meira Filho, B. A. Callander, N. Harris, A. Kattenberg, and K. Maskell, pp. 285–358, Cambridge University Press, Cambridge, U.K., 1996.

Kelly, P. M., and M. T, L. Wigley, The influence of solar forcing on global mean temperature since 1861, *Nature, 347*, 460–462, 1990.

Kelly, P. M., and T. M. L. Wigley, Solar cycle length, greenhouse forcing and global climate, *Nature, 360*, 328–330, 1992.

Kim, H. H., and N. E. Huang, Newly found evidence of Sun-climate relationships, pp. 37, NASA, Greenbelt, MD, 1993.

Labitzke, K., On interannual variability of the middle stratosphere during northern winters, *J. Meteorol. Soc. Japan, 60*, 124–139, 1982.

Labitzke, K., Sunspots, the QBO, and the stratospheric temperature in the north polar region, *Geophys. Res. Lett., 14*, 535–537, 1987.

Labitzke, K., and H. van Loon, Associations between the 11-year solar cycle, the QBO and the atmosphere. Part I: The troposphere and stratosphere in the northern hemisphere in winter, *J. Atmos. Terr. Physics, 50*, 197–206, 1988.

Labitzke, K., and H. van Loon, Associations between the 11-year solar cycle, the QBO and the atmosphere. Part III: Aspects of associations, *J. Climate, 2*, 554–565, 1989.

Labitzke, K., and H. van Loon, Associations between the 11-year solar cycle and the atmosphere. Part V: Summer, *J. Clim., 5*, 240–251, 1992.

Labitzke, K., and H. van Loon, Some recent studies of probable connections between solar and atmospheric variability, *Ann. Geophysicae, 11*, 1084–1094, 1993.

Labitzke, K., and H. van Loon, The signal of the 11-year sunspot cycle in the upper troposphere-lower stratosphere, *Space Sci. Rev., 80*, 393–410, 1997.

Lean, J., Reconstructions of Past Solar Variability, in *Climatic Variations and Forcing Mechanisms of the Last 2,000 Years*, pp. 519–532, Springer-Verlag, Berlin, 1996.

Lean, J., and D. Rind, The sun and climate, *Consequences, 2* (1), 27–36, 1996.

Lean, J., J. Beer, and R. Bradley, Reconstruction of solar irradiance since 1610: Implications for climate change, *Geophys. Res. Lett., 22* (23), 3195–3198, 1995.

Lempert, R. J., and M. E. Schlesinger, Robust Strategies for Abating Climate Change, *Climatic Change, 45* (3–4), 387–401, 2000.

Lempert, R. J., and M. E. Schlesinger, Adaptive Strategies for Climate Change, in *Innovative Energy Systems for CO_2 Stabilization*, edited by R.G. Watts, Cambridge University Press, Cambridge, 2001.

Lempert, R. J., M. E. Schlesinger, and S. C. Bankes, When we don't know the costs or the benefits: Adaptive strategies for abating climate change, *Climatic Change, 33*, 235–274, 1996.

Lempert, R. J., M. E. Schlesinger, S. C. Bankes, and N. G. Andronova, The impacts of climate variability on near-term policy choices and the value of information, *Climatic Change, 45* (1), 129–161, 2000.

Lempert, R. J., M. E. Schlesinger, and J. K. Hammitt, The impact of potential abrupt climate changes on near-term policy choices, *Climatic Change, 26*, 351–376, 1994.

Marshall, Institute, Global Warming Update: Recent Scientific Findings, pp. 32, Marshall Institute, Washington, D.C., 1992.

Marshall, S., R. J. Oglesby, J. W. Larson, and B. Saltzman, A comparison of GCM sensitivity to changes in CO_2 and solar luminosity, *Geophys. Res. Lett.*, *21* (23), 2487–2490, 1994.

Mass, C., and S. H. Schneider, Statistical evidence on the influence of sunspots and volcanic dust on long-term temperature records, *J. Atmos. Sci.*, *34*, 1995–2004, 1977.

Mendoza, B., R. Garduño, and J. Adem, The impact of solar irradiance on the Maunder Minimum climate, *J. Geomag. Geoelectr.*, *49*, 957–964, 1997.

Mitchell, J. M., Evidence of a 22-year rhythm of drought in the western United States related to the Hale solar cycle since the 17th century, in *Solar-Terrestrial Influences on Weather and Climate*, edited by B.M. McCormac, and T. A. Seliga, pp. 125–143, D. Reidel, Dordrecht, 1979.

Myhre, G., E. J. Highwood, K. P. Shine, and F. Stordal, New estimates of radiative forcing due to well mixed greenhouse gases, *Geophs. Res. Lett.*, *25* (14), 2715–2718, 1998.

Nastrom, G. D., and A. D. Belmont, Evidence for a solar cycle signal in tropospheric winds, *J. Geophys. Res.*, *85* (C1), 443–452, 1980.

Nesme-Ribes, E., E. N. Ferreira, R. Sadourny, H. L. Treut, and Z. X. Li, Solar dynamics and its impact on solar irradiance and the terrestrial climate, *J. Geophys. Res.*, 1993.

Nicolis, C., The effect of solar output, infrared cooling and latitudinal heat transport on the evolution of the earth's climate, *Tellus*, *31*, 193–198, 1979.

North, G. R., J. G. Mengel, and D. A. Short, Climatic response to a time varying solar constant, in *Weather and Climate Responses to Solar Variations*, edited by B.M. McCormac, pp. 243–256, Colorado Associated University Press, Boulder, 1983.

Otto-Bliesner, B. L., and D. D. Houghton, Sensitivity of the seasonal climate of a general circulation model to ocean surface conditions and solar forcing, *J. Geophys. Res.*, *91* (D6), 6682–6694, 1986.

Pap, J. M., Total solar and spectral irradiance variations from near-UV to infrared, in *The variable shape of the sun: Astrophysical consequences*, edited by J.P. Rozelot, Springer-Verlag, 2001.

Parker, D. E., P. D. Jones, C. K. Folland, and A. Bevan, Interdecadal changes of surface temperature since the late nineteenth century, *J. Geophys. Res.*, *99*, 14,373–14,399, 1994.

Peck, S. C., and T. J. Teisberg, Global Warming Uncertainties and the Value of Information: An Analysis Using CETA, *Resource and Energy Economics*, *15*, 71–97, 1993.

Peng, L., M.-D. Chou, and A. Arking, Climate studies with a multi-layer energy balance model. Part I: Model description and sensitivity to the solar constant, *J. Atmos. Sci.*, *39* (12), 2639–2656, 1982.

Ramankutty, N., An Empirical Estimate of Climate Sensitivity, M. S. thesis, University of Illinois at Urbana-Champaign, Urbana, IL, 1994.

Reid, G. C., Influence of solar variability on global sea surface temperature, *Nature*, *329*, 142–143, 1987.

Reid, G. C., Solar total irradiance variations and the global sea surface temperature record, *J. Geophys. Res.*, *96* (D2), 2835–2844, 1991.

Rind, D., J. Lean, and R. Healy, Simulated time-dependent climate response to solar radiative forcing since 1600, *J. Geophys. Res.*, *104* (D2), 1973–1990, 1999.

Rind, D., and J. Overpeck, Hypothesized causes of decade-to-century climate variability: Climate model results, *Quaternary Sci. Rev.*, *12* (6), 357–374, 1993.

Robock, A., Internally and externally caused climate changes, *J. Atmos. Sci.*, *35*, 1111–1122, 1978.

Rozanov, E. V., M. E. Schlesinger, T. A. Egorova, B. Li, N. Andronova, and V. A. Zubov, Atmospheric response to the observed increase of solar UV radiation from solar minimum to solar maximum simulated by the University of Illinois at Urbana-Champaign climate-chemistry model, *J. Geophys. Res.*, *109*, D01110, doi: 10.1029/2003JD003796, 2004.

Rozanov, E., M. E. Schlesinger, and V. A. Zubov, The University of Illinois, Urbana-Champaign Three-dimensional Stratosphere-Troposphere General Circulation Model with Interactive Ozone Photochemistry: Fifteen-year Control Run Climatology, *J. Geophys. Res.*, *106* (D21), 27,233–27,254, 2001.

Santer, B. D., K. E. Taylor, T. M. L. Wigley, T. C. Johns, P. D. Jones, D. J. Karoly, J. F. B. Mitchell, A. H. Oort, J. E. Penner, V. Ramaswamy, M. D. Schwarzkopf, R. J. Stouffer, and S. Tett, A search for human influences on the thermal structure of the atmosphere, *Nature*, *382*, 39–46, 1996.

Santer, B. D., K. E. Taylor, T. M. L. Wigley, J. E. Penner, and P. D. Jones, Towards the detection and attribution of an anthropogenic effect on climate, *Climate Dynamics*, *12*, 77–100, 1995.

Santer, B. D., T. M. L. Wigley, and P. D. Jones, Correlation methods in fingerprint detection studies, *Climate Dynamics*, *8*, 265–276, 1993.

Schlesinger, M. E., Greenhouse policy, *National Geographic Research & Exploration*, *9* (2), 159–172, 1993.

Schlesinger, M. E., and N. G. Andronova, Climate Sensitivity, in *Encyclopedia of Global Environmental Change*, pp. 301–308, Wiley, London, 2001.

Schlesinger, M. E., N. G. Andronova, B. Entwistle, A. Ghanem, N. Ramankutty, W. Wang, and F. Yang, Modeling and simulation of climate and climate change, in *Past and Present Variability of the Solar-Terrestrial System: Measurement, Data Analysis and Theoretical Models. Proceedings of the International School of Physics "Enrico Fermi" CXXXIII*, edited by G. Cini Castagnoli, and A. Provenzale, pp. 389–429, IOS Press, Amsterdam, 1997.

Schlesinger, M. E., and X. Jiang, Revised projection of future greenhouse warming, *Nature*, *350*, 219–221, 1991.

Schlesinger, M. E., X. Jiang, and R. J. Charlson, Implication of anthropogenic atmospheric sulphate for the sensitivity of the climate system, in *Climate Change and Energy Policy: Proceedings of the International Conference on Global Climate Change: Its Mitigation Through Improved Production and Use of Energy*, edited by L. Rosen, and R. Glasser, pp. 75–108, American Institute of Physics, New York, 1992.

Schlesinger, M. E., and N. Ramankutty, Implications for global warming of intercycle solar-irradiance variations, *Nature*, *360*, 330–333, 1992.

Schlesinger, M. E., and N. Ramankutty, Low-frequency oscillation. Reply, *Nature*, *372*, 508–509, 1994a.

Schlesinger, M. E., and N. Ramankutty, An oscillation in the global climate system of period 65–70 years, *Nature*, *367*, 723–726, 1994b.

Schlesinger, M. E., and N. Ramankutty, Is the recently reported 65–70 year surface-temperature oscillation the result of climatic noise?, *J. Geophys. Res.*, *100*, 13,767–13,774, 1995.

Schneider, S. H., and C. Mass, Volcanic dust, sunspots, and temperature trends, *Science*, *198*, 741–746, 1975.

Schönwiese, C., Analysis and prediction of global climate temperature change based on multiforced observational statistics, *Environ. Pollution*, *83*, 149–154, 1994a.

Schönwiese, C.-D., R. Ullrich, F. Beck, and J. Rapp, Solar signals in global climatic change, *Climatic Change*, *27* (259–281), 1994b.

Schuurmans, C. J. E., Influence of solar activity on solar temperatures, *Climatic Change*, *1*, 231–237, 1978.

Seitz, F., Global Warming and Ozone Hole Controversies: A Challenge to Scientific Judgment, pp. 33, George C. Marshall Institute, Washington, D.C., 1994.

Sellers, W. D., A global climate model based on the energy balance of the earth-atmosphere system, *J. Appl. Meteor.*, *8*, 392–400, 1969.

Sellers, W. D., The effect of a solar perturbation on a global climate model, *J. Climate Appl. Meteorol.*, *24*, 770–776, 1985.

Shindell, D., D. Rind, N. Balachandran, J. Lean, and P. Lonergan, Solar cycle variability, ozone, and climate, *Science*, *284*, 305–308, 1999.

Shindell, D. T., G. A. Schmidt, M. E. Mann, D. Rind, and A. Waple, Solar forcing of regional climate change during the Maunder Minimum, *Science*, *294*, 2149–2152, 2001.

Sofia, S., and L. H. Li, Solar variability and climate, *J. Geophys. Res.*, *106*, 12,969–12,974, 2001.

Solanki, S. K., and M. Fligge, A reconstruction of total solar irradiance since 1700, *Geophys. Res. Lett.*, *26* (16), 2465–2468, 1999.

Solow, A. R., Bootstrapping correlated data, *Mathematical Geology*, *17*, 769–775, 1985.

Soon, W., E. Posmentier, and S. Baliunas, Climate hypersensitivity to solar forcing?, *Ann. Geophys, Atm. Hydr.*, *18* (5), 583–588, 2000.

Stevenson, D. S., C. E. Johnson, W. J. Collins, R. G. Derwent, K. P. Shine, and J. M. Edwards, Evolution of tropospheric ozone radiative forcing, *Geophys. Res. Lett.*, *25* (20), 3819–3822, 1998.

Stott, P. A., S. F. B. Tett, G. S. Jones, M. R. Allen, W. J. Ingram, and J. F. B. Mitchell, Attribution of twentieth century temperature change to natural and anthropogenic causes, *Climate Dynamics*, *17*, 1–21, 2001.

Syktus, J., H. Gordon, and J. Chappell, Sensitivity of a coupled atmosphere-dynamic upper ocean GCM to variations of CO2, solar constant, and orbital forcing, *Geophys. Res. Lett.*, *21*, 1599–1602, 1994.

Tett, S. F. B., P. A. Stott, M. R. Allen, W. J. Ingram, and J. F. B. Mitchell, Causes of Earth's near-surface temperature change in the twentieth century, *Nature*, *399*, 569–572, 1999.

Tol, R. S. J., and P. Vellinga, Climate change, the enhanced greenhouse effect and the influence of the sun: A statistical analysis, *Theor. Appl. Climatol.*, *61* (1–2), 1–7, 1998.

Wetherald, R. T., and S. Manabe, The effects of changing the solar constant on the climate of a general circulation model, *J. Atmos. Sci.*, *32*, 2044–2059, 1975.

White, W. B., D. R. Cayan, and J. Lean, Global upper ocean heat storage response to radiative forcing from changing solar irradiance and increasing greenhouse gas/aerosol concentrations, *J. Geophys. Res.*, *2* (C10), 21,355–21,366, 1998.

Wigley, T. M. L., The climate of the past 10,000 years and the role of the sun, in *Secular Solar and Geomagnetic Variations in the Last 10,000 Years*, edited by F.R. Stephenson, and A.W. Wolfendale, pp. 209–224, Kluwer, 1988.

Wigley, T. M. L., and S. C. B. Raper, Climate change due to solar irradiance changes, *Geophys. Res. Lett.*, *17* (12), 2169–2172, 1990.

Willson, R. C., Irradiance observations of SMM, Spacelab 1, UARS and Atlas experiments, in *The Sun as a Variable Star*, edited by J. Pap, X. Fröhlich, H. Hudson, and K. Solanki, pp. 54–62, Cambridge Univ. Press, New York, 1994.

Natalia G. Andronova, Climate Research Group, Department of Atmospheric Sciences, 105 S. Gregory St., Urbana, IL 61801, U.S.A.

Michael E. Schlesinger, Climate Research Group, Department of Atmospheric Science, 105 S. Gregory St., Urbana, IL 61801, U.S.A.

Effects of Solar UV Variability on the Stratosphere

Lon L. Hood

Lunar and Planetary Laboratory, University of Arizona, Tucson, Arizona

Previously thought to produce only relatively minor changes in ozone concentration, radiative heating, and zonal circulation in the upper stratosphere, solar ultraviolet (UV) variations at wavelengths near 200 nm are increasingly recognized as a significant source of decadal variability throughout the stratosphere. On the time scale of the 27-day solar rotation period, UV variations produce a stratospheric ozone response at low latitudes that agrees approximately with current photochemical model predictions. In addition, statistical studies suggest an unmodeled dynamical component of the 27-day response that extends to the low and middle stratosphere. On the time scale of the 11-year solar cycle, the ozone response derived from available data is characterized by a strong maximum in the upper stratosphere, a negligible response in the middle stratosphere, and a second strong maximum in the tropical lower stratosphere. The 11-year temperature response derived from NCEP/CPC data is characterized by a similar altitude dependence. However, in the middle and upper stratosphere, disagreements exist between analyses of alternate temperature data sets and further work is needed to establish more accurately the 11-year temperature response. In the lower stratosphere, in contrast to most model predictions, relatively large-amplitude, apparent solar cycle variations of geopotential height, ozone, and temperature are observed primarily at tropical and subtropical latitudes. As shown by the original work of *Labitzke and van Loon* [1988], additional large responses can be detected in the polar winter lower stratosphere if the data are separated according to the phase of the equatorial quasi-biennial wind oscillation. A possible explanation for the unexpectedly large lower stratospheric responses indicated by observational studies is that solar UV forcing in the upper stratosphere may influence the selection of preferred internal circulation modes in the winter stratosphere.

1. INTRODUCTION

A component of long-term climate change that remains poorly understood at present is that associated with solar variability. Although statistical models suggest that solar variability (together with anthropogenic and volcanic forcing) is a significant contributor to climate change during the last one to three centuries (see, e.g., *Mitchell et al.* [2001]), the precise nature of the solar forcing and how it results in tropospheric climate change is not yet well established.

One mechanism by which solar variability may influence tropospheric climate involves changes in solar ultraviolet (UV) flux at wavelengths that affect ozone production and radiative heating in the stratosphere. Photodissociation of molecular oxygen (leading to ozone production) occurs in

Solar Variability and its Effects on Climate
Geophysical Monograph 141
10.1029/141GM20

the Schumann-Runge bands and in the Herzberg continuum at wavelengths less than 242 nm (e.g., *Brasseur and Solomon* [1984]). Radiation at these wavelengths (190–240 nm) can penetrate into the upper stratosphere (30 to 50 km; *Herzberg* [1965]) where the increased atmospheric number density results in the formation of the ozone layer. Absorption of solar radiation by ozone at wavelengths between 200 and 300 nm is, in turn, responsible for radiative heating of the stratosphere. The most obvious consequences of this radiative heating are the positive temperature gradient with altitude and the increased static stability of the stratosphere. The latitudinal gradient of radiative heating drives a zonal circulation in the midlatitude upper stratosphere with maximum westerly amplitude near the time of winter solstice in both hemispheres. Solar-induced changes in upper stratospheric ozone and radiative heating will therefore perturb the zonal wind field [*Kodera and Yamazaki*, 1990; *Hood et al.*, 1993].

About 30 years ago, it was speculated that solar-induced changes in upper stratospheric winds may modify the properties of upwardly propagating planetary waves, thereby leading to secondary effects on circulation in the lower stratosphere and troposphere [*Hines*, 1974]. This proposed mechanism is mainly operative only in the winter hemisphere since planetary wave activity is effectively trapped by summertime easterlies [*Charney and Drazin*, 1961]. More recently, it has been established theoretically that angular momentum transfer to the zonal wind field by breaking Rossby and gravity waves at a given level can indeed significantly influence the evolution of the wind field at lower levels [*Haynes et al.*, 1991]. Nevertheless, in view of the small amplitude of solar UV forcing in the upper stratosphere, it has generally been considered unlikely that major changes of circulation in the lower stratosphere and troposphere would result from relatively weak solar UV forcing.

During the early 1990's, however, analyses of ozone and meteorological data began to suggest a substantial solar cycle variation of ozone, temperature, and zonal wind in both the upper and lower stratosphere. Consistent with the solar UV forcing mechanism, the observed decadal variations of upper stratospheric ozone and temperature were in phase with one another and the largest variation of upper stratospheric zonal wind occurred at middle latitudes near winter solstice in both hemispheres [*Kodera and Yamazaki*, 1990; *Hood et al.*, 1993; *Chandra and McPeters*, 1994]. In the lower stratosphere, however, observational studies indicated an unexpectedly large variation of circulation and column ozone occurring on the time scale of the 11-year solar cycle [*Labitzke and van Loon*, 1988; *Hood*, 1997; *Zerefos et al.*, 1997; *van Loon and Labitzke*, 1998; *van Loon and Shea*, 2000; *Labitzke*, 2001].

In this chapter, a review is presented of recent efforts to understand and quantify both the direct effect of solar ultraviolet spectral irradiance variations in the upper stratosphere and their subsequent indirect effects on circulation in the lower stratosphere and troposphere. Because of the limited lengths of available data records, emphasis is placed here on solar UV forcing occurring on the time scales of the solar rotation period and the 11-year solar activity cycle. However, it should be realized that other, possibly larger changes in solar UV spectral irradiance may occur on decadal and century time scales [*Lean et al.*, 1997; *Lean*, 2001]. In section 2, the variability of solar UV radiation relevant to the stratosphere on the 27-day and 11-year time scales is briefly summarized (see the chapter by G. Rottman for a more complete description). In section 3, stratospheric effects of solar UV variations on the 27-day time scale are briefly discussed. (For a more detailed description, see the reviews by *Hood* [1987; 1999].) In section 4, observations of the response of stratospheric ozone, temperature, geopotential height, and zonal wind to solar cycle UV variations are summarized. In section 5, a conceptual model for how direct upper stratospheric effects of solar cycle UV variations can indirectly modify the circulation in the lower stratosphere and troposphere is described. In support of this model, recent evidence is summarized indicating that a decadal modulation of the tropical branch of the Brewer-Dobson circulation, which is largely forced by planetary wave driving in the extratropical middle and upper stratosphere, may be mainly responsible for solar cycle variations of ozone and temperature in the tropical lower stratosphere. Finally, the status of current efforts to understand the altitude dependence of the solar cycle variation of ozone in the middle and upper stratosphere is reviewed. Conclusions and possible directions for future work are given in section 6.

2. SOLAR UV SPECTRAL IRRADIANCE VARIABILITY

The change in solar flux over a solar rotation or solar cycle increases rapidly with decreasing wavelength (see the chapter by G. Rottman, this volume). At wavelengths shortward of the Al I edge at 205 nm, changes exceeding 5–6% occur on both the solar rotation and solar cycle time scales. At wavelengths between 205 nm and 242 nm, changes on these time scales are approximately half as large as at 205 nm. Amplitudes gradually decrease to nearly zero by 300 nm. The ratio of the change in flux at any wavelength in the UV band to that at 205 nm appears to be nearly constant on different time scales at least up to that of the solar cycle (e.g., *Donnelly et al.* [1988]; *Donnelly* [1991]). Therefore, direct satellite-based measurements of the solar flux near

205 nm as a function of time together with a flux ratio table would ideally provide much of the information needed by stratospheric photochemical and radiative heating models. Unfortunately, the tendency for the calibration of satellite-based instruments to change with time combined with the limited durations and lack of complete intercalibration between different instruments has not allowed such a precise time series of the solar 205 nm flux to be constructed.

In lieu of a direct measurement of long-term changes in the 205 nm flux, an index of solar radiation near this wavelength has been developed that makes use of radiation within the core of the Mg II line at 280 nm. At this wavelength, the radiation originates from nearly the same level in the solar photosphere as does the 205 nm flux. Consequently, the variability at 280 nm is nearly the same as that near 205 nm. The ratio of the flux at 280 nm to that on the wings of the Mg II line (known as the Mg II core-to-wing ratio) is nearly independent of wavelength-dependent instrument calibration changes [*Heath and Schlesinger*, 1986]. This flux ratio is therefore commonly used as a reliable proxy for long-term solar flux changes at wavelengths near 205 nm. Figure 1 is a plot of the Mg II core-to-wing ratio (L. Puga, R. Viereck, NOAA Space Environment Center, Boulder, Colorado) for the period during which satellite spectral irradiance data are available. Note that the amplitudes of 27-day variations are comparable to those of the 11-year solar cycle under solar maximum conditions. The amplitudes of solar cycles 21 and 22 were ~6% while that of cycle 23 had not yet reached its maximum when Figure 1 was prepared.

The penetration depth into the stratosphere of solar UV radiation is also a strong function of wavelength (e.g., *Herzberg* [1965]). At the 121.5 nm solar Lyman α wavelength, the penetration altitude is approximately 70 km. In the Schumann-Runge bands (~180 - 200 nm) radiation penetrates to an altitude range between about 40 and 60 km. At 205 nm, the altitude of penetration approaches a minimum of 30 km. At wavelengths between 205 nm and 300 nm, where solar spectral irradiance variability decreases gradually to zero, the altitude of penetration varies between 30 and 35 km. Thus, the maximum depth of penetration occurs near 205 nm where UV flux changes as large as 6% are observed on both the solar cycle and solar rotation time scales.

3. STRATOSPHERIC RESPONSES ON THE SOLAR ROTATION TIME SCALE

During the interval surrounding the maximum of solar cycle 21, instruments on the Nimbus 7 spacecraft were obtaining the first complete daily global measurements of stratospheric composition and thermal structure. In particu-

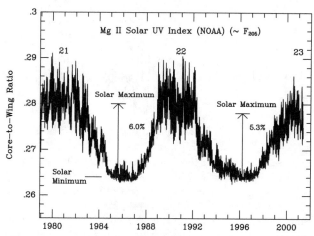

Figure 1. Time series of daily Mg II solar ultraviolet flux index (L. Puga, private communication, 2001). The cycle numbers 21, 22, and 23 are indicated.

lar, the Solar Backscattered Ultraviolet (SBUV) and Limb Infrared Monitor of the Stratosphere (LIMS) instruments were measuring stratospheric ozone mixing ratios in unprecedented detail while the Stratosphere and Mesosphere Sounder (SAMS) and LIMS instruments were also obtaining initial global measurements of stratospheric temperature profiles. These data presented an ideal opportunity to investigate, for the first time, the existence and properties of stratospheric responses to solar ultraviolet flux changes occurring on the time scale of the 27-day solar rotation period. Although evidence for the detectability of ozone responses to 27-day solar variations was also obtained using Nimbus 4 Backscattered Ultraviolet data [*Hood*, 1984], these responses were more easily detected using the improved Nimbus 7 data at a time of high-amplitude solar rotation variations [*Gille et al.*, 1984; *Heath and Schlesinger*, 1985; *Hood*, 1986; *Keating et al.*, 1987]. Because of the relatively large number of 27-day cycles that are available for analysis, observed responses on this time scale can be measured with greater confidence and therefore impose firm constraints on stratospheric models.

Shown in Figure 2 (from *Hood* [1986]) is a comparison of Nimbus 7 SBUV ozone mixing ratio data at 1.5 hPa and at low latitudes (30°S to 30°N) to direct measurements of the 205 nm solar flux by the same instrument over a 22-month period in 1979 and 1980 (from *Hood* [1986]). To facilitate the detection of ozone responses occurring on the 27-day time scale (and to eliminate instrument-related drifts in the solar flux data), both time series were filtered to minimize variations with periods greater than approximately 35 days. The time series were also smoothed with a 7-day running average and linear interpolation algorithm. The residuals correlate with one another especially during intervals of

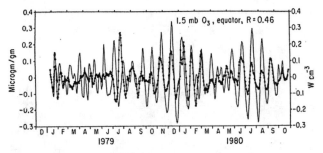

Figure 2. The thin solid line is the solar ultraviolet flux at 205 nm measured by the Nimbus 7 SBUV instrument. The dots connected by a heavy line represent measurements of the equatorial ozone mixing ratio near 1.5 hPa obtained by the same SBUV instrument. Both time series have been filtered to minimize fluctuations with periods greater than 35 days and less than 7 days (from *Hood* [1986]).

strong UV variations (i.e., Nov. 1979 to Feb. 1980 and June to Sept. 1980). Although the overall correlation coefficient is only 0.46, the value of this coefficient increases to 0.7 or 0.8 during the latter intervals. Also, the correlation coefficient was found to be significantly larger when direct measures of the solar 205 nm flux were used (rather than indirect measures such as the 10.7 cm radio flux or geomagnetic indices). It is therefore established that the observed ozone responses shown in Figure 2 are produced directly (through photochemistry) and/or indirectly (through dynamical feedbacks) by solar UV flux changes. Finally, ozone responses as a function of altitude in the stratosphere were estimated by regression using two independent time intervals during the maximum of solar cycle 21 [*Hood and Cantrell*, 1988]. Results from these two time intervals were consistent with one another, further confirming that the response could be directly measured.

The ozone response amplitude, or sensitivity, is defined as the percent change in ozone mixing ratio at a given pressure level that occurs for a 1% change in the solar 205 nm flux. On the 27-day time scale, tropical ozone response amplitudes reach a maximum of 0.4 to 0.5 in the 2 to 3 hPa pressure range. Response amplitudes decrease with increasing altitude to ~0.3 by 1 hPa and also with decreasing altitude to ~0.1 by 10 hPa (see, e.g., Figure 2 of *Hood* [1999]). The phase lag of the ozone response relative to the UV flux maximum increases continuously with decreasing altitude and ranges from -2.4 ± 0.6 days at 1 hPa to +3.2 ± 0.5 days at 10 hPa. A number of photochemical model calculations have indicated that the observed response amplitudes and phase lags on the solar rotation time scale are approximately consistent with theoretical expectations [*Brasseur*, 1993; *Fleming et al.*, 1995; *Chen et al.*, 1997; *Williams et al.*,

2001]. In fact, accurate measurements of ozone response amplitudes and phase lags can be applied to test the validity of photochemical models of the upper stratosphere [*Hood and Douglass*, 1988].

Since the 1980's when the most detailed measurements of 27-day ozone responses were made, several other studies have confirmed the basic character of these responses using alternate data sets, i.e. those acquired using the Microwave Limb Sounder (MLS) on the Upper Atmosphere Research Satellite [*Waters*, 1993] and the NOAA 11 SBUV/2 instrument [*Lienesch et al.*, 1996]. Specifically, it was shown that ozone responses derived from MLS data agreed with those estimated from Nimbus 7 SBUV data when differences in local times of measurement were accounted for [*Hood and Zhou*, 1998]. To investigate whether the ozone profile response differed significantly during the solar cycle 22 maximum period (around 1990) as compared with the solar cycle 21 maximum period (around 1980), two separate 4-year intervals (1979–1982 and 1989–1992) of daily Nimbus 7 SBUV and/or NOAA 11 SBUV/2 data for tropical latitudes were analyzed using cross-spectral and regression methods [*Hood and Zhou*, 1999]. Results showed that ozone profile sensitivities and phase lags are in agreement between these two intervals when statistical uncertainties and differences in data processing algorithms were considered. Most recently, *Zhou et al.* [2000] have reported a partial correlation analysis of the ozone response to 27-day solar UV variations after first removing dynamically induced ozone variations. The latter result mainly from the temperature dependence of reaction rates controlling the ozone balance. It was found that the ozone sensitivity to solar UV was again independent of time periods used in the analysis. However, the ozone sensitivity to temperature at 1–2 hPa increased significantly from solar cycle 21 to solar cycle 22. It was speculated that the latter might reflect long-term changes in the composition of the upper stratosphere.

In addition to the observed ozone response to 27-day solar UV forcing, there is some evidence for an associated temperature response based on the analysis of Nimbus 7 Stratosphere and Mesosphere Sounder (SAMS) data (for a more detailed review, see *Hood* [1987]). The temperatures derived from the SAMS data have a vertical resolution of 8–10 km [*Rodgers et al.*, 1984] and are based on limb scans obtained in the afternoon and early evening hours of each day, i.e. shortly after the local times when SBUV ozone profiles were obtained. Using the SAMS data, the estimated temperature response at low latitudes (30°S to 30°N) has a maximum amplitude of about 0.06% (~0.16 K) per 1% change in the 205 nm flux near the 1 hPa level [*Hood*, 1986]. The phase lag of the temperature response relative to the UV flux maximum was estimated as about +6 days at 1 hPa. The reality of this weak temperature response was supported by detailed time-progressive cross correlation analyses (see, e.g., Figure 9 of *Hood* [1987]) and by analyses of separate 22-month time intervals [*Hood and Cantrell* [1988]. Using an analytic model, it was also shown that the observed mean temperature responses have approximately the correct amplitudes and phase lags needed to explain the negative ozone phase lags that are measured above 2 hPa [*Hood,* 1987].

Finally, there are several indications that the upper stratospheric response to solar UV variations on the 27-day time scale has a significant dynamical component. First, the derived positive temperature phase lags ~6 days at 1 hPa are about twice as large as would be predicted by one-dimensional models that consider only radiative-photochemical processes. Apparently, the radiative temperature response is significantly modified by dynamical processes. Second, there is some provisional empirical evidence that latitudinal ozone and temperature oscillations during the northern hemisphere winter may be modulated by 27-day solar UV variability. As shown in Figure 3 (from *Hood and Jirikowic* [1991]), at low latitudes, 205 nm UV variations correlate approximately with 1.5 hPa ozone perturbations (see also Figure 2) while temperature variations lag the ozone perturbations by many days. Moving to higher latitudes in the northern hemisphere, one sees from the figure that the ozone and temperature oscillations change phase at about 40°N and are opposite in phase to the lower-latitude oscillations by about 60°N. Many of the negative ozone perturbations at 60°N appear to correlate with positive solar UV maxima. Out-of-phase ozone and temperature oscillations at low and high latitudes are a common characteristic of the northern winter and are associated with occasional increases in absorption of upward propagating planetary-scale waves by the zonal wind field (e.g., *Chandra* [1985; 1986]; *Holton and Mass* [1976]; *Kodera and Kuroda* [2000]). In the most extreme cases, wave absorption results in a strong warming of the polar stratosphere and an associated weaker cooling in the tropics, i.e., "stratospheric warmings" [*Matsuno,* 1971]. In the upper stratosphere, where ozone is nearly in photochemical equilibrium, the temperature dependences of the dominant reaction rates controlling the ozone balance result in ozone changes that are opposite in sign to the temperature perturbations at the same latitudes. Since these ozone and temperature oscillations tend to be in phase with UV forcing during periods of strong 27-day solar UV variations, it may be speculated that planetary wave absorption (and resulting latitudinal oscillations of ozone and temperature) is modulated in part by solar UV induced

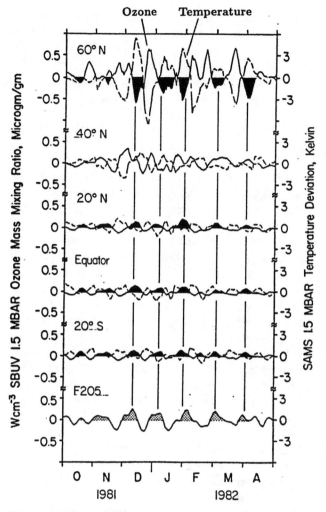

Figure 3. Latitude dependence of zonal mean Nimbus 7 SBUV ozone and SAMS temperature deviations for a 7-month period centered on the 1981-82 Northern Hemisphere winter. The deviation from the 35-day running mean of the SBUV solar 205-nm flux is shown in the lower panel for comparison. The ozone deviations are emphasized by dark shading in the upper plots (from *Hood and Jirikowic* [1991]).

changes in the stratosphere [*Hood,* 1987; *Hood and Jirikowic,* 1991].

In agreement with the above interpretation, a number of observational studies have reported significant spectral coherency between solar UV flux (as represented by the 10.7 cm radio flux) and harmonic components of planetary waves at levels between 50 and 10 hPa and at frequencies near the solar rotation period [*Ebel and Batz,* 1977; *Ebel et al.,* 1981; 1986; *Dameris et al.,* 1986; *Soukharev and Labitzke,* 2000a,b]. These results are broadly consistent with the mechanism of *Hines* [1974] (noted in section 1), which proposes that solar induced variations in stratospheric and mesospheric zonal winds may alter the transmission-reflection properties of upwardly propagating planetary waves. However, a detailed theoretical model describing this solar-induced modulation of planetary waves on the 27-day time scale has not yet been developed.

4. STRATOSPHERIC RESPONSES TO THE 11-YEAR SOLAR CYCLE

Beginning in the early 1990's, the lengths of continuous global satellite measurements of basic stratospheric parameters (composition, temperature) began to exceed one solar cycle. In the upper stratosphere (1–3 hPa), measurements of latitudinally averaged and deseasonalized ozone and temperature at low latitudes during solar cycle 21 showed a large decadal variation that correlated approximately with long-term changes in the Mg II UV index (Figure 1) and with the solar 10.7 cm radio flux. In the tropical middle stratosphere (5–10 hPa), no significant decadal variation was present. Although one decade of measurements is hardly sufficient to demonstrate a solar cycle variation, preliminary analyses of data acquired during the last decade (1990 to 2000) support the possibility that the decadal variations observed in the 1980's were also present in the 1990's (*Hood and Soukharev* [2000]; see also *McCormack et al.* [1997]).

In further support of the existence of an 11-year signal in the stratosphere, analyses of lower stratospheric meteorological data for the northern hemisphere have indicated a solar-correlated response occurring over a period of 4 solar cycles [*Labitzke and van Loon*, 1988; 1993; *Labitzke*, 2001]. Also, analyses of ground-based total ozone data have indicated a significant response over a similar period (e.g., *Zerefos et al.* [1997]). The basic characteristics of the 11-year signal in both the upper stratosphere and the lower stratosphere are summarized below.

4.1 Middle and Upper Stratospheric Ozone

Initial estimates for the solar cycle variation of ozone mixing ratio in the middle and upper stratosphere have been reported by a number of analysts [*Chandra*, 1991; *Hood et al.*, 1993; *Chandra and McPeters*, 1994; *McCormack and Hood*, 1996; *Wang et al.*, 1996; *Hood and Soukharev*, 2000]. Figure 4 compares the observationally inferred solar cycle variation of stratospheric ozone in the tropics (35°S to 35°N) with the predictions of several recent stratospheric models. In the middle and upper stratosphere, the observed ozone changes are estimated by applying a multiple regression sta-

tistical model to Nimbus 7 SBUV and NOAA 11 SBUV/2 ozone profile data over the 1979 to 1994 period [*McCormack and Hood*, 1996]. These estimates have been supported by independent analyses of Stratospheric Aerosol and Gas Experiment (SAGE) I and II data over the 1979 to 1996 period (*Wang et al.*, 1996; see Fig. 3.5 of *Stolarski et al.* [1998]). In the lower stratosphere, the estimated change is based on analyses of global satellite column ozone data after correcting for the contribution from the middle and upper stratosphere [*Hood*, 1997]. Specifically, a combined total ozone record consisting of Nimbus 7 Total Ozone Mapping Spectrometer (TOMS), Meteor 3 TOMS, NOAA 11 SBUV, and Earth Probe TOMS data [*Stolarski et al.*, 2000] was analyzed using a multiple regression statistical model that includes seasonal, quasi-biennial oscillation, solar cycle, and volcanic aerosol terms (e.g., *McCormack et al.* [1997]). The solar cycle variation of total ozone at low latitudes is in the range of 1–2% (see also *Zerefos et al.* [1997]); in contrast to model predictions, most of this column ozone response occurs in the lower stratosphere [*Hood*, 1997].

As shown in Figure 4, in terms of percent change, the observed mean low-latitude response is largest in the upper stratosphere (1-3 hPa) and amounts to about $4 \pm 1\%$ from solar minimum to maximum. Unexpectedly, the response decreases to nearly zero in the middle stratosphere (5–10 hPa) before increasing again in the lower stratosphere. If regression analyses are performed at individual latitudes and pressure levels, the results are as shown in Figure 5a. At

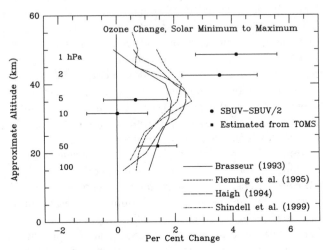

Figure 4. The filled circles with 2σ error bars represent the solar cycle variation of ozone in the low-latitude stratosphere as estimated from Nimbus 7 SBUV and NOAA 11 SBUV/2 data over a 16-year period. For comparison, the lines show the variation estimated by a series of two- and three-dimensional stratospheric models.

the equator, the observed annually averaged response is actually negative near 10 hPa although the statistical significance is marginal. As shown in Figure 4, most two- and three-dimensional stratospheric models have predicted a different altitude dependence for the solar cycle variation of ozone. Specifically, most models predict that solar cycle per cent ozone changes should be largest in the middle stratosphere ~5 hPa) and should be much less in the upper and lower stratosphere [*Brasseur,* 1993; *Fleming et al.,* 1995; *Haigh,* 1994; *Shindell et al.,* 1999]. More detailed, three-dimensional simulations using the CCSR/NIES coupled chemistry-climate model [*Takigawa et al.,* 1999] and the Meteorological Office model UMETRAC [*Austin,* 2001] have not resulted in significantly better agreement, as shown recently by *Labitzke et al.* [2001].

4.2 Middle and Upper Stratospheric Temperature

As reviewed by *Chanin et al.* [1998], data sets available for analyzing long-term variations in stratospheric temperature at pressure levels above ~10 hPa consist of (a) satellite remote sensing data (1979–present); (b) lidar soundings (1979–present); and (c) rocketsonde data (time intervals varying with station location, but most beginning in the 1960's and ending by the mid-1990's). The satellite data that are relevant to the middle and upper stratosphere consist primarily of that acquired by the Stratospheric Sounding Units (SSU) on the National Oceanic and Atmospheric Administration (NOAA) polar orbiting operational satellites [*Nash and Forrester,* 1986; *Scaife et al.,* 2000]. Lidar soundings that are useful for long-term variability studies are presently limited primarily to measurements from a single station, Haute Provence Observatory in southern France (e.g., *Keckhut et al.* [1995]). Rocketsonde data are available from > 10 stations since the 1960's distributed globally but concentrated at low to middle northern latitudes (see, e.g., *Keckhut et al.* [1999]; *Dunkerton et al.* [1998]).

In the case of the SSU data, it must be emphasized that estimation of long-term temperature variability is hindered by the deterioration of the radiometers during their lifetimes and by changes of satellite platforms, which result in differences in orbital characteristics and in the effective weighting functions for the various SSU channels. In addition, changes in the procedures that are adopted for analysis of these data (e.g., for inclusion in the NOAA National Centers for Environmental Prediction, or NCEP, data set) have, in some cases, introduced effective changes in calibration. Nevertheless, methods have been developed that attempt to correct for these effects [*Nash and Edge,* 1989; *Gelman et al.,* 1986; 1994; *Finger et al.,* 1993; *Scaife et al.,* 2000]. For

example, the method of *Finger et al.* [1993] includes a comparison of the derived SSU temperatures with co-located rocketsonde and lidar observations to detect systematic biases.

Several analyses of the SSU data have indicated evidence for a solar cycle component of middle and upper stratospheric temperature [*Hood et al.,* 1993; *McCormack and Hood,* 1996; *Scaife et al.,* 2000]. Using adjustment procedures recommended by *Gelman et al.* [1986] and *Finger et al.* [1993], *Hood et al.* [1993] carried out an initial analysis of 12.4 years of SSU temperature data as compiled at the NCEP Climate Prediction Center (NCEP/CPC; formerly, NMC/CAC) and found evidence for an apparent solar cycle variation in the upper stratosphere. The analysis was later extended to 14 years by *McCormack and Hood* [1996] who also investigated latitude and seasonal dependences of the temperature response. It was found that the decadal variation of temperature near 1 hPa was approximately in phase with the decadal variation of upper stratospheric ozone. The altitude dependence of the temperature response at low latitudes was also estimated (maximum near the stratopause and minimum in the 5–10 hPa range). This altitude dependence was approximately consistent with that of the estimated ozone response.

Since the study of *McCormack and Hood* [1996], the NCEP data have been reanalyzed [*Kalnay et al.,* 1996] and temperature data above 10 hPa have been reprocessed at the NCEP/CPC using an extension of the methods of *Gelman et al.* [1986] and *Finger et al.* [1993] (S. Zhou, private communication, 1998). Figure 5b shows the estimated solar cycle variation of stratospheric temperature as derived from the NCEP reanalysis data as a function of latitude and altitude over the 1980 to 1997 period. Like the ozone profile data, the NCEP/CPC temperature data also yield a relatively large variation in the tropical upper stratosphere while the variation in the tropical middle stratosphere (5–10 hPa) is much less or is even slightly negative. An increased temperature variation from solar minimum to maximum is seen in the tropical lower stratosphere where the ozone response also increases. It has been argued by, e.g., *Hood et al.* [1993] that the occurrence of both ozone and temperature increases at low latitudes from solar minimum to maximum near the 1 hPa level is supportive of an origin involving photochemical and radiative effects of solar cycle changes in ultraviolet radiation. The main alternative of an internal dynamical variation with a period ~10 years can be excluded since the resulting temperature variation near the stratopause would drive ozone variations of opposite sign, unlike the in-phase variation of both temperature and ozone that is observed.

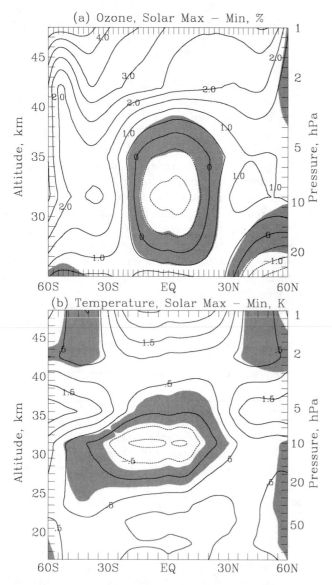

Figure 5. Annually averaged solar cycle variation of stratospheric ozone (a) and temperature (b) as derived using a multiple regression statistical model (see the text). Shaded regions are not significantly different from zero at the 2σ level.

Recently, alternate estimates for the solar cycle component of stratospheric temperature variations have been derived from the SSU data [*Scaife et al.*, 2000] (see also Figure 5-19 of *Chanin et al.* [1998]). These authors analyzed the SSU data over the 1979 to 1997 period by applying adjustment procedures developed by *Nash and Edge* [1989] to the measured SSU radiances. Although evidence for a solar cycle variation in the upper stratosphere was obtained, the derived amplitude and altitude dependence

were different from those shown in Figure 5b. Specifically, their estimated temperature change from solar minimum to maximum has a reduced maximum amplitude of only ~0.8 K near 4 hPa in the tropics. The estimated variation near 1 hPa is ~0.4 K. For comparison, as shown in Figure 5b, the maximum amplitude derived from the NCEP reanalysis data near 1 hPa is >2 K and no significant response is present in the middle stratosphere. As suggested by *Chanin et al.* [1998] (see their Figure 5-18), after application of the adjustments recommended by *Finger et al.* [1993], the SSU data as compiled at the NCEP Climate Prediction Center (CPC) may still exhibit significant offsets near the times of satellite transitions. *Chanin et al.* therefore conclude that the time series analyzed by *Scaife et al.* [2000] is more nearly free of instrument-related errors.

One independent estimate for the solar cycle variation of temperature in the middle and upper stratosphere has been made using rocketsonde data (*Dunkerton et al.* [1998]; see also *Keckhut et al.* [1999]). The time interval of the analysis was from 1962 to 1991, or about three solar cycles. Using station average measurements over an altitude range of 28 to 56 km and at latitudes from 9°S to 38°N, a mean amplitude of 1.1 K from solar minimum to maximum was estimated (see Figure 4 of *Dunkerton et al.* [1998]). For comparison, averaging the NCEP reanalysis results shown in Figure 5b over roughly the same altitude and latitude range yields a mean amplitude of around 1 K, in approximate agreement with the results of *Dunkerton et al.* However, averaging the results of *Scaife et al.* [2000] (see also Figure 5-18 of *Chanin et al.* [1998]) over the same altitude and latitude range yields a mean amplitude of no more than 0.6 K, about a factor of two less than indicated by the rocketsonde data. As reviewed by *Chanin et al.* [1998], uncertainties also characterize the rocketsonde data (see also *Keckhut et al.* [1999]). In any case, the absence of a solar signal near 1 hPa in the SSU results published by Scaife et al. disagrees with both the rocket data and the NCEP/CPC data. Also, virtually all models show that ozone and temperature increase at 1 hPa as solar UV increases. This is therefore a major, fundamental disagreement and suggests that more work is needed to evaluate uncertainties and adjustment procedures for the SSU data.

4.3 Lower Stratospheric Geopotential, Ozone, and Temperature

A 10-12 year oscillation of the heights of constant-pressure surfaces in the northern lower stratosphere was first reported by *Labitzke and van Loon* [1988]. As shown in Figure 6, correlations of 30 hPa geopotential height with the

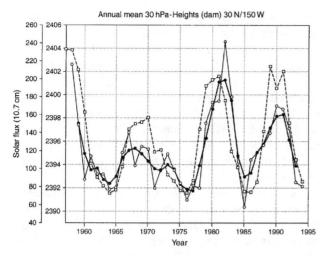

Figure 6. Time series of annual mean 10.7 cm solar flux (open squares), annual mean 30 hPa height at 30°N, 150°W (open circles), and a 3-year running mean of the 30 hPa heights for each year (heavy solid line) (K. Labitzke, private communication). The heights are in decameters. (Updated from *Labitzke and van Loon* [1995])

solar 10.7 cm radio flux can be fairly high at certain longitudes near 30°N latitude. A similar correlation is present in the Southern Hemisphere with maximum amplitude near 30°S latitude in winter [*van Loon and Labitzke,* 1998]. Although the available time record (~4 decades) is still too short to prove statistically that the decadal oscillation of Figure 6 is solar in origin, this is the leading hypothesis at the present time. Because geopotential height is determined in part by the temperature at lower levels, the data of Figure 6 imply an apparent solar cycle variation in temperature records for the lower stratosphere and upper troposphere. At high latitudes in winter, a solar cycle signal in lower stratospheric meteorological parameters can be found only if the data are first separated according to the phase of the equatorial quasi-biennial wind oscillation (QBO) [*Labitzke and van Loon,* 1988; *Gray et al.,* 2001a]. At low latitudes (<50°), however, division of the data according to QBO phase is unnecessary to detect an apparent solar response although the QBO is still a major source of interannual variability.

In addition to the quasi-decadal oscillation of lower stratospheric geopotential height, a decadal oscillation of total column ozone is also present with a global mean amplitude of 1.5–2.0% [*Hood,* 1997; *Zerefos et al.,* 1997] (see, for example, Figure 10a). Most of this ozone column oscillation occurs in the lower stratosphere, as verified by analyses of ozone profile data (see Table 1 of *Hood* [1997]). Because the ozone lifetime in the lower stratosphere is

much longer than dynamical time scales, total ozone is a sensitive indicator of circulation changes in this altitude range. Like the oscillation in geopotential height, the quasi-decadal oscillation of total ozone is approximately in phase with the solar cycle and maximum amplitudes of the ozone oscillation occur near 30° latitude in both hemispheres [*McCormack et al.,* 1997].

Observational evidence for a quasi-decadal variation of lower stratospheric temperature has been reported in several studies. Using reprocessed NCEP Reanalysis data [*Kalnay et al.,* 1996], *Labitzke* [2001] has recently found evidence for significant correlations and temperature differences between solar maxima and minima near the tropical tropopause (30°S to 30°N); ~100 hPa). Annually and zonally averaged 100 hPa temperature differences of more than 1.5 K were estimated near 20°S (see Figure 3 of *Labitzke* [2001]). However, although the NCEP reanalyses have been reprocessed to minimize artificial errors associated with satellite calibration differences or changes in processing methods, significant errors still remain. It is therefore desirable to consider alternative data sources.

One alternative to the NCEP reanalysis temperature data in the tropical lower stratosphere is provided by measurements from Channel 4 of the Microwave Sounding Units on the TIROS-N NOAA operational satellites [*Spencer et al.,* 1990; *Spencer and Christy,* 1993]. Channel 4 yields brightness temperature measurements at a series of frequencies near the 60-GHz oxygen absorption band. These represent weighted mean atmospheric temperatures in a layer with peak power at 75 hPa and half-power at 40 and 120 hPa, i.e. mainly in the lowermost stratosphere but including a significant contribution from the upper troposphere in the tropics. Standard errors of measurement for 5-day means at individual grid points at 2.5° horizontal resolution are less than 0.25° for most of the globe and less than 0.15° for the tropics. These microwave soundings are not affected by the presence of aerosols. Correlations between MSU4 and radiosonde profiles for monthly anomalies are usually above 0.94 and approach unity in some cases. Monthly signal-to-noise ratios are usually over 500 at individual grid points. Unlike the NCEP reanalysis data near the tropical tropopause, MSU Channel 4 temperature deviation time series at individual latitudes in the tropics show no evidence for intercalibration offsets.

Plots of the deviation from the long-term monthly mean of the MSU4 data for the tropics are characterized by a long-term, approximately decadal variation (see, e.g., Figure 10b). This variation appears to be separate and distinct from the positive temperature anomalies resulting from the El Chichon and Pinatubo volcanic aerosol injection events. Specifically, excluding periods following these two major eruptions, the mean temperature deviation during solar minima is 0.5–0.8 K lower than during solar maxima. Application of a standard multiple regression statistical model to the tropical MSU4 data yields an estimated solar cycle variation of 0.70 ± 0.18 K. This amplitude is significantly less than estimated from the NCEP Reanalysis data. Although part of this difference may reflect the wider spread of altitudes sampled by the MSU instrument, a major part of the difference likely reflects reduced calibration errors in the MSU data. During the past several years as the solar UV flux has increased toward the next solar maximum, the mean tropical temperature has also increased significantly by more than 0.5 K. An increase of similar magnitude occurred between 1986 and the 1990 solar maximum. These temperature increases cannot be attributed easily to volcanic influences and are consistent with a solar origin. However, extension of the data record for several more years into the current solar maximum (in the absence of a major new volcanic aerosol injection event) would allow a firmer test of this provisional conclusion.

4.4 Lower Stratospheric QBO

In a recent paper, *Salby and Callaghan* [2000] reported statistical evidence for an 11-year solar cycle modulation of the equatorial quasi-biennial wind oscillation (QBO). For example, they find that the duration of the westerly QBO phase (i.e., the time period when equatorial winds are from the west) in the middle stratosphere varies with a systematic pattern resembling the curve of the 11-year solar cycle. *Soukharev and Hood* [2001] have reported further statistical investigation using both equatorial wind data from 50 to 1 hPa and long-term proxy solar ultraviolet flux time series (10.7 cm solar radio flux and sunspot numbers). Spectral analysis of the solar time series yielded evidence for a significant spectral peak at periods between 25 and 30 months, approximately equivalent to the mean QBO period, as had also been noted by earlier authors [*Shapiro and Ward,* 1962]. Cross spectral analysis of the 10.7 cm solar flux and equatorial zonal wind time series showed significant coherency at the QBO period at all available pressure levels. The phase lag of the wind data relative to the solar flux at the QBO period ranges from 0–1 months near the stratopause (1 hPa) to 20–24 months in the lower stratosphere (50 hPa). The nearly in-phase relationship near the stratopause suggests a possible modulation of the QBO at this level by the radiative and photochemical effects of solar ultraviolet variations.

Some evidence for a solar modulation of the equatorial QBO can be extracted from NCEP reanalysis data. (Note that zonal winds are calculated from latitudinal gradients in geopotential height and are therefore less susceptible to calibration offsets such as those discussed in the previous section.) Figure 7a plots the length of the QBO westerly phase versus the 10.7 cm solar flux using NCEP 50 hPa equatorial wind data for a 40-year period. The solid line is a regression line. Figure 7b plots the maximum easterly wind speed during 18 QBO cycles versus the 10.7 cm flux. As can be seen, both the duration of the westerly phase and the amplitude of the easterly phase tend to be larger under solar minimum conditions.

Some insight into how the QBO may be modulated by solar UV variability can be obtained by considering the relationship between the QBO in the equatorial lower stratosphere and the SAO in the upper stratosphere. Near 50 km, where the maximum response of ozone, temperature, and zonal wind to solar UV flux changes occurs, the SAO is the dominant mode of variability. At this level, there is both observational and theoretical evidence for a coupling

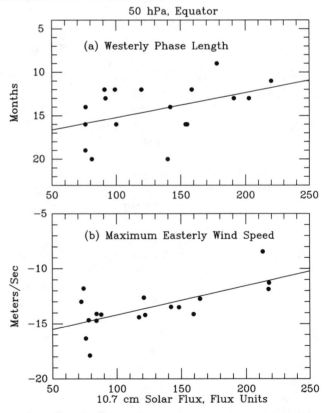

Figure 7. (a) Duration of the westerly phase of the quasi-biennial wind oscillation (QBO) at 50 hPa based on NCEP 50 hPa equatorial wind data for a 40-year period plotted against the 10.7 cm solar flux (see the text). (b) Maximum easterly wind speed during each QBO cycle plotted against the 10.7 cm flux. The solid lines in each diagram are regression lines (from *Soukharev and Hood* [2001]).

between the QBO and the SAO. Statistical analyses of radiosonde-derived zonal wind data show that the westerly phase of the QBO forms preferentially during the westerly phase of the SAO [*Baldwin et al.*, 2001]. Model calculations have also shown a similar tendency [*Gray and Pyle*, 1989]. A positive feedback may exist since the strength of the easterly phase of the SAO in the mesosphere is apparently modulated by the phase of the stratospheric QBO through the filtering of upward-propagating gravity waves [*Burrage et al.*, 1996; *Garcia et al.*, 1997]. It is therefore possible that solar UV-driven changes in upper stratospheric and lower mesospheric wind fields may propagate downward through coupling between the SAO and the QBO. Additional quasi-decadal variability in the extratropical lower stratosphere could be ascribable to coupling of the QBO with the annual cycle in the mean meridional circulation. Clearly, further work is needed to test the validity of this hypothesized mechanism.

5. PHYSICAL INTERPRETATIONS AND MODEL SIMULATIONS

The observational evidence summarized in section 4 for solar cycle variations of upper stratospheric ozone and temperature at low latitudes is qualitatively consistent with the hypothesis of direct photochemical and radiative forcing by solar ultraviolet variations. However, the relatively large apparent solar cycle variations of ozone, temperature, geopotential height, and zonal wind observed in the lower stratosphere (subsections 4.3 and 4.4) are not directly explicable in terms of these processes alone. Also, the unexpected altitude dependence of the estimated ozone and temperature responses (Figures 4 and 5) is not predicted by most existing models.

In this section, a working hypothesis or "conceptual model" for explaining how the direct upper stratospheric effects of solar UV variations may lead to unexpectedly large solar cycle signals in the lower stratosphere is described. Some recent observational evidence that appears to support this hypothesis is also summarized. In addition, observational evidence that solar-induced circulation anomalies in the lower stratosphere can produce detectable effects on tropospheric circulation is reviewed. Finally, current efforts to understand and simulate the unexpected altitude dependence of the apparent solar cycle ozone variation in the middle and upper stratosphere are discussed.

5.1 Downward Propagation From the Upper to the Lower Stratosphere

During the 1990's, a conceptual model for the downward propagation of direct solar UV-induced effects from the upper stratosphere to the lower stratosphere was developed by a number of authors [*Kodera*, 1995; *Hood et al.*, 1993; *Rind and Balachandran*, 1995; *Holton*, 1994] (for recent updates, see also *Kuroda and Kodera* [2002], *Kodera* [2002], and *Kodera and Kuroda* [2002]). This model is at least qualitatively consistent with the original suggestion of *Hines* [1974] and is based on a combination of both observational analyses and general circulation model simulations. The model is also supported by earlier theoretical work showing that angular momentum transfer to the zonal wind field by breaking Rossby and gravity waves at a given level can control the evolution of the wind field at lower levels [*Haynes et al.*, 1991].

To provide some background, in the simplest description, there are two preferred internal modes, or types of circulation, in the winter stratosphere [*Labitzke*, 1982]. In one mode, the polar night jet is relatively strong, polar lower stratospheric temperatures are relatively cold, and the winter is characterized by few major stratospheric warmings. In the other mode, opposite conditions dominate. This dichotomy and the tendency of the stratosphere to stay in one or the other of these modes for most of the winter has been described in terms of several closely related atmospheric circulation patterns in the recent literature, i.e., the northern (or southern) annular mode [*Thompson and Wallace*, 2000] and the polar night jet (PNJ) oscillation [*Kuroda and Kodera*, 2002]. In the troposphere, the annular mode in the Northern Hemisphere (NH) is called the "Arctic Oscillation" (AO) [*Thompson and Wallace*, 1998], also referred to as the North Atlantic Oscillation (NAO) in the Atlantic / European sector [*Hurrell et al.*, 2001]. The tendency for the stratosphere to occupy one or the other of these two states is considered to be an internal mode of variability; no external forcing is required to initiate and maintain a given state. In other words, the energy for the generation and maintenance of these modes is entirely internal. However, if an external forcing exists during the critical initiation phase of the early winter, then this external forcing can determine which mode is realized. Possible external forcings that could influence the selection of preferred modes in the winter stratosphere include the equatorial QBO [*Holton and Tan*, 1980], volcanic aerosols (e.g., *Kodera* [1995]), global warming [*Perlwitz and Graf*, 1995], the El Nino-Southern Oscillation (ENSO), and solar UV forcing. If solar UV forcing is stronger than other forcings at the beginning of a given winter, then this forcing can dominantly determine which internal mode is selected.

As shown in Figure 8 (from Kodera [1995]), there is observational evidence that, on average, the positive annular (or strong polar night jet mode) tends to be selected in the northern winter under solar maximum conditions. The

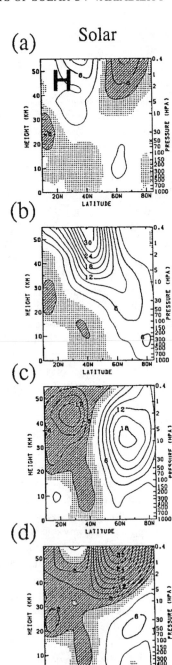

Figure 8. Approximate mean evolution of the zonal wind in the Northern winter stratosphere during solar maximum years from November (a) to February (d). The initial solar radiative forcing is indicated schematically by the symbol "H". Although the initial wind anomaly is larger (more than 9 m/s) than expected from solar UV forcing alone, this anomaly may be amplified through wave-mean-flow interactions, possibly involving the onset of the westerly phase of the SAO (see the text). (From *Kodera* [1995])

figure shows the results of a composite analysis in which Northern Hemispheric winters over a 12-year period were divided into two groups according to whether the sun was near solar maximum (1980, 1981, 1982, 1989, 1990, 1991, 1992) or solar minimum (1985, 1986, 1987, 1988). Subtracting monthly mean zonal wind fields for solar minimum from those for solar maximum yielded the difference wind fields (solar maximum minus solar minimum) shown in the figure for the months of (a) November; (b) December; (c) January; and (d) February. In Figure 8a, the symbol "H" indicates the initial forcing of radiative heating in the equatorial upper stratosphere in late fall / early winter. Comparing Figures 8a and 8b, a westerly zonal wind anomaly with maximum amplitude develops near the mid-latitude stratopause by December. Analysis of simultaneous planetary wave fluxes (not shown here) indicates that the upward flux in the stratosphere is weakened as the zonal wind anomaly amplifies. This coupled behavior of zonal wind anomalies and planetary wave fluxes strongly suggests that the evolution of the system is controlled by wave-mean flow interactions (e.g., *Andrews et al.* [1987]). In January and February (Figure 8c,d), the zonal wind anomaly shifts poleward and downward with time. It should be emphasized, however, that interannual variability (especially that associated with the QBO) is large so that any single winter near solar maximum is unlikely to exhibit the precise behavior shown in Figure 8.

Recently, *Kuroda and Kodera* [2002] have reported an extension of the analysis of *Kodera* [1995] in time to 1998 (19 years) and in coverage to include the Southern Hemisphere (SH). Results show that the strong polar night jet mode is also favored in the SH and that the early winter zonal wind anomalies are similar in both hemispheres. However, the time evolution of the wind field differs between hemispheres; this difference was attributed to known differences in the annular (or PNJ) mode in the SH as compared to the NH.

As indicated schematically in Figure 9a (from *Kodera and Kuroda* [2002]), the observations of Figure 8 are consistent with the following simplified interpretation. First, direct solar UV forcing, consisting of increases in equatorial upper stratospheric ozone and radiative heating, produces, through the thermal wind relationship, an enhancement of the zonal wind (U) in the subtropics of the winter hemisphere. Second, apparently via wave-mean-flow interactions, this initial solar-induced zonal wind anomaly in the subtropical upper stratosphere propagates downward and poleward.

In support of the possibility that solar UV forcing can influence the selection of preferred winter modes, it is known theoretically that differences in UV heating will produce a larger deviation of the zonal wind in the upper stratosphere during early winter near the times of winter solstice (e.g., *Hood et al.* [1993]). Near this time in both the Northern and Southern Hemispheres (December and June/July, respectively), statistical analyses of NCEP zonal wind data near the mid-latitude stratopause yield the largest positive solar regression coefficients [*Kodera and Yamazaki*, 1990; *Hood et al.*, 1993; Kodera, 1995]. However, the observationally derived wind changes from solar minimum to maximum are much larger than calculated by most existing models. It is therefore still not well understood how solar UV forcing can yield wind field changes as large as those shown in Figure 8.

From a purely empirical standpoint, there is recent observational evidence that the equatorial upper stratosphere can be more effective in determining the evolution of the northern winter circulation than the equatorial lower stratosphere. Specifically, an analysis of rocketsonde zonal wind data extending to 58 km shows that the largest correlation between northern polar temperature in January / February and equatorial wind occurs in the upper stratosphere in September / October near the onset of the westerly phase of the semi-annual wind oscillation (SAO) [*Gray et al.*, 2001a]. The SAO is dominant in the upper stratosphere and is influenced by the QBO in the lower stratosphere through the effect of the latter on the vertical propagation of waves that drive the SAO.

Supporting model simulations demonstrate that forcing of a stratosphere / mesosphere model by observed winds in the upper stratosphere (32–58 km) as well as in the lower stratosphere is necessary to reproduce the observed dependence of polar winter temperatures on the phase of the QBO [*Gray et al.*, 2001b]. Since solar cycle UV flux changes produce zonal wind anomalies in the upper stratosphere, it may be hypothesized that these zonal wind changes will also modify the development of the westerly regime of the SAO. This would help to explain how weak UV heating differences could result in relatively large zonal wind changes observed near the mid-latitude stratopause between solar minimum and maximum [*Kodera and Yamazaki*, 1990; *Hood et al.*, 1993]. Clearly, this possibility must be tested further using future equatorial wind data and model simulations.

5.2 Coupling to the Tropical Lower Stratosphere

A possible mechanism by which solar-induced changes in the upper stratosphere may couple to the tropical lower stratosphere is illustrated schematically in Figure 9b (from *Kodera and Kuroda* [2002]). In the simplest description, changes in wave absorption that are associated with the development of a stronger polar vortex under solar maximum conditions (averaging over other sources of interannual variability) also modify the mean meridional or "Brewer-Dobson" circulation. The predicted anomaly in the Brewer-Dobson circulation is downward in the tropics (less upwelling) under solar maximum conditions.

In support of the hypothesis of Figure 9b, Figure 10 (from *Hood and Soukharev* [2002]) compares a time series of extratropical eddy heat flux (a measure of the rate of wave absorption) with tropical column ozone and MSU4 temperature records. The (a) and (b) parts of the figure represent deviations from long-term monthly means of tropical column ozone and vertically sampled (40–120 hPa) lower stratospheric temperature, respectively, averaged in the tropics (20°S to 20°N). The ozone record is that compiled

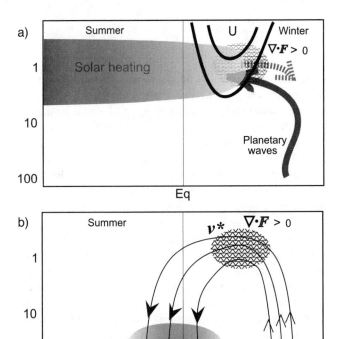

Figure 9. Schematic illustration of the effect of solar UV variability on the lower stratosphere (from Kodera and Kuroda [2002]). (a) Increased ozone production and radiative heating in the summer hemisphere enhances the zonal wind in the subtropical upper stratosphere of the winter hemisphere. Upward propagating planetary waves are deflected poleward (dashed arrow), decreasing the rate of planetary wave absorption ($\nabla \cdot \mathbf{F} > 0$). (b) The reduced wave absorption in the extratropical upper stratosphere induces an equatorward anomaly in the Brewer-Dobson circulation. This, in turn, produces a warm anomaly in the tropical lower stratosphere.

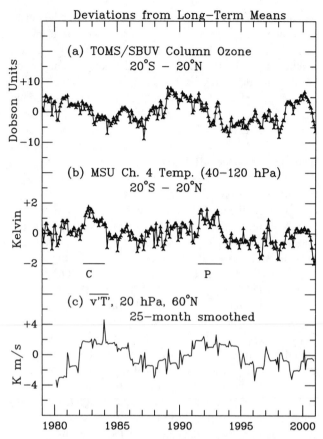

Figure 10. Comparisons of decadal variations present in (a) tropical column ozone; (b) tropical lower stratospheric temperature; and (c) extratropical meridional eddy heat flux, a proxy for the rate of planetary wave absorption by the zonal flow. (See the text)

by *Stolarski et al.* [2000] from a combination of Nimbus 7 Total Ozone Mapping Spectrometer (TOMS), NOAA 9/11 Solar Backscattered Ultraviolet (SBUV), and Earth Probe TOMS data. The temperature time series is calculated from the MSU Channel 4 record [*Spencer et al.*, 1990; *Spencer and Christy*, 1993]. Shown in Figure 10c is the meridional eddy heat flux at 20 hPa, 60°N after smoothing with a 25-month boxcar filter. The eddy heat flux is defined as $v'T'$, the zonally averaged product of the deviations of meridional velocity and temperature from the zonal mean at a given pressure level and latitude. $v'T'$ is nearly proportional to the vertical component of the planetary wave flux and is a good proxy for the rate of wave absorption (see, e.g., Coy et al. [1997]).

In Figure 10, the $v'T'$ series, exhibits a decadal variation with maxima in ~ 1984 and ~ 1993. The tropical ozone series exhibits decadal maxima at ~ 1980, ~ 1990, and ~ 2000, approximately in phase with the solar cycle. As was discussed in section 4.3, the tropical MSU4 temperature

record is characterized by pronounced heating episodes associated with aerosol injections following the El Chichon (C) and Pinatubo (P) volcanic eruptions. However, outside of these volcanically perturbed intervals, a decadal temperature variation is present that correlates approximately with the ozone record.

The $v'T'$ decadal variation shown in Figure 10c has approximately the correct phase to be responsible for the decadal tropical ozone and temperature variations shown in Figure 10a,b. To see that this is the case, one may consider the following simplified forms of the ozone continuity and thermodynamic energy equations:

$$\frac{\partial \overline{\mu}}{\partial t} \approx -\overline{w}^* \frac{\partial \overline{\mu}}{\partial z} - (\overline{\mu} - \overline{\mu}_{eq})/\tau_c \tag{1}$$

$$\frac{\partial \overline{T}}{\partial t} \approx -\overline{w}^* \left[\frac{\partial \overline{T}}{\partial z} + g/c_p \right] - (\overline{T} - \overline{T}_{eq})/\tau_r \tag{2}$$

where $\overline{\mu}$ is the zonal mean ozone mixing ratio, $\overline{\mu}_{eq}$ is an equilibrium value for $\overline{\mu}$, τ_c is the odd oxygen lifetime, $\overline{T}$ is the zonal mean temperature, g/c_p is the adiabatic lapse rate, τ_r is the radiative lifetime, and $\overline{w}^*$ is the vertical velocity in the transformed Eulerian mean formalism (e.g., *Andrews et al.* [1987]). In writing (1) and (2), we have assumed that vertical ozone and heat transports dominate, for the situation in question, over other transport processes for ozone and diabatic heating for temperature. If $\overline{w}^*$ in the tropical lower stratosphere is at least partly driven by extratropical wave forcing, then we expect that $\overline{w}^*$ will increase when $v'T'$ in the extratropical stratosphere increases. For example, during a sudden warming event, $v'T'$ in the extratropical lower stratosphere increases, the polar night jet decelerates, and the induced meridional circulation anomaly is such as to increase $\overline{w}^*$ in the tropical lower stratosphere. If so, then, for time scales comparable to or shorter than τ_c and τ_r, it follows from (1) and (2) that $v'T'$ will correlate inversely with the *tendency* of the tropical ozone and temperature change. This relationship is observed on short time scales during stratospheric warming events [*Randel*, 1993]: Maxima in $v'T'$ in the extratropical lower stratosphere near 60°N occur at the same time as rapid negative rates of change of column ozone in the tropics and summer hemisphere. Consistent with this expectation, the decadal peaks in $v'T'$ shown in Figure 10c occur during periods following solar maxima at times when both column ozone and temperature were declining at maximum rates.

To investigate whether the amplitude of the observed decadal variation of $v'T'$ in Figure 10c is sufficient to explain the amplitudes of the ozone and temperature variations shown in Figure 10a,b, one may consider observed relationships between these quantities on short time scales (e.g., during sudden warmings), assuming that the same relationships will hold approximately on long time scales. Specifically, examinations of daily data show that increases of $v'T'$ at 20 hPa, 60°N of ~ 60 K-m/s are associated with tropical column ozone and MSU4 temperature tendencies of roughly ~ -0.2 DU/day and ~ -0.04 K/day, respectively (see, e.g., *Randel* [1993] in the case of column ozone). From Figure 10c, we see that the mean anomaly of $v'T'$ has a peak amplitude of ~ 2 K-m/s during years following solar maxima. From the daily data results, we would therefore expect corresponding ozone and temperature tendencies of ~ -0.2 DU/month and ~ -0.04 K/month. For comparison, ozone and temperature tendencies derived directly from the data of Figure 10a,b are ~ -0.2 DU/month and ~ -0.03 K/month during years following solar maxima. Although a more rigorous analysis is clearly needed, we can provisionally conclude that extratropical wave driving (as gauged by the time series of $v'T'$ shown in Figure 10c) is probably a likely cause of the observed decadal ozone and temperature variations in the tropical lower stratosphere.

5.3 GCM Simulations

As reviewed by *Haigh* [1999b], GCM simulations have shown that the calculated response of lower stratospheric circulation to solar ultraviolet flux changes is sensitively dependent on whether the model includes changes in stratospheric ozone. Early experiments considered only the radiative effects of solar UV changes on the upper stratosphere and employed unrealistically large flux changes to produce significant impacts on the lower stratosphere [*Kodera*, 1991; *Kodera et al.*, 1991; *Balachandran and Rind*, 1995; *Rind and Balachandran*, 1995; *Jackman et al.*, 1996; *Balachandran et al.*, 1999]. Nevertheless, these studies were valuable for demonstrating the downward propagation of solar UV induced perturbations from the upper to the lower stratosphere. Later experiments showed that lower stratospheric effects of solar UV variations are sensitive to the photochemical response of ozone in the middle and upper stratosphere [*Haigh*, 1994; 1996; *Shindell et al.*, 1999; *Haigh*, 1999a; *Rozanov et al.*, 2000; *Gray et al.*, 2001b]. *Shindell et al.* [1999] applied a GCM including an interactive parameterization of stratospheric chemistry to show how upper stratospheric ozone changes may amplify observed 11-year irradiance changes to affect circulation in the lower stratosphere and troposphere. They found that the model could simulate qualitatively aspects of the observed response, including 30 hPa geopotential height variations in the northern winter hemisphere. However, for those runs

that adopted realistic solar flux and stratospheric ozone changes, the amplitudes of the calculated solar-induced changes in the lower stratosphere were generally less than those that have been inferred observationally. Similarly, the simulations of *Haigh* [1999a] showed a broadening of the tropical Hadley circulation, subtropical warming, and an associated increase in geopotential heights that are in qualitative agreement with the observations of *Labitzke* [2001]. However, the amplitudes of the height changes ~10 m) were much less than those estimated observationally (75–100 m).

5.4 Tropospheric Consequences

In general, composite analyses of large stratospheric circulation anomalies, including realizations of the annular and PNJ modes, have shown that anomalies appearing initially above 50 km altitude often descend to the lowermost stratosphere and tend to be followed by specific tropospheric weather regimes during a 10–60 day period [*Baldwin and Dunkerton*, 1999; 2001]. As emphasized by the latter authors, the downward phase propagation of these anomalies does not necessarily imply a downward influence. A better interpretation is that waves originating in the troposphere can modify the stratospheric conditions for wave propagation, thereby leading to the movement of zonal wind anomalies poleward and downward [*Dunkerton*, 2000]. In the equatorial stratosphere, an analogous interaction of the mean flow with waves originating in the troposphere occurs during the transition from one phase of the QBO to the other.

However, as reviewed above, observations indicate that solar cycle UV variations can initiate the development of a positive annular (or PNJ) mode in the winter upper stratosphere. According to the statistical results of *Baldwin and Dunkerton* [1999; 2001], such a change in stratospheric circulation would inevitably have tropospheric consequences. In this sense, solar UV variability, while directly affecting only the upper stratosphere, can lead to significant differences in tropospheric circulation between solar minimum and maximum. In particular, according to Figure 8c,d, an enhanced zonal wind anomaly at middle latitudes in the NH troposphere (representing a positive annular or AO mode) appears by January under solar maximum conditions [*Kodera*, 1995; *Kuroda and Kodera*, 2002]. To first order, the main effect of increased solar UV flux on tropospheric circulation is therefore a positive increase in the Arctic Oscillation index. Statistical studies of NCEP Reanalysis data over a 40-year period [*Ruzmaikin and Feynman*, 2002] confirm that the annular mode in the NH (NAM) is positively influenced by solar UV flux and that the extent of the influence depends on the phase of the QBO. The latter authors specifically find that tropospheric effects of solar UV flux changes are seen mainly in early winter for the west QBO phase and in late winter for the east QBO phase.

5.5 The Response of Middle and Upper Stratospheric Ozone

Several possible explanations for the differences indicated in Figure 4 between ozone observations and models have been suggested. First, as noted already in section 4.1 above, it must be admitted that the available data are limited in duration. The possibility can therefore not be completely excluded that the ozone measurements themselves are biased because of intercalibration uncertainties and the limited period for which satellite ozone observations are available ~2 solar cycles). However, comparisons of measurements obtained from independent satellite profile records (i.e., SAGE I/II and SBUV-SBUV/2) as well as comparisons of measurements made in the two separate solar cycles for which data are available suggest that observational biases may not be capable of explaining the magnitude of the differences shown in Figure 4 [Hood and Soukharev, 2000].

Another possibility is that there may be limitations in the statistical methods used to determine the ozone response to the solar cycle. In particular, linear regression models may not account for possible non-linear interactions between the seasonal cycle, the QBO, and volcanic signals. Further work to evaluate this source of uncertainty (e.g., by using artificial data) is therefore desirable.

Turning to possible physical explanations of the differences seen in Figure 4, it should first be noted that, at the altitudes of interest, ozone is nearly in photochemical equilibrium. Hence, viable explanations most probably require differences in chemical production/loss rates compared to those that are predicted in nominal photochemical models. In this regard, it has been pointed out by Siskind et al. [1998] that transport-induced decadal variations of CH_4 and H_2O may have occurred during the last few decades. The CH_4 concentration effectively determines the abundance of reactive chlorine while that of H_2O controls the odd hydrogen abundance. Such decadal variations could either be independent of solar forcing (resulting, for example, from changes in tropical upwelling associated with volcanic aerosol injections) or could be a consequence of solar forcing in a manner that is not predicted by existing models. In either case, significant modifications of the effective ozone response to solar cycle UV variations would be expected. Observational evidence for decadal-scale variations in CH_4 and H_2O in the middle and upper stratosphere have been

reported based on data from the Halogen Occultation Experiment (HALOE) on the Upper Atmosphere Research Satellite [Nedoluha et al. 1998a,b; Randel et al., 1999].

In addition to unmodeled decadal variations in methane and water vapor, it has been suggested by Callis et al. [2000] that the discrepancy could be due to long-term (2–10 year) fluctuations in stratospheric odd nitrogen (NO_y). Such long-term fluctuations could originate from one of several mechanisms, including changes in photochemical production or transport. Callis et al. specifically proposed that decadal variations in NO_y could be driven by solar cycle changes in energetic electron precipitation, which can directly modulate the production of NO_y in the lower thermosphere and mesosphere. They argued that this upper atmospheric NO_y can be transported downward and equatorward into the mid-latitude stratosphere where it could influence global ozone abundances. On the other hand, the efficiency of odd nitrogen transport from the polar upper atmosphere to the mid-latitude stratosphere is still controversial and other models have not predicted a large contribution from this source (e.g., Jackman et al. [1995]; Siskind et al. [1997]).

6. SUMMARY AND CONCLUSIONS

Stratospheric responses to solar ultraviolet variations have now been observed on both the solar rotation and the solar cycle time scales. Although the observed responses are qualitatively consistent with theoretical expectations, there are important quantitative differences. This is especially true on the solar cycle time scale in the lower stratosphere where observed responses are much larger than expected from existing models.

As reviewed in section 3, the 27-day ozone response in the tropics has been accurately measured and is in good agreement with recent stratospheric model calculations. There is also good evidence for a response of tropical upper stratospheric temperature on this time scale. However, the derived positive temperature phase lags are significantly larger than expected from models that consider only radiative and photochemical processes. It is therefore inferred that a dynamical component of the response exists. This inference is supported by observational evidence that latitudinal ozone and temperature oscillations during the Northern Hemisphere winter may be modulated by 27-day solar UV variability.

As reviewed in section 4, despite the short lengths of available data records, substantial evidence exists for a significant solar cycle variation in both the upper and the lower stratosphere. The observationally inferred ozone response in the middle and upper stratosphere exhibits a quite different dependence on altitude (maximizing in the upper stratosphere, approaching zero or negative values in the middle stratosphere) than that which is expected from current radiative-photochemical models. It is unlikely that this difference can be explained by observational biases alone. As reviewed in section 5.5, other suggested explanations have focused mainly on possible decadal variations of trace constituents that affect the ozone loss rate but that may not be accounted for in current models. Of most interest in this regard are unmodeled decadal variations of CH_4, H_2O, and NO_y. It is possible that unmodeled differences in transport of these trace constituents between solar minimum and maximum are responsible for much of the discrepancy between the observed and modeled ozone response.

As reviewed in section 5.1, the unexpectedly large solar cycle signal in the lower stratosphere may be caused by the apparent ability of weak solar UV forcing in the upper stratosphere to influence the selection of preferred internal modes, or types of circulation, in the winter stratosphere. Specifically, there is increasing observational evidence that the positive phase of the annular mode (or polar night jet mode) tends to be selected under solar maximum conditions in both the Northern and the Southern Hemispheres [*Kodera*, 1995; *Kuroda and Kodera*, 2002; *Ruzmaikin and Feynman*, 2002]. However, it must be emphasized that this tendency is superposed on large interannual variability driven by other forcing mechanisms (e.g., the QBO). On average, the initial solar forcing appears to occur near the time of winter solstice in both hemispheres when relatively large westerly zonal wind increases are observed near the midlatitude stratopause under solar maximum conditions [*Kodera and Yamazaki*, 1990; *Hood et al.*, 1993]. These large zonal wind increases may be a result of modification of the development of the westerly regime of the semi-annual oscillation in the fall and early winter by solar UV forcing near the stratopause.

As discussed in section 5.2, coupling of the upper stratospheric response on the 11-year time scale to the tropical lower stratosphere most probably occurs via the Brewer-Dobson circulation (Figure 9b). The apparent preferred selection of the positive phase of the annular mode in winter (stronger polar night jet) under solar maximum conditions is associated with a reduction of extratropical wave absorption ($\nabla \cdot \mathbf{F} > 0$; $v'T' < 0$) approaching solar maxima. Because of the importance of extratropical wave absorption in forcing the Brewer-Dobson circulation (*Holton et al.* [1995]), this leads to reduced upwelling in the tropical lower stratosphere near solar maxima. Observational evidence that apparent solar cycle variations of ozone and tem-

perature in the tropical lower stratosphere may be driven by decadal variations in extratropical wave forcing was described in relation to Figure 10.

As discussed in section 5.4, the existence of tropospheric consequences of solar UV forcing is supported by recent statistical evidence [*Ruzmaikin and Feynman*, 2002]. Specifically, anomalous tropospheric weather regimes (regardless of origin) are often preceded by stratospheric circulation anomalies appearing initially above 50 km altitude [*Baldwin and Dunkerton*, 1999; 2001]. It follows that those stratospheric circulation anomalies that are of solar origin (i.e., some positive phases of the annular mode in the winter hemisphere occurring near solar maxima), will be associated with later tropospheric circulation anomalies.

In conclusion, some progress has been made during the last decade toward understanding the response of the stratosphere to solar cycle changes in UV flux. However, remaining differences between observations and model simulations indicate that further work in both observational and modeling areas is needed before a full understanding will be achieved. By implication, current general circulation model simulations of the effect of solar UV variability on climate change must be regarded as provisional. More realistic simulations must await a more complete knowledge of the processes that lead to the observed stratospheric effects and how these effects are transmitted to the troposphere.

Acknowledgments. Several colleagues, especially Kuni Kodera, John McCormack, and Boris Soukharev, contributed significant ideas and work that are contained herein. An anonymous reviewer and K. Labitzke also provided insightful criticisms that improved the quality of the final manuscript. Special thanks are due to instrument and data processing teams who are responsible for the production of stratospheric remote sensing data discussed in this review. Preparation of the review was supported by grants from the NASA Atmospheric Chemistry Modeling and Analysis Program and the NASA Solar Influences Research Program.

REFERENCES

Andrews, D. G., J. R. Holton and C. B. Leovy, *Middle Atmosphere Dynamics*, Academic Press, 489 pp., 1987.

Austin, J., A three-dimensional coupled chemistry-climate model simulation of past stratospheric trends, *J. Atmos. Sci.*, In press, 2001.

Balachandran, N. K. and D. Rind, Modeling the effects of UV variability and the QBO on the troposphere / stratosphere system. Part I: The middle atmosphere, *J. Climate*, 8, 2058-2079, 1995.

Balachandran, N. K., D. Rind, P. Lonergan, and D. Shindell, Effects of solar cycle variability on the lower stratosphere and the troposphere, *J. Geophys. Res.*, 104, 27321-27339, 1999.

Baldwin, M. P., et al., The quasi-biennial oscillation, *Rev. Geophys., 39*, 179-229, 2001.

Baldwin, M. P. and T. J. Dunkerton, Propagation of the Arctic Oscillation from the stratosphere to the troposphere, *J. Geophys. Res., 104*, 30937-30945, 1999.

Baldwin, M. P. and T. J. Dunkerton, Stratospheric harbingers of anomalous weather regimes, *Science, 294*, 581-584, 2001.

Brasseur, G., The response of the middle atmosphere to long-term and short-term solar variability: A two-dimensional model, *J. Geophys. Res., 98*, 23079-23090, 1993.

Brasseur, G., and S. Solomon, *Aeronomy of the Middle Atmosphere*, 441 pp., D. Reidel, Hingham, Mass., 1984.

Burrage, M.D., R.A. Vincent, H.G. Mayr, W.R. Skinner, N.F. Arnold, and P.B. Hays, Long-term variability in the equatorial middle atmosphere zonal wind, *J. Geophys. Res., 101*, 12847-12854, 1996.

Callis, L. B., M. Natarajan, and J. Lambeth, Calculated upper stratospheric effects of solar UV flux and NO_y variations during the 11-year solar cycle, *Geophys. Res. Lett., 27*, 3869-3872, 2000.

Chandra, S., Solar-induced oscillations in the stratosphere: A myth or reality, *J. Geophys. Res., 90*, 2331-2339, 1985.

Chandra, S., The solar and dynamically induced oscillations in the stratosphere, *J. Geophys. Res., 91*, 2719-2734, 1986.

Chandra, S., The solar UV related changes in total ozone from a solar rotation to a solar cycle, *Geophys. Res. Lett., 18*, 837-840, 1991.

Chandra, S. and R. D. McPeters, The solar cycle variation of ozone in the stratosphere inferred from Nimbus 7 and NOAA 11 satellites, J. Geophys. Res., 99, 20665-20671, 1994.

Chanin, M.-L., V. Ramaswamy, D. J. Gaffen, W. J. Randel, R. B. Rood, and M. Shiotani, Trends in stratospheric temperatures, in *Scientific Assessment of Ozone Depletion: 1998*, Global Ozone Research and Monitoring Project - Report No. 44, p. 5.1-5.59, World Meteorological Organization, Geneva, Switzerland, 1998.

Charney, J. G., and P. G. Drazin, Propagation of planetary-scale disturbances from the lower into the upper atmosphere, *J. Geophys. Res., 66*, 83-109, 1961.

Chen, Li, J. London, and G. Brasseur, Middle atmospheric ozone and temperature responses to solar irradiance variations over 27-day periods, *J. Geophys. Res., 102*, 29957-29979, 1997.

Coy, L., E. Nash, and P. Newman, Meteorology of the polar vortex: Spring 1997, *Geophys. Res. Lett., 24*, 2693-2696, 1997.

Dameris, M., A. Ebel, and H. J. Jakobs, Three-dimensional simulation of quasiperiodic perturbations attributed to solar activity effects in the middle atmosphere, *Annal. Geophysicae, 4, A*, 287-296, 1986.

Donnelly, R. F., Solar UV spectral irradiance variations, *J. Geomagn. Geoelect., 43*, Supplement, Part 2, 835-842, 1991.

Donnelly, R. F., D. E. Stevens, J. Barrett, and K. Pfendt, Short-term temporal variations of Nimbus 7 measurements of the solar UV spectral irradiance, *Tech. Memo. ERL ARL-154*, Natl. Oceanic and Atmos. Admin., Air Resources Lab., Silver Spring, MD, 1987.

Dunkerton, T., Midwinter deceleration of the subtropical mesospheric jet and interannual variability of the high-latitude flow in UKMO Analyses, *J. Atmos. Sci., 57*, 3838-3855, 2000.

Dunkerton, T., D. Delisi, and M. Baldwin, Middle atmosphere cooling trend in historical rocketsonde data, *Geophys. Res. Lett., 25*, 3371-3374, 1998.

Ebel, A. and W. Batz, Response of stratospheric circulation at 10 mbar to solar activity oscillations resulting from the sun's rotation, *Tellus, 29*, 41-47, 1977.

Ebel, A., M. Dameris, H. Hass, A. H. Manson, C. W. Meek, and K. Petzoldt, Vertical change of the response to solar activity oscillations with periods around 13.6 and 27 days in the middle atmosphere, *Ann. Geophys., 4, A*, 271-280, 1986.

Ebel, A., B. Schwister, and K. Labitzke, Planetary waves and solar activity in the stratosphere between 50 and 10 mbar, *J. Geophys. Res., 86*, 9729-9738, 1981.

Finger, F., R. Nagatani, M. Gelman, C. Long, and A. J. Miller, Consistency between variations of ozone and temperature in the stratosphere, *Geophys. Res. Lett., 22*, 3477-3480, 1995.

Fleming, E. L., S. Chandra, C. H. Jackman, D. B. Considine, and A. R. Douglass, The middle atmospheric response to short and long term solar UV variations: Analysis of observations and 2D model results, J. *Atmos. Terr. Phys., 57*, 333-365, 1995.

Gelman, M., A. J. Miller, K. W. Johnson, and R. M. Nagatani, Mean zonal wind and temperature structure during the PMP-1 winter periods, *Adv. Space Res., 6*, 17-26, 1986.

Gelman, M., A. J. Miller, R. Nagatani, and C. S. Long, Use of UARS data in the NOAA Stratospheric Monitoring Program, *Adv. Space Res., 14*, (9)21-(9)31, 1994.

Garcia, R. R., T. J. Dunkerton, R. S. Lieberman, and R. A. Vincent, Climatology of the semi-annual oscillation of the tropical middle atmosphere, *J. Geophys. Res., 102*, 26,019 -26,032, 1997.

Gille, J., C. M. Smythe, and D. F. Heath, Observed ozone response to variations in solar ultraviolet radiation, *Science, 225*, 315-317, 1984.

Gray, L. J., and J. A. Pyle, A two-dimensional model of the quasi-biennial oscillation of ozone, *J. Atmo. Sci, 46*, 203-220, 1989.

Gray, L. J., S. J. Phipps, T. J. Dunkerton, M. P. Baldwin, E. F. Drysdale, and M. R. Allen, A data study of the influence of the equatorial upper stratosphere on Northern Hemisphere stratospheric sudden warmings, *Q. J. R. Meteorol. Soc., 127*, 19 pp., 2001a.

Gray, L. J., E. F. Drysdale, T. J. Dunkerton, and B. N. Lawrence, Model studies of the inter annual variability of the Northern Hemisphere winter circulation: The role of the quasi-biennial oscillation, *Q. J. R. Meteorol. Soc., 127*, 29 pp., 2001b.

Haigh, J. D., The role of stratospheric ozone in modulating the solar radiative forcing of climate, *Nature, 370*, 544-546, 1994.

Haigh, J. D., On the impact of solar variability on climate, *Science, 272*, 981-984, 1996.

Haigh, J. D., A GCM study of climate change in response to the 11-year solar cycle, *Q. J. R. Meteorol. Soc., 125*, 871-892, 1999a.

Haigh, J. D., Modelling the impact of solar variability on climate, *J. Atmos. Solar-Terr. Phys., 61*, 63-72, 1999b.

Haynes, P. H., C. J. Marks, M. McIntyre, T. Shepherd, and K. Shine, On the "downward control" of extratropical diabatic circulations by eddy-induced mean zonal forces, *J. Atmos. Sci., 48*, 651-678, 1991.

Heath, D. F. and B. M. Schlesinger, The global response of stratospheric ozone to ultraviolet solar flux variations, in *Atmospheric Ozone*, Proceedings of the Quadrennial Ozone Symposium, Halkidiki, Greece, edited by C. S. Zerefos and A. Ghazi, pp. 666-670, D. Reidel, Hingham, Mass., 1985.

Heath, D. and B. Schlesinger, The Mg 280 nm doublet as a monitor of changes in solar ultraviolet irradiance, *J. Geophys. Res., 91*, 8672-8682, 1986.

Herzberg, L., in *Physics of the Earth's Upper Atmosphere*, edited by C. Hines, I. Paghis, T. R. Hartz, and J. A. Fejer, Prentice Hall, Englewood Cliffs, N. J., 1965.

Hines, C. O., A possible mechanism for the production of Sun-weather correlations, *J. Atmos. Sci., 31*, 589-591, 1974.

Holton, J. R., The quasi-biennial oscillation in the Earth's atmosphere and its links to longer period variability, in *The Solar Engine and Its Influence on Terrestrial Atmosphere and Climate*, NATO ASI Series, Vol. 125, E. Nesme-Ribes, ed., pp. 259-273, Springer-Verlag, Berlin, Heidelberg, 1994.

Holton, J. R., P. Haynes, M. McIntyre, A. Douglass, R. Rood, and L. Pfister, Stratosphere-troposphere exchange, *Rev. Geophys., 33*, 403-439, 1995.

Holton, J. R., and C. Mass, Stratospheric vacillation cycles, *J. Atmos. Sci., 33*, 2218-2225, 1976.

Holton, J. R., and H.-C. Tan, The influence of the equatorial quasi-biennial oscillation on the global circulation at 50 mb, *J. Atmos. Sci., 37*, 2200-2208, 1980.

Hood, L. L., The temporal behavior of upper stratospheric ozone at low latitudes: Evidence from Nimbus 4 BUV data for short-term responses to solar ultraviolet variations, *J. Geophys. Res., 89*, 9557-9568, 1984.

Hood, L. L., Coupled stratospheric ozone and temperature responses to short-term changes in solar ultraviolet flux: An analysis of Nimbus 7 SBUV and SAMS data, *J. Geophys. Res., 91*, 5264-5276, 1986.

Hood, L. L., Solar ultraviolet radiation induced variations in the stratosphere and mesosphere, *J. Geophys. Res., 92*, 876-888, 1987.

Hood, L. L., The solar cycle variation of total ozone: Dynamical forcing in the lower stratosphere, *J. Geophys. Res., 102*, 1355-1370, 1997.

Hood, L. L., Effects of short-term solar UV variability on the stratosphere, *J. Atmos. Solar-Terr. Phys., 61*, 45-51, 1999.

Hood, L. and S. Cantrell, Stratospheric ozone and temperature responses to short-term solar ultraviolet variations: Reproducibility of low-latitude response measurements, *Ann. Geophys., 6*, 525-530, 1988.

Hood, L. L. and A. R. Douglass, Stratospheric responses to solar ultraviolet variations: Comparisons with photochemical models, *J. Geophys. Res., 93*, 3905-3911, 1988.

Hood, L. L. and J. Jirikowic, Stratospheric dynamical effects of solar ultraviolet variations: Evidence from zonal mean ozone and temperature data, *J. Geophys. Res., 96*, 7565-7577, 1991.

Hood, L. L., J. Jirikowic, and J. P. McCormack, Quasi-decadal variability of the stratosphere: Influence of long-term solar ultraviolet variations, *J. Atmos. Sci., 50*, 3941-3958, 1993.

Hood, L. and B. Soukharev, The solar component of long-term stratospheric variability: Observations, model comparisons, and possible mechanisms (extended abstract), SPARC 2000 Symposium, Mar del Plata, Argentina, November, 2000.

Hood, L. and B. Soukharev, Quasi-decadal variability of the tropical lower stratosphere: The role of extratropical wave forcing, *J. Atmos. Sci.*, submitted, 2002.

Hood, L. L. and S. Zhou, Stratospheric effects of 27-day solar ultraviolet variations: An analysis of UARS MLS ozone and temperature data, *J. Geophys. Res., 103*, 3629-3638, 1998.

Hood, L. and S. Zhou, Stratospheric effects of 27-day solar ultraviolet variations: The column ozone response and comparisons of solar cycles 21 and 22, *J. Geophys. Res., 104*, 26473-26479, 1999.

Hurrell, J., Y. Kurshnir, and M. Visbeck, The North Atlantic Oscillation, *Science, 291*, 603-605, 2001.

Jackman, C. H., E. L. Fleming, S. Chandra, D. B. Considine, and J. E. Rosenfield, Past, present, and future modeled ozone trends with comparisons to observed trends, *J. Geophys. Res., 101*, 28753-28767, 1996.

Jackman, C. H., F. M. Vitt, D. B. Considine, and E. L. Fleming, Energetic particle precipitation effects on odd nitrogen and ozone over the solar cycle time scale, in *The Solar Cycle Variation of the Stratosphere: A STEP Working Group 5 Report*, edited by L. Hood, University of Arizona, Tucson, 120 pp., 1995.

Kalnay, E. et al., The NCEP/NCAR 40 Year reanalysis project, *Bull. Am. Meteorol. Soc., 77*, 437-471, 1996.

Keating, G., J. Nicholson III, D. F. Young, G. Brasseur, and A. De Rudder, Response of middle atmosphere to short-term solar ultraviolet variations, 1, Observations, *J. Geophys. Res., 92*, 889-902, 1987.

Keckhut, P., A. Hauchecorne, and M.-L. Chanin, Midlatitude long-term variability of the middle atmosphere: Trends and cyclic and episodic changes, *J. Geophys. Res., 100*, 18887-18897, 1995.

Keckhut, P., F. J. Schmidlin, A. Hauchecorne, and M.-L. Chanin, Trend estimates from rocketsondes at low latitude stations (8S-34N), taking into account instrumental changes and natural variability, *J. Atmos. Solar-Terr. Phys., 61*, 447-459, 1999.

Kodera, K., The solar and equatorial QBO influences on the stratospheric circulation during the early northern hemisphere winter, *Geophys. Res. Lett., 18*, 1023-1026, 1991.

Kodera, K., On the origin and nature of the interannual variability of the winter stratospheric circulation in the northern hemisphere, *J. Geophys. Res., 100*, 14077-14087, 1995.

Kodera, K., Dynamical response to the solar cycle I: Winter stratopause, *J. Geophys. Res.*, submitted, 2002.

Kodera, K., M. Chiba, and K. Shibata, A general circulation model study of the solar and QBO modulation of the stratospheric circulation during the northern hemisphere winter, *Geophys. Res. Lett., 18}*, 1209-1212, 1991.

Kodera, K. and Y. Kuroda, A mechanistic model study of slowly propagating coupled stratosphere-troposphere variability, *J. Geophys. Res., 105*, 12361-12370, 2000.

Kodera, K. and Y. Kuroda, Dynamical response to the solar cycle II: Lower stratosphere and the influence of the QBO, *J. Geophys. Res.*, submitted, 2002.

Kodera, K., and K. Yamazaki, Long-term variation of upper stratospheric circulation in northern hemisphere in December, *J. Meteor. Soc. Japan, 68*, 101-105, 1990.

Kuroda, Y. and K. Kodera, Effect of solar activity on the polar-night jet oscillation in the Northern and Southern hemisphere winter, *J. Meteorol. Soc. Japan*, in press, 2002.

Labitzke, K., The global signal of the 11-year sunspot cycle in the stratosphere: Differences between solar maxima and minima, *Meteorologische Zeitschrift, 10*, 901-908, 2001.

Labitzke, K., On the interannual variability of the middle stratosphere during the northern winters, *J. Met. Soc. Japan, 60*, 124-139, 1982.

Labitzke, K., J. Austin, N. Butchart, J. Knight, Masaaki Takahashi, Miwa Nakamoto, Tatsuya Nagashima, J. Haigh, and V. Williams, The global signal of the 11-year solar cycle in the stratosphere: observations and model results, *J. Atmos. Solar Terrestrial Physics*, in press, 2001.

Labitzke, K., and H. van Loon, Associations between the 11-year solar cycle, the QBO and the atmosphere: I, the troposphere and stratosphere in the northern hemisphere in winter, *J. Atmos. Terr. Phys., 50*, 197-206, 1988.

Labitzke, K., and H. van Loon, Some recent studies of probable connections between solar and atmospheric variability, *Ann. Geophysicae, 11*, 1084-1094, 1993.

Labitzke, K., and H. van Loon, Connection between the troposphere and stratosphere on a decadal scale, *Tellus, 47A*, 275-286, 1995.

Lean, J. L., Solar irradiance and climate forcing in the near future, *Geophys. Res. Lett., 28*, 4119-4122, 2001.

Lean, J. L., G. J. Rottman, H. L. Kyle, T. N. Woods, J. R. Hickey, and L. C. Puga, Detection and parameterization of variations in solar mid- and near-ultraviolet radiation (200-400 nm), *J. Geophys. Res., 102*, 29,939-29,956, 1997.

Lienesch, J. H., W. Planet, M. T. DeLand, K. Laaman, R. P. Cebula, E. Hilsenrath, and K. Horvath, Validation of NOAA-9 SBUV/2 total ozone measurements during the 1994 Antarctic ozone hole, *Geophys. Res. Lett., 23*, 2593-2596, 1996.

Matsuno, T., A dynamical model of the stratospheric sudden warming, *J. Atmos. Sci., 28*, 1479-1494, 1971.

McCormack, J., and L. Hood, Apparent solar cycle variations of upper stratospheric ozone and temperature: Latitude and seasonal dependences, *J. Geophys. Res., 101*, 20933-20944, 1996.

McCormack, J. P., L. L. Hood, R. Nagatani, A. J. Miller, W. Planet, and R. McPeters, Approximate separation of volcanic and 11-year signals in the SBUV-SBUV/2 total ozone record over the 1979-1995 period, *Geophys. Res. Lett., 24*, 2729-2732, 1997.

Mitchell, J. F. B., D. J. Karoly, G. C. Hegerl, F. W. Zwiers, M. R. Allen, and J. Marengo, Detection of climate change and attribution of causes, in *Climate Change 2001: The Scientific Basis. Contribution of Working Group 1 to the Third Assessment Report of the Intergovernmental Panel on Climate Change*, Houghton, J. T., Y. Ding, D. J. Griggs, M. Noguer, P. J. van der

Linden, X. Dai, K. Maskell, and C. A. Johnson, eds., Cambridge University Press, Cambridge, United Kingdom and New York, U.S.A., 881 pp., 2001.

Nash, J. and P. R. Edge, Temperature changes in the stratosphere and lower mesosphere 1979-1988 inferred from TOVS radiance observations, *Adv. Space Res., 9*, 333-341, 1989.

Nash, J. and G. F. Forrester, Long-term monitoring of stratospheric temperature trends using radiance measurements obtained by the TIROS-N series of NOAA spacecraft, *Adv. Space Res., 6*, 37-44, 1986.

Nedoluha, G. E., R. M. Bevilacqua, R. M. Gomez, D. E. Siskind, and B. C. Hicks, Increases in middle atmospheric water vapor as observed by the Halogen Occultation Experiment and the ground-based Water Vapor Millimeter-wave Spectrometer from 1991 to 1997, *J. Geophys. Res., 103*, 3531-3543, 1998a.

Nedoluha, G. E., D. E. Siskind, J. T. Bacmeister, R. M. Bevilacqua, and J. M. Russell III, Changes in upper stratospheric CH_4 and NO_2 as measured by HALOE and implications for changes in transport, *Geophys. Res. Lett., 25*, 987-990, 1998b.

Perlwitz, J. and H. Graf, The statistical connection between tropospheric and stratospheric circulation of the Northern Hemisphere in winter, *J. Climate, 8*, 2281-2295, 1995.

Randel, W., Global variations of zonal mean ozone during stratospheric warming events, *J. Atmos. Sci., 50*, 3308-3321, 1993.

Randel, W., F. Wu, J. Russell III, and J. Waters, Space-time patterns of trends in stratospheric constituents derived from UARS measurements, *J. Geophys. Res., 104*, 3711-3727, 1999.

Rind, D., and N. K. Balachandran, Modeling the effects of UV variability and the QBO on the troposphere / stratosphere system, Part II: The troposphere, *J. Climate, 8*, 2080-2095, 1995.

Rodgers, C. D., R. L. Jones, and J. J. Barnett, Retrieval of temperature and composition from Nimbus 7 SAMS measurements, *J. Geophys. Res., 89*, 5280-5286, 1984.

Rozanov, E., M. E. Schlesinger, F. Yang, S. Malyshev, N. Andronova, V. Zubov, and T. Egorova, Sensitivity of the UIUC stratosphere / troposphere GCM with interactive photochemistry to the observed increase of solar UV radiation (extended abstract), SPARC 2000 Symposium, Mar del Plata, Argentina, November, 2000.

Ruzmaikin, A. and J. Feynman, Solar influence on a major mode of atmospheric variability, *J. Geophys. Res.*, in press, 2002.

Salby, M., and P. Callaghan, Connection between the solar cycle and the QBO: The missing link, *J. Clim., 13*, 2652-2662, 2000.

Scaife, A., J. Austin, N. Butchart, S. Pawson, M. Keil, J. Nash, and I. N. James, Seasonal and interannual variability of the stratosphere diagnosed from UKMO TOVS analyses, *Q. J. R. Meteorol. Soc., 126*, 2585-2604, 2000.

Shapiro, R., and F. Ward, A neglected cycle in sunspot numbers? *J. Atmos. Sci., 19*, 506-508, 1962.

Shindell, D., D. Rind, N. Balachandran, J. Lean, and P. Lonergan, Solar cycle variability, ozone, and climate, *Science, 284*, 305-308, 1999.

Siskind, D. E., and J. T. Bacmeister, M. E. Summers, J. M. Russell III, Two dimensional model calculations of nitric oxide transport in the middle atmosphere and comparison with Halogen

Occultation Experiment data, *J. Geophys. Res., 102*, 3,527-3,545, 1997.

Siskind, D. E., L. Froidevaux, J. M. Russell, and J. Lean, Implications of upper stratospheric trace constituent changes observed by HALOE for O_3 and ClO from 1992 to 1995, *Geophys. Res. Lett., 25*, 3513-3516, 1998.

Soukharev, B. and K. Labitzke, The 11-year solar cycle, the sun's rotation, and the middle stratosphere in winter Part I: Response of zonal means, *J. Atmos. Solar-Terr. Phys., 62*, 335-346, 2000a.

Soukharev, B. and K. Labitzke, The 11-year solar cycle, the sun's rotation, and the middle atmosphere in winter. Part II: Response of planetary waves, *J. Atmos. Solar-Terr. Phys., 63*, 1931-1939, 2000b.

Soukharev, B. and L. Hood, Possible solar modulation of the equatorial quasi-biennial oscillation: Additional statistical evidence, *J. Geophys. Res., 106*, 14855-14868, 2001.

Spencer, R. W. and J. R. Christy, Precision lower stratospheric temperature monitoring with the MSU: Technique, validation, and results 1979-1991, *J. Climate, 6*, 1194-1204, 1993.

Spencer, R. W., J. Christy, and N. Grody, Global atmospheric temperature monitoring with satellite microwave measurements: Method and results 1979-84, *J. Climate, 3*, 1111-1128, 1990.

Stolarski, R., R. McPeters, G. Labow, S. Hollandsworth Frith, and L. Flynn, On the long-term calibration of the TOMS total ozone record, In *Proc. Quad. Ozone Symposium*, edited by R. Bojkov and S. Kazuo, Hokkaido University, Sapporo, Japan, p. 33-34, 2000.

Stolarski, R., W. Randel, L. Bishop, D. Cunnold, D. deMuer, S. Godin, D. Hofmann, S. Hollandsworth, L. Hood, J. Logan, J. Miller, M. Newchurch, S. Oltmans, O. Uchino, and R. Wang, Ozone change as a function of altitude, in *SPARC/IOC/GAW Assessment of Trends in the Vertical Distribution of Ozone*, SPARC Report No. 1, WMO Ozone Research and Monitoring Project Report No. 43, Harris, N., R. Hudson, and C. Phillips, eds., SPARC Office, BP3, 91371 Verrieres le Buisson Cedex, France, 289 pp., 1998.

Takigawa, M., Takahashi, M. and Akiyoshi, H., Simulation of ozone and other chemical species using a Center for Climate System Research/National Institute for Environmental Studies atmospheric GCM with coupled stratospheric chemistry, *J. Geophys. Res., 104*, 14,003-14,018, 1999.

Thompson, D. W. J., and J. M. Wallace, The Arctic Oscillation signature in the wintertime geopotential height and temperature fields, *Geophys. Res. Lett., 25*, 1297-1300, 1998.

Thompson, D. W. J., and J. M. Wallace, Annular modes in the extratropical circulation. Part I: Month-to-month variability, *J. Climate, 13*, 1000-1016, 2000.

van Loon, H., and K. Labitzke, The global range of the stratospheric decadal wave, Part I: Its association with the sunspot cycle in summer and in the annual mean, and with the troposphere, *J. Climate, 11*, 1529-1537, 1998.

van Loon, H. and D. Shea, The global 11-year signal in July-August, *Geophys. Res. Lett., 27*, 2965-2968, 2000.

Wang, H. J., D. M. Cunnold, and X. Bao, A critical analysis of SAGE ozone trends, *J. Geophys. Res., 101*, 12495-12514, 1996.

Waters, J., Microwave limb sounding, chap. 8, in *Atmospheric Remote Sensing by Microwave Radiometry*, edited by M. A. Janssen, John Wiley, New York, 1993.

Williams, V., J. Austin, and J. D. Haigh, Model simulations of the impact of the 27-day solar rotation period on stratospheric ozone and temperature, *Adv. Space Res.*, in press, 2001.

Zerefos, C. W., K. Tourpali, B. R. Bojkov, D. S. Balis, B. Rognerund, and I. S. A. Isaksen, Solar activity-total ozone relationships: Observations and model studies with heterogeneous chemistry, *J. Geophys. Res., 102*, 1561-1569, 1997.

Zhou, S., A. J. Miller, and L. L. Hood, A partial correlation analysis of the stratospheric ozone response to 27-day solar UV variations with temperature effect removed, *J. Geophys. Res., 105*, 4491-4500, 2000.

Lon L. Hood, Lunar and Planetary Laboratory, University of Arizona, Tucson, Arizona 85721.

The Effect of Solar Proton Events on Ozone and Other Constituents

NASA Goddard Space Flight Center, Greenbelt, Maryland.

Solar proton events (SPEs) can cause changes in constituents in the Earth's middle atmosphere. The highly energetic protons cause ionizations, excitations, dissociations, and dissociative ionizations of the background constituents. Complicated ion chemistry leads to HO_x production and dissociation of N_2 leads to NO_y production. Both the HO_x and NO_y increases can result in changes to ozone in the stratosphere and mesosphere. The HO_x increases lead to short-lived ozone decreases in the mesosphere and upper stratosphere due to the short lifetimes of the HO_x constituents. The NO_y increases lead to long-lived stratospheric ozone changes because of the long lifetime of NO_y constituents in this region. The NO_y-induced ozone changes are generally decreases, however, the NO_y constituents can interfere with chlorine and bromine radicals in the lowest part of the stratosphere and cause ozone increases. Temperature changes have been predicted to occur as a result of the larger SPEs. Atmospheric changes have been observed as a result of eleven SPEs since 1969. Neutral wind variations were measured shortly after the July 1982 and April 1984 SPEs. The recent July 2000 SPE caused NO_x increases that lasted for two months past the event. The two periods of largest SPEs (August 1972 and October 1989) caused ozone decreases that lasted for several weeks past the events.

1. INTRODUCTION

Periodically, the Sun erupts in a solar flare and an associated coronal mass ejection (CME) that results in an intense flux of solar particles in interplanetary space. The solar particles from a CME can impact the Earth's magnetosphere if the location of the Earth is aligned relative to the solar flare and the solar magnetic field (see Figure 1). This is known as a solar proton event (SPE). The sunward side of the magnetosphere is further flattened from the norm and the tail is additionally elongated. Most particles are then drawn in on the far side of the magnetosphere and are carried into the polar cap regions (generally >60° geomagnetic latitude).

The eruptions of solar protons, which are more frequent near solar maximum, can produce ionizations, dissociations, dissociative ionizations, and excitations in the middle atmosphere. The very important middle atmospheric families of HO_x (H, OH, HO_2) and NO_y (N, NO, NO_2, NO_3, N_2O_5, HNO_3, HO_2NO_2, $ClONO_2$, $BrONO_2$) are produced either directly or through a photochemical sequence. As a result, the chemistry of the polar middle atmosphere (stratosphere and mesosphere) can be dramatically altered by large solar proton events.

There has been proof that the Sun can produce significant fluxes of highly energetic particles for about 60 years [*Forbush*, 1946]. A large SPE that occurred over 30 years ago in November 1969 gave the first evidence that ozone could be depleted by these solar eruptions [*Weeks et al.*, 1972]. The influence of SPEs on the atmosphere has matured over the years and evidence of impacts by at least

Solar Variability and its Effects on Climate
Geophysical Monograph 141
Published in 2004 by the American Geophysical Union
10.1029/141GM21

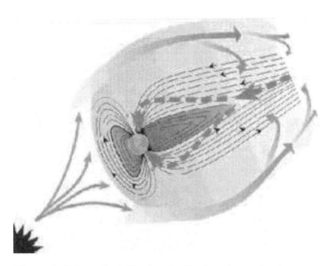

Figure 1. Schematic indicating the Earth and associated magnetosphere interacting with solar particles from a CME. The CME resulted from an explosion on the Sun (lower left corner) and caused further flattening of the sunward side of the magnetosphere and additional elongation of the tail. This was taken from an image on the CDROM entitled "The Dynamic Sun 4.0" from the NASA/ESA Solar and Heliospheric Observatory (SOHO) Mission.

11 large events (see Table 1) is well documented in the refereed literature and will be discussed below.

Other energetic charged particles, besides solar protons, also have an influence on the Earth's atmosphere. Auroral electrons primarily influence the thermosphere and upper part of the mesosphere with associated bremsstrahlung penetrating to the upper and middle stratosphere. Relativistic electrons primarily influence the mesosphere and upper part of the stratosphere, although their associated bremsstrahlung can reach the middle to lower stratosphere. These electrons are influenced by the Earth's magnetic field and affect the subauroral region (geomagnetic latitudes between 60° and 70°). Galactic cosmic rays (GCRs) deposit most of their energy in the lower stratosphere and upper troposphere. These very high energy particles are not as affected by the Earth's magnetic field and the higher energy GCRs can influence all the way to tropical latitudes.

Solar protons, which primarily impact the mesosphere and stratosphere, are affected by the Earth's magnetic field and deposit most of their energy in the polar cap regions (>60° geomagnetic latitude). Like electron precipitation events, the solar proton influences can be sporadic and impulsive. The very highest energy solar protons (>100 MeV) can influence the middle to lower stratosphere, a region also affected by the lower energy GCRs. There are similarities in the atmospheric influences among all precip-

itating charged particles, e.g., most of the energy deposited in the atmosphere results in ionizations. There are also differences in the temporal characteristics of the particular events and the spatial influences (both altitude and latitude regions). The details of the differences among these precipitating charged particles is beyond the scope of this review and we will only focus on the effects of the solar protons.

This paper is divided into seven primary sections, including the introduction. We discuss the very important solar proton measurements in section 2. Knowledge of the solar proton flux is crucial in computing the atmospheric impacts. The solar protons produce both HO_x and NO_y constituents, which are discussed in sections 3 and 4, respectively. The impact of the SPEs on ozone is discussed in section 5 and the temperature and dynamical changes from SPEs are discussed in section 6. Finally, the conclusions are given in section 7.

2. SOLAR PROTON MEASUREMENTS

Solar protons have been measured by several satellites over the past few decades. The Interplanetary Monitoring Platform (IMP) series of satellites has provided measurements from 1963 to the present. The IMP satellites are in orbits that position them in interplanetary space most of the time. The last of these satellites, IMP-8, launched on October 26, 1973, is in a near circular, 35 Earth radii, 12-day orbit. The NOAA Geostationary Operational Environmental Satellites (GOES) have provided measurements since 1975. The GOES series of satellites are in geosynchronous orbits at about 35,800 km above the Earth. These IMP and GOES satellite orbits, far from the Earth's atmosphere, provide reliable measurements of incoming solar protons.

The solar proton flux can change dramatically within a matter of minutes and primarily influences the polar regions. The temporal characteristics and spatial influence of the solar protons will be discussed in the next two subsections.

2.1. Temporal Characteristics

The solar proton flux is generally quite small for protons with sufficient energy to penetrate into the middle atmosphere. However, SPEs can bring significant fluxes of protons in the near Earth environment, which can influence the middle atmosphere. Although SPEs can occur at anytime during the eleven year solar cycle, more SPEs happen near solar maximum. The October 1989 SPEs, which produced the highest flux of solar protons in the past thirty years, showvery large changes in proton fluxes over a short period

Table 1. List of solar proton events (SPEs) and measured atmospheric constituent influence documented in the literature. The effects on ozone were all decreases and the effects on NO and NO_2 were all increases. This is not an exhaustive list of publications and other references are given later for particular SPEs.

Date of SPEs	Effects of SPEs	Publication(s)
November 1969	Ozone	*Weeks et al.* [1972]
January 1971	Ozone	*McPeters et al.* [1981]
September 1971	Ozone	*McPeters et al.* [1981]
August 1972	Ozone	*Heath et al.* [1977], also others
June 1979	Ozone	*McPeters and Jackman* [1985]
August 1979	Ozone	*McPeters and Jackman* [1985]
October 1981	Ozone	*McPeters and Jackman* [1985]
July 1982	Ozone,	Ozone in *Thomas et al.* [1983], also others;
	NO	NO in *McPeters* [1986]
December 1982	Ozone	*McPeters and Jackman* [1985]
October 1989	Ozone,	Ozone in *Reid et al.* [1991], also others;
	NO,	NO in *Zadorozhny et al.* [1992];
	NO_2	NO_2 in *Jackman et al.* [1995]
July 2000	Ozone,	Ozone in *Jackman et al.* [2001];
	NO, NO_2	NO & NO_2 during event in *Jackman et al.* [2001];
		NO & NO_2 after event in *Randall et al.* [2001]

of time. The protons with energies from 4.2 to 8.7 MeV, which deposit most of their energy between about 75 and 85 km, increased their flux by about a thousand in minutes (see Figure 2). Ultimately, the increase in proton flux was five orders of magnitude above background in a couple of hours.

2.2. Spatial Influence

Higher energy protons deposit their energy lower in the atmosphere. The precipitating protons primarily lose their energy in the creation of ion pairs in the atmosphere. An ion pair is created when a precipitating proton removes an electron (called a secondary electron) from the neutral molecule or atom, leaving behind a positive ion. The protons impart energy on the secondary electrons and these freed charged particles also cause further ionizations in the atmosphere. A computation of the rate of ionizations by incoming mono-energetic protons and their associated electrons is given in Figure 3. These protons have a flux of 1 cm^{-2} s^{-1} ster^{-1} and are isotropic in distribution. Approximately 35 eV is expended in the production of an ion pair [*Porter et al.*, 1976]. The total ionization rate for an SPE is computed using the actual particle spectrum multiplied by the individual monoenergetic deposition values.

Because of the guiding influence of the Earth's magnetic field, the effects are generally confined to the polar regions and have their largest influence on the regions of the polar caps, defined as the northern and southern areas with latitudes greater than about 60° geomagnetic. The impact of the

very large July 2000 SPE on mesospheric ozone is illustrated in Plate 1. This plate shows the NOAA 14 SBUV/2 ozone measurements for the 0.5 hPa (~55 km) level [*Jackman et al.*, 2001]. The plot on the left indicates ozone amounts on July 13, before the SPE. The plot on the right indicates ozone amounts on July 14-15, during the maximum intensity of the SPE. The polar cap (>60° geomagnetic) is outlined by the thick white oval. This plate illustrates the very significant changes in ozone during the SPE that are generally confined to the polar cap. The ozone reduction slightly outside the polar cap near 90°E longitude is probably caused by the Earth's magnetic field being perturbed somewhat during this very large solar disturbance. The ozone in the area outside the polar cap region did not change significantly from July 13 to July 14/15.

The atmospheric influence of a SPE is dependent on the energy spectrum of the solar protons and the absolute flux levels at the particular energies. The August 1972 SPEs were the second largest in the past 30 years and caused very significant atmospheric effects (discussed later). The computed ionization rates for these very large SPEs are given in Figure 4.

The largest values of ionization rate were computed to occur within several hours after the onset of these SPEs, and were caused by the highest energy protons. These SPEs were very "hard spectrum" events with a huge flux of very energetic protons. These fast moving protons (with energies greater than 30 MeV) arrived first and dominated the first several hours of these SPEs (Figure 4). These particular

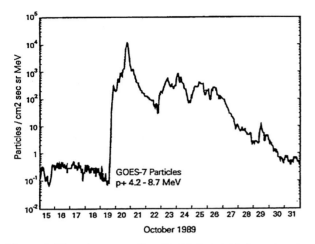

Figure 2. The solar proton flux (E=4.2-8.7 MeV) from GOES-7 for the second half of October 1989. Reprinted from *J. Atmos. Terr. Phys.*, 54, A. M. Zadorozhny, G. A. Tuchkov, V. N. Kikhtenko, J. Lastovicka, J. Boska, and A. Novak, Nitric oxide and lower ionosphere quantities during solar particle events of October 1989 after rocket and ground-based measurements, 183-192, Fig. 1, Copyright(1992), with permission from Elsevier Science.

SPEs had computed ionization rates of over 1000 cm^{-3} s^{-1} for over a day with peak ion rates over 40000 cm^{-3} s^{-1} in the stratosphere (~10-50 km). Not all SPEs cause such large ion pair production in the stratosphere and, in general, most large SPEs induce atmospheric changes primarily in the mesosphere (~50-90 km).

3. PRODUCTION OF HO_x CONSTITUENTS BY SPEs –MODEL COMPUTATIONS

Precipitating protons also produce HO_x constituents. The basic theory for the HO_x production by these particles has been known for the past thirty years. *Solomon et al.* [1981] provide a clear analysis of the current understanding of this process. Some earlier papers that discussed this process are *Swider and Keneshea* [1973] and *Frederick* [1976]. The production of HO_x relies on complicated ion chemistry that takes place after the initial formation of ion pairs.

The production of the original positive ions N_2^+, O_2^+, N^+, and O^+ depends on the efficiency for production from both the primary protons and their associated electrons. The O_2^+ constituent has a smaller ionization potential than N_2^+, N^+, or O^+, thus most ions rapidly charge exchange to form it. NO^+, also a potential source of odd hydrogen [*Solomon et al.*, 1981], is formed when N^+ and O^+ atoms react with O_2 and N_2, respectively. The oxonium ion (O_4^+) is generally formed from O_2^+ via the path

$$O_2^+ + O_2 + M \rightarrow O_4^+ + M$$

Water cluster ions are then formed via

$$O_4^+ + H_2O \rightarrow O_2^+ \cdot H_2O + O_2.$$

Further clustering follows this reaction with an ultimate recombination between the positive ion and an electron or negative ion. The shortest path leading to HO_x production is given by:

$$O_2^+ \cdot H_2O + H_2O \rightarrow H_3O^+ \cdot OH + O_2$$
$$H_3O^+ \cdot OH + e^- \rightarrow H + OH + H_2O$$
$$Net: H_2O \rightarrow H + OH.$$

Solomon et al. [1981] discuss other paths for $O_2^+ \cdot H_2O$, which lead to HO_x production through recombination with free electrons. The negative ions (especially NO_3^-) become more abundant than the free electrons below about 70 km. A possible path in this region is:

$$O_2^+ \cdot H_2O + H_2O \rightarrow H_3O^+ \cdot OH + O_2$$
$$H_3O^+ \cdot OH + H_2O \rightarrow H^+ \cdot (H_2O)_2 + OH$$
$$H^+ \cdot (H_2O)_2 + NO_3^- \rightarrow HNO_3 + H_2O + H_2O$$
$$HNO_3 + h\nu \rightarrow OH + NO_2$$
$$Net: H_2O + NO_3 \rightarrow OH + OH + NO_2.$$

Either this path or a similar one takes place in the region where negative ions prevail. The number of HO_x constituents produced per ionization as a function of altitude and ionization rate is shown in Figure 5 for daytime, polar

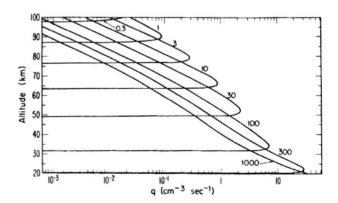

Figure 3. Rate of production of ion pairs in the atmosphere by monoenergetic fluxes of protons incident isotropically over the upward-looking hemisphere with a flux of 1 proton cm^{-2} s^{-1} $ster^{-1}$. Curves are for energies from 0.3 MeV to 1000 MeV. Reprinted from Kluwer Academic Publishers book *Physics of the Sun, Vol. III*, edited by P. A. Sturrock, pp. 251-278, Chapter 12 – Solar energetic particles and their effects on the terrestrial environment, G. C. Reid, Fig. 1, 1986, with kind permission from Kluwer Academic Publishers.

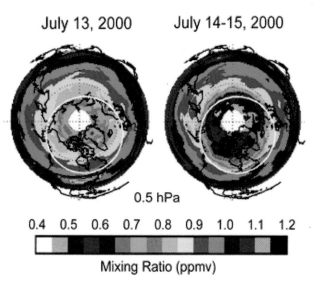

Plate 1. Taken from Fig. 1 of *Jackman et al.* [2001]. NOAA 14 SBUV/2 Northern Hemisphere polar ozone in ppmv before (July 13, 2000) and during (July 14/15) the solar proton event period at 0.5 hPa (~55 km). The white circle indicates the boundary at 60°N geomagnetic, above which solar protons can penetrate to the Earth's atmosphere.

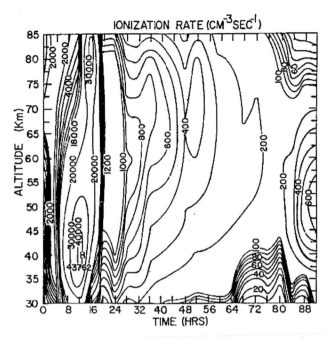

Figure 4. Contours of ionization rate in units of ionizations cm⁻³ s⁻¹ for the first four days of the August 1972 proton events. Time 0 is August 4, 1972, at 339 U.T. Reprinted from *Planet. Space Sci.*, *29*, Rusch, D. W., J.-C. Gerard, S. Solomon, P. J. Crutzen, and G. C. Reid, The effect of particle precipitation events on the neutral and ion chemistry of the middle atmosphere, 1, Odd nitrogen, 767-774, Fig. 4, Copyright(1981), with permission from Elsevier Science.

summer conditions of temperature, air density, and solar zenith angle. Each ion pair produces about two HO_x constituents up to an altitude of approximately 70 km. Above 70 km, the production is less than two HO_x constituents per ion pair [*Solomon et al.*, 1981]. As the ionization rate increases the computed electron density increases and the probability of the recombination of the intermediate positive ions with the electrons or negative ions increases. Thus at all altitudes the number of HO_x constituents produced per ion pair is seen to decrease with ionization rate. The production of the HO_x constituents produced per ion pair is not very dependent on solar zenith angle and will be as effective at night as during the daytime.

The HO_x constituents react fairly quickly to destroy each other through reactions like

$$OH + HO_2 \rightarrow H_2O + O_2.$$

Therefore, the lifetime of HO_x is only on the order of hours in the upper stratosphere and mesosphere. Any corresponding change in the atmosphere caused by the HO_x species would only last for a couple of hours past any SPE (see section 5.1).

Large changes in HO_x constituents have been computed to result from SPEs. For example, during the July 2000 SPE, HO_x was computed to increase by over 100% [*Jackman et al.*, 2001] at sunrise. To the best of our knowledge there are no published measurements of HO_x changes during SPEs. Such measurements would be extremely useful to validate the theory of the HO_x production during SPEs.

4. PRODUCTION OF NO_y CONSTITUENTS BY SPES

All SPEs produce NO_y, however, the SPE-caused NO_y enhancement has only been measured as a result of the three very large SPEs that occurred in July 1982, October 1989, and July 2000 (Table 1). These SPEs all showed substantial NO and/or NO_2 increases. *McPeters* [1986] used Nimbus 7 SBUV measurements to derive NO increases during the July 1982 SPE. *Zadorozhny et al.* [1992] used rockets for NO and *Jackman et al.* [1995] employed SAGE II measurements for NO_2 to derive the influences of the October 1989

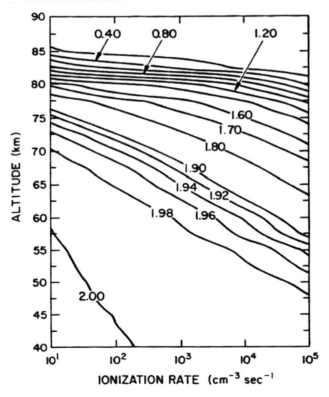

Figure 5. Contours of odd hydrogen particles produced per ionization as a function of altitude and ionization rate for daytime summer polar conditions. Reprinted from *Planet. Space Sci.*, *29*, S. Solomon, D. W. Rusch, J.-C. Gerard, G. C. Reid, and P. J. Crutzen, The effect of particle precipitation events on the neutral and ion chemistry of the middle atmosphere, 2, Odd hydrogen, 885-892, Fig. 2, Copyright(1981), with permission from Elsevier Science.

SPEs on NO_y. *Jackman et al.* [2001] and *Randall et al.* [2001] used the Upper Atmosphere Research Satellite (UARS) HALOE instrument data to establish the July 2000 SPE increases in NO_x ($NO+NO_2$). The influence of SPEs on NO_y in the atmosphere through both measurements and model simulations will be discussed in this section.

4.1. Measurements and Model Computations of SPE-caused NO_y

The precipitating protons and associated secondary electrons also produce atomic nitrogen (N) via dissociations, predissociations, or dissociative ionizations in collisions with N_2. Estimates of the number of NO_y constituents created per ion pair range from 0.33 [*Warneck*, 1972] up to 2.5 [*Fabian et al.*, 1979]. Recent publications show only small differences and range from 1.25 [*Jackman et al.*, 1990] up to 1.3 [*Reid et al.*, 1991] NO_y constituents produced per ion pair.

The production of NO_y by SPEs has been predicted since the mid-1970s [*Crutzen et al.*, 1975]. The NO increase after the July 1982 SPE was inferred from the Nimbus 7 SBUV instrument to be about 6×10^{14} NO molecules cm^{-2} above 1 hPa at polar latitudes [*McPeters*, 1986]. *Jackman et al.* [1990] computed an NO increase of 7×10^{14} above 1 hPa from the July SPE using an assumption that all the NO_y produced during this SPE resulted in the production of NO. *Zadorozhny et al.* [1992] measured NO enhancements of 2.6×10^{15} cm^{-2} between 50 and 90 km at southern polar latitudes with a rocket-borne instrument as a result of the October 1989 SPEs. *Jackman et al.* [1995] computed a production of NO in that altitude range of 3.0×10^{15} cm^{-2}, again assuming all the NO_y produced during these October 1989 SPEs resulted in the production of NO. Both of these computations are in reasonable agreement but higher than the measurements because some of the NO would be destroyed in the daytime through the reactions

$$NO + h\nu \rightarrow N + O$$
followed by $N + NO \rightarrow N_2 + O.$

This mechanism is the primary method whereby odd nitrogen (NO_y) is returned to even nitrogen (N_2).

4.2. Model Computations– Yearly Production of NO_y

There are other sources of NO_y in the atmosphere and the importance of SPEs in producing NO_y becomes clear only when comparing to these other sources. A summary of the computations of the production of NO_y molecules per year is given in Figure 6 for the northern polar stratosphere (lati-

tudes >50°N) over the 1955 to 1993 time period [*Jackman et al.*, 2000]. The sources include SPEs, GCRs, "in situ" oxidation of N_2O, and horizontal transport from lower latitudes.

The full histograms indicate the SPE-caused NO_y production for both the stratosphere and mesosphere, while the light gray areas of the histograms indicate the SPE-caused stratospheric NO_y production only. The annual production rates of NO_y from GCRs, "in situ" oxidation of N_2O, and horizontal transport from lower latitudes are represented in Figure 6 by the solid, dash-dot and dashed lines, respectively. There is only about a 10% variation in the "in situ" oxidation of N_2O and horizontal transport from lower latitudes, caused by the solar cycle ultraviolet flux variation which feeds into each of these sources [*Vitt and Jackman*, 1996]. This minor variation is hardly noticeable on the log scale used for the ordinate in the figure and the constant values of 1×10^{33} and 9×10^{33} NO_y molecules/year are assumed for the "in situ" oxidation of N_2O and horizontal transport of NO_y from lower latitudes, respectively. The GCR contribution to NO_y was computed to vary from 6.9 to 9.6×10^{32} molecules/year.

There is another source of NO_y in the polar middle atmosphere, energetic electrons. Unfortunately, we do not have a very reliable estimate of this source of NO_y on an annual

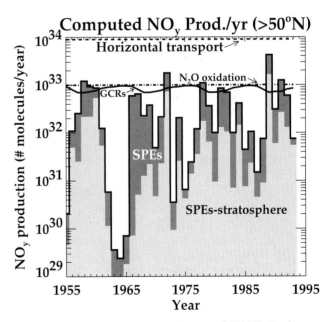

Figure 6. Taken from Fig. 1 of *Jackman et al.* [2000]. Total number of NO_y molecules produced per year in the northern polar stratosphere by SPEs (histogram indicating both the total and the stratospheric contributions), galactic cosmic rays (GCRs) indicated by solid line, N_2O oxidation in the polar region only (dash-dot line), and horizontal transport of NO_y from lower latitudes into this region (dashed line).

basis. The higher energy electrons and associated bremsstrahlung provide a direct middle atmospheric source of NO_y and the lower energy electrons provide an indirect source that is dependent on the downward transport of the NO_y to the middle atmosphere during polar night.

The oxidation of N_2O at lower latitudes and subsequent horizontal transport of NO_y into the polar regions is about an order of magnitude larger than the NO_y sources of GCRs and "in situ" oxidation of N_2O. The NO_y production from SPEs for the periods 1955-1973 and 1974-1993 was taken from *Jackman et al.* [1980] and *Vitt and Jackman* [1996], respectively. This SPE-generated annual production of NO_y is computed to vary by orders of magnitude over the 1955-1993 time period. There is a general solar cycle dependence with more (less) SPE-produced NO_y near solar maximum (minimum). The annual SPE source of NO_y in the Northern Hemisphere was larger than 1.2×10^{33} NO_y molecules/year during only 2 years (1972 and 1989). These were years in which very large fluxes of extremely energetic protons accompanied the SPEs in August 1972 and October 1989.

4.3. Measurements and Model Computations of July 2000 SPE-produced NO_y

The recent SPE of July 2000 also produced substantial changes in NO_y. Measurements of HALOE NO and NO_2 during [*Jackman et al.*, 2001] and a couple of months after the event [*Randall et al.*, 2001] indicated very large increases as a result of this event. Figure 7a shows the increases in HALOE Version 19 NO_x (NO+NO_2) above background. The HALOE measurements were taken at different longitudes and slightly different latitudes (65.4°N to 68.5°N) over the July 12-15 period, however, the NO_x changes resulting from the July 2000 SPE dominated any of the geographically-caused differences. The apparent enhancements in HALOE NO_x above 0.03 hPa during the first half of July 14 are probably related to other natural variabilities and not due to the SPE. Enhancements of NO_x greater than 50 ppbv and 200 ppbv are observed at 0.3 hPa and 0.01 hPa, respectively, near the end of July 15. Background NO_x levels are typically 1-5 ppbv at 0.3 hPa and 20-60 ppbv at 0.01 hPa, thus the NO_x increases caused by this SPE were very large.

Model computed NO_x changes for 65°N are given in Figure 7b. The predicted increase in the NO_x constituents from the SPE are similar to the measurements in the stratosphere and lower mesosphere up to ~0.3 hPa. However, in the middle and upper mesosphere (<0.3 hPa), the model predicts larger increases in the NO_x than observed. It is possible that some of these model/measurement differences may be related to the fraction of atomic N produced by the protons and associated electrons that end up in excited

states (e.g., $N(^2D)$) or the ground state ($N(^4S)$). The family modeling approach of *Jackman et al.* [2001] assumed that virtually all the NO_y produced above 0.3 hPa by the July 2000 SPE was in the form of NO, implying a nearly 100% production of $N(^2D)$ by the precipitating particles. *Rusch et al.* [1981] showed that there are huge differences in the final results of model computations of NO_y enhancements from SPEs that depend strongly on the branching ratios of the N atoms produced. Further study of these model/measurement

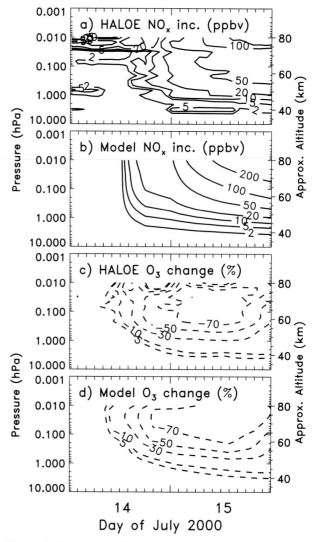

Figure 7. Taken from Fig. 2 of *Jackman et al.* [2001]. Polar Northern Hemisphere pressure versus time cross sections during the SPE period (July 14-15, 2000) for a) HALOE NO_x and b) model NO_x increases, both for contour levels 2, 5, 10, 20, 50, 100, and 200 ppbv; and c) HALOE ozone and d) model ozone decreases, both for contour levels −5, −10, −30, −50, and −70%. The HALOE NO_x and ozone changes were computed by comparing to the background average of the July 12-13 observations.

differences is needed to sort out the discrepancies.

Randall et al. [2001] showed evidence that the large NO_x enhancements from the July 2000 SPE lasted for at least two months past the event. They showed HALOE measurements in early September 2000 in the Southern Hemisphere polar vortex, indicating substantial enhancements of NO_x in the middle stratosphere. These large increases were almost certainly caused by the July 2000 SPE. The NO_x peak in the southern polar middle stratosphere occurred near 32 km and was probably a result of descent of the SPE-created NO_x from the mesosphere and upper stratosphere during the winter [*Randall et al.*, 2001].

5. OZONE EFFECTS BY SPES

Ozone is influenced by SPEs through the production of both HO_x and NO_y constituents. There are measured influences on atmospheric constituents during eleven large SPEs (Table 1). These SPEs all showed ozone decreases and three of the SPEs had NO or NO_2 increases that were measured as well as discussed in the previous section. A number of papers over a span of nearly 30 years have provided the evidence of these SPE influences on ozone in the atmosphere.

Solar cycle 20 provided the first evidence of atmospheric effects by SPEs. *Weeks et al.* [1972] described large mesospheric ozone decreases associated with the November 1969 SPE in this solar cycle. The ozone decreases caused by the August 1972 SPEs, the second largest episode of SPEs in the past 30 years, were originally shown in the Nimbus 4 satellite BUV instrument data in *Heath et al.* [1977] and were reanalyzed by *McPeters et al.* [1981], who was able to subtract the direct effects of the high energy protons on the BUV instrument. *Reagan et al.* [1981] showed the spatial extent of the ozone depletion and *Jackman and McPeters* [1987] were able to provide ozone depletion information for nearly two months after the August 1972 SPEs, again using BUV data. *McPeters et al.* [1981] was also able to provide reliable information about substantial short-lived ozone depletion at 1 mbar (~50 km) and above as a result of the January and September 1971 SPEs using BUV data.

Both the Solar Mesosphere Explorer (SME) satellite [*Thomas et al.*, 1983] and the SBUV instrument aboard Nimbus 7 [*McPeters and Jackman*, 1985] showed large mesospheric depletions during the large July 1982 SPE. *McPeters and Jackman* [1985] studied other SPEs in solar cycle 21 and provided evidence of ozone depletion in June and August 1979, October 1981, and December 1982.

The largest SPEs of the past 30 years occurred in solar cycle 22 in October 1989. Measurements by NOAA-11 SBUV/2 [*Jackman et al.*, 1993] and SAGE II [*Jackman et al.*, 1995] indicated very substantial long-term ozone depletion caused by this event.

A more recent SPE in July 2000 during solar cycle 23 caused ozone depletion that was measured by the NOAA-14 SBUV/2 and UARS HALOE instruments [*Jackman et al.*, 2001]. This SPE was the third largest in the past 30 years and provided good information about the spatial and temporal extent of the ozone changes caused by this atmospheric disturbance. Other large SPEs in this solar cycle that probably caused ozone decreases occurred in November 2000 and November 2001.

All SPEs produce both HO_x and NO_y constituents, both of which influence ozone. Some of these SPE-related ozone influences are short-term, caused by the HO_x increases, whereas others are longer-term, caused by the NO_y increases. These different ozone effects will be discussed below.

5.1. Short-term Effects from HO_x Constituents

The depletion of ozone by the SPE-enhanced HO_x constituents has been simulated in several papers [e.g., *Swider and Keneshea*, 1973; *Frederick*, 1976; *Swider et al.*, 1978; *Solomon et al.*, 1981; *Jackman and McPeters*, 1985]. All the SPEs given in Table 1 have a HO_x-induced ozone depletion associated with them. As noted in section 3, the HO_x constituents do not last more than a couple of hours after an SPE. However, the HO_x constituents can have very significant influences on ozone, especially in the mesosphere. There are several catalytic processes through which the HO_x constituents destroy ozone.

An important catalytic process in the stratosphere (~10-50 km) which leads to ozone destruction is:

$$OH + O_3 \rightarrow HO_2 + O_2$$
$$O + HO_2 \rightarrow OH + O_2$$
$$\text{Net: } O_3 + O \rightarrow 2O_2.$$

At mesospheric altitudes (~50-90 km) the catalytic process

$$H + O_3 \rightarrow OH + O_2$$
$$O + OH \rightarrow H + O_2$$
$$\text{Net: } O + O_3 \rightarrow 2O_2$$

is a dominant mechanism for ozone destruction.

The influence of a SPE on ozone through the HO_x constituents is very dependent on the solar zenith angle. The ambient HO_x level is maintained by the following two mechanisms

$$H_2O + h\nu \rightarrow H + OH \text{ (dominant above 70 km)}$$

and $O(^1D) + H_2O \rightarrow OH + OH$ (dominant below 70 km).

Both of these reactions are only active in the sunlight. The first mechanism is directly dependent on the sun and the second mechanism is dependent on the production of $O(^1D)$, a very short-lived constituent, which is produced by the photolysis of O_3. Since the HO_x production by the SPEs is not dependent on sunlight (see section 3), the SPE-produced HO_x will have a larger impact on ozone at larger solar zenith angles, where the ambient HO_x production is less. *Solomon et al.* [1983] and *Jackman and McPeters* [1985] provide further details about this solar zenith angle dependence of the SPE-produced HO_x impact on ozone.

This dependence of the ozone depletion caused by SPE-produced HO_x on solar zenith angle can be seen quite well in Figure 8 taken from *Solomon et al.* [1983]. The figure shows both model computations and measurements during different portions of the Solar Mesosphere Explorer (SME) orbit. The model computations are close to the observations during the AM portions of the SME orbit, the AM portion of the orbit being at the very large solar zenith angles. There is less agreement between the model and observations during the PM portions of the SME orbit. Notice that the measurements show larger amounts of ozone depletion at the larger solar zenith angles, consistent with theoretical predictions.

The ozone depletion during the July 2000 SPE was also mostly caused by the HO_x production (shown in Figure 7). The percentage decreases in the HALOE sunrise ozone measurements are given in Figure 7c where the July 14-15 values are compared to the background average of the July 12-13 observations. The HALOE observed ozone reductions started on July 14 and reached over 70% during most of July 15 in the middle mesosphere between 0.3 and 0.01 hPa. The model computations in Figure 7d are fairly similar to the HALOE measurements during most of the event.

Although there appears to be reasonable agreement between the model and HALOE measurements, *Jackman et al.* [2001] did point out some discrepancies between the model and SBUV/2 measurements at 1-2 hPa (~45-50 km). The SBUV/2 measurements indicated a larger ozone depletion than predicted. Reaction rate adjustments for HO_x constituents discussed in *Conway et al.* [2000] did not help the model/measurement agreement. These discrepancies need to be investigated further.

5.2. Long-term Effects from NO_y Constituents

The influence of the SPE-enhanced NO_y constituents on ozone has been understood nearly as long as the impact of

the SPE-enhanced HO_x constituents. *Crutzen et al.* [1975] predicted that the nitric oxide (NO) produced during three large SPEs between 1960 and 1972 would probably have been enough to cause an ozone change. The *Heath et al.* [1977] results showed that there were large stratospheric ozone reductions apparent in the Nimbus-4 BUV instrument data up to 19 days past the August 1972 events and were probably caused by the NO_y enhancements. Several other papers, including *Fabian et al.* [1979], *Maeda and Heath* [1980/1981], *Reagan et al.* [1981], *Solomon and Crutzen* [1981], *Rusch et al.* [1981], and *Jackman et al.* [1990, 1995, 2000], studied various aspects of NO_y influence on stratos-

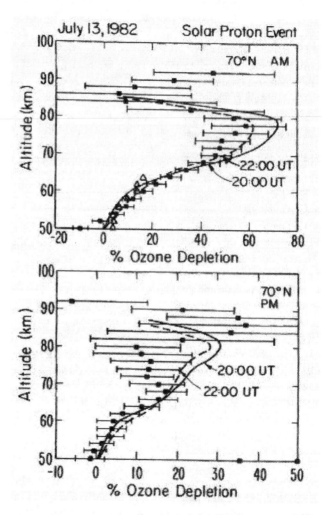

Figure 8. Taken from Fig. 3 of *Solomon et al.* [1983]. Observed ozone depletion on July 13, 1982 at 70°N latitude on the AM and PM portions of the Solar Mesosphere Explorer (SME) orbit (each point represents a mean of three orbits on July 13, 1982 near 1830, 2120, and 2206 UT). Triangles denote data from the UV spectrometer. Model calculated profiles for 2000 and 2200 UT are shown.

pheric ozone. The primary catalytic cycle for NO_y destruction of ozone is

$$NO + O_3 \rightarrow NO_2 + O_2$$
$$NO_2 + O \rightarrow NO + O_2$$
$$\text{Net: } O_3 + O \rightarrow 2O_2.$$

A comparison of the Nimbus 4 BUV measurements and model predictions of ozone destruction caused by the extremely large August 1972 SPEs is given in Figure 9. Virtually all the ozone destruction observed beyond day 222 was caused by the NO_y enhancements from this SPE. Both the measurements and model computations indicate ozone depletions of over 20% in the upper stratosphere during the SPE with depletions of over 15% persisting for about two months after the SPE. The major difference between the measurements and model results is in the upper stratospheric and lower mesospheric region (near 50 km), where the model indicates a faster recovery from the initial SPE-caused ozone depletion than is indicated in the measurements.

The long lifetime of the NO_y constituents allows the influence on ozone to last for a number of months past the event. Figure 10 shows the model predicted temporal behavior of profile ozone and NO_y for 1989-1992 at 75°N. Predicted upper stratospheric increases in NO_y over 100% during and shortly after the extremely large October 1989 SPEs produced significant upper stratospheric ozone decreases (>10%). The downward transport in the late fall and winter caused the very large enhancements of NO_y in the upper stratosphere to be moved to lower stratospheric levels with a corresponding ozone decrease. NO_y enhancements of over 10% accompanied by ozone decreases of greater than 2% persisted for over a year past these events. Ozone decreases down to 100 mbar (~16 km) were predicted by the spring of 1990 as a result of these events in 1989.

Although most recent theoretical studies have been completed with two-dimensional (2D) atmospheric models, three-dimensional (3D) atmospheric models have also played a role in our understanding of the influence of SPEs on the atmosphere. *Jackman et al.* [1993] was able to simulate the large Northern and Southern Hemisphere measured differences in ozone response from the October 1989 SPEs with a 3D model over the October-December 1989 time period, whereas they were unable to simulate these interhemispheric differences with a 2D model. *Jackman et al.* [1995] used a 3D model to simulate the time period October 1989 to April 1990 to study the longer term influences of the October 1989 SPEs. They found a reasonable simulation of the SAGE II NO_2 observations. The *Jackman et al.* [1995] simulations found less good agreement with SAGE II ozone changes and suggested that other factors, such as heteroge-

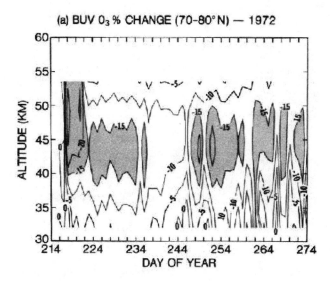

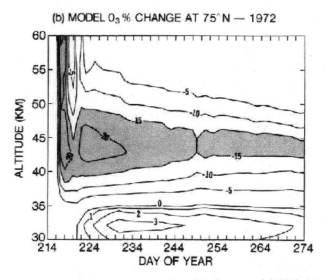

Figure 9. Taken from Figs. 6a and 7a of *Jackman et al.* [1990]. (a) Nimbus 4 BUV measured ozone percentage change as a function of day of year in 1972 for 70°-80°N band. (b) Model predicted ozone percentage change as a function of day of year in 1972 for 75°N. The contour levels of –30, –20, –15, –10, –5, 0, +1, +2, and +3%.

neous chemistry, might also be influencing constituents of this region.

5.3. Self-healing and Halogen Interference Leading to Ozone Increases

Ozone is usually depleted by SPE enhancements of HO_x and NO_y, but it can also be increased in certain regions. *Jackman and McPeters* [1985] showed that ozone "self-healing" should also be associated with SPEs. When ozone is decreased at a higher altitude, increased ultraviolet (UV)

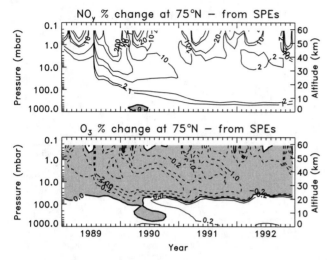

Figure 10. Taken from Fig. 6 of *Jackman et al.* [2000]. Model-computed percentage changes in NO_y and O_3 at 75°N for 1989-1992 resulting from SPEs. Contour intervals for NO_y are 0, 1, 2, 10, 20, 100, and 200%. Contour intervals for O_3 are +0.2, 0, –0.2, –1, –2, –10, and –20%.

radiation penetrates to lower altitudes. The increased UV radiation at very low altitudes (or extremely large solar zenith angles) can lead to an ozone production due to an increase in O_2 dissociation

$$O_2 + h\nu \; (<242 \text{ nm}) \rightarrow O + O$$

followed by

$$O + O_2 + M \rightarrow O_3$$
$$\text{and } O + O_2 + M \rightarrow O_3.$$

Predicted ozone increases below about 40 mbar (~22 km) in 1989 in Figure 10 (lower graph) are the result of self-healing. Such effects are small and the ozone self-healing only partially compensates for the higher altitude depletion of ozone caused by the SPE-enhanced NO_y.

The ozone enhancements in the lowest part of the stratosphere (below ~22 km) in late 1990, 1991, and 1992 shown in Figure 10 (lower graph) were caused by the interference of downward transported NO_y with chlorine and bromine constituents [*Jackman et al.*, 2000]. More ClO and BrO radicals were tied up into the reservoir species $ClONO_2$ and $BrONO_2$ by the enhanced NO_2 in the lower stratosphere through

$$ClO + NO_2 + M \rightarrow ClONO_2 + M$$
$$\text{and } BrO + NO_2 + M \rightarrow BrONO_2 + M.$$

Since the chlorine and bromine radicals are very important in the control of ozone in the lower stratosphere, the

interference production of the reservoir species $ClONO_2$ and $BrONO_2$ resulted in a net decrease in the catalytic loss of ozone. This process then led to a predicted ozone increase, which is especially important in periods of high chlorine and bromine loading such as the present time [*Jackman et al.*, 2000].

5.4. Total Column Ozone Changes

The predicted influence of the SPEs in October 1989 on total column ozone is illustrated in Figure 11. This largest period of SPEs of the past 30 years was predicted to have caused a maximum total ozone depletion of about 2%. The largest predicted ozone depletions due to these SPEs were in late 1989 and 1990. Total ozone depletions were less in 1991 and were almost non-existent or positive in 1992. These impacts on total column ozone are fairly small and would be difficult to discern in measurements because of large daily fluctuations at polar latitudes due to meteorological and seasonal variations [e.g., see *WMO*, 1999]. *Marin and Lastovicka* [1998] were not able to find any signal for SPEs in a Dobson total ozone record for central Europe (50°N).

The influence of SPEs on global total ozone variations was compared to the other natural (solar cycle ultraviolet variations and large volcanic eruptions) fluctuations as well as the anthropogenically-caused ozone variations in *Jackman et al.* [1996]. The annually averaged almost global total ozone, computed between 65°S and 65°N, was predicted in *Jackman et al.* [1996] to: 1) decrease ~4% from 1979 to 1995 due to anthropogenically-induced increases of chlorine and bromine; 2) decrease a maximum of ~2.8% in 1992 as a result of the very large Mt. Pinatubo volcanic

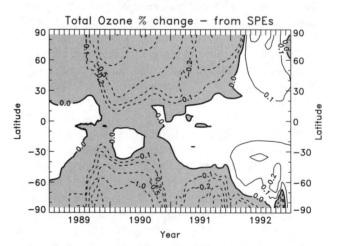

Figure 11. Taken from Fig. 3 of *Jackman et al.* [2000]. Model-computed percentage total ozone changes from 1989-1992 resulting from SPEs. Contour intervals are +0.2, +0.1, 0, –0.1, –0.2, –0.5, –1, and –2%.

eruption; 3) vary by ~1.2% as a result of solar ultraviolet flux variations over a solar cycle; 4) decrease by ~0.2% as a result of the extremely large SPEs in October 1989; and 5) vary by ~0.02% as a result of GCRs over a solar cycle. The influence of SPEs on global total ozone variations is thus rather small compared to other natural and humankind-caused fluctuations or trends.

6. SPE-CAUSED TEMPERATURE AND DYNAMICAL VARIATIONS

Large SPEs can also influence temperature and dynamics in the middle atmosphere. The temperature change associated with SPEs has been discussed in previous studies [e.g., *Banks*, 1979; *Reagan et al.*, 1981; *Jackman and McPeters*, 1985; *Roble et al.*, 1987; *Reid et al.*, 1991; and *Jackman et al.*, 1995]. Other studies [*Rottger*, 1992; *Johnson and Luhmann*, 1993] have documented changes in long-period gravity waves and neutral winds in the polar mesosphere as a result of SPEs.

SPEs can cause both heating and cooling of the middle atmosphere. Direct atmospheric heating is caused by the precipitating particles through Joule heating. *Banks* [1979] computed Joule dissipation temperature rate changes of 1° to 10°K per day in the mesosphere as a result of the very large August 1972 SPEs. *Roble et al.* [1987] computed similar mesospheric temperature rate changes during the July 1982 SPE. The July 1982 SPE was comparable to the August 1972 SPEs in the mesosphere. These mesospheric temperature rate forcings from Joule heating are comparable to or larger than the solar ultraviolet heating in this region and could cause small short-term temperature increases. Negligible temperature rate changes via Joule heating were computed in the stratosphere as a result of these SPEs.

Since ozone is one of the primary radiative absorbing (heating) gases of the middle atmosphere, a decrease in ozone as a result of SPEs could be expected to result in lower temperatures. *Reagan et al.* [1981] calculated a temperature decrease of 2.2°K at 50 km during the extremely large August 1972 SPEs. *Jackman and McPeters* [1985] computed a temperature decrease of a maximum of 1.1°K at 50 km due to the July 1982 SPE. The July 1982 SPE was smaller than the August 1972 SPEs in the stratosphere. *Reid et al.* [1991] and *Jackman et al.* [1995] calculated maximum temperature decreases of about 3°K at 75°S in the upper stratosphere (40-50 km) as a result of the very large October 1989 SPEs. These temperature changes are longer-lived than those associated with the Joule heating and would last as long as the SPE-caused ozone decreases, which could be for several months in the stratosphere.

Rottger [1992] discussed large wave amplitudes in the meridional wind (long-period gravity waves) in the polar lower mesosphere during SPEs in October and December of 1989. These changes may be caused by the Joule heating as well as the heating rate changes generated from ozone depletions, both effects from SPEs. Such SPE-caused atmospheric variations follow a certain temporal and spatial variation and could produce vertical wavelengths of about 10 km and northward velocities up to ± 40 m s^{-1} due to the wave [*Rottger*, 1992].

Consistent with these observations by *Rottger* [1992], *Johnson and Luhmann* [1993] showed observations of variations in upper mesospheric neutral winds accompanying the July 1982 and April 1984 SPEs. As a result of the July 1982 SPE, the mean wind shifted to the west-southwest by 5-30 m s^{-1} over the 81-90 km range, which was approximately twice the normal standard deviation. Similar mean wind shifts were observed as a result of the April 1984 SPE, although this occurred primarily at lower altitudes (65-75 km). *Johnson and Luhmann* [1993] speculate that the Joule heating and ozone depletion probably played a role in the SPE-related wind changes.

7. CONCLUSIONS

The polar middle atmosphere is readily influenced by large SPEs. SPEs can cause huge changes in the constituents in the polar middle atmosphere and can be used to study the natural variations in ozone. HO$_x$ increases of greater than 100% at high solar zenith angles were computed in the mesosphere as a result of large SPEs. The NO$_y$-induced stratospheric increases from SPEs are more persistent and are computed to be greater than 10% for over a year past the extremely large October 1989 SPEs.

The ozone depletion in eleven SPEs was documented in the refereed literature in the last 32 years. All these SPEs showed ozone depletions during the events, which were primarily HO$_x$-induced. The HO$_x$-induced ozone depletions from the larger SPEs were computed and measured at high solar zenith angles to be quite large (>50%) in the middle to upper mesosphere. Three of these SPEs had lingering ozone depletions past the events, which were primarily NO$_y$-induced. Ozone depletions of 15% from the extremely large August 1972 SPEs persisted for two months after these events in the upper stratosphere.

Polar total column ozone was computed to be decreased by a maximum of about 2% as a result of the extremely large October 1989 SPEs. Annually averaged global total ozone decreases were calculated to be 0.2% in 1990 as a result of these events, relatively small compared to other natural and humankind related changes.

SPEs may also change the temperature and dynamics of the middle atmosphere either directly through Joule heating or indirectly through an associated ozone depletion. Joule heating from very large SPEs was comparable to or larger in the mesosphere to the solar ultraviolet heating. Polar mesospheric temperatures were predicted to be lowered through ozone depletion in the upper stratosphere by 1-3°K as a result of very large SPEs Gravity waves and winds in the mesosphere were observed to vary as a result of these changed heating rates from the SPEs.

These solar events input quantifiable amounts of HO_x and NO_y into the atmosphere and can be used to test the current understanding of the atmospheric chemistry and dynamics. Although there is a general agreement between model predictions and measurements concerning SPE-induced ozone influences, some differences remain. The largest differences between the measured and modeled ozone depletions caused by SPEs occur in the upper stratosphere, both for the HO_x- and the NO_y-induced changes. This region should be further scrutinized during large SPEs to resolve these model/measurement inconsistencies.

Measurements of HO_x induced changes due to SPEs would be extremely useful to check our understanding of the mesospheric and upper stratospheric influences from the SPEs. Such HO_x measurements would be particularly useful during the nighttime when other HO_x production mechanisms are at a minimum [e.g., *Brinksma et al.*, 2001]. It may be possible to measure the temperature influences of SPEs in the future, if the measurements have a precision of less than 1° K. We anticipate further progress in our understanding of the influence of SPEs on constituents in the middle atmosphere in the next several years.

Acknowledgments. We thank NASA Headquarters Atmospheric Chemistry Modeling and Analysis Program for support during the time that this manuscript was written. The paper was improved by clarifications required to address topical editor John McCormack's and an anonymous referee's several good comments. We thank Judi Bordeaux, who provided assistance with scanning several of the other figures from various papers. We thank Steele Hill, who provided a publication quality image for Figure 1.

REFERENCES

Banks, P. M., Joule heating in the high-latitude mesosphere, *J. Geophys. Res.*, *84*, 6709-6712, 1979.

Brinksma, E. J., Y. J. Meijer, I. S. McDermid, R. P. Cageao, J. B. Bergwerff, D. P. J. Swart, W. Ubachs, W. A. Matthews, W. Hogervorst, and J. W. Hovenier, First lidar observations of mesospheric hydroxyl, *Geophys. Res. Lett.*, *25*, 51-54, 1998.

Conway, R. R., M. E. Summers, M. H. Stevens, J. G. Cardon, P. Preusse, and D. Offermann, Satellite observations of upper stratospheric and mesospheric OH: the HO_x dilemma, *Geophys. Res. Lett.*, *27*, 2613-2616, 2000.

Crutzen, P. J., I. S. A. Isaksen, and G. C. Reid, Solar proton events: Stratospheric sources of nitric oxide, *Science*, *189*, 457-458 1975.

Fabian, P., J. A. Pyle, and R. J. Wells, The August 1972 solar proton event and the atmospheric ozone layer, *Nature*, *277*, 458-460, 1979.

Forbush, S. E., Three unusual cosmic-ray increases possibly due to charged particles from the Sun, *Phys. Rev.*, *70*, 771-772, 1946.

Frederick, J. E., Solar corpuscular emission and neutral chemistry in the Earth's middle atmosphere, *J. Geophys. Res.*, *81*, 3179-3186, 1976.

Heath, D. F., A. J. Krueger, and P. J. Crutzen, Solar proton event: influence on stratospheric ozone, *Science*, *197*, 886-889, 1977.

Jackman, C. H., J. E. Frederick, and R. S. Stolarski, Production of odd nitrogen in the stratosphere and mesosphere: An intercomparison of source strengths, *J. Geophys. Res.*, *85*, 7495-7505, 1980.

Jackman, C. H., and R. D. McPeters, The response of ozone to solar proton events during solar cycle 21: A theoretical interpretation, *J. Geophys. Res.*, *90*, 7955-7966, 1985.

Jackman, C. H., and R. D. McPeters, Solar proton events as tests for the fidelity of middle atmosphere models, *Physica Scripta*, *T18*, 309-316, 1987.

Jackman, C. H., A. R. Douglass, R. B. Rood, R. D. McPeters, and P. E. Meade, Effect of solar proton events on the middle atmosphere during the past two solar cycles as computed using a two-dimensional model, *J. Geophys. Res.*, *95*, 7417-7428, 1990.

Jackman, C. H., J. E. Nielsen, D. J. Allen, M. C. Cerniglia, R. D. McPeters, A. R. Douglass, and R. B. Rood, The effects of the October 1989 solar proton events on the stratosphere as computed using a three-dimensional model, *Geophys. Res. Lett.*, *20*, 459-462, 1993.

Jackman, C. H., M. C. Cerniglia, J. E. Nielsen, D. J. Allen, J. M. Zawodny, R. D. McPeters, A. R. Douglass, J. E. Rosenfield, and R. B. Rood, Two-dimensional and three-dimensional model simulations, measurements, and interpretation of the influence of the October 1989 solar proton events on the middle atmosphere, *J. Geophys. Res.*, *100*, 11,641-11,660, 1995.

Jackman, C. H., E. L. Fleming, S. Chandra, D. B. Considine, and J. E. Rosenfield, Past, present and future modeled ozone trends with comparisons to observed trends, *J. Geophys. Res.*, *101*, 28,753-28,767, 1996.

Jackman, C. H., E. L. Fleming, and F. M. Vitt, Influence of extremely large solar proton events in a changing stratosphere, *J. Geophys. Res.*, *105*, 11659-11670, 2000.

Jackman, C. H., R. D. McPeters, G. J. Labow, E. L. Fleming, C. J. Praderas, and J. M. Russell, Northern hemisphere atmospheric effects due to the July 2000 solar proton event, *Geophys. Res. Lett.*, *28*, 2883-2886, 2001.

Johnson, R. M., and J. G. Luhmann, Poker Flat MST radar observations of high latitude neutral winds at the mesopause during and after solar proton events, *J. Atmos. Terr. Phys.*, *55*, 1203-1218, 1993.

Maeda, K., and D. F. Heath, Stratospheric ozone response to a solar proton event: Hemispheric asymmetries, *Pure Appl. Geophys.*, *119*, 1-8, 1980/1981.

Marin, D., and J. Lastovicka, Do solar flares affect total ozone?, *Stud. Geophys. Geod. 42*, 533-539, 1998.

McPeters, R. D., A nitric oxide increase observed following the July 1982 solar proton event, *Geophys. Res. Lett., 13*, 667-670, 1986.

McPeters, R. D., and C. H. Jackman, The response of ozone to solar proton events during solar cycle 21: the observations, *J. Geophys. Res., 90*, 7945-7954, 1985.

McPeters, R. D., C. H. Jackman, and E. G. Stassinopoulos, Observations of ozone depletion associated with solar proton events, *J. Geophys. Res., 86*, 12,071-12,081, 1981.

Porter, H. S., C. H. Jackman, and A. E. S. Green, Efficiencies for production of atomic nitrogen and oxygen by relativistic proton impact in air, *J. Chem. Phys., 65*, 154-167, 1976.

Randall, C. E., D. E. Siskind, and R. M. Bevilacqua, Stratospheric NO_x enhancements in the southern hemisphere polar vortex in winter and spring of 2000, *Geophys. Res. Lett,, 28*, 2385-2388, 2001.

Reagan, J. B., R. E. Meyerott, R. W. Nightingale, R. C. Gunton, R. G. Johnson, J. E. Evans, W. L. Imhof, D. F. Heath, and A. J. Krueger, Effects of the August 1972 solar particle events on stratospheric ozone, *J. Geophys. Res., 86*, 1473-1494, 1981.

Reid, G. C., Chapter 12 – Solar energetic particles and their effects on the terrestrial environment, in *Physics of the Sun, Vol. III*, edited by P. A. Sturrock, pp. 251-278, D. Reidel Publishing Company, 1986.

Reid, G. C., S. Solomon, and R. R. Garcia, Response of the middle atmosphere to the solar proton events of August-December 1989, *Geophys. Res. Lett., 18*, 1019-1022, 1991.

Roble, R. G., B. A. Emery, T. L. Killeen, G. C. Reid, S. Solomon, R. R. Garcia, D. S. Evans, G. R. Carignan, R. A. Heelis, W. B. Hanson, D. J. Winningham, N. W. Spencer, and L. H. Brace, Joule heating in the mesosphere and thermosphere during the July 13, 1982, solar proton event, *J. Geophys. Res., 92*, 6083-6090, 1987.

Rottger, J., Solar proton events: A source for long-period gravity waves in the polar mesosphere?, *COSPAR Colloquia Series, Vol. 5, Solar Terrestrial Energy Program*, 473-476, Proceedings 1992 STEP Symposium/5th COSPAR Colloquium, Laurel, USA, 24-28 Aug. 1992, 1992.

Rusch, D. W., J.-C. Gerard, S. Solomon, P. J. Crutzen, and G. C. Reid, The effect of particle precipitation events on the neutral and ion chemistry of the middle atmosphere, 1, Odd nitrogen, *Planet. Space Sci., 29*, 767-774, 1981.

Solomon, S., and P. J. Crutzen, Analysis of the August 1972 solar proton event including chlorine chemistry, *J. Geophys. Res., 86*, 1140-1146, 1981.

Solomon, S., D. W. Rusch, J.-C. Gerard, G. C. Reid, and P. J. Crutzen, The effect of particle precipitation events on the neutral and ion chemistry of the middle atmosphere, 2, Odd hydrogen, *Planet. Space Sci., 29*, 885-892, 1981.

Solomon, S., G. C. Reid, D. W. Rusch, and R. J. Thomas, Mesospheric ozone depletion during the solar proton event of July 13, 1982, 2, Comparison between theory and measurements, *Geophys. Res. Lett., 10*, 257-260, 1983.

Swider, W., and T. J. Keneshea, Decrease of ozone and atomic oxygen in the lower mesosphere during a PCA event, *Planet. Space Sci., 21*, 1969-1973, 1973.

Swider, W., T. J. Keneshea, and C. I. Foley, An SPE disturbed D-region model, *Planet. Space Sci., 26*, 883-892, 1978.

Thomas, R. J., C. A. Barth, G. J. Rottman, D. W. Rusch, G. H. Mount, G. M. Lawrence, R. W. Sanders, G. E. Thomas, and L. E. Clemens, Mesospheric ozone depletion during the solar proton event of July 13, 1982, 1, Measurments, *Geophys. Res. Lett., 10*, 257-260, 1983.

Vitt, F. M., and C. H. Jackman, A comparison of sources of odd nitrogen production from 1974 through 1993 in the Earth's middle atmosphere as calculated using a two-dimensional model, *J. Geophys. Res., 101*, 6729-6739, 1996.

Warneck, P., Cosmic radiation as a source of odd nitrogen in the stratosphere, *J. Geophys. Res., 77*, 6589-6591, 1972.

Weeks, L. H., R. S. CuiKay, and J. R. Corbin, Ozone measurements in the mesosphere during the solar proton event of 2 November 1969, *J. Atmos. Sci., 29*, 1138-1142, 1972.

World Meteorological Organization (WMO), Scientific Assessment of Ozone Depletion: 1998, *Rep. 44*, Global Ozone Res. And Monit. Proj., Geneva, 1999.

Zadorozhny, A. M., G. A. Tuchkov, V. N. Kikhtenko, J. Lastovicka, J. Boska, and A. Novak, Nitric oxide and lower ionosphere quantities during solar particle events of October 1989 after rocket and ground-based measurements, *J. Atmos. Terr. Phys., 54*, 183-192, 1992.

Charles H. Jackman and Richard D. McPeters, both at: Code 916, NASA Goddard Space Flight Center, Greenbelt, MD, 20771.

Atmospheric Ionization and Clouds as Links Between Solar Activity and Climate

Brian A. Tinsley

University of Texas at Dallas, Richardson, Texas

Fangqun Yu

State University of New York at Albany, New York

Observations of changes in cloud properties that correlate with the 11-year cycles observed in space particle fluxes are reviewed. The correlations can be understood in terms of one or both of two microphysical processes; ion mediated nucleation (IMN) and electroscavenging. IMN relies on the presence of ions to provide the condensation sites for sulfuric acid and water vapors to produce new aerosol particles, which, under certain conditions, might grow into sizes that can be activated as cloud condensation nuclei (CCN). Electroscavenging depends on the buildup of space charge at the tops and bottoms of clouds as the vertical current density (J_z) in the global electric circuit encounters the increased electrical resistivity of the clouds. Space charge is electrostatic charge density due to a difference between the concentrations of positive and negative ions. Calculations indicate that this electrostatic charge on aerosol particles can enhance the rate at which they are scavenged by cloud droplets. The aerosol particles for which scavenging is important are those that act as in-situ ice forming nuclei (IFN) and CCN. Both IMN and electroscavenging depend on the presence of atmospheric ions that are generated, in regions of the atmosphere relevant for effects on clouds, by galactic cosmic rays (GCR). The space charge depends, in addition, on the magnitude of J_z. The magnitude of J_z depends not only on the GCR flux, but also on the fluxes of MeV electrons from the radiation belts, and the ionospheric potentials generated by the solar wind, that can vary independently of the GCR flux.

1. INTRODUCTION

Clouds play a key role in the energy budget of Earth's surface and lower atmosphere, and are probably the largest contributor to the uncertainty concerning the global climate change [*IPCC*, 1996]. Small modifications of the amount, distribution, or radiative properties of clouds can have significant impacts on the predicted climate [*Hartmann*, 1993]. The suggestion that space particle fluxes, as modulated by solar activity and the solar wind, affect clouds and climate has a long history. Space particle fluxes include galactic cosmic rays (GCR); MeV electrons from the radiation belts; occasional energetic solar proton events; and the bulk solar wind plasma itself. Following a note by *Ney* [1959], *McDonald and Roberts* [1960] and *Roberts and Olsen* [1973], speculated that their correlations between short-term solar variability and intensification of cyclones in the Gulf of Alaska might involve stratospheric ionization

Solar Variability and its Effects on Climate
Geophysical Monograph 141

changes, produced by solar particle precipitation, affecting clouds. A more in-depth discussion of theoretical aspects of such a mechanism was made by *Dickinson* [1975]. He pointed out that while direct condensation of water vapor in the ionization produced by GCR would not occur in the atmosphere (it requires supersaturations of several hundred percent) the presence of sulfuric acid vapor in the atmosphere allowed condensation on the ions of H_2SO_4 molecules together with H_2O molecules. Provided that these can grow large enough to act as cloud condensation nuclei (CCN), they could affect the particle size distribution and lifetime of clouds, and thus the radiative properties affecting climate. This process is now termed ion-mediated nucleation [*Yu and Turco*, 2000, 2001] , and in later sections we will discuss it in more detail. *Svensmark and Friis-Christensen* [1997] and *Svensmark* [1998] demonstrated correlations of cloud cover with GCR flux, and speculated that ionization processes could affect nucleation or the phase transitions of water vapor. *Marsh and Svensmark* [2000] first noted the potential importance of IMN for such GCR-cloud links.

A more indirect effect of GCR flux changes was suggested by *Herman and Goldberg* [1978] and *Markson and Muir* [1980], in that GCR and other space particle fluxes modulate the current flow in the global electric circuit [*Israël*, 1973; *Tinsley*, 1996; *Bering et al.*, 1998], and the passage of this current density through clouds affects their initial electrification. There are a number of ways in which weak electrification can affect the microphysics of clouds, with consequences for cloud lifetime, radiative properties, and precipitation efficiency. It was suggested by *Tinsley and Deen* [1991] that electrical processes in clouds containing supercooled water lead to enhanced production of ice, which directly leads to enhanced precipitation by the Wegener-Bergeron-Findeisen process. Recent work [*Tinsley et al.*, 2000, 2001] has focused on a process called electroscavenging as an electrical link leading to precipitation. In this case the climatic effect may not just be due to changes in cloud cover and the atmospheric radiation balance, but also to dynamical effects on the storm system that is undergoing precipitation, due to the latent heat transfer involved. An increase in precipitation entails a decrease in the amount of cloud water that is available for evaporation into air being entrained into the air mass of a storm, or an increase in condensation in the entrained air. The reduction of this diabatic (non-adiabatic) cooling or equivalent diabatic heating increases uplift and redistributes vorticity towards the center of the storm (non-conservation of potential vorticity). For a winter cyclone, that is drawing vorticity from the latitudinal shear in the winter circulation, the vorticity redistribution is likely to enhance feedback processes, and increase the overall strength of the storm.

The effects of concentration and overall increase of vorticity appear to be responsible for the observed correlations of tropospheric dynamics and temperature with particle fluxes on the day-to-day timescale [*Tinsley*, 2000]. There is an increase in meridional transport of heat and momentum associated with increased vorticity in winter storms, that is concentrated in the cyclogenesis longitudes, and the cumulative effects over a winter season have the potential for significant feedback on the circulation itself, e. g., on Rossby wave amplitude and regional changes in temperature and storm track latitude.

Thus *Tinsley and Deen* [1991] suggested that shifts in winter storm track latitude in the North Atlantic, that had been observed to correlate with sunspot number for six solar cycles, were a consequence of cosmic ray induced freezing of supercooled water in storm clouds. The climatic effects of these changes in Rossby wave amplitude are much more regional than global. The regional changes show differences in sign, so that the global mean changes, although not negligible, represent relatively small residuals. The greatest societal impact would appear to be from regional changes on the decadal and longer timescales.

In addition to proposed effects of solar modulation of atmospheric electricity on clouds and climate there are changes in solar spectral and total irradiance that affect ozone and have been proposed to affect climate, as discussed in this volume [*Haigh, Hood*] and by *Shindell et al.* [2001]. On the decadal and longer timescales one cannot easily distinguish between effects on climate of particle flux changes and effects of irradiance changes. However, the temporal signatures are completely different on the day-to-day timescale. The review by *Tinsley* [2000] describes a long history of studies of day-to-day changes in atmospheric vorticity and temperature that correlate unambiguously with short-term space particle flux variations. The latitudinal and seasonal variations are fully consistent with the hypothesis of electroscavenging inducing ice production and precipitation from clouds, affecting atmospheric temperature and dynamics. The same processes must operate on the decadal and longer timescales; however the relative importance compared to irradiance effects on those timescales remains to be determined.

In this review we will first discuss observations supporting the two hypotheses—ion mediated nucleation and electroscavenging—as links between solar activity and effects on clouds and climate. The influence of solar variability on atmospheric ionization will then be reviewed, followed by outlines of the ion-mediated nucleation and electroscavenging mechanisms. The consequences for precipitation, atmospheric dynamics and radiation of the cloud changes associated with electroscavenging and ion-mediated nucleation will then be discussed. Thereafter, the extent to which the correlations with atmospheric changes support one or the other or both of the mechanisms will be examined.

2. CORRELATIONS OF CLOUD PROPERTIES WITH DECADAL GCR AND SUNSPOT VARIATIONS

Among the many reported decadal timescale correlations of meteorological parameters with solar activity, one of the least ambiguous as an effect of space particle fluxes on clouds is that shown in Figure 1. This is a correlation of precipitation and precipitation efficiency with GCR flux in the Southern Ocean that is greatest at the highest geomagnetic latitudes, where the amplitude of the GCR flux variations and the associated vertical current density (J_z) variations are greatest [*Kniveton and Todd*, 2001]. The location of the geomagnetic pole is marked by an X. The precipitation data were from the Climate Prediction Center Merged Analysis of Precipitation (CMAP) product. The amplitudes of the precipitation and precipitation efficiency variations were 7–9% at 65–75° geomagnetic latitudes and at those latitudes the GCR flux and the vertical current density J_z vary by 15–20% over the solar cycle. The statistical significance of the correlation with GCR flux is better than 95% over a large oceanic region as shown in Figure 1. There is a tendency for reversed correlation at lower latitudes.

Another decadal variation that is very consistent with cloud forcing by space particle fluxes is the correlation of cloud cover with cosmic ray flux that was noted earlier.

Svensmark and Friis-Christensen [1997] and *Svensmark* [1998] analyzed total cloud cover data from the C2 data set of the International Satellite Cloud Climatology Project (ISCCP). Later analyses of the altitude-resolved ISCCP-D2 data set by *Marsh and Svensmark* [2000a,b] and *Palle and Butler* [2000] showed a significant positive correlation with the frequency of low clouds, below about 3.2 km, but not with clouds at higher altitudes. The correlation is strongest at low latitudes (between 45°N and 45°S) with the intertropical convergence zone excluded. The global average amplitude of the cloud cover change is 1.5–2% as illustrated in Figure 2.

In addition to the high correlation between low cloud cover and cosmic ray fluxes, *Marsh and Svensmark* [2000a,b] find the cosmic ray flux is also strongly correlated with the cloud-top temperatures of low clouds. In this case, a band of significantly high correlation is centered around the tropics where stratocumulus and marine stratus clouds are dominant. Again there is no obvious correlation for middle and high clouds.

Inferred cloud cover changes over Russia that correlate with the 11-year cycle of cosmic ray flux changes have been reported by *Veretenenko and Pudovkin* [1999, 2000]. The

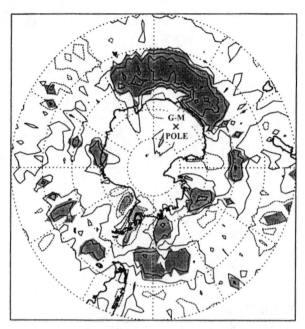

Figure 1. Correlation coefficients between 12-month moving averaged cosmic ray flux and precipitation. Positive and negative correlation coefficients are shown as solid and broken lines respectively, with a contour interval of 0.2. Shading indicates areas with locally significant correlations at the 95% level, where the amplitude of the response is 7–9% of the mean value. From *Kniveton and Todd* [2001].

cloud cover changes were inferred from 1961–1986 observations by a network of actinometric stations, and were negatively correlated with GCR flux below about 57° geographic latitude (about 50° geomagnetic latitude) and positively correlated at higher latitudes.

Cloud cover changes over the USA that correlate with the 11-year solar cycle have been reported by *Udelhofen and Cess* [2001] from cloud and actinometric observations in 1900–1987. Relative to GCR variations the correlations were negative below about 42° geographic latitude (about 50° geomagnetic), with a tendency to positive correlations at higher latitudes. Thus, the continental-low-latitude cloud cover variations over the USA and Russia have the same phase relative to GCR variations, and about the same geomagnetic latitude at which the correlation reverses. These continental-low-latitude variations have opposite phase to the mainly oceanic low-latitude cloud cover observations of *Svensmark and Friis-Christensen* [1997].

Changes in cloud cover on the day-to-day timescale that correlate with Forbush decreases of GCR have been reported by *Todd and Kniveton* [2001]. We do not discuss them here as the present focus is on climate change on the decadal and longer timescales.

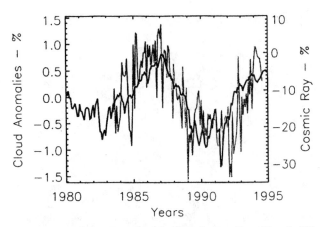

Figure 2. Monthly mean values for global anomalies of low (>680 hPa) cloud cover (grey), and GCR fluxes from Climax (black, normalized to May 1965). The global average temporal mean low cloud cover was 28%. From *Marsh and Svensmark* [2000a].

3. THE INFLUENCE OF SOLAR VARIABILITY ON ATMOSPHERIC IONIZATION

3.1. Space Particle Fluxes

The ionization at cloud level that determines the rate of electroscavenging and IMN is influenced by space particle fluxes as illustrated in Figure 3. These inputs are in the form of GCR; MeV electrons precipitating from the radiation belts with associated X-ray bremsstrahlung; the bulk solar wind plasma with its embedded magnetic fields that determines the horizontal distribution of potential across the polar cap ionospheres; and occasional energetic solar proton events. The latter occur too infrequently to have a significant effect on climate, and are not illustrated.

The GCR flux is responsible for almost all of the production of ionization below 15 km altitude, that determines the conductivity in that region. The MeV electrons and their associated X-rays produce ionization in the stratosphere, and affect the conductivity there. The current flow in the global electric circuit is generated mainly by charge separation in deep convective clouds in the tropics, and maintains the global ionosphere at a potential of about 250 kV. Variations above and below this value occur in the high latitude regions due to solar wind—magnetosphere—ionosphere coupling processes. The current density J_z varies horizontally due to variations in the local vertical column resistance (this is affected by the GCR and MeV electron fluxes) and by variations in the local ionospheric potential (especially to those in the high latitude regions). Because J_z flowing through clouds in the troposphere responds to conductivity and potential changes occurring all the way up to 120 km altitude, it is a very effective coupling agent for

linking inputs in the stratosphere and ionosphere with cloud levels.

3.2. Ion Concentrations

The GCR flux in the troposphere produces positive air ions of concentration n^+, and negative air ions of concentration n^-, and controls the variations of the total ion concentration $n_t = n^+ + n^-$. Both electroscavenging and IMN depend on these ion concentrations, with IMN production rates, in current theory, dependent on n_t. For electroscavenging it is the difference between positive and negative ion concentrations, that gives the space charge density $\rho = (n^+ - n^-)e$, that is more important than the total ion concentration. The ions are taken as being singly charged, with charges of $\pm e$, where e is the elementary charge.

The space charge density is dependent on J_z, that creates vertical electric fields $E_z = J_z / \sigma (z)$, where $\sigma (z)$ is the local conductivity at altitude z that depends on both n_t and the ion mobility. Where there are local vertical gradients in conductivity there will be gradients in E_z and space charge will be generated according to Poisson's equation:

$$\rho = \varepsilon_o \nabla \cdot E = \varepsilon_o J_z \, d/dz(1/\sigma) \qquad (1)$$

Strong conductivity gradients covering a significant horizontal extent are found for example, at the upper and lower boundaries of stratus, stratocumulus, and cirrus clouds, stratified dust and aerosol layers, and at the top of the mixing layer [*Sagalyn and Burke*, 1985]. The space charge density variations, along with the ion production variations, are modulated by solar activity influencing the space particle fluxes on the day-to-day and decadal timescales with relative amplitude 3–20%, as discussed below. The amplitude variations increase to a factor of two or more on the century timescale. They are ubiquitous throughout the troposphere, with greater amplitude at high latitudes. They constitute the solar forcing agent with the largest relative amplitude of any that might affect climate.

3.3. GCR and Ion Production Variations

The GCR flux is modulated in interplanetary space by changes in magnetic fields embedded in the solar wind, with larger modulation amplitude for the lower energy component (less than 5 GeV/nucleon). The Earth's magnetic field confines the low energy particles to higher geomagnetic latitudes, so that only GCR of order 10 GeV/nucleon penetrate into the atmosphere at the magnetic equator. The filtering effect of the geomagnetic field is also variable in time, with the magnetospheric currents that grow during periods of

FORCINGS BY SPACE PARTICLE FLUXES

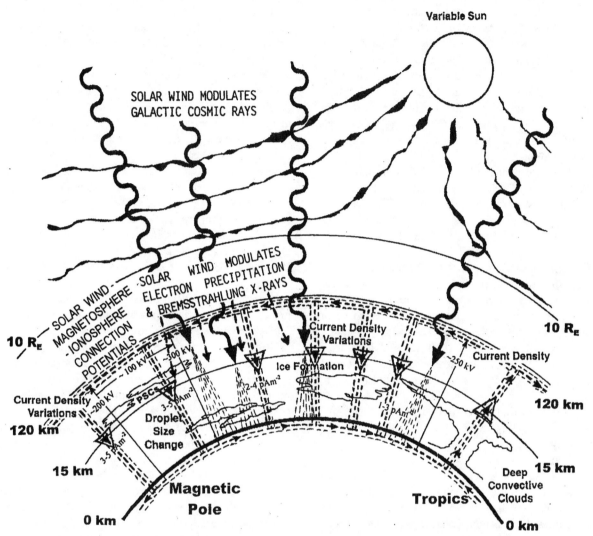

Figure 3. Solar wind variations modulate the fluxes of GeV galactic cosmic rays and the MeV electrons coming into the atmosphere, and the ionospheric potential in the polar caps. The fluxes of energetic particles change the vertical column resistance between the ionosphere and the surface, particularly at middle and high latitudes, and this together with the variations in ionospheric potential, change the electrical currents flowing down from the ionosphere into clouds.

magnetic activity allowing particles of a given energy to penetrate to lower latitudes [*Flückiger et al.*, 1987].

The latitude and time variations of GCR flux into the atmosphere, together with altitude profiles of the resulting ion production, have been reviewed by *Herman and Goldberg* [1978]; *Bazilevskaya* [2000]; and *Bazilevskaya et al.* [2000]. The altitude profiles of ion concentration peak at about 12 km, where the bulk of the primary cosmic ray flux, mainly protons of energy 1–5 GeV, is stopped. The ion pair production rates at 12 km vary from about 20 to 50 ion pairs $cm^{-3}s^{-1}$, with the higher production at higher latitudes and

weaker solar activity. Below the 15 km level the production is by cascades of secondary particles, particularly μ mesons. Atmospheric attenuation of secondaries reduces the production rate to about 10 $cm^{-3}s^{-1}$ at 5 km and 2 $cm^{-3}s^{-1}$ near sea level. Over continents the ion production in the lowest few km is several times greater, due to surface radioactivity [*Hoppel et al.*, 1986].

At geomagnetic latitudes of about 40° (e.g, Climax Colorado) the amplitude of the solar cycle modulation of the GCR flux above about 1GeV/nucleon, as measured by neutron monitors, is about 20%. At equatorial latitudes the modulation is only 3–5% because of the weaker modulation of the 10 GeV/nucleon particles. A similar decrease of modulation amplitude with decreasing latitude applies to Forbush decreases. These are short-term reductions in the GCR flux, with onset time usually less than a day and recovery time about a week, that occur when coronal mass ejections pass over the Earth. So both for the solar cycle variations and the Forbush decreases, the ion pair production rates in the stratosphere and troposphere and their effects on n_i and J_z and r follow the GCR variations, and are consequently also modulated most strongly at the higher geomagnetic latitudes.

In the polar stratospheres the ion production is mainly by lower energy (0.1–0.5 MeV) GCR that are modulated more strongly in interplanetary space than the 1–5GeV component. the balloon measurements of *Bazilevskaya et al.* [2000] show a factor of two modulation of the 500 MeV component at 25–30 km altitude on the 11-yr solar cycle. On the century time scale the GCR flux changes in the 1–5 GeV component are considerably larger than the 11-yr cycle variations. *McCracken and McDonald* [2001] review ¹⁰Be data from polar ice cores and find increases, compared to present solar cycle averages, of a factor of two for epochs of low solar activity near 1900 and 1800 AD, and a factor of three near 1700 AD. They suggest that for more extended solar minima the flux could rise an additional 60%.

4. ION-MEDIATED NUCLEATION

4.1. General

The effects of aerosols on chemistry and climate are sensitive to particle size and concentration, which are influenced significantly by nucleation processes that are not well understood. It has been demonstrated recently that air ions may play an important role in the production of new particles under typical tropospheric conditions [*Yu and Turco,* 2000, 2001]. Essentially, the production of H_2SO_4 in the gas phase by photochemical reactions in the atmosphere results in supersaturated H_2SO_4 vapor. There is competition between deposition of H_2SO_4 on pre-existing aerosols and on the ions. The competition results in a complicated dependence of the nucleation rates on the concentrations of pre-existing aerosols, ions, and H_2SO_4 vapor. The charged molecular clusters, condensing around ions, are much more stable and can grow significantly faster than corresponding neutral clusters, and thus can preferentially achieve stable, observable sizes. The proposed ion-mediated nucleation (IMN) theory can physically explain the enhanced growth rate (a factor of ~ 10) of sub-nanometer clusters as observed by *Weber et al.* [1997], and seems to account consistently

for ultrafine aerosol formation in jet plumes [*Yu and Turco,* 1997, *Yu et al.,* 1998, 1999; *Kärcher et al.,* 1998], in motor vehicle wake [*Yu,* 2001a], in marine boundary layer [*Yu and Turco,* 2001], in clean continental air as well as for the diurnal variation in the atmospheric mobility spectrum [*Yu and Turco,* 2000].

The IMN theory adds another important parameter—ion concentration (n_i) dependent on the ionization rate (Q)—to the list of known parameters controlling the nucleation rate of atmospheric particles [*Yu,* 2001b, 2002]. Galactic cosmic rays (GCRs) are the dominant sources of ionization in the oceanic troposphere, and in the continental troposphere except in the boundary layer where ionization by radioactive nuclides from the soil usually dominates in the lowest 1 km [*Reiter,* 1992]. Recently, the hypothesis that IMN affects cloud properties has received increasing attention, as a result of the high correlation noted earlier between total cloud cover over midlatitude ocean and GCR intensity [*Svensmark and Friis-Christensen,* 1997], and the development of the IMN theory [*Marsh and Svensmark,* 2000a; *Yu and Turco,* 2000, 2001]. More recent results, indicating the presence of a significant positive correlation between GCR intensity and the frequency of low clouds, below about 3.2 km but none with clouds at higher altitudes [*Marsh and Svensmark,* 2000a,b; *Palle and Butler,* 2000] is at first sight surprising, since the solar modulation of the cosmic ray intensity is a maximum around the tropopause and decreases with decreasing altitudes. However, IMN theory does predict strong altitude dependencies of ultrafine aerosol production, as we now discuss.

Figure 4 shows three key steps involved in the IMN-Cloud hypothesis, which can also be called the CGR-CN-CCN-cloud hypothesis. If confirmed, they might offer a physically-based mechanism connecting GCR fluxes and global cloud properties. First, the modulation of galactic cosmic radiation by the solar cycle will cause an observable variation in aerosol production and condensation nuclei (CN) population in the lower atmosphere. The term CN is an alternative name for ultrafine particles, or stable clusters of diameter a few nanometers that can act as condensation nuclei in cloud chambers where the supersaturation is several hundred percent. Second, to act as cloud condensation nuclei (CCN) in the atmosphere, where the supersaturation nearly always less than 2%, they must grow in diameter by a factor of about 10 to act as condensation nuclei. If the number of CCN thus produced is comparable to the total CCN abundance, a systematic change in the ultrafine particle production rate will affect this abundance. Third, a change in CCN abundance will affect the cloud properties.

Yu and Turco [2001] demonstrated that systematic variations in ionization levels due to the modulation of galactic

GCR-CN-CCN-Cloud Hypothesis

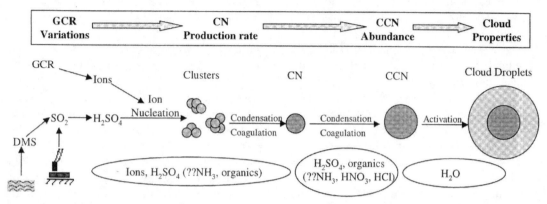

Figure 4. Schematic illustrating of GCR-CN-CCN-Cloud hypothesis that, if confirmed, might explain the correlation between variations of GCR flux and low cloud cover. The possible dominating species involved in the different phases of CN formation and growth processes are also indicated. The organics species may play an important role in growing the CN into the size of CCN. [From *Yu*, 2002].

cosmic radiation by the solar cycle are sufficient to cause an observable variation in condensation nuclei (CN) production in the marine boundary layer. Since the dominating number of cloud condensation nuclei (CCN) over ocean evolves from newly formed ultrafine particles [e.g., *Fitzgerald*, 1991], it is physically plausible that a systematic increase in ultrafine production rate will increase the CCN abundance, though the magnitude of such effect is currently unknown. It is well known that the CCN abundance affects cloud properties [*Twomey*, 1977, 1991; *Albrecht*, 1989; *Hobbs*, 1993]. Clouds that form in air containing high CCN concentrations tend to have high droplet concentrations, which lead to an increase in both albedo and absorption. Increases in the CCN concentration also inhibit rainfall and therefore increase cloud lifetimes (cloud coverage). These effects—which are due to more, smaller droplets at fixed liquid water content—are particularly significant in marine air, where the CCN concentrations are generally quite low.

In order to assess the magnitude of the effect of GCR variations on global cloudiness properties, all three steps have to be quantitatively understood. This is not an easy job and requires much further study. However, as far as the sign of the effect (e.g., the correlation between GCR variations and cloudiness) is concerned, the first step is critical. An increase in CN production is expected to increase the CCN abundance and cloud cover, but an increase in GCR fluxes does not always lead to an increase in CN production [Yu, 2001b, 2002]. The dependence of CN production on Q is a complex function of Q and $[H_2SO_4]$, as well as ambient conditions (T, RH, surface area of pre-existing particles, etc). Since Q, $[H_2SO_4]$, and ambient conditions vary significant-

ly with altitudes, it is of interest to understand how systematic change of ionization rates as a result of solar activity will affect the particle production at different altitudes.

4.2. Altitude and Solar Cycle Dependencies

Ambient ions are continuously generated by galactic cosmic rays as noted earlier, with the magnitude of the ionization rate variations being a function of latitude and altitude. During a solar cycle, the values of Q vary by ~20–25% in the upper troposphere and ~5–10% in the lower troposphere for high latitudes, and by ~4–7% in the upper troposphere and ~3–5% in the lower troposphere for low latitudes [*Ney*, 1959]. The effect of such systematic change in ionization rate on the altitude profile for the production of ultrafine particles has been studied by *Yu* [2002].

The advanced particle microphysics (APM) model which *Yu* [2002] employed for the study simulates a size-resolved multi-component aerosol system via a unified collisional mechanism involving both neutral and charged particles down to molecular sizes. The size-resolved ion-ion recombination coefficients, ion-neutral collision kernels, and neutral-neutral interaction coefficients calculated in the model are physically consistent and naturally altitude (temperature T, pressure P, and relative humidity RH) dependent [*Yu and Turco*, 2001]. The baseline values of Q at different altitudes are from observations [*Millikan et al.*, 1944; *Neher*, 1971], and the temperature and pressure are according to the US standard atmosphere. In Yu's study, the $[H_2SO_4]$ and RH were parameterized in a way so that they are constant in the lowest 2 km of atmosphere ($2 \times 10^7/cm^3$ and 90%, respectively) and gradually decrease with altitude above 2 km.

These parameterizations are reasonable and are within the range of the observed values in various field campaigns where significant nucleation events have been identified [*Weber et al.,* 1999; *Clarke et al.,* 1999]. The ion concentration is initialized as $\sqrt{Q/\alpha}$ where α is ion-ion recombination coefficient. The pre-existing particles are initialized as two log-normal modes with total number densities of 19.5/cm³ and 0.6/cm³, median dry diameters of 0.09 mm and 0.3 μm, and standard deviations of 1.6 and 1.5, respectively. This gives an initial wet surface area of ~ 4.2 μm²/cm³ at 90% relative humidity, corresponding to a cloud-processed clean air mass where typical significant aerosol nucleation has been observed.

Figure 5 shows the total condensation nuclei bigger than 3nm ($N_{d>3\ nm}$) after three hours of simulations at different altitudes. The line with open circles is for the baseline Q values while the line with filled circles is for Q values 20% over the corresponding baseline values. The shaded areas in Figure 5 are low, middle, and high cloud regions as defined in ISCCP cloud data. [H_2SO_4], Q, T, and RH at each altitude are fixed during the three-hour simulations. The production rates of ultrafine particles are sensitive to both [H_2SO_4] and n_t (or Q). The growth rate of the ion clusters is controlled by [H_2SO_4], while n_t determines the lifetime of charged clusters as well as the availability of ions. The neutralization by ion-ion recombination will make the growing charged clusters lose their growth advantage and the resulting neutral clusters may dissociate if smaller than the critical size. At typical [H_2SO_4] where significant nucleation has been observed, for very low Q most of the ion clusters have sufficient time to reach the larger stable sizes prior to recombination and the nucleation rate is limited by Q. As Q (or altitude) increases, ion concentration increases but the lifetime of ions decreases and hence the fraction of ions having sufficient time to grow to the stable sizes decreases. As a result, the total number of particles nucleated first increases rapidly but later on decreases as Q (or altitude) increases. The altitude of the turning point is around 4 km under the vertical profiles assumed in this study.

It is clear from Figure 5 that an increase in GCR ionization rate associated with solar activity leads to an increase in the ultrafine production rate (i.e., $dN/dQ > 0$) in the lower troposphere (as indicated by the arrows) but a decrease in the ultrafine production rate (i.e., $dN/dQ < 0$) in the upper troposphere (as indicated by the arrows). In the middle troposphere, dN/dQ changes sign and the average value of dN/dQ is small compared to that of lower and upper troposphere.

As we pointed out earlier, the magnitudes of ionization rate variations during one solar cycle are smaller than 20% for low latitudes. If a smaller value of ionization rate varia-

tion was used, the difference between two curves shown in Figure 5 would be reduced but the altitude-dependent behavior of changes in CN production would not change. Based on the GCR-CN-CCN-Cloud hypothesis and the influence of GCR ionization change on particle formation rate at different altitudes as shown in Figure 5, we can expect that if GCR variations have any impact on cloudiness, they should correlate positively with low cloud amount and negatively with high cloud amount. For middle clouds, such a correlation (if any) is likely to be weak.

We conclude that solar-modulated GCR fluxes can affect the CN abundance and such effect is likely altitude-dependent. Thus, the first key process (i.e, influence of GCR variations on nucleation and CN abundance) in the proposed GCR-CN-CCN-Cloud hypothesis seems to be valid. While, the second and third processes in the proposed hypothesis are logical and physically plausible, we currently do not know how much the GCR produced CN will affect the CCN abundance and cloud properties. Laboratory and field measurements, as well as theoretical studies are needed to validate the predicted dependent-behaviors of nucleation on ionization rates at different altitudes, to investigate the

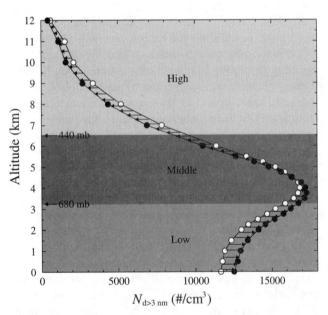

Figure 5. Simulated concentrations of total condensation nuclei larger than 3nm after three hours of simulations at different altitudes. The open circles are for the baseline Q values while the closed circles are for Q values 20% over corresponding baseline values. Thus the arrows indicate the changes in $N_{d>3\ nm}$ as ionization rates increase by 20%. The shaded areas are the ranges of Q corresponding to low (>680 hPa), middle (440–680 hPa), and high (<440 hPa) cloud regions as defined in ISCCP cloud data. [From *Yu,* 2002].

effect of GCR variations on CCN abundance, and to clarify the complex microphysics of aerosol/cloud interactions.

As discussed earlier, the ion-mediated nucleation is sensitive to the ion concentration. Since the space charge density variation as a result of the ionosphere-earth current density change is due to a difference in the concentrations of positive and negative ions, it may also have an impact on ion-mediated nucleation. Charge of mainly one sign in the interaction regions at the boundaries of clouds, as opposed to equal numbers of charges of both signs away from the clouds in clear air, may actually favor the ion-mediated nucleation for clusters of that one sign, because the lifetime of those ion clusters will increase due to the reduced recombination. Recently, significant nucleation has been observed in the top boundaries of clouds [e.g., *Keil and Wendisch*, 2001; *Weber et al.*, 2001]. Thus, ion-mediated nucleation may also respond to the ionosphere-earth current density changes. Further research on this issue is obviously required.

5. THE GLOBAL ELECTRIC CIRCUIT AND ELECTROSCAVENGING

5.1. Modulation of J_z in the Global Circuit

The global electric circuit was illustrated pictorially in Figure 3, and a schematic circuit diagram is given in Figure 6. General properties of the circuit have been reviewed by *Bering et al.*, [1998]. Earlier comprehensive reviews have been given by *NAS* [1986] and *Israël* [1973]. The polar potential pattern is superimposed on the thunderstorm-generated potentials. In a given high latitude region the overhead ionospheric potential, V_i is the sum of the thunderstorm-generated potential and the superimposed magnetosphere-ionosphere generated potential for that geomagnetic latitude and geomagnetic local time. During magnetic storms the changes in V_i from the mean can be as high as 30% within regions extending up to 30° of latitude out from the geomagnetic poles [*Tinsley et al.* 1998].

The vertical current density J_z is determined both by the local ionospheric potential V_i and by the local ionosphere-to-Earth-surface column resistance R, with $J_z = V_i /R$, and R given by $R = \int_0^{120km}(1/\sigma(z))dz$. In Figure 6 the tropospheric contributions to the vertical column resistances at given latitudes are designated T_E, T_L, T_H, and T_P, for equatorial, low, high and polar geomagnetic latitudes, with corresponding designations S_E, S_L, S_H, and S_P for the stratospheric contributions. The conductivities in the mesosphere and lower thermosphere are considered to be so high that their contribution to the column resistance can be neglected. At any latitude the current density is given to a good approximation by $J_z = V_i /(T+S)$.

The GCR flux modulates T and S continuously. The more intermittent fluxes of MeV electrons precipitating into the stratosphere with their associated X-ray Bremsstrahlung modulate S_P and S_H, and the occasional solar proton events modulate S_P. The flux of MeV electrons in the magnetosphere is strongly correlated with solar wind velocity, as discussed by *Li et al.* [20001a,b] who also describe the time and the latitude variation of the precipitating flux. The S_P and S_H contributions to R are normally small, on account of the relatively high conductivity in the stratosphere under normal conditions. But their contribution can be significant when the general stratospheric conductivity has been greatly reduced by H_2SO_4 liquid aerosol particles and vapor following volcanic eruptions [*Tinsley*, 2000]. Polar stratospheric clouds may have the same effect. Measured potential gradients (proportional to J_z) near the surface at the South Pole have been analyzed by *Tinsley et al.* [1998] and show the expected correlation with the solar wind-driven changes in the overhead ionospheric potential. Measured potential gradients and associated J_z variations responding to MeV electron flux changes are discussed by *Tinsley* [2000].

The schematic shown in Figure 6 is for one quarter of a meridional section through the three dimensional global circuit. Comprehensive numerical models of the circuit have been constructed by *Hays and Roble* [1979] and *Roble and Hays* [1979]. A model by *Sapkota and Vareshneya* [1990] includes in addition a first attempt to take into account the effect of volcanic H_2SO_4 aerosols and vapor on the stratospheric column resistance, as well as a more detailed version of the latitude variation of GCR modulation.

As indicated in Figure 6, horizontal potential differences of order 100 kV are generated, high on the dawn side and low on the dusk side, producing corresponding changes in V_i and J_z. The dawn-dusk potential difference has a strong dependency on the product of the solar wind velocity, v_{sw}, and the Bz(GSM) north-south solar wind magnetic field component [*Boyle et al.*, 1997]. Figure 6 includes not only a horizontal potential generator in the northern polar cap, but a solar wind-driven generator of a potential difference between the northern polar cap ionosphere and the southern polar cap ionosphere. This potential depends on the product of v_{sw} and By(GSM), the east-west solar wind magnetic field component. These products are measures of the east-west and north-south components of the electric field in the solar wind. The electric field is due to the motion of the solar wind with its embedded magnetic field relative to the Earth. The resulting potential differences across the magnetosphere are partially coupled onto magnetospheric field lines and conducted by them down into the polar ionospheres.

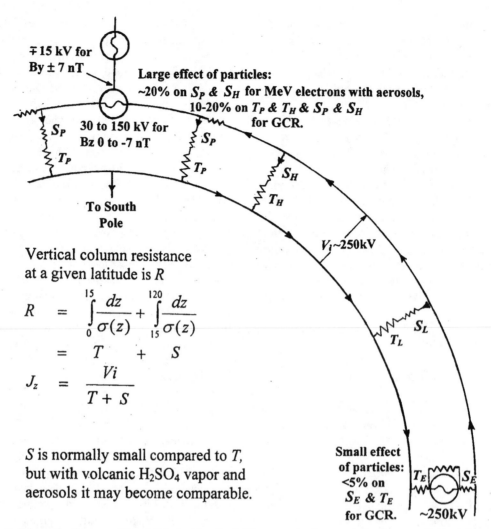

Figure 6. Schematic diagram of the global electric circuit. The geometry is essentially plane-parallel and the equatorial and low latitudes have much smaller column conductivity changes, due to energetic particle influx, than high or polar latitudes. The main generator is tropical thunderstorms, producing an ionospheric potential of order 250 kV over most of the globe, except in the polar regions, where superimposed dawn-to-dusk potentials and a pole-to-pole potential are produced by solar wind-magnetosphere-ionosphere interactions.

5.2. Variation With Latitude of J_z Responses to GCR Changes

The latitude variation of J_z responses to GCR changes is important because models show that it changes sign between high and low latitudes, and therefore any cloud response to GCR, that is linked via J_z changes, would also be expected to change sign.

Roble and Hays [1979] showed that during a Forbush decrease the resulting conductivity decrease at high latitudes (increase of R) results in a diversion of the (assumed constant) current output of the tropical thunderstorm generators from high latitudes to low. This is the case even with

conductivity decreases at middle and low latitudes as well, since these decreases are smaller than at high latitudes.

The treatment of this effect by *Sapkota and Varshneya* [1990] is illustrated in Figure 7, for a relatively large decrease of GCR flux, of 35% at high latitudes and 12% at equatorial latitudes. The values of ΔJ_z and ΔE_z, the changes in J_z and E_z from values before the GCR flux change, are plotted against latitude for longitude 72.5°E. The effect of changes in surface altitude (orography) and thus in near-surface resistivity with latitude and longitude cause the departures from symmetry about the equator. The effects are especially important for E_z that is proportional to the altitude-dependent near-surface resistivity.

5.3. Charges on Droplets and Aerosol Particles

As noted earlier, a region of space charge develops at the tops and bottoms of clouds because of the flow of current through the gradients of conductivity there. In the limiting case of the cloud conductivity being zero and the cloud of very large horizontal extent the full ionosphere-earth potential difference of ~250 kV would appear between the top and bottom of the cloud. The gradients of conductivity at cloud boundaries arise because of the decrease in the concentration of the high-mobility light ions, due to the presence droplets and condensation on nuclei (forming haze particles) that increase the surface area for attachment and recombination of the ions. Also the humidity approaching 100% ensures that water molecules cluster on ions, reducing their mobility. The conductivity changes have been discussed by *Anderson and Trent* [1996], *Reiter* [1992], and *Dolezalek* [1963]. Observations and theory have been compared by *Pruppacher and Klett* [1997], who give a range of between a factor of 3 and 40 for the reduction of conductivity within clouds. *Rust and Moore* [1974] found a factor of 10 reduction in clouds compared to clear air at the same altitude.

To estimate the approximate space charge density ρ for cloud tops we assume that the conductivity σ decreases by a factor of 10 from outside the cloud at 5 km, where $\sigma \sim 10^{-13}$ $\Omega^{-1}m^{-1}$ [*Gringel et al.*, 1986], to inside where $\sigma \sim 10^{-14}$ $\Omega^{-1}m^{-1}$. If the width of the interface is 1 m, then application of equation (1) gives $\rho \sim 10^4 e$ cm^{-3} where e is the elementary charge. If the width is 10 m thick then $\rho \sim 10^3 e$ cm^{-3}. The layer of charge density in either case amounts to a surface charge of 1.6×10^{-9} C m^{-2} that is supplied by J_z in a time of order 15 minutes.

The turbulence at the interface between clear and cloudy air mixes parcels containing charge density, whether on air ions or aerosol particles or droplets, both inwards and outwards. On a longer timescale convection and mixing within the cloud will move parcels containing negative charge from cloud base to cloud top, and parcels containing positive charge from cloud top to cloud base. In general the amounts of positive and negative charge distributed in the cloud will scale as J_z. This applies also to space charge convected into the cloud from other regions in the troposphere where there are conductivity gradients, e. g., on dust layers or from near the surface. However, if ice is eventually produced by processes other than electroscavenging, thunderstorm processes involving ice particle collisions may provide a great deal of additional charge independently of J_z variations.

Many measurements have been made of droplet charges, and a summary is given in Figure 18.1 of *Pruppacher and Klett* [1997]. For clouds in which no ice has formed, and for droplets of radius about 10 mm, charges of about 100e per droplet are typical. The charge would initially be deposited as space charge in interface layers at the tops and bottoms of clouds, by the current J_z flowing through the conductivity gradients in those locations. The charges migrate from air ions to the droplets, which are the largest objects present. The space charge deposited is positive at cloud top and negative at cloud base, and becomes mixed through the cloud depending on the amount of convection and turbulence present.

Charges on aerosol particles are also highly variable. Charge is transferred from air ions to particles, but the greatest charge is present on the residues of charged droplets, immediately after the droplets have evaporated, when they retain essentially all of the charge of the original droplets. As discussed by *Tinsley et al.* [2000] they lose this charge with an initial decay time constant estimated to be about 15 minutes for mid level clouds. The charge does not decay to zero, but asymptotically approaches an equilibrium value that varies with the amount of space charge present, or more precisely with the ratio n^+/n^-. For example, in the interface layer at cloud top with a moderate amount of mixing and evaporation a droplet with a charge of say 100e creates a charged aerosol particle with 100e that can be scavenged by other droplets. For a typical aerosol particle of radius 0.3μm and with $n^+/n^- = 10$ the particle charge decays to an equilibrium value of 10e.

5.4. Electroscavenging

Small electric charges on aerosol particles induce the 'scavenging', or collection of such particles by falling droplets. Figure 8 illustrates the movement across streamlines caused by the electrical forces. Aerosol particles, including CCN, will be retained by the droplet after making contact, and these will then be unavailable for further

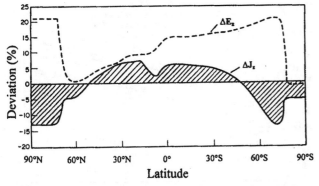

Figure 7. Percentage changes in J_z and near-surface E_z for a Forbush decrease of 35% (polar) to 12% (equatorial) for longitude 72.5°E. The departures from symmetry about the equator are due to the variations in surface altitude, that have greatest effect on the near-surface E_z. Adapted from *Sapkota and Varshneya* [1990].

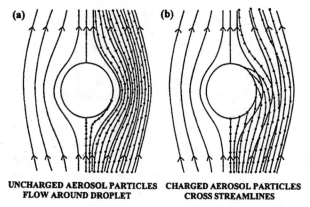

UNCHARGED AEROSOL PARTICLES FLOW AROUND DROPLET

CHARGED AEROSOL PARTICLES CROSS STREAMLINES

Figure 8. (a) Schematic of aerosol flow around a falling droplet in the absence of electrical forces. (b) Schematic of effect of electrical forces in moving aerosol particles across streamlines.

condensation cycles. If the scavenged particle is an ice-forming nucleus, (IFN), and the droplet is supercooled (at a temperature below freezing but still liquid) the droplet is likely to freeze when contact is made.

Electroscavenging rates are determined by calculating trajectories of charged aerosol particles relative to falling droplets, under the influence of electrical and other forces. Such trajectories have been calculated by *Tinsley et al.* [2000, 2001]. The simulations take into account the previously neglected force between the particle charge and its image charge on the droplet, and the results show that this is very important. Collision efficiencies are calculated, being defined as the fraction of those particles in a cylindrical volume swept out by a falling droplet, that actually make contact with the droplet. The tendency for particles to be carried around the droplet by the flow results in collision efficiencies for uncharged particles in the range 10^{-2} to less than 10^{-3} for particles of radius a in the range 0.1 μm to 1 μm in the path of typical cloud droplets.

For particles in this size range, which is the most important one for contact ice nucleation, the electrical effects were found in most cases to dominate the scavenging, even with charges of the same sign on the droplet. Although there is a long-range repulsion between charges of the same sign, the flow carries these larger particles against the repulsion close to the droplet, so that the short-range attractive force, due to the attraction between the particle charge and its image charge, ensures particle collection. Figure 9 shows the variation of collection rate with droplet radius for droplet charges of 100e and particle charges of 20e of the same sign, where e is the elementary charge of 1.6 x 10^{-19} C. The rapid increase of collection rate with droplet radius is due to the volume swept out in a given time varying approximately as the fourth power of the droplet radius (both fall velocity and cross sectional area increase as radius

squared), although the collection efficiencies actually decrease with increasing droplet radius for the larger particles. This variation of collection rate means that electroscavenging processes are likely to be more important for clouds where the droplet size distributions include an appreciable fraction with droplet radii greater than about 10 μm. The low concentrations of aerosol particles and condensation nuclei over the oceans generally results in larger average droplet sizes there, as observed by *Bréon and Colzy* [2000]. Thus, electroscavenging will be more important for oceanic than for continental type clouds.

5.5. Electroscavenging and Ice Formation

When clouds cool below 0°C there is usually little production of ice until temperatures of –10° to –20°C are reached [*Rogers and Yau*, 1989] and for very pure air and water it is necessary to cool below –40°C before droplets freeze (homogeneous nucleation). In the middle and lower troposphere with temperatures above –40°C the freezing of these supercooled droplets is due to a variety of heterogeneous processes. Contact ice nucleation occurs when an aerosol particle with a suitable surface, making it an IFN, makes contact with a supercooled droplet. The formation of ice is an important first step in the production of precipitation in supercooled clouds. It is only necessary that the supercooling be in the region near the cloud top, as the frozen droplets grow by absorbing vapor there, and when they are large enough to fall below that region they will continue to accrete liquid water while in the cloud, even if they melt and continue to do so as droplets.

The clouds where contact ice nucleation is likely to be the most important are those where slow uplift has recently produced cloud top temperatures below, but not greatly below

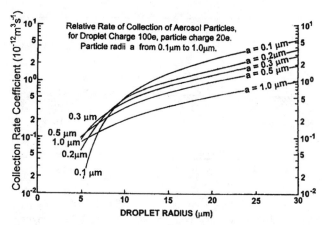

Figure 9. Collection rate coefficients as a function of droplet radius for particle radii a as labeled. The particle charges are all 20e and the droplet charges are 100e, where e = 1.6 x 10^{-19} C.

0°C, and where a moderate amount of mixing is occurring (as in marine stratocululus) that results in the evaporation of charged droplets. Mixing at the interface between the cloudy and clear air at cloud top provides an optimum location for electroscavenging, leading to contact ice nucleation and enhanced precipitation. This interface region is where droplets become charged from the space charge deposited in the conductivity gradient, and where evaporation of these charged droplets is occurring, and where there is further mixing, so that other supercooled droplets can fall through the air parcels containing charged evaporation residues and electroscavenge them.

5.6. Electroscavenging and CCN Removal

A consequence of electroscavenging is that the scavenged aerosol particles (both cloud condensation nuclei and ice forming nuclei) become incorporated in the droplets, and the remaining particle size distributions become depleted in the component removed. If the droplets precipitate then these CCN and IFN will be removed from the air mass, and if the droplets evaporate a number of individual scavenged particles are likely to be consolidated into larger sized particles, although the volatile components will tend to evaporate and be available for further cycles of IMN. There is a tendency for preferential removal of the smaller condensation nuclei and generation of giant nuclei.

In maritime clouds the observed concentrations of CCN are considerably lower than over continents as noted, and so the recycling of evaporation residues as CCN is likely to be greater than for continental clouds. Repeated cycles of cloud formation and evaporation occur, so that removal by electroscavenging of CCN during evaporation is followed by effects on the size distribution of the droplets that are formed during the next condensation cycle. The size range of interest for these hygroscopic particles extends down to roughly 0.02 μm (20 nm) radius, below which they would not grow large enough by absorbing water to be activated for droplet formation. Because of uncertainty in CCN size distributions and surface coatings we can only estimate roughly the time required for a significant reduction in CCN population by electroscavenging.

Using the data of Figure 9 and the six droplet size distributions typical of oceanic clouds discussed by *Tinsley* [2001] the estimated lifetime of condensation nuclei against scavenging ranges from 0.5 hr to 5.5 hr. The estimate quoted earlier for the initial time constant for decay of the charges on newly formed evaporation residues is about 15 minutes, and this leaves insufficient time for significant effects on the CCN population, on account of the decay of the particle charge. However, more time is provided by the repeated cycles of condensation and evaporation. *Pruppacher and Jaenicke* [1995] reviewed the subject of aerosol processing by clouds, and pointed out that most clouds evaporate and do not precipitate, and that aerosol particles and water in the atmosphere typically undergo at least three cycles of condensation and evaporation before being removed. Also, the probability of significant electroscavenging of CCN is increased by particles having a smaller charge than 20e, but a finite value maintained for the lifetime of the cloud in regions of finite space charge. Clearly, there is a need for time-dependent modeling that includes charging, mixing, and evaporation, to arrive at more quantitative evaluations of electroscavenging effects on CCN.

6. EFFECTS ON PRECIPITATION AND LIFETIMES OF CLOUDS

The macroscopic properties of clouds such as albedos, lifetimes, fractional cloud cover and amount of precipitation are affected by the microphysical processes, such as the addition or removal of cloud condensation nuclei by ion-mediated nucleation or electroscavenging respectively, or from the production of ice by contact ice nucleation due to electroscavenging. The importance of the effects depends on cloud temperature, thickness, and type.

For cold clouds (those with tops colder than 0°C) there are two effects by which electroscavenging can increase precipitation. The first is the production of ice as described previously. The frozen droplets grow by the deposition of vapor until they reach a radius of ~25 μm, at which stage they will fall fast enough to grow more rapidly by collision and coalescence with smaller droplets and ice crystals in their path [*Pruppacher and Klett,* 1997]. If the cloud depth is great enough they attain radii of ~100 μm and have large enough fall speeds to leave the cloud within a typical cloud lifetime. The timescale for electroscavenging and contact ice nucleation to increase the precipitation in cold clouds, following a change in ionosphere-earth current density J_z, is hours or less. This is because the time constant for the space charge ρ to develop on clouds and on droplets that evaporate to generate charged evaporation residues is less than one hour, and the mixing processes and contact ice nucleation processes and timescales for growth of ice particles have similar short timescales.

The second way that electroscavenging can increase precipitation in cold clouds is by the removal of CCN as described earlier, that leads to an increase in the average droplet size. With larger droplets there is a higher rate of electroscavenging, as can be seen from the increase in collection rate coefficient with droplet size in Figure 9. This increases the rate of contact ice nucleation that initiates precipitation as described above. The time scale for removal of CCN by electroscavenging is likely to be that of days rather than hours, as repeated cycles of cloud formation and evaporation would appear to be necessary to significantly change the CCN concentrations.

In cold clouds as well as in warm clouds ion-mediated nucleation can affect the CCN concentration. Increases in the CCN concentration producing decreases in the average droplet size were shown by *Twomey* [1977] to produce an increase in cloud albedo. This effect is well known as a result of anthropogenic production of CCN. From the results reported by *Schwartz* [1996] the maximum increase in cloud top reflectance is for clouds of about 300 meters thickness, where there is an increase in cloud top reflectance by about 5% (from 50% to 55%) for a doubling of the droplet concentration from 100 cm^{-3} to 200 cm^{-3}. This would suggest a quite small effect for ~5% change in droplet concentration.

In warm clouds another effect due to changes in CCN concentration, that would appear to be more important than albedo effects, is to change the probability of precipitation. The smaller average droplet size with increased CCN concentration reduces the droplet fall speeds and increases the time necessary for collision and coalescence; processes that are required for the production of precipitation in thick clouds and drizzle in thinner clouds [*Beard and Ochs, 1993*]. With less drizzle production the cloud lifetime and cloud cover are likely to increase.

In thick storm clouds changes in the heavy precipitation that contributes significantly to latent (diabatic) heat transfer can occur as a result of either electroscavenging or IMN changes. We discuss the consequences of this in the next section. Again, there is a need for time dependent cloud modeling to arrive at more quantitative evaluations of the effects of IMN and electroscavenging on CCN concentrations, ice production, precipitation rates, cloud lifetimes and cloud cover.

7. DYNAMICAL AND RADIATIVE EFFECTS OF CLOUD CHANGES

7.1. Effects of Precipitation on Diabatic Heating and Dynamical Changes

From the foregoing theoretical considerations it is apparent that an increase in electroscavenging can lead to increased precipitation, and an increase in ion-mediated nucleation can lead to decreased precipitation in certain types of clouds and air masses. When precipitation removes

water from a cloud there is less cloud water available to cool the air mass by evaporation when the cloud eventually dissipates. In winter storms where there is continuing uplift of air from lower levels the continuing precipitation entails a reduction of the evaporation of water into unsaturated air being entrained into the storm in updrafts and in the mixing at cloud tops. Again, there is less cooling of the air mass than would otherwise have been the case without precipitation. The effect of precipitation is thus one of diabatic (non-adiabatic) heating of the air mass, and this can have important effects on the dynamics of the air mass, (nonconservation of potential vorticity) because of the large latent heat content of moist air in comparison with the specific heat of the same mass of dry air.

In a winter cyclone the primary driver of the dynamics is the baroclinic instability in the winter circulation, with the storm extracting vorticity from the latitudinal shear in the circulation, and converting it to the vorticity of the cyclone. The effective diabatic heating associated with precipitation and reduced cooling of entrained air amounts to an increase in potential vorticity and uplift in the air mass, and is likely to concentrate the vorticity near the cyclone center. In addition, by enhancing the feedback processes inherent in the baroclinic instability, it can increase the overall vorticity of the cyclone. It has been demonstrated analytically by *van Delden* [1989] and from numerical storm simulations by *Zimmerman et al.* [1989] and *Mallet et al.* [1999] that a positive feedback exists between the storm dynamical configuration and the diabatic processes. Thus precipitation changes explain the many reported examples of correlations of the vorticity area index (VAI) with GCR flux change and J_z reviewed by *Tinsley* [2000].

Winter cyclones grow in the presence of a well-developed baroclinic winter circulation, and their growth tends to reduce the baroclinicity, by dissipating momentum and by transporting heat poleward. These effects are concentrated in cyclogenesis longitudes and downstream from them. The changes associated with external forcing and reflected in the VAI variations, when sustained over a season, have the potential for significant effects on the circulation itself. Such changes should appear as changes in Rossby wave amplitude, and in regional variations in average storm track latitude.

Figure 10 compares observed changes in (a) GCR flux above about 1 GeV measured by the Climax, Colorado neutron monitor [*NGDC*, 1989] and above about 500 MeV measured by balloons at 18 km altitude in the Arctic [see *Bazilevskaya*, 2000]; (b) the number of winter storms in the western North Atlantic crossing 60°W between 35°N and 65°N in the west phase of the QBO [from *Labitzke and van*

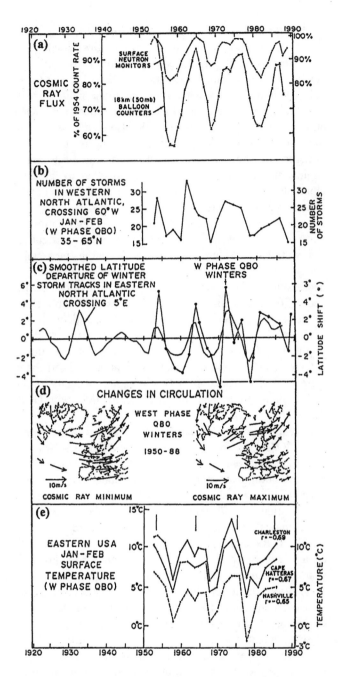

Figure 10. Comparison of GCR variability with tropospheric dynamics and temperature in the North Atlantic-Europe region. (a) GCR flux above about 1 GeV/nucleon at Climax, Colorado and above about 500MeV/nucleon in the Arctic Stratosphere; (b) storm frequencies in the western North Atlantic; (c) storm track latitudes in the eastern North Atlantic; (d) 700 hPa wind speeds and directions in the North Atlantic and European regions for GCR maximum and GCR minimum winters; (e) winter temperatures for three eastern USA locations. For data sources and details see *Tinsley and Deen* [1991].

Loon, 1989]; (c) the latitude shift in storm tracks in the eastern North Atlantic crossing 5°E above 50°N for both QBO phases (smoothed) and for the west QBO phase [see *Tinsley and Deen,* 1991]; (d) the 700 hPa wind speed and direction in the North Atlantic and European regions, for west QBO phase, and cosmic ray minimum winters compared with cosmic ray maximum winters 1950–1988 [from *Venne and Dartt,* 1990]; (e) the mean Jan-Feb. temperatures for three eastern USA stations for west QBO phase winters [*van Loon and Labitzke,* 1988]. These correlated decadal variations were postulated by *Tinsley and Deen* [1991] to be due to increased cyclogenesis in the western north Atlantic responding to increased cosmic ray flux and cloud responses at solar minimum relative to those at solar maximum, in accordance with the general scenario that we have now described in more detail above. This general scenario does not imply that the effects of cosmic ray flux on clouds are fully responsible for the observed decadal atmospheric variations, as such variations at these longitudes have been attributed to an internal coupled atmosphere-ocean oscillation known as the North Atlantic Oscillation [*Walker and Bliss,* 1932; *van Loon and Rogers,* 1978]. The influence of the GCR effect on clouds need only be strong enough to phase lock the NAO to the solar cycle, during periods of stronger solar activity, to account for the six consecutive cycles of storm track latitude that have remained in phase with the solar cycle in the mid-20th century.

An effect of increased cyclogenesis is to increase the meridional transport of heat and momentum and weaken the prevailing westerly zonal winds, as we see in Figure 10(d) for central Western Europe. Thus increased cyclogenesis in the North Atlantic due to greater GCR flux during the Maunder Minimum may have contributed to the reduced zonal winds and colder winters in Europe at that time [*Luterbacher et al.,* 2001]. The main effect has been attributed by *Shindell et al.* [2001] to reduced solar UV during the Maunder Minimum.

The quantitative study of the dynamical consequences of externally forced cloud and diabatic heating changes has received little attention so far, but the physical theory needed for numerically modeling such effects seems to be reasonably well established. Such models would provide a good test of the general scenario describe qualitatively above.

7.2. Effects of Cloud Cover Changes on Radiative Balance

Clouds play a key role in the energy budget of Earth's surface and lower atmosphere. The net cloud radiative forcing depends on the altitude and optical thickness of clouds.

Optically thin clouds at high and middle altitudes tend to warm due to their relatively high transparency at short wavelengths but significant opacity in long wavelengths, while optically thick clouds tend to cool as a result of the dominance of the increased albedo of shortwave solar radiation. Overall, clouds reflect more solar radiation than they trap, leading to a net cooling of ~27.7 W/m² from the mean global cloud cover of ~63.3% [*Hartmann*, 1993].

As discussed in the earlier sections, particle flux induced changes in CCN abundance (via electroscavenging and/or ion-mediated nucleation) may affect the cloud albedo, opacity, and lifetime (hence cloud cover fraction). An increase in CCN tends to increase the cloud albedo and lifetime. Based on the average low cloud radiative forcing of $-16.7 W/m^2$, *Marsh and Svensmark* [2000a] estimated a change in net low cloud forcing of ~1.2W/m² associated with a ~2% absolute change (~7% relative change) in low cloud cover over a solar cycle. These cloud cover changes are sufficient to change the atmospheric heating profile, as discussed by *Yu*, [2002]. Associated with the radiative effects of such large-scale cloud cover changes are likely to be changes in atmospheric circulation. *Dickinson* [1975] pointed out that a change in upper cloud opacity by 20% would produce heating rates in the column below of order 0.1°C per day, which as a differential across a zone 15° in latitude would lead to changes in zonal winds at the tropopause of order 2 m s⁻¹.

8. CONSISTENCY OF CORRELATIONS WITH MECHANISMS

While the energy inputs into the troposphere from the space particle fluxes are orders of magnitude less than those in the correlated atmospheric variations, the free energy available in the supercooled liquid water and supersaturated water vapor are adequate to provide the needed energy amplification, and this would be supplemented by modulation of solar and terrestrial radiation. There is adequate amplitude, relative to their means, of variations in the particle fluxes to drive linear amplitude changes.

At high latitudes both ion production Q and space charge ρ are positively correlated with both GCR changes and J_z changes. (Although parameter $\sigma(z)$ is in the denominator of equation (1), it does not change this result. It applies to the tropospheric cloud level, and varies by a smaller percentage with GCR changes than does J_z. This is because J_z is determined by the reciprocal of R, and at higher altitudes the contributions to R vary by considerably larger percentages.) At low latitudes Q is positively correlated with GCR changes whereas J_z and ρ are negatively correlated, as shown in Figure 7.

Thus the decadal low latitude and low altitude oceanic cloud cover changes [*Marsh and Svensmark*, 2000a] are consistent with ion-mediated nucleation responding to changes in Q. In addition, the anticorrelation of J_z with GCR at low latitudes means that at solar minimum, while there is a maximum of production of CCN by ion-mediated nucleation due to Q changes, there would also be a minimum of removal of CCN by electroscavenging due to opposite ρ changes. So both ion-mediated nucleation and electroscavenging mechanisms could be acting together to increase cloud albedo (and possibly absorption), cloud lifetime, and cloud cover at low latitudes.

The decadal precipitation changes at high latitudes observed by *Kniveton and Todd* [2001] are consistent in sign with J_z and ρ increasing electroscavenging and precipitation at solar minimum, but not with Q and IMN changes decreasing it. The tendency for negative correlations below about 50° geomagnetic latitude is consistent with electroscavenging responding to the low latitude anticorrelation of J_z with GCR flux, as well as with the effects of IMN, as in the low latitude low altitude cloud cover variations. Effects of IMN could contribute for high latitudes if there were sufficient reduction of H_2SO_4 vapor concentration at high latitudes to give an anticorrelation of IMN with GCR flux there.

Recent papers by *Ram and Stolz* [1999] and *Donarummo et al.* [2002] have found 11 and 22-yr and longer solar periods in dust concentrations in a high latitude (Greenland) ice core extending over more than 90,000 annual layers. They attribute the changes in dust concentration to changes in precipitation and soil moisture. The presence of the 22-yr cycle, that is a feature of neutron monitor records, suggested that GCR changes, rather than total or spectral irradiance changes were responsible. Using only data for the last 400 years, *Donarummo et al.* [2002] compared the phase of the 11-yr variation to that of the sunspot cycle. The phase showed a tendency to reverse during periods following explosive volcanic eruptions, when the upper tropospheric H_2SO_4 content was greatly increased. They attributed the 11-yr dust variations in years without volcanic H_2SO_4 to increases (at solar minimum) of electroscavenging, ice production, precipitation and soil moisture. They attributed the 11-yr variations with opposite phase, in years with high volcanic H_2SO_4 to a greater effect of enhanced IMN that (at solar minimum) inhibited precipitation, with resulting low soil moisture.

The continental cloud cover variations over Russia and the USA have opposite variations with GCR flux to those over the oceans, that we attribute to ion-mediated nucleation and electroscavenging. It is of interest that both *Udelhofen and Cess* [2001] and *Veretenenko and Pudovkin* [1999] suggested that circulation changes rather than direct micro-physical effects could be responsible. As discussed above, circulation changes are to be expected as a result of ocean-ic cloud changes, which are present over a considerably larger fraction of the globe.

9. CONCLUSIONS

Effects of space particle flux variations on clouds and cli-mate have been reviewed in the context of two proposed mechanisms—electroscavenging and ion-mediated nucle-ation. The particle flux changes directly affect atmospheric concentrations of ions, and indirectly affect the density of space charge in the troposphere via modulation of current flow in the global electric circuit. The two proposed mech-anisms respond to changes in space charge and/or ion con-centration, and are likely to affect properties of clouds such as droplet size distributions, cloud albedo, infrared opacity, precipitation rates, cloud lifetimes, and hence fractional cloud cover. These cloud and precipitation changes in turn affect atmospheric temperature and dynamics. The effec-tiveness of the two mechanisms depend on latitude as well as altitude, cloud type and atmospheric sulfate content.

Considered together, the two proposed mechanisms have the following strengths and weaknesses. Strengths:- the observations of the decadal variations of clouds and atmos-pheric dynamics are consistent in a number of details (such as variation with latitude, and effects related to the presence of atmospheric sulfate) with one or other or both mecha-nisms. There is adequate available energy and relative amplitude to drive the observed atmospheric changes. Weaknesses: —there is no decisive result at present to deter-mine how much of the observed decadal variations are due to particle flux inputs as compared to total or spectral irra-diance changes. (However, there is no such ambiguity con-cerning the correlations of atmospheric dynamics with par-ticle fluxes on the day-to-day timescale.) Quantitative eval-uations of various aspects of the cloud microphysics have not yet been made. For example, for IMN the growth of the ultrafine particles to CCN size has not yet been modeled, nor has the effect of changes in droplet size on eventual cloud cover. For electroscavenging, the cloud processes linking ice production rates with precipitation rates for var-ious types of clouds has not yet been modeled. Field meas-urements on the global circuit as well as on cloud micro-physical parameters are needed, together with laboratory measurements of IMN and electroscavenging.

Clearly there is a great deal of model development that is needed in order to provide quantitative relationships between atmospheric ionization and macroscopic clouds properties. However, evaluations of the radiative and dynamical consequences for climate of estimated precipita-tion and cloud cover changes could be made with present modeling capabilities. Improved cloud cover and precipita-tion data covering more solar cycles would be useful for validating the present observational results, and as more accurate inputs into global climate models.

Acknowledgments. We wish to acknowledge support from NSF under grants ATM 9903424 and 0104966.

REFERENCES

Anderson R. V. and E. M. Trent, Evaluation of the use of atmos-pheric-electricity recordings in fog forecasting, *Naval Research Laboratory Report, No. 6427,* 20 pp, 1966.

Albrecht, B. A., Aerosols, cloud microphysics, and fractional cloudiness, *Science, 245,* 1227-1230, 1989.

Bazilevskaya, G. A., Observations of variability in cosmic rays, *Space Sci. Rev., 94,* 25-38, 2000.

Bazilevskaya, G. A., M. B. Krainev, and V. S. Makhmutov, Effects of cosmic rays on the Earth's environment, *J. Atmos. Solar Terr. Phys., 62,* 1577-1586, 2000.

Beard, K. V., and H. T. Ochs, Charging mechanisms in clouds and thunderstorms, in *The Earth's Electrical Environment*, pp. 114-130, National Academy Press, Washington, D.C., 1986.

Bering, E. A. III, A. A. Few, and J. R. Benbrook, The global elec-tric circuit, *Phys. Today, 51,* 24-30, 1998.

Boyle, C. B., P. F. Reiff, and M. R. Hairston, Empirical polar cap potentials, *J. Geophys. Res., 102,* 111-125, 1997.

Bréon, F.-M., and S. Colzy, Global distribution of cloud droplet effective radius from POLDER polarization measurements, *Geophys. Res. Lett., 27,* 4065-4068, 2000.

Clarke, A. D., V. N. Kapustin, F. L Eisele, R. J. Weber, and P. H. McMurry, Particle production near marine clouds: Sulfuric acid and predictions from classical binary nucleation, *Geophys. Res. Lett., 26,* 2425-2428, 1999.

Dickinson, R. E., Solar variability and the lower atmosphere, *Bull. Am. Meteorol. Soc., 56,* 1240-1248, 1975.

Dolezalek, H., The atmospheric fog effect, *Rev. Geophys., 1,* 231-282, 1963.

Donarummo, J. Jr., M. Ram and M. R. Stolz, Sun/dust correlations and volcanic interference, *Geophys. Res. Lett., 29(9),* 75-1 - 75-4, 2002.

Fitzgerald, J. W., Marine aerosols - a review, *Atmos. Environ., 25A,* 533-545, 1991.

Flückiger, E. O., D. F. Smart and M. A. Shea, A procedure for estimating the changes in cosmic ray cutoff rigidities and asymptotic directions at low and middle altitudes during periods of enhanced geomagnetic activity, *J. Geophys. Res., 91,* 7925-7930, 1986.

Gringel, W., J. M. Rosen, and D. J. Hoffman, Electrical structure from 0 to 30 kilometers, pp. 166-182, in *The Earth's Electrical Environment*, NAS Press, Washington, D.C., 1986.

Haigh, J. P., Fundamentals of the Earth's Atmosphere and Climate, *this volume.*

Hays, P. B. and R. G. Roble, A quasi-static model of global atmospheric electricity, I. The lower atmosphere, *J. Geophys. Res., 84,* 3291-3305, 1979.

Hartmann, D.L. Radiative effects of clouds on Earth's climate, in *Aerosol-Cloud-Climate Interactions,* edited by P.V. Hobbs, Academic Press Inc., San Diego, 1993.

Herman, J. R. and R. A. Goldberg, *Sun, Weather, and Climate,* NASA , SP-426, Washington, D.C., 1978.

Hobbs, P.V., Aerosol-cloud interactions, in *Aerosol-cloud-climate interactions,* edited by P.V. Hobbs, Academic Press Inc., San Diego, 1993.

Hood, L. l., Effects of Solar UV variability on the stratosphere, *this volume.*

Hoppel, W. A., R. V. Anderson, and J. C. Willett, Atmospheric electricity in the planetary boundary layer, pp. 149-165 in *The Earth's Electrical Environment,* NAS Press, Washington, D.C., 1986.

Israël, H., *Atmospheric Electricity,* vol. II, translated from German, Israel Program for Scientific Translations, Jerusalem, 1973.

IPCC, *Climate Change 1995: The Science of Climate Change,* Intergovernmental Panel on Climate Change, edited by J. T. Houghton et al., Cambridge Univ. Press, New York, 1996.

Kärcher, B., F. Yu, F. P. Schroeder and R. P. Turco, Ultrafine aerosol particles in aircraft plumes: Analysis of growth mechanisms, *Geophys. Res. Lett., 25,* 2793-2796, 1998.

Keil, A., and M. Wendisch, Bursts of Aitken mode and ultrafine particles observed at the top of continental boundary layer clouds, *J. Aerosol Sci., 32,* 649-660, 2001.

Kniveton, D. R., and M. C. Todd, On the relationship of cosmic ray flux and precipitation, *Geophys. Res. Lett. 28,* 1527-1530, 2001.

Labitzke, K., and H. van Loon, Association between the 11-year solar cycle, the QBO, and the atmosphere, III, Aspects of the association, *J. Clim., 2,* 554-565, 1989.

Li, X., M. Temerin, D. N. Baker, G. D. Reeves and D. Larson, Quantitative prediction of radiation belt electrons at geostationary orbit based on solar wind measurements, *Geophys. Res., Lett., 28,* 1887-1890, 2001a.

Li, X., D. N. Baker, S. G. Kanekal, M. Looper and M. Teremin, Long term measurements of radiation belts by SAMPEX and their variations, *Geophys. Res. Lett., 28,* 3827-3830, 2001b.

Luterbacher, J., R. Rickli, E. Xoplaki, C. Tinguely, C. Beck, C. Pfister and H. Wanner, The late Maunder Minimum (1675–1715)—A key period for studying decadal climate change in Europe, *Climate Change, 49,* 441-462, 2001.

Mallet, I., J.-P Cammas, P. Mascart and P. Bechtold, Effects of cloud diabatic heating on the early development of the FASTEX IOP17 cyclone, *Q. J. Roy. Meteorol. Soc., 125,* 3439-3467, 1999.

Markson, R., and M. Muir, Solar wind control of the Earth's electric field, *Science, 206,* 979, 1980.

Marsh, N., and H. Svensmark, Cosmic rays, clouds, and climate, *Space Sci. Rev., 94,* 215-230, 2000a.

Marsh, N., and H. Svensmark, Low cloud properties influenced by cosmic rays, *Phys. Rev. Lett., 85,* 5004-5007, 2000b.

McCracken, K. G., and F. B. McDonald, The long term modulation of the galactic cosmic radiation, 1500-2000, *Proceedings of the International Cosmic Ray Conference, 2001,* Copernicus Gesellschaft, 2001.

McDonald, N. J., and W. O. Roberts, Further evidence of a solar corpuscular influence on large scale circulation at 300 mb, *J. Geophys. Res., 65,* 529-534, 1960.

Millikan, R. A., H. V. Neher, and W. H. Pickering, Further studies on the origin of cosmic rays, *Phys. Rev., 66,* 295-302, 1944.

NAS, *The Earth's Electrical Environment,* Geophysics Study Committee, NAS, Nat. Acad. Press, Washington, D.C., 1986.

Neher, H.V., Cosmic-Rays at high latitudes and altitudes covering four solar maxima, *J. Geophys. Res., 76,* 1637-1651, 1971.

Ney, E. P., Cosmic radiation and the weather, *Nature, 183,* 451, 1959.

NGDC, *Cosmic Ray Hourly Count Rates 1953-1987,* National Geophysical Data Center, NOAA, Boulder, Colo., 1989.

Palle, E., and C. J. Butler, Sunshine, clouds and cosmic rays, in *Proceedings of the first SOLSPA Euroconference,* Canary Islands, September, 2000.

Pruppacher, H. R., and Klett, J. D., *Microphysics of Clouds and Precipitation,* 2nd rev. ed., Kluwer, Dordrecht, 1997.

Pruppacher, H. R., and R. Jaenicke, The processing of water vapor and aerosols by atmospheric clouds, a global estimate, *Atmospheric Res., 38,* 283-295, 1995.

Ram, M., and M. R. Stolz, Possible solar influences on the dust profile of the GISP2 ice core from central Greenland, *Geophys. Res. Lett., 26,* 1043-1046, 1999.

Reiter, R., *Phenomena in Atmospheric and Environmental Electricity,* Elseiver, Amsterdam, 1992.

Roberts, W. O., and Olson, R. H., Geomagnetic storms and wintertime 300 mb trough development in the North Pacific–North America area, *J. Atmos. Sci., 30,* 135-140, 1973.

Roble, R. G. and P. B. Hays, A quasi-static model of global atmospheric electricity, II. Electrical coupling between the upper and the lower atmosphere, *J. Geophys. Res., 84,* 7247-7256, 1979.

Rogers, R. R., and M. K. Yau, *A Short Course in Cloud Physics,* 3rd. ed., Pergamon Press, Oxford, 1989.

Rust, W. D., and C. B. Moore, Electrical conditions near the bases of thunderclouds over New Mexico, *Q. J. R. Meteorol. Soc., 100,* 450-468, 1974.

Sagalyn, R. C. and H. K. Burke, Atmospheric Electricity, in *Handbook of Geophysics and the Space Environment,* edited by A. S. Jursa, Air Force Geophysics Lab., Bedford, MA., 1985.

Sapkota, B. K. and N. C. Varshneya, On the global atmospheric electrical circuit, *J. Atmos. Terr. Phys., 52,* 1-20, 1990.

Schwartz, S. E., Cloud droplet nucleation and its connection to aerosol properties, in *Nucleation and Atmospheric Aerosols,* Edited by M. Kulmala and P. E. Wagner, pp. 770-779, Pergamon Press, Oxford, 1996.

Shindell, D. T., G. A. Schmidt, M. E. Mann, D. Rind and A. Waple, Solar Forcing of Regional Climate Change During the Maunder Minimum, *Science,* 294, 2149-2152, 2001

Svensmark, H., Influence of cosmic rays on Earth's climate, *Phys. Rev. Lett., 81,* 5027-5030, 1998.

Svensmark, H., and E. Friis-Christensen, Variation of cosmic ray flux and global cloud coverage—a missing link in solar climate relations, *J. Atmos. Solar Terr. Phys., 59*, 1225-1232, 1997.

Tinsley, B. A., Correlations of atmospheric dynamics with solar wind-induced changes of air-earth current density into cloud tops, *J. Geophys. Res., 101*, 29,701-29,714, 1996.

Tinsley, B. A., Influence of the solar wind on the global electric circuit, and inferred effects on cloud microphysics, temperature, and dynamics of the troposphere, *Space Sci. Rev., 94*, 231-258, 2000.

Tinsley, B. A., and G. W. Deen, Apparent tropospheric response to MeV-GeV particle flux variations: A connection via electrofreezing of supercooled water in high level clouds? *J. Geophys. Res., 96*, 22283-22296, 1991.

Tinsley, B. A., W. Liu, R. P. Rohrbaugh, and M. Kirkland, South pole electric field responses to overhead ionospheric convection, *J. Geophys. Res., 103*, 26,137-26146, 1998.

Tinsley, B. A., Rohrbaugh, R. P., Hei, M., and Beard, K. V.: 2000, Effects of image charges on the scavenging of aerosol particles by cloud droplets, and on droplet charging and possible ice nucleation processes, *J. Atmos. Sci., 57*, 2118-2134, 2000.

Tinsley, B. A., R. P. Rohrbaugh and M. Hei., Electroscavenging in clouds with broad droplet size distributions and weak electrification, *Atmosph. Res., 59-60*, 115-135, 2001.

Todd, M. C., and D. R. Kniveton, Changes in cloud cover associated with Forbush decreases of galactic cosmic rays, *J. Geophys. Res., 106*, 32,031-32,041, 2001.

Twomey, S., The influence of pollution on the shortwave albedo of clouds, *J. Atmos. Sci., 34*, 1149-1152, 1977.

Twomey, S., Aerosols, clouds and radiation, *Atmos. Environ., 25A*, 2435-2442, 1991.

Udelhofen, P. M., and R. Cess, Cloud cover variations over the United States: An influence of cosmic rays or solar variability?, *Geophys. Res. Lett., 28*, 2617-2620, 2001.

van Delden, A., On the deepening and filling of balanced cyclones by diabatic heating, *Meteor. Atmos. Phys., 41*, 127, 1989.

van Loon, H., and K. Labitzke, Association between the 11 year solar cycle, the QBO, and the atmosphere, Part II, surface and 700 mb on the northern hemisphere in winter, *J. Clim., 1*, 905-920, 1988.

van Loon, H., and J. C. Rogers, The seesaw in winter temperatures between Greenland and Northern Europe, part I, General description, *Mon. Weather Rev., 106*, 296-310, 1978.

Venne, D. E., and D. G. Dartt, An examination of possible solar cycle QBO effects in the northern hemisphere troposphere, *J. Clim., 3*, 272-281, 1990.

Vereteenenko, S. V., and M. I. Pudovkin, Variations in solar radiation input to the lower atmosphere associated with different helio/geophysical factors, *J. Atmos. Solar Terr. Phys., 61*, 521-529, 1999.

Vereteenenko, S. V., and M. I. Pudovkin, Latitudinal dependence of helio/geophysical effects on the solar radiation input to the lower atmosphere, *J. Atmos. Solar Terr. Phys., 62*, 567-571, 2000.

Walker, G. T., and E. W. Bliss. World Weather V, *Mem. Roy. Meteorol. Soc., 4*, 53-84, 1932.

Weber, R. J., et al., Measurement of new particle formation and ultrafine particle growth rates at a clean continental site, *J. Geophys. Res., 102*, 4375-4385, 1997.

Weber, R. J., et al., New particle formation in the remote troposphere: A comparison of observations at various sites, *Geophys. Res. Lett., 26*, 307-310, 1999.

Weber, R. J., et al., Measurements of enhanced H_2SO_4 and 3-4 nm particles near a frontal cloud during the First Aerosol Characterization Experiment (ACE 1), *J. Geophys. Res., 106*, (D20), 24, 107, 2000JD000109, 2001.

Yu, F., Chemiions and nanoparticle formation in diesel engine exhaust, *Geophys. Res. Lett., 28*, 4191-4194, 2001a.

Yu, F., On the mechanism controlling atmospheric particle formation, *J. Aerosol Sci., 32* (S1), S603-604, 2001b.

Yu, F., Altitude variations of cosmic ray induced production of aerosols: Implications for global cloudiness and climate, *J. Geophy. Res., 107 (A7)*, SIA, 8-1 - 8-10, 2002.

Yu, F., and R. P. Turco, The role of ions in the formation and evolution of particles in aircraft plumes, *Geophys. Res. Lett., 24*, 1927-1930, 1997.

Yu, F., and R. P. Turco, Ultrafine aerosol formation via ion-mediated nucleation, *Geophys. Res. Lett., 27*, 883-886, 2000.

Yu, F. and R. P. Turco, From molecular clusters to nanoparticles: The role of ambient ionization in tropospheric aerosol formation, *J. Geophys. Res., 106*, 4797-4814, 2001.

Yu, F., R. P. Turco, B. Kärcher, and F. P. Schröder, On the mechanisms controlling the formation and properties of volatile particles in aircraft wakes, *Geophys. Res. Lett., 25*, 3839-3842, 1998.

Yu, F., R. P. Turco and B. Kärcher, The possible role of organics in the formation and evolution of ultrafine aircraft particles, *J. Geophys. Res., 104*, 4079-4087, 1999.

Zimmerman, J. E., P. J. Smith and D. R. Smith, The role of latent heat release in the evolution of a weak extratropical cyclone, *Mon. Wea. Rev., 117*, 1039-1057, 1989.

Impact of Solar EUV, XUV, and X-Ray Variations on Earth's Atmosphere

Tim Fuller-Rowell

CIRES University of Colorado and NOAA Space Environment Center, Boulder, Colorado

Stan Solomon and Ray Roble

High Altitude Observatory, NCAR, Boulder, Colorado

Rodney Viereck

NOAA Space Environment Center, Boulder, Colorado

Solar extreme ultraviolet (EUV and XUV) radiation (1–103 nm) is the main source of energy and ionization of the upper atmosphere. Changes in EUV flux over the solar cycle and solar rotation time-scales give rise to large changes in neutral temperature and density, and ion density. Thermal expansion of the upper atmosphere over the solar cycle changes the drag on low-Earth orbiting spacecraft by a factor of ten. The increase in plasma density impacts propagation of high-frequency (HF) radio communication signals, and introduces positioning errors by increasing the phase delay in single frequency Global Positioning System navigation signals. Reasonable agreement between observed and globally averaged models of the energy budget and ionospheric production indicates a high level of maturity in our understanding of aeronomic processes initiated by solar photon impact of Earth's atmosphere. Much of our current understanding is based on satellite data from over twenty years ago, but higher resolution observations at EUV and XUV wavelengths may shed light on some of the remaining discrepancies in the energy budget. On time-scales less than a day, the flux at X-ray wavelengths (0.1-0.8 nm) are highly variable, and can increase by three orders of magnitude during a major solar flare. During these events, D-region ion density increases and follows the time history of the flare, causing absorption of HF radio signals and disruption of low frequency navigation systems. EUV flares have also been observed on these same time-scales and can change EUV flux by ~50% for a short period (less than an hour) and have a modest impact on neutral and plasma density in the upper atmosphere. Ionization by soft X-rays (XUV; 1-30 nm) are an important source of plasma in the E-region ionosphere, contribute significantly to the higher energies in the photoelectron spectrum, and are dominant in the time-varying production of low-latitude nitric oxide in the lower thermosphere. Recent measurements of the flux of soft X-rays by SNOE have largely resolved discrepancies in the observed and modeled photoelectron spectrum and E-region electron density profile. It remains to be seen if modeled solar produced NO densities can now be brought into agreement with observations at low latitudes, including the diurnal and latitude structure and the relative contribution of molecular ions.

Solar Variability and its Effects on Climate
Geophysical Monograph 141
Copyright 2004 by the American Geophysical Union
10.1029/141GM23

1. INTRODUCTION

Solar extreme ultraviolet (EUV and XUV; $\lambda < 102.7$ nm) radiation is the primary agent responsible for the creation of Earth's ionosphere. EUV, together with solar far ultraviolet (FUV; $102.7 < \lambda < 200$ nm) wavelengths, also plays a fundamental role in the maintenance of the thermal and composition structure of the thermosphere and ionosphere. The main purpose of this section is to describe the impact of photons shortward of 102.7 nm, including EUV, XUV and X-rays, which are all sufficiently energetic to ionize the major species in the upper atmosphere. Reference will also be made to the longer FUV wavelengths including the Schumann-Runge continuum and bands ($130 < \lambda < 175$ nm) due to their importance in heating and dissociating the upper atmosphere. A separation will also be made of the affect at the shorter wavelengths, with $\lambda < 30$ nm, which are often referred to as soft X-rays or XUV. The XUV solar flux is important in the generation of the E-region ionosphere, the photoelectron spectrum, and in the production of minor species in the lower thermosphere, particularly those derived from the odd-nitrogen chemical processes, such as nitric oxide, NO. The intensity of soft X-rays and production of NO has had an interesting history [Barth et al., 1988; 1999].

Although solar radiation is the dominant source of energy in the upper atmosphere under "quiet" geomagnetic conditions, additional heating, ionization, and dissociation can arise from the magnetosphere. At high latitudes, auroral particles precipitating from the magnetosphere can ionize the atmosphere and enhance the D- and E-region ion densities, and locally can have a much greater impact than solar photons. The other major source of energy from the magnetosphere is the ohmic dissipation of ionospheric currents (also known as Joule heating), driven by the convection electric field mapped from the magnetosphere. Joule heating normally provides significantly less heat than solar radiation for the global upper atmosphere, but during a large geomagnetic storm it can exceed solar photon heating for several hours. Ions in the high latitude upper atmosphere can also be driven to high velocities by the imposed magnetospheric convection electric field, and are a significant momentum source for the upper atmosphere. Tidal, planetary, and gravity waves propagating from the lower atmosphere are the third major source of energy and momentum for the upper atmosphere. Neither the magnetospheric nor the lower atmospheric sources will be considered in detail in this Chapter.

2. UPPER ATMOSPHERE RESPONSE TO EUV

The global mean structure of the Earth's thermosphere is maintained primarily by the highly variable EUV and FUV radiation. Current models and measurements of the radiative flux, at these wavelengths, have been covered comprehensively by Woods et al. (this volume) They show that the integrated EUV energy flux increases by about a factor of three over the solar cycle. Although the processes that establish the overall structure of the thermosphere and ionosphere had been studied for many years, Roble et al. [1987] provided the first comprehensive, quantitative model assessment of the globally averaged structure of the thermosphere from low to high solar activity. Many of the observed global mean characteristics of the upper atmosphere could be captured in the model using estimates of the radiative output from the sun from Hinteregger [1981] and Hinteregger et al. [1981] together with knowledge of the numerous aeronomic processes that operate in the atmosphere above 100 km. Figure 1 is a schematic of the major pathways within the thermosphere that establishes its global mean behavior, including direct ionization of atmospheric species, the creation of photoelectrons, the dissociation of molecular oxygen and nitrogen.

The source of the photoelectrons is the photoionization process, which produced a gas of free electrons. Momentum conservation implies that much of the excess photoionization energy is imparted to the lighter electrons [Rees, 1989] producing a spectrum of electrons with a Maxwellian temperature significantly hotter than the surrounding ambient medium. Subsequent collisions of this energetic electron gas with neutral particles can cause secondary ionization and further heating of the neutral gas.

At lower altitudes (<180 km), frequent collisions of ions with the neutral atmosphere limits diffusive transport, so that the ionosphere is close to chemical equilibrium. At high altitudes, molecular diffusion becomes more important in controlling the vertical distribution of ion density and species concentration. Molecular diffusion has a larger impact on the vertical distribution of the neutral species.

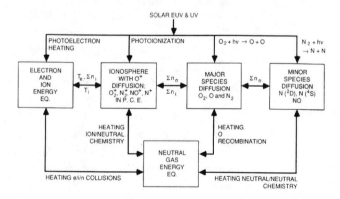

Figure 1. Schematic of the major processes within the thermosphere that establish its global mean structure.

Most of the loss of atomic oxygen occurs below 110 km by three-body recombination process. As a result, the globally averaged vertical profiles of the neutral species are close to diffusive equilibrium in the upper atmosphere.

2.1. Neutral Gas Heating

The neutral gas heating rate consists of the following ten component processes (1) absorption of solar UV radiation in the O_2 Schumann-Runge (S-R) continuum region, (2) the same in the S-R bands, (3) heating by neutral-neutral chemical reactions as given in Table 1, (4) heating by ion-neutral chemical reactions as given in Table 2, (5) heating by colli-

Table 1. Major ion-neutral reactions in the upper atmosphere.

$$N_2^+ + O \rightarrow N(^2D) + NO^+$$
$$NO^+ + e^- \rightarrow N(^2D) + O \ (80\%)$$
$$\rightarrow N(^4S) + O \ (20\%)$$
$$N_2^+ + e^- \rightarrow N(^4S) + N(^4S) \ (10\%)$$
$$\rightarrow N(^4S) + N(^2D) \ (90\%)$$
$$N^+ + O_2 \rightarrow NO^+ + O$$
$$N^+ + O_2 \rightarrow O_2^+ + N(^4S)$$
$$N^+ + O \rightarrow O^+ + N$$
$$O_2^+ + e^- \rightarrow O + O$$
$$O_2^+ + N(^4S) \rightarrow NO^+ + O$$
$$O_2^+ + NO \rightarrow NO^+ + O_2$$
$$O^+ + O_2 \rightarrow O_2^+ + O$$
$$O^+ + N_2 \rightarrow NO^+ + N(^4S)$$
$$NO + h\nu_{Ly\,\alpha} \rightarrow NO^+ + e^-$$

Table 2. Major neutral-neutral reactions in the upper atmosphere

$$O_2 + h\nu \rightarrow O + O$$
$$O + O + M \rightarrow O_2 + M$$
$$O + OH \rightarrow O_2 + H$$
$$O + HO_2 \rightarrow O_2 + OH$$
$$O + O \rightarrow O_2 + h\nu$$
$$N(^2D) + O_2 \rightarrow NO + O$$
$$N(^4S) + O_2 \rightarrow NO + O$$
$$N(^2D) + O \rightarrow N(^4S) + O$$
$$N(^4S) + NO \rightarrow N_2 + O$$
$$N(^2D) + e^- \rightarrow N(^4S) + e^-$$
$$N(^2D) + NO \rightarrow N_2 + O$$
$$N(^2D) \rightarrow N(^4S) + h\nu$$
$$NO + h\nu \rightarrow N(^4S) + O$$
$$NO + h\nu_{Ly\,\alpha} \rightarrow NO^+ + e^-$$

sions between ambient electrons, ions, and neutrals, (6) quenching of $O(^1D)$ by N_2 and O_2, (7) atomic oxygen recombination, (8) absorption of solar UV energy in the O_3 Hartley bands, (9) absorption of solar Lyman-α radiation, and (10) heating by fast photoelectrons.

The absorption of solar radiation by the atmospheric species is given by:

$$\bar{Q}(Z) = \sum_j F\left(\lambda_j\right)\sigma_j E_2\left(\sum_i N_{i\perp}\sigma_{ij}\right)\left(h\nu_j\varepsilon_j\right), \qquad (1)$$

where F is the solar flux, σ_j is the absorption cross section at wavelength λ_j, $N_{i\perp}$ is the perpendicular column number density of absorbing species i, and σ_{ij} is the absorption cross section of species i at wavelength j; $h\nu_j$ is the absorbed solar energy, ε_j is a heating efficiency, and E_2 is the exponential integral. A similar expression can be used to describe photoionization production rates of N_2^+, N^+, O_2^+, O^+, and NO^+, and the photodissociation rates of O_2, N_2, O_3, and NO. Effective absorption and ionization cross-sections for the major atmospheric species were parameterized by *Torr et al.* (1979) into 37 EUV wavelength intervals for use in expressions such as Equation 1. Absorption cross-sections of O_2 over eight wavelength bands in the S-R continuum were also prepared by *Torr et al.* [1980].

Figure 2 shows calculated profiles from *Roble et al.* [1987] of the ten component heating processes at low and high solar activity; the estimated 10.7 cm flux values were 67 and 243 units for the low and high solar activity conditions, respectively. The total neutral heating rate per unit mass peaks in the upper thermosphere, and increases from 15 to 46 Jkg⁻¹s⁻¹ over the two levels of solar activity. Below 150 km, the dominant heating process is O_2 absorption in the S-R continuum; in the middle thermosphere exothermic neutral-neutral and ion-neutral chemical reactions are the main contributors, and at the high levels, the heating is largely from collisions between electrons, ions, and neutrals. The resulting change in global exospheric temperature with solar activity is shown in Figure 3, with values increasing from 700 K at low solar activity to over 1200 K at solar maximum. The solar UV and EUV heating processes are balanced by vertical heat conduction and infrared radiative cooling mainly from NO at 5.3 μm and CO_2 at 15 μm. Details of the production of NO will be covered in a Section 4.

As the thermosphere is heated over the solar cycle, the neutral gas expands to higher altitudes, in such a way that a given pressure level will rise. Within one scale height of the heat source, the local expansion of the gas causes a decrease in density. At all higher altitudes, above the isopycnic

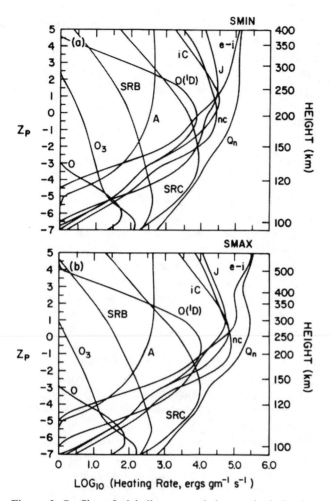

2.2. Ionization and Dissociation

In addition to heating the atmosphere, EUV is also responsible for ionizing and dissociating the neutral species. Similar expressions to Equation 1 describe the photoionization production rates of N_2^+, N^+, O_2^+, O^+, and NO^+, and photodissociation rates of O_2, N_2, and NO. The energy required to ionize the major species are 13.62 eV for O, 12.06 eV for O_2, and 15.58 eV for N_2, corresponding to EUV wavelengths of 91.03 nm, 102.8 nm, and 79.58 nm respectively. NO has a lower ionization threshold energy of 9.264 eV equivalent to a wavelength of 133.8 nm. Photon energies in excess of the ionization threshold are capable of either ionizing the neutral species or, if sufficiently energetic, can also dissociatively ionize O_2 and N_2. Dissociation of O_2, without ionization, requires significantly less energy so that a large fraction of the production of O can come from the larger fluxes at S-R continuum wavelengths. A comprehensive description of the various pathways and their branching ratios can be found in *Roble et al.* [1987]. The shorter wavelength become increasingly more important for the production of the energetic photoelectron spectrum, which can further ionize and dissociate atmospheric species. A more detailed account of its importance will be covered in Section 4.

Figure 2. Profiles of globally averaged thermospheric heating rates from various processes originating from the impact of solar radiation on Earth's upper atmosphere at low (a) and high (b) solar activity, from *Roble et al.* [1987]. The total neutral gas heating rate (Q_n) consists of the following component processes (1) absorption of solar UV radiation in the O_2 Schumann-Runge (S-R) continuum region (SRC), (2) the same in the S-R bands (SRB), (3) heating by neutral-neutral chemical reactions as given in Table 1 (nc), (4) heating by ion-neutral chemical reactions as given in Table 2 (iC), (5) heating by collisions between ambient electrons, ions, and neutrals (e-i), (6) quenching of $O(^1D)$ by N_2 and O_2 ($O(^1D)$), (7) atomic oxygen recombination (O), (8) absorption of solar UV energy in the O_3 Hartley bands (O_3). Also shown are estimates of the globally averaged Joule (J) and auroral (A) heating.

[*Rishbeth and Garriott,* 1969] the rise of the atmospheric column more than compensates for the local expansion. At a fixed height in the upper thermosphere, i.e., at 400 km, the neutral density increases by about a factor of 10 over the solar cycle because of the thermal expansion of the gas. Over a solar rotation, EUV flux can rise by about 50% caus-

ing neutral density to change by about a factor of two to three at 400 km altitude.

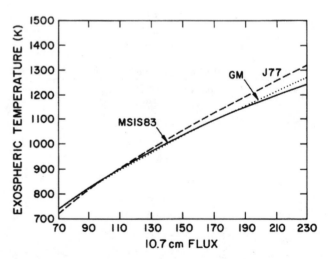

Figure 3. Global mean exospheric temperature variation with solar F10.7 cm flux. Curves representing predictions calculated by Jacchia (1977) empirical model (J77) and by MSIS-83 and the calculated global mean (GM) curves from Roble et al. (1987), are shown.

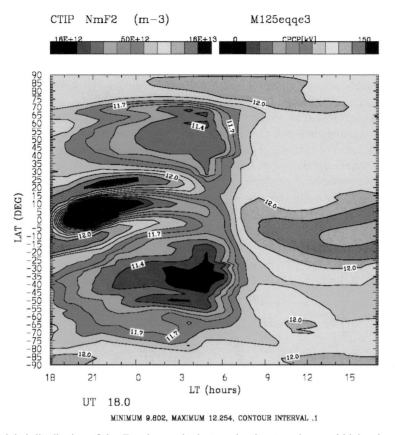

Plate 1. The global distribution of the *F*-region peak electron density at equinox and high solar activity (F10.7=180) from a coupled thermosphere ionosphere plasmasphere electrodynamic model (CTIPe).

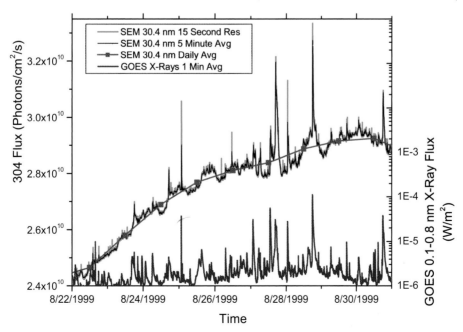

Plate 2. The solar EUV and soft X-ray fluxes for a 9-day period in August 1999. EUV flux is shown as both 15 second and daily-averaged data.

The ion-neutral chemical interactions within the ionosphere are complex especially if one considers the metastable species [*Torr and Torr,* 1978; 1982; *Torr,* 1985]. However, *Torr et al.* [1979] demonstrated that the major reactions in the *F* region can be reduced to a simpler set for modeling global ionospheric processes. The major ion-neutral chemical reactions are shown in Table 1. In addition to direct solar photon dissociation, ion-neutral and neutral-neutral reactions in Tables 1 and 2 also break the O_2 chemical bond. These reactions are driven primarily by solar EUV processes so they have a strong solar cycle dependence, significantly greater than the solar cycle variation in the S-R wavelengths.

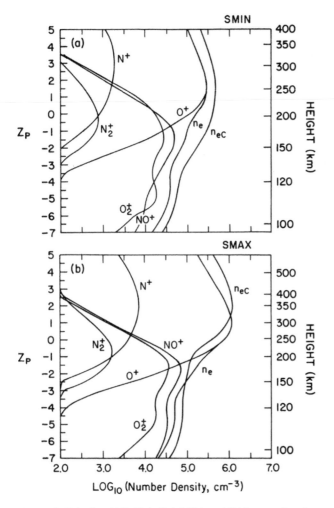

Figure 4. Calculated N^+, N_2^+, O_2^+, NO^+, and O^+ ion number density profiles, and electron density n_e profiles (cm^{-3}) from the global mean model of Roble et al. (1987), and the globally averaged electron density profile from Chiu (1975) empirical model n_{eC} for (a) solar minimum and (b) solar maximum conditions.

Calculated global mean electron density profiles and ion number density profiles are shown in Figure 4 for low and high solar activity [from *Roble et al.,* 1987]. Molecular ions O_2^+ and NO^+ dominate in the lower thermosphere up to an altitude of about 180 km, and O^+ dominates above. The transition altitude varies with solar activity mainly due to thermal expansion rather than substantial changes in the vertical profiles of EUV production. Nitrogen ions N^+ and N_2^+ are minor constituents throughout the ionosphere. The calculated globally averaged peak electron density at solar minimum is 2.8×10^{11} m^{-3} at 255 km compared to the solar maximum value of 1.3×10^{12} m^{-3} at 320 km.

The global distribution of the *F*-region peak electron density at equinox and high solar activity (F10.7=180) is shown in Plate 1. The plate is from a simulation of a coupled thermosphere, ionosphere, plasmasphere, physical model [*Fuller-Rowell et al.,* 1996; *Millward et al.,* 1996], where the solar EUV flux is derived from the Hinteregger [1981] reference spectra. Maximum solar photoionization occurs at the geographic equator on the dayside. The actual peaks in electron density, however, occur either side of the magnetic equator, the so-called equatorial ionization anomaly, due to the redistribution of plasma by low latitude electric fields [*Anderson,* 1981]. The mid and low latitude electrodynamic drifts are driven by the neutral wind dynamo, from both tidal winds propagating from the lower atmosphere into the lower thermosphere and EUV driven winds at higher altitudes.

2.3. EUV Flares

The major changes in EUV intensity are expected to occur over solar cycle and solar rotation timescales. There have been indications of short period variability (minutes to hours) at EUV wavelengths [*Knecht and Davies,* 1961; *Donnelly,* 1971] based on the observed response of the ionosphere. However, definitive interpretation of the observations was not possible due to lack of high cadence EUV observations of the solar flux. Recently, however, measurements from the Solar EUV Monitor (SEM) instrument on the SOHO spacecraft provide confirmation. Based on earlier studies [*Kane and Donnelley,* 1971; *Wood et al.,* 1972] and on preliminary analysis of the SOHO SEM data, the increase in the disk-integrated solar EUV flux can be more than 50% and last for tens of minutes. The estimates of the EUV flux enhancements ranged from 1.5 to 10 depending on the EUV wavelength and the flare magnitude [*Kane and Donnelly,* 1971; *Wood et al.,* 1972].

More recent observations from the SOHO Solar EUV Monitor (SEM) further quantify the solar EUV flux enhancements associated with solar flares [*Judge et al.,* 1998]. In particular, the SEM 30.4 nm channel (26–34 nm) shows significant correlation with the GOES XRS soft X-ray data. In Plate 2, the EUV flux in the SEM 30.4 nm channel is shown as 15 second, 5 minute, and daily-averaged values. For each of the X-ray flares seen in the lower curve, the EUV flux is also enhanced. The enhancement of the X-rays is typically a factor of 10-100 while the EUV flux enhancement is of the order 5–20%. The impact of the X-rays themselves will be discussed in Section 5.

During a particularly large X6-class X-ray flare in July 2000 (the Bastille Day event), the SEM 30.4 nm channel observed a peak flare enhancement in excess of 50% with a 40% enhancement lasting about 30 minutes. For comparison, fluxes of the GOES soft X-ray and SOHO SEM EUV channels are shown in Table 3. Note that the bands are different widths so absolute comparison is less meaningful than the flare enhancement factor.

Early studies by *Donnelly* [1968, 1970] found that radio waves reflected off the ionosphere would undergo Sudden Frequency Deviations (SFDs) and it was hypothesized that these were the results of solar EUV flares. Later the correlation was made between SFDs and solar EUV flares [*Donnelly,* 1971]. The 50% enhancement in EUV flux during the Bastille Day event coincided with a 50% increase in both the N_2 LBH airglow and the OI 135.6 emission, as observed by both the FUV instrument on the IMAGE spacecraft and the SSULI instrument on the ARGOS spacecraft [*private communication, Nicholas et al.,* 2001]. This level of airglow response is consistent with earlier observations by *Opal* [1973]. Unfortunately, to our knowledge, there are no measurements of the Earth's dayside ionosphere at low latitudes at cadences and frequencies needed to observe the response to the Bastille Day event.

Table 3. Soft X-ray and EUV solar flux enhancement for the Bastille Day solar flare on 14 July 2000.

Instrument (Peak λ Band)	Initial Flux (W/m²)	Peak Flux (W/m²)	Flare Enhancement
GOES XRS (0.05–0.4 nm)	1×10^{-7}	2×10^{-4}	2000
GOES XRS (0.1–0.8 nm)	2×10^{-6}	6×10^{-4}	300
SOHO SEM (1–5 nm)	6.5×10^{-3}	1.5×10^{-2}	2.3
SOHO SEM (26–34 nm)	5.2×10^{-4}	7.9×10^{-4}	1.5

To further investigate EUV flares, a coupled thermosphere ionosphere model [*CTIM, Fuller-Rowell et al.,* 1996] has been used to simulate the possible atmospheric and ionospheric response. The model is global and solves the fluid equations for momentum, energy, continuity, major species composition, and odd-nitrogen chemistry for the neutral atmosphere. For the ionosphere, it solves for production, loss, and transport of O^+ and H^+ and assumes chemical equilibrium for the atomic and molecular ions N_2^+, N^+, O_2^+, O^+, and NO^+.

The CTIM model uses the F10.7 cm radio flux as a proxy for solar EUV after [*Heroux and Hinteregger* 1978]. The initial EUV flare studies involved simply scaling the F10.7 cm flux to provide a short enhancement in the solar flux. While it is known that the EUV spectrum of a flare is different from that of the quiet sun, scaling F10.7 will produce an enhancement similar to a flare at least in the EUV portion of the spectrum. However, it will not produce the soft X-ray enhancement described above.

The F10.7 cm flux was set at 220 as the initial value and for the control run, it was left constant. For the flare run, the F10.7 cm flux was increased from 220 to 350 over a 12-minute period (10.2 to 10.4 hours) and then allowed to exponentially decay back to the original 220 value with about a 1-hour time constant (see Plate 3). Note that at the end of the 2-hour run, the flux levels were down to about 250. These levels were designed to mimic the EUV changes expected during an X-class solar X-ray flare such as the Bastille Day event. The modeled response of the low latitude ionosphere, neutral density, and NO concentration are shown in Plate 3.

The electron density responded quite quickly with a 20% enhancement in the E-region and a 50% enhancement on the bottom side of the F-region at about 200 km. The F-region peak changed by less than 10% in response to the simulated EUV flare so it would be difficult to detect in ionosonde observations. Cursory examination of the peak F-region electron density from some suitably placed ionosondes did not reveal a clear flare signature above the normal geophysical variability, which is usually around 25%. The neutral temperature and density responded more slowly, with a 20% enhancement in density above 600 km which was still increasing at the end of the 2-hour run, illustrating the much longer time constant for the thermal response than the short duration of the flare. In addition, the NO density was found to increase by just over 10% at about 150 km; with the peak NO normally near 110 km altitude the maximum response at 150 km is on the topside of the profile. As with the electron density enhancement, the NO density responded quickly to the solar flux levels and peaked very

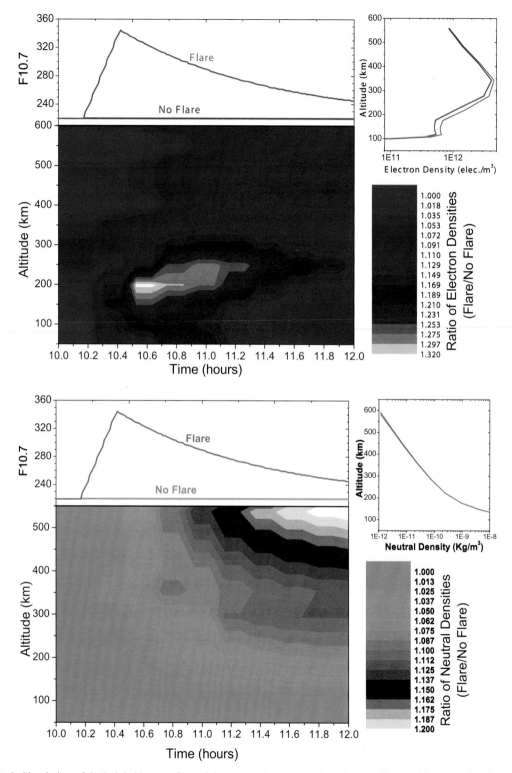

Plate 3. Simulation of the height/time profiles of the expected response of a) electron density, b) neutral density, and c) nitric oxide number density, at low latitude on the dayside of the Earth to the EUV flare on the Bastille Day event using a coupled thermosphere ionosphere model.

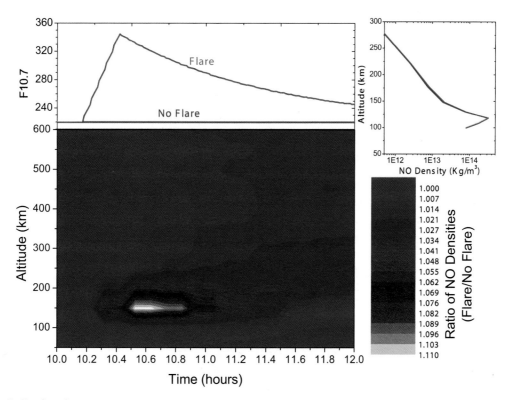

Plate 3. Continued.

close in time to the peak in solar flux. At 150 km NO is closer to chemical equilibrium than at the peak at 110 km.

Since the N_2 LBH and OI airglow emissions are the product of photoelectrons on N_2 and O respectively, it is presumed that the airglow response would be directly proportional to the electron enhancement. Therefore, the airglow response should follow the EUV flux without significant delay. The modeled electron density increase of 50% is in reasonable agreement with the 50% enhancement observed in the N_2 LBH and OI airglow emissions. It should also be noted that the 30% enhancement in the 600 km neutral density could have a noticeable effect on the orbital parameters of satellites but, as yet, no high-cadence satellite drag observations have been made during a large EUV flare. A more detailed investigation into the response of the upper atmosphere to EUV flares is ongoing.

3. UPPER ATMOSPHERE RESPONSE TO SOFT X-RAYS (XUV)

The soft X-ray region of the solar spectrum from 1–30 nm, designated XUV, has been difficult to quantify and has been the source of considerable discussion. Solar XUV photons are few in number, but they are energetic, variable, and are very important for production of photoelectrons, E-region ionization, the lower thermosphere odd-nitrogen balance, and daytime airglow emissions. They deposit their energy throughout the thermosphere, but especially in the altitude range from 100 to 200 km. Although the reference spectrum obtained from the Atmosphere Explorer (AE) program and from rocket measurements by *Hinteregger et al.* [1981] has been considered an acceptable standard for the extreme ultraviolet (EUV) from 30–103 nm, there has been considerable doubt concerning its validity shortward of 25 nm. The problem dates back to comparisons with AE photoelectron measurements by *Richards and Torr* [1984], which has been reviewed by *Solomon* [1991] and *Bailey et al.* [2000]. *Richards and Torr* proposed a factor ~3 increase in the *Hinteregger et al.* solar XUV flux below 25 nm, which enabled theoretical estimates of the photoelectron spectrum to be generally consistent with data from the Dynamics Explorer–2 satellite [*Winningham et al.*, 1989]. Several other lines of evidence pointed in the same direction, including airglow analyses, calculation of nitric oxide densities, and modeling of the lower ionosphere. *Barth et al.* [1988] and *Siskind et al.* [1990; 1995] invoked increases in the soft X-ray 1.8 – 5.0 nm solar flux in order to account for NO densities from a range of satellite and ground-based observations. *Buonsanto* [1990] and

Buonsanto et al. [1992; 1995] found a significant shortfall in modeled electron densities of the E and F_1 region ionosphere between 100 and 200 km, when compared to incoherent scatter radar and ionosonde measurements. Because XUV photons produce additional ionization through generation of photoelectrons and subsequent photoelectron impact ionization of neutral species, the photoelectron and ion density problems are intertwined in the lower ionosphere.

New measurements of solar irradiance in the soft X-ray region of the spectrum from the SNOE satellite [*Bailey et al.*, 2000] show that the full-disk solar irradiance in the ~2–20 nm region is higher than a standard model [*Hinteregger et al.*, 1981] by about a factor of four at all levels of solar activity. This confirms contentions developed from several lines of evidence, [*e.g.*, *Richards et al.*, 1984; 1994] but to a larger degree than previously suspected. This finding has many implications for the thermosphere and ionosphere [*Solomon et al.*, 2001].

3.1. The Photoelectron Spectrum

The confirmation of the increase in XUV flux from the SNOE observations has largely resolved the issues regarding the photoelectron spectrum. *Solomon et al.* [2001] was able to demonstrate excellent agreement between measured photoelectron fluxes and those modeled by GLOW [*Solomon et al.*, 1988; *Solomon and Abreu*, 1989] using the new XUV fluxes. A photoelectron energy spectrum meas-

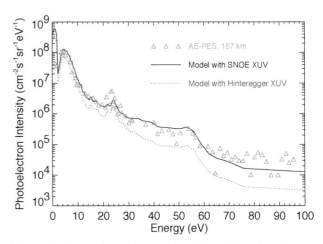

Figure 5. Comparison between the AE satellite photoelectron spectrum measurement and two runs of the GLOW model. In the first run, the Hinteregger *et al.* (1981) model solar spectrum was employed. In the second, the same solar spectrum is used, but multiplied by a factor of 4 shortward of 25 nm to approximate the SNOE findings.

ured by the AE Photoelectron Spectrometer (PES) was selected for comparison from those published by *Lee et al.* [1980]. It was obtained near 167 km altitude at low solar activity, early in the AE-E mission. Figure 5 shows the comparison between the AE–PES measurement and two runs of the GLOW model. In the first run, the *Hinteregger et al.* [1981] model solar spectrum was employed. In the second, the same solar spectrum is used, but multiplied by a factor of 4 shortward of 25 nm to approximate the SNOE findings, as discussed above. The model with increased solar XUV is a much better match for the data. Some discrepancies remain, for instance at the features near 25 eV caused by the intense solar He II line at 30.4 nm. Above 70 eV, the measurements are noisy and probably unreliable, so the lack of fit there should not be considered meaningful. Nevertheless, a further increase in the solar spectrum shortward of 10 nm could be supported by both the solar and photoelectron measurements.

3.2. Nitric Oxide Production

Nitric Oxide (NO) is an important minor species in the thermosphere and ionosphere. It was first suggested by *Kockarts* [1980] that NO infrared cooling by the emission of the 5.3 mm fundamental band in the terrestrial thermosphere is an important mechanism above 120 km altitude. This was confirmed with the globally averaged model of *Roble et al.* [1987]. There are a number of pathways for the production and loss of NO (see Table 1 and 2), but the primary source of NO is the reaction of excited atomic nitrogen, $N(^2D)$, with molecular oxygen. The atomic nitrogen itself is created by a number of the ion-neutral reactions and by direct dissociation of molecular nitrogen by photons and photoelectrons. The substantial shortfall in modeled NO density found by many theoretical calculations has usually been attributed to an underestimate of the soft X-ray solar flux by the reference models.

With the simultaneous measurements of thermospheric NO density and solar soft X-ray irradiances on the SNOE satellite a number of the problems were resolved. *Barth et al.* [1999] showed a strong correlation between solar soft X-ray flux and NO density in the tropics at low geomagnetic activity. Photochemical model calculations that used the new measurements of soft X-ray irradiances as input parameters adequately reproduced the magnitude of the time-varying component of the thermospheric nitric oxide in the tropics. Although the model calculations were able to capture the changes in NO, the calculations also implied that there was an additional source of nitric oxide that did not vary with the time-period of the solar rotation. Figure 6

shows a more recent illustration of the strong relationship between soft X-rays and NO at 110 km over a longer time period, indicating that the residual NO density is now small when extrapolating the linear relationship to zero irradiance. Modeling is still required to determine if the latest measurements of soft X-ray flux bring mid and low latitude NO concentrations, and their diurnal variation, into agreement with observations.

3.3. E-Region Electron Density

Using contemporary solar reference spectra, *Buonsanto* [1990] and *Buonsanto et al.* [1992; 1995] found a significant shortfall in modeled E and F_1 region ionospheric density between 100 and 200 km, when compared to incoherent scatter radar and ionosonde measurements. The modeled electron density profiles were compared with measurements from the Haystack Observatory incoherent scatter radar at Millstone Hill, intercalibrated with ionosonde measurements. With the new observations of solar soft X-ray flux from SNOE, which indicated fluxes four times the previous reference spectra, *Solomon et al.* [2001] has been able to resolve the issue. Figure 7a, shows a comparison of incoherent scatter measurements of the electron density profile near solar minimum to that modeled using the *Hinteregger et al.* [1981] solar spectrum, and using the fourfold increase shortward of 25 nm. The increased XUV irradiance obtains a much better match to the observed electron density profile. The overall magnitude is similar, and the "valley" region is much less prominent. There are two aspects of the observations that are not replicated by the model: the wave-like vertical structure, and the increase in density above 180 km. The former may be due to neutral atmosphere or ionos-

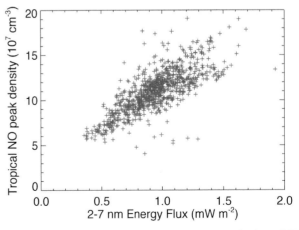

Figure 6. Correlation between SNOE observations of solar soft X-ray flux and NO density in the tropics at low geomagnetic activity.

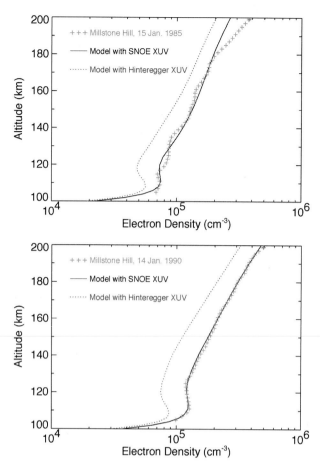

Figure 7. Comparison of incoherent scatter measurements of the electron density profile near a) solar minimum, and b) solar maximum, to that modeled using the Hinteregger *et al.* (1981) solar spectrum, and using the fourfold increase shortward of 25 nm.

pheric structure that is not captured by a semi-empirical model of atmospheric climatology or by a photochemical equilibrium model. The latter is evidence of diffusive transport of O^+, which starts to become important at altitudes slightly below 200 km at low solar activity, and is not accounted for in this model formulation. Figure 7b shows the same comparison at moderately high solar activity. Excellent agreement is again obtained throughout the altitude range when increased solar XUV irradiance is employed in the model.

4. IMPACT OF SOLAR X-RAYS

Solar X-rays penetrate deeply into the atmosphere so their heating effect is negligible. They are, however, a significant source of ions in the D-region, below about 100 km attitude. Historically, the significance of X-rays and D-region ion-

ization has been related to absorption of high frequency radio waves. Solar X-rays have been monitored by the XRS instrument on the NOAA-GOES satellites for more than two decades. The GOES XRS data led to the C-M-X flare classification system (see below) and enabled the connection between solar activity and terrestrial effects to become established. Long-range communications using high frequency (HF) radio waves (3–30 MHz) depend on reflection of the signals in the ionosphere. Radio waves are typically reflected near the peak of the F_2 layer (~300 km altitude), but along the path to and from the F_2 peak the radio wave signal suffers attenuation due to absorption by the intervening ionosphere. Most of this absorption occurs in the D-region and so is very dependent on the intensity of the solar X-ray flux.

Absorption is the process by which the energy of radio waves are converted into heat and electromagnetic (EM) noise through interactions between the radio wave, ionospheric electrons, and the neutral atmosphere (See *Davies* [1990] for a more extensive description of the absorption process). Most of the absorption occurs in the ionospheric D region (50 to 90 km altitude) because here, the product of the electron density and the electron/neutral collision frequency, attains a maximum. Within this region, the neutral density is relatively constant over time, so variations in the local electron density drive the total amount of absorption. The electron density is a function of many parameters and normally varies with local time, latitude, season, and over the solar cycle. These "natural" changes are predictable, and affect absorption only moderately at the lowest HF frequencies. Much more significant changes to the electron density, and therefore the absorption strength, are seen as a result of solar X-ray flares (the classic short wave fade).

Although recent SOHO observations have clearly demonstrated an EUV flare component, traditionally, solar flares were associated with an increase in the X-ray flux in the 0.1 to 0.8 nm wavelength range. The flares, which can last from a few minutes to several hours, are rated C, M, or X according to the 0.1–0.8 nm flux as measured by instruments on the GOES satellites. To qualify as a C-class flare the flux, F, must fall within the range $10^{-6} \le F < 10^{-5}$ W m^{-2}, for M-class $10^{-5} \le F < 10^{-4}$, and X-class $F \ge 10^{-4}$. In standard notation the letters act as multipliers, for example C3.2 equates to a flux of 3.2×10^{-6} W m^{-2}.

The C, M, and X classification is based on the full-disk X-ray emission from the sun. During periods of high solar activity, such as solar maximum, the background flux may increase to C-class levels for days at a time, even without flare activity. The D region electron density is directly driven by the total X-ray flux regardless of the source, so these periods of high background flux are equally important to radio absorption. Due to geometric effects, D region ionization is greatest at the sub-solar point, where the sun is directly overhead. The amount of ionization and absorption falls with distance away from the sub-solar point, reaching zero at the day/night terminator. The night-side of the Earth is unaffected due to the rapid ion recombination at D-region altitudes.

5. SUMMARY AND CONCLUSIONS

Solar EUV and XUV are clearly the dominant source of heating in the upper thermosphere and are the primary cause of the global ionosphere. Only at high latitudes or during a major geomagnetic storm is this not the case. Substantial agreement has been demonstrated between global models and observations indicating a mature level of scientific understanding. A substantially new aspect has come from the recent SOHO observations of the solar flux during EUV flares. Although the flares do not have a dramatic impact on the thermosphere and ionosphere, the response is measurable particularly in the changes in airglow. Estimates in the EUV flare spectrum remain to be quantified, as does a more comprehensive comparison of the modeled and observed upper atmosphere response.

The improvement in the XUV reference spectrum from the SNOE observations has been shown to have significant implications for the thermosphere and ionosphere. The new models of the photoelectron spectrum and the E and F1-region electron density profiles are now in excellent agreement with data. The relationship between soft X-rays and low latitude nitric oxide concentrations is now clear, but it remains to be seen if modeled solar produced NO densities can now be brought into agreement with observations at low latitudes, including the diurnal and latitude structure and issues on the relative contribution of molecular ions NO^+ and O_2^+.

The revised XUV fluxes have an addition impact on the globally averaged exospheric temperature. The model of *Roble et al.* (1987), using the original *Hinteregger et al.* [1981] solar flux, showed that the heating effect of the 15 - 20 nm wavelength band contributed nearly 50 and 20 K to globally averaged exospheric temperature at high and low solar activity, respectively. With a factor of four increase in the XUV flux, the implication is that the thermospheric temperature will rise by an additional 150 K at high solar activity. At the same time, the increase in XUV would also increase the globally averaged nitric oxide, which presumably will also increase the 5.3 μm infrared cooling by nitric oxide. With the changes in the NO radiative cooling and the need for higher CO_2 15 μm cooling rate coefficients [*Sharma and Roble,* 2001], it remains to be seen if the combination of increased heating at XUV wavelengths is balanced by increased infrared cooling, to bring globally averaged temperature in reasonable agreement with the empirical models, such as MSIS.

REFERENCES

Anderson, D. N., Modeling of the ambient, low-latitude *F*-region ionosphere—A review, *J. Atmos. Terr. Phys.* 43, 753, 1981.

Bailey, S. M., T. N. Woods, C. A. Barth, S. C. Solomon, L. R. Canfield, and R. Korde, Measurements of the solar soft X-ray irradiance from the Student Nitric Oxide Explorer: First analysis and underflight calibrations, *J. Geophys. Res.* 105, 27179, 2000.

Barth, C.A., W.K. Tobiska, D.E. Siskind, and D.D. Cleary, Solar terrestrial coupling: Low latitude thermospheric nitric oxide, *Geophys. Res. Lett.*, 15, 92-94, 1988.

Barth, C.A., S.M. Bailey, and S.C. Solomon, Solar terrestrial coupling: Solar soft X-rays and thermospheric nitric oxide, *Geophys. Res. Lett.*, 26, 1251-1254, 1999.

Buonsanto, M. J., A study of the daytime $E–F_1$ region ionosphere at mid-latitudes, *J. Geophys. Res.* 95, 7735, 1990.

Buonsanto, M. J., S. C. Solomon, and W. K. Tobiska, Comparison of measured and modeled solar EUV flux and its effect on the E-F1 region of the ionosphere, *J. Geophys. Res.* 97, 10,513, 1992.

Buonsanto, M. J., P. G. Richards, W. K. Tobiska, S. C. Solomon, Y.-K. Tung, and J. A. Fennelly, Ionospheric electron densities calculated using different EUV flux models and cross sections: Comparison with radar data, *J. Geophys. Res.* 100, 14569, 1995.

Davies, K., *Ionospheric Radio,* Peter Peregrinus LTD Publisher, London, 247-249, 1990.

Donnelly, R. F., Early detection of a solar flare: A study of X-ray, extreme ultraviolet, H-alpha, and solar radio emission from solar flares, *ESSA Tech Report, ERL 81-DSL 2,* 1968

Donnelly, R. F., Extreme ultraviolet flashes of solar flares observed via sudden frequency deviations, *ESSA Tech. Report, ERL 169-SDL 14,* 1970.

Donnelly, R. F., Extreme ultraviolet flashes of solar flares observed via sudden frequency deviations: experimental results, *Solar Phys.* 20, 188–203, 1971.

Fuller-Rowell, T. J., R. J. Moffett, S. Quegan, D. Rees, M. V. Codrescu, and G. H. Millward, A Coupled Thermosphere-Ionosphere Model (CTIM), *STEP Handbook on Ionospheric Models,* R. W. Schunk (ed.), Utah State Univ., Logan, UT, 1996.

Heroux, L. and H. E. Hinteregger, Aeronomical reference spectrum for solar UV below 2000 Å, *J. Geophys. Res.* 83, 5305–5308, 1978

Hinteregger, H. E., Representations of solar EUV fluxes for aeronomical applications, *Adv. Space Res.* 1, 39, 1981.

Hinteregger, H. E., K. Fukui, and B. R. Gilson, Observational, reference, and model data on solar EUV from measurements on AE-E, *Geophys. Res. Lett.* 8, 1147–1150, 1981.

Judge, D. L., D. R., McMullin, H. S. Ogawa, D. Hovestadt, B. Klecker, M. Hilchenbach, E. Mobius, L. R. Canfield, R. E. Vest, R. Watts, C. Tarrio, M. Kuhne, P. Wurz, First solar EUV irradiances obtained from SOHO by the CELIAS/SEM, *Solar Phys.* 177, 161–163, 1998.

Kane, S. R. and R. F. Donnelly, Impulsive hard X-ray and ultraviolet emissions during solar flares, *Astrophys. J.* 164, 1971

Knecht, R. W. and K. Davies, Solar flare effects in the *F*-region of the ionosphere, *Nature* 190, 797–798, 1961.

Kockarts, G., Nitric oxide cooling in the terrestrial thermosphere, *Geophys. Res. Lett.* 7, 137–140, 1980.

Lee, J. S., J. P. Doering, T. A. Potemra, and L. H. Brace, Measurements of the ambient photoelectron spectrum from Atmosphere Explorer, 1, AE-E measurements below 300 km during solar minimum conditions, *Planet. Space. Sci.* 28, 947, 1980.

Millward, G. H., H. Rishbeth, T. J. Fuller-Rowell, A. D. Aylward, S. Quegan, and R. J. Moffett, Ionospheric F_2 layer seasonal and semiannual variations, *J. Geophys. Res.* 101, 5149, 1996.

Nicholas, A. C., *et al., private communication*, 2001

Opal, C. B., Enhancements of the photoelectron-excited dayglow during solar flares, *Space Research XIII* 13, 797–802, 1973.

Rees, M.H., *Physics and Chemistry of the Upper Atmosphere*, Cambridge University Press, Cambridge, 1989.

Richards, P. G. and D. G. Torr, An investigation of the consistency of the ionospheric measurements of the photoelectron flux and solar EUV flux, *J. Geophys. Res.* 89, 5625, 1984.

Richards, P. G., M. R. Torr, and D. G. Torr, Photo-dissociation of N_2: A significant source of atomic nitrogen, *J. Geophys. Res.* 86, 1495–1498, 1984.

Richards, P. G., J. A. Fennelly, and D. G. Torr, EUVAC: A solar EUV flux model for aeronomic calculations, *J. Geophys. Res.* 99, 8981–8992, 1994.

Rishbeth, H. and O. K. Garriott, *Introduction to Ionospheric Physics,* Academic Press, New York, 1969.

Roble, R. G., E. C. Ridley, and R. E. Dickinson, On the global mean structure of the thermosphere, *J. Geophys. Res.* 92, 8745–8758, 1987.

Siskind, D.E., D.J. Strickland, R.R. Meier, T. Majeed, and F.G. Eparvier, On the relationship between the solar soft X-ray flux and thermospheric nitric oxide: An update with an improved photoelectron model, *J. Geophys. Res.,* 100, 19687-19694, 1995.

Siskind, D.E., C.A. Barth, and D.D. Cleary, The possible effect of solar soft X-rays on thermospheric nitric oxide, *J. Geophys. Res.,* 95, 4311, 4317, 1990.

Sharma, R.D. and R.G. Roble, Impact of the new rate coefficient for the O atom vibrational deactivation and photodissociation of NO on the temperature and density structure of the terrestrial atmosphere. J. Geophys. Res., 106, 21343-21350, 2001.

Solomon, S. C., Optical aeronomy, *U.S. Natl. Rep. Intl. Union Geod. Geophys. 1987–1990, Rev. Geophys.* 29, 1089, 1991.

Solomon, S. C. and V. J. Abreu, The 630 nm dayglow, *J. Geophys. Res.* 94, 6817, 1989.

Solomon, S. C., P. B. Hays, and V. J. Abreu, The auroral 6300 Å emission: observations and modeling, *J. Geophys. Res.* 93, 9867, 1988.

Solomon, S. C., S. M. Bailey, and T. N. Woods, Effect of solar soft X-rays on the lower ionosphere, *Geophys. Res. Lett.,* 11, 2149, 2001.

Torr, D. G., The photochemistry of the upper atmosphere, in *The Photochemistry of Atmospheres,* J. S. Levine (ed.), Academic Press, Inc., 165–278, 1985.

Torr, D. G. and M. R. Torr, Review of rate coefficients of ionic reactions determined from measurements made by Atmosphere Explorer Satellites, *Rev. Geophys.* 16, 327, 1978.

Torr, M. R. and D. G. Torr, The role of metastable species in the thermosphere, *Rev. Geophys.* 20, 91, 1982.

Torr, M. R., D. G. Torr, and H. E. Hinteregger, Solar flux variability in the Schumann-Runge continuum as a function of solar cycle 21, *J. Geophys. Res.* 85, 6063–6068, 1980.

Torr, M. R., D. G. Torr, R. A. Ong, and H. E. Hinteregger, Ionization frequencies for major thermospheric constituents as a function of solar cycle 21, *Geophys. Res. Lett.* 6, 771–774, 1979.

Winningham, J. D., D. T. Decker, J. U. Kozyra, J. R. Jasperse, and A. F. Nagy, Energetic (>60 eV) atmospheric photoelectrons, *J. Geophys. Res.* 94, 15335, 1989.

Wood, A. T., R. W. Noyes, A. K. Dupree, M. C E. Huber, W. H. Parkinson, E. M. Reeves, and G. L. Withbroe, Solar Flares in the Extreme Ultraviolet, *Solar Phys.* 24, 169-179, 1972.

Woods, T. N., G. J. Rottman, S. M. Bailey, S. C. Solomon, and J. R. Worden, Solar extreme ultraviolet irradiance measured during solar cycle 22, *Solar Phys.* 177, 133, 1998.

Woods, T.N., Acton, L.W., Bailey, S., Eparvier, F., Garcia, H., Judge, D., Lean, J., Mariska, J.T., McMullin, D., Schmidtke, G., Solomon, S.C., Tobiska, W.K., Warren, H.P., and Viereck, R., Solar Extreme Ultraviolet and X-ray Irradiance Variations, *this issue.*

Tim Fuller-Rowell, Space Environment Center, 325 Broadway, Boulder, CO 80305

Raymond G. Roble, High Altitude Observatory, National Center for Atmospheric Research, Boulder, CO 80303

Stan C. Solomon, LASP, University of Colorado, Boulder, CO 80309

Rodney Viereck, Space Environment Center, 325 Broadway, Boulder, CO 80305

Section 4

Science Requirements
and Required Future Measurements

Science Requirements and Required Future Measurements

William Sprigg

Institute for the Study of Planet Earth, University of Arizona, Tucson, Arizona

Judit M. Pap

Goddard Earth Sciences and Technology Center, University of Maryland, Baltimore County, Baltimore, Maryland

The role of solar variability in climate variability and change has been debated for a long time. Now, new results from various space experiments monitoring the radiative and particle emissions from the Sun together with detailed studies of their terrestrial impacts have opened an exciting new era in both solar and atmospheric physics. Being so close, the Sun is the only star where we have a chance to identify and study in detail the processes responsible for changes in irradiance on time scales from minutes to decades—the longest time scale over which high precision data are available. High-resolution spatial and temporal observations conducted in various space and ground-based experiments demonstrate that the surface of the Sun and its outer atmosphere are highly variable on almost all spatial scales, and that many of the observed changes are linked to interior processes taking place in the Sun's convective zone or below. The broad collection of the material in this Monograph clearly shows that the variable solar energy output affects the Earth's atmosphere and climate in many fundamental ways. However, a quantitative understanding of all the involved processes and their relationship to the climate system and its response remains elusive. Based on the current database and knowledge, it remains to be seen what role solar forcing will play in future climate.

1. SUMMARY

One of the most exciting and important challenges in science today is to understand climate variability and to make reliable predictions. The Earth's climate is a complex system driven by external and internal forces. Climate can vary over a large range of time scales as a consequence of natural variability or anthropogenic influence, or both. The dominant driving force of the climate system is the Sun. Solar energy at various wavelengths is absorbed in the Earth's atmosphere, oceans and land and redistributed by the climate system. Interaction of solar radiation with the atmosphere results in complex ionization, radiative, chemical, and dynamical processes. Therefore, accurate knowledge of the solar radiation received by Earth at various wavelengths and from energetic particles with varying intensities, as well as a better knowledge of the solar-terrestrial interactions and their temporal and spatial variability are crucial to quantify the solar influence on climate and to distinguish between natural and anthropogenic influences.

The basic issue in climate research today is global warming and its consequences. Observations of steadily increas-

Solar Variability and its Effects on Climate
Geophysical Monograph 141
Copyright 2004 by the American Geophysical Union
10.1029/141GM24

ing concentrations of greenhouse gases—primarily manmade—in the Earth's atmosphere have led to an expectation of global warming during the coming decades. However, the greenhouse effect competes with other climate forcing mechanisms, such as solar variability, cosmic ray flux changes, desertification, deforestation, and changes in natural and manmade atmospheric aerosols. Indeed, the climate is always changing, and has forever been so, including periods before the industrial era began [*Bradley and Jones*, 1993; *Muesheler et al.*, this volume].

Another important issue in the field of climate research is the degree to which various forcing factors affect the climate. Although the Sun supplies the overwhelming part of the energy in the Earth's climate system, the measured 0.1% change in total irradiance over the 11-year solar cycle is estimated too small to cause direct changes in the Earth's climate above its intrinsic noise. However, larger variations on longer time scales cannot be ruled out and feedback mechanisms within the climate system may significantly amplify weak forcing. Several papers [*Haigh*; *Hood*; *Schlesinger and Andronova*, this volume and references therein], show that variations in UV irradiance and the response of the stratosphere and mesosphere to these UV changes may have a comparable or larger effect on climate through some type of amplification mechanism.

Less is known about the possible effect of variations in the visible and IR spectral bands, simply because long-term and high-precision space-based measurements are lacking —despite the fact that the thermal balance of the Earth's atmosphere is largely determined by the 300 to 10,000 nm spectral region. This spectral band contains 99% of the total solar radiation. Roughly 30% of the incoming radiation in these spectral regions is reflected or scattered back to space, the remaining 70% is absorbed by the atmosphere, ocean and land surfaces. Small but persistent variations could have a major impact if they exist, since the Earth's oceans, land surfaces, and atmosphere, all respond differently to heating by different wavelengths. At the extremes, for example, visible light at the blue end of the spectrum penetrates the entire mixed layer of the ocean, while infrared radiation is absorbed by the top few millimeters of skin. Transfer of heat from the ocean to the atmosphere is likely to be affected significantly by the spectral distribution of irradiance changes [e.g., *Reid*, 1997; 1999].

Current atmosphere-ocean-land general circulation models (GCMs) have been improved significantly during recent years. However, they still lack the spatial resolution and some of the processes that are necessary to account for all the effects induced by solar changes. Due to their complexity, GCMs are too slow to simulate climatic processes on

time scales of ten to hundreds of thousands of years. These are the time scales on which the largest variations in climate have occurred, influenced by orbital forcing known as the Milankovich effect. The Milankovich effect is a well-known climate driver, revealed in many various climate data sets [see *Muscheler et al.*, this volume].

Attempts to associate past climate changes with solar variability are described by *Muscheler et al.*; *Damon and Perystikh*; *and North et al.* [in this Monograph]. Additional evidence is presented by *Hansen and Lacis* [1990]; *Reid* [1991, 1997; 1999]; *Lean et al.* [1995]; *Lean* [1997]; *White et al.* [1997]; *Solanki and Fligge* [1998; 1999]. Theoretical considerations and study of variable solar-type stars do not rule out the possibility that long-term irradiance variations are significantly larger, in the range of 0.4–0.7%, than the 0.1% changes observed over the last two and half solar cycles [e.g. *Nesme-Ribes et al.*, 1994; *Zhang et al.*, 1994; IPCC 1995 Report; *Radick*, this volume].

In addition to the above studies, several papers [*Wigley* 1999; *Reid*, 1999; *Schlesinger and Andronova*, this volume and references therein], show that greenhouse gases and aerosols cannot explain all the measured temperature changes within the last 130 years. This is not surprising considering the complexity of the climate system. Adding solar variability to the greenhouse gas and aerosol effects improves agreement between modeled and observed temperature variations. However, the lack of understanding of the underlying mechanisms of irradiance variability is the largest obstacle to revealing the role of solar variability in climate change. Direct space irradiance observations cover only two and a half solar cycles. Thus, physical understanding of solar irradiance variability is essential if decade to century irradiance time series are needed. Without this understanding, reliance on statistical "trendsetting" attempts is necessary [e.g., *Lean et al.*, 1995; *Solanki and Fligge*, 1998; 1999; *Lean, 2001*; *Fox*, this volume].

Current empirical irradiance models based on magnetic surrogates (sunspots and faculae) have served as statistical proxies for about 90% of the short-term irradiance variations, and they are correlated with only about 70% of the long-term changes over the solar cycle. Detailed observations of these surface features show that such correlative modeling is physically flawed but useful for statistical predictions [e.g. *Kuhn and Armstrong*, this volume]. Physical understanding of irradiance variability is a difficult problem since global effects, like changes in the radius, photospheric temperature or large-scale convective cells [*Kuhn et al.*, 1988, 1998; *Ribes et al.*, 1992; *Pap et al.*, 2001; *Kuhn and Armstrong*; *Sofia and Li*, this volume, and references therein] all can contribute to irradiance variations. Indeed, old

[*Kuhn et al.*, 1988] and new measurements by *Gray and Livingston* [1997] show that a measured 1.5K variation in the photospheric temperature may also explain the observed long-term irradiance variations, and these changes are not part of the simple faculae and sunspot correlative irradiance descriptions.

Another irradiance modeling problem is to explain the breakdown of the linear relationship between the amplitude of solar irradiance variations and that of the solar cycle at the time of the maximum of solar cycle 23. While current irradiance models used in climate studies assume that solar irradiance varies in a fashion similar to the sunspot number —showing larger cycle-to-cycle variations during high activity cycles and smaller variations during weaker activity cycles, current observations have led to different results. Detailed measurements of facular and sunspot irradiance contributions on the Sun [*Kuhn and Libbrecht*, 1991] show that magnetic features peaked at different times during the last solar cycle. Furthermore, it has been found that both the averaged full disk magnetic field strength as well as sunspot indices (sunspot number and total sunspot area) had lower values during solar cycle 23 than during previous solar cycles [*Chapman et al.*, 2001; *Pap et al.*, 2002] In contrast, the long-term variation of total irradiance and the Mg II h & k core-to-wing ratio is rather symmetrical over the last two and a half solar cycles [see papers by *Fröhlich*; and *Rottman et al.*, this volume]—showing that the maximum level of solar irradiance is higher during solar cycle 23 than the maximum of magnetic indices [*Pap et al.*, 2002].

The climate implication of these results is especially important. Some statistical irradiance models assume a linear relation between the long-term changes in the number of sunspots and total irradiance [e.g., *Lean et al.*, 1995; *Solanki and Fligge, 1998*]. However, these models may not adequately represent long-term irradiance variations over several solar cycles, and back to the Maunder Minimum. These long-term irradiance models also assume that irradiance variations are direct consequences of sunspot blocking and facular brightening [*Foukal and Lean*, 1988]. Another long-term irradiance model, which is extended back to the time of the Maunder Minimum and is widely used in climate studies [*Lean et al.*, 1995], assumes that the distribution of the Ca K emissions of solar-type of stars is pertinent to the Sun, using contemporary solar and stellar Ca II K measurements.

However, these assumptions may not be correct [*Rast et al.*, 1999; *Kuhn and Armstrong*, this volume]. Current stellar observations [*Radick*, this volume] do not necessarily confirm the Ca K emission distribution of solar-type stars found by *Baliunas and Jastrow* [1990] though this work is still in progress. On the other hand, as shown by the results presented in this Monograph [see papers by *Sofia and Li*; *Kuhn and Armstrong*; and *Fox*], there is growing evidence that the observed irradiance variations that are based solely on surface manifestations of solar magnetic activity cannot be explained simply. Indeed, hints from precise photometry (like the SOHO/MDI and VIRGO measurements) and helioseismology indicate that solar irradiance and luminosity vary in response to global solar perturbations. Hence, understanding changes in the solar interior will likely lead to forecasting of solar irradiance variability and, subsequently, to the climatic consequences. Therefore, it is imperative to develop appropriate, interdisciplinary, coordinated research and observation strategies aimed at these insights.

2. FUTURE REQUIREMENTS AND JUSTIFICATION

2.1. Measurement Requirements

Accurate long-term total and spectrally resolved solar irradiance measurements are required for full understanding of the response of Earth's atmosphere and climate to irradiance changes. Space-based irradiance observations over the last two and a half solar cycles span a time interval too short to reveal secular changes and/or to establish conclusively whether there are significant changes in the amplitude or the character of irradiance variations on longer time scales. Since the time period of interest far exceeds the lifespan of any single experiment, continuous measurement programs must be formulated to compile composite irradiance time series from data of several experiments. Because the absolute accuracy of the current measurements is limited (±0.2% in case of total irradiance and about ±5% for UV irradiance), overlapping and redundant measurements are needed to ensure that the resulting composite data sets represent the "true" solar behavior. Use of experiments utilizing different design concepts also contributes to this goal. Construction of such a composite normally includes adjustments of the component individual data sets to the same irradiance level [*Fröhlich*, this volume; *Fröhlich and Lean*, 1998; and *Pap and Fröhlich*, 1999]. That such a measurement program will be needed is amply illustrated by the experience in building composite total solar irradiance data sets derived from redundant and overlapping measurements [*Fröhlich*, this volume] and from intercomparisons of existing UV measurements [*Rottman et al.*, this volume].

2.1.1. Total irradiance. Multiple instruments for measuring total and spectral irradiance from UV to IR are especially important because the measured variations and the absolute accuracy of the measurements are about the same. *Gaps in the data and/or instrumental drifts would make it difficult, if not impossible, to adjust various data sets provided by only one experiment.* The largest obstacle in creating the current long-term total irradiance composite [see *Fröhlich*, this volume] is the two-year gap between the SMM/ACRIM I and UARS/ACRIM II measurements. Adjustment of the ACRIM I and ACRIM II data now must be made through the Nimbus-7/ERB and/or the ERBE measurements. As shown by several papers [*Willson*, 1997; *Fröhlich and Lean*, 1998; *Pap and Fröhlich*, 1999; *Fröhlich*, 2000], different approaches of moving the two ACRIM time series to the same level have led to different conclusions, due especially to the instrumental drifts in the Nimbus-7 data after 1989 [see *Lee et al.*, 1995 and *Fröhlich*, this volume].

Continuous daily total irradiance measurements are currently provided by two experiments: SOHO/VIRGO [see details by *Fröhlich et al.*, 1997; *Fröhlich*, 2000; and this volume], and ACRIM III on ACRIMSAT [*Willson*, 2001]. The "Solar Radiation and Climate Experiment" (SORCE) was successfully launched on January 25, 2003 [see details by *Woods et al.*, 2000]. SORCE carries four instruments, including the "Total Irradiance Monitor" (TIM) radiometer [see details by *Lawrence et al.*, 2000]. SORCE—overlaps with both VIRGO and ACRIM III to extend the present long-term total irradiance database to 2007. Although short duration measurements from the "International Space Station" and the French PICARD experiment will provide total irradiance data in the 2006–2008 time frame, no experiment has yet been approved for the time interval of 2008 and 2012. Total irradiance measurements from the "National Polar Orbiting Operational Environmental Satellite System" (NPOESS) are planned to begin only around 2012.

Lack of an additional total irradiance experiment between SORCE and NPOESS will lead, at best, to a "one radiometer measurement" strategy between 2007–2008 or no measurement period between 2007–2008 and 2012. Since current empirical models cannot replace direct radiometric space measurements and no other alternative data exist to fill voids between consecutive measurements, the possible gap between SORCE and NPOESS virtually assures that a long-term irradiance database for climate studies will not materialize. In keeping with the community's ongoing recommendation of continuous measurements and the important results presented in this Monograph, a program should be established to ensure the continuity of total irradiance measurements between SORCE and NPOESS.

2.1.2. Spectral irradiance from near-UV to IR. The SunPhotometers (SPMs) on SOHO/VIRGO provided the first long-term spectral irradiance measurements in three selected wavelengths with 5 nm bands at 402, 500, and 862 nm from the minimum to the maximum of solar cycle 23 [*Fröhlich et al.* 1997; *Pap et al.*, 1999; 2002; *Fröhlich*, this volume]. However, because of instrumental effects, the degradation of the SPM instrument cannot be corrected by intercomparing the primary and back-up instruments. The degradation effect in the SPM spectral data, which masks most of the solar variations, can be removed only by means of statistical methods [see *Pap et al.*, 1999; *Fröhlich*, this volume]. Unfortunately, these statistical methods remove both the instrumental effects and the solar-cycle-related long-term trends. Thus, after removing instrumental effects the SPM data provide information only about the short-term variations of spectral irradiance.

As shown by *Fröhlich* [this volume], the VIRGO results confirm that active regions modulate total and spectral irradiance on time scales of days to weeks in a similar manner, but they clearly indicate that this modulation cannot be explained only in terms of sunspot flux blocking and facular enhancements [*Fröhlich et al.*, 1997; *Wehrli et at.*, 1998; *Pap et al.*, 1999]. The VIRGO results demonstrate that small active regions can cause significant events in total and spectral irradiance; however, the observed changes are different in the various spectral bands. Although the VIRGO/SPM instrument has provided important new results about the differences in the short-term variations in spectral irradiance, the VIRGO/SPM instrument gives only "snapshots" of the solar spectrum, yet the spectral distribution of total irradiance variations at various near-UV, visible and IR wavelengths is extremely important for both solar physics and climate research.

The "Spectral Irradiance Monitor" (SIM) on SORCE is the first experiment to detail the spectral distribution of irradiance variations in the entire 200 nm to 2 μm wavelength interval over which the instrument's resolution varies from 1 nm to 34 nm [*Lawrence et al.*, 1998; *Harder et al.*, 2000]. The SIM instrument, similar to TIM, employs the basic concept of electrical-substitution radiometers along with improved materials and modern digital signal processing techniques. Operation of an additional SIM between the SORCE and NPOESS eras is essential for obtaining a spectral irradiance dataset of sufficient duration, covering at least a full 11-year solar cycle in the visible and infrared. Until high precision and wavelength resolution spectral measurements are available in the visible and infrared, the climate impact of these variations will remain unclear. It must be emphasized that even small, but consistent variations in these spectral bands (especially in the near-IR) may be important.

2.1.3. UV/EUV/XUV Variations. In contrast to the climate effect of irradiance variations where most studies focus on variability over decades to centuries, the Earth's upper atmosphere responds significantly to solar activity as manifested by energetic radiations and particles over time scales from days to a few years [*Fuller-Rowell et al.;* and *Jackman and McPeters,* this Monograph]. Although solar radiation below 300 nm represents only about 1% of the Sun's total electromagnetic output, variations in this spectral region are especially important because this energy is entirely absorbed by the Earth's atmosphere. Consequently, these radiations play a significant role in heating the Earth's atmosphere and establishing its chemical composition through photodissociation and photoionization processes. Variations in this portion of the solar spectrum are closely connected to the Sun's magnetic activity, and the magnitude of the spectral changes generally increases with decreasing wavelengths. Variations in the UV and EUV affect a number of processes in Earth's middle and upper atmosphere, such as ozone concentrations, the ionization of the E and F regions of the Earth's ionosphere, and the thermospheric molecular and atomic densities [e.g. *Brasseur,* 1993; *Rottman et al.;* and *Woods et al.;* this volume]. The latter controls the lifetime of Earth-orbiting satellites [*White et al.,* 1994].

Long-term continuous UV irradiance measurements are available from late 1978 in the spectral range of about 160 nm to 400 nm [see *Rottman et al.,* this volume]. One particular set of these UV measurements on various NOAA satellites was produced as a by-product of the SBUV and SBUV2 ozone measurements [*DeLand and Cebula,* 1998]. SBUV2 aboard NOAA-11 was calibrated via comparisons with several flights of a similar instrument aboard the space shuttle [*Cebula and DeLand,* 1998]. Later, the two UARS experiments, SOLSTICE and SUSIM, were the first to carry their own means of calibration, the former by bright blue stars and the latter by onboard deuterium lamps and redundant optical channels [*Rottman et al.,* 1993; *Brueckner et al.,* 1993]. In the near term, the UARS experiments overlap with SORCE, which carries the next generation of self-calibrating instruments and was launched on January 25, 2003. Although new NOAA satellites continue to be launched every two years, their solar UV instrumentations are not capable of maintaining calibration. Thus, only short-term variations on time scales of weeks to months can reliably and accurately be measured. If the SORCE mission does not continue past its planned completion in 2007, a gap may

appear in the measurement record since follow-on solar UV monitoring by NPOESS is not contemplated before 2012. As noted previously, measurement gaps in the irradiance record make difficult the construction of composite UV irradiance data sets needed for atmospheric and climate studies [*Haigh; Hood,* this volume]. Furthermore, NPOESS UV measurements, as currently planned, are not expected to extend below 200 nm—despite their importance for climate (and aeronomical) studies.

Although variations in EUV or XUV wavelengths have no direct effect on climate, the amplitude of these variations are relatively large and they significantly affect the Earth's thermosphere and ionosphere [*Woods et al.; Fuller-Rowell et al.,* this volume]. The ionized medium within the ionosphere also affects radio waves and a variety of phenomena related to every day human activities. The very existence and general properties of the ionosphere allow ionospheric radio propagation and determine the band of radio frequencies that can be used. Any change in ionospheric conditions has considerable effect on telecommunications—among the world's largest industries, comprising telephony and telegraphy, broadcasting, navigational aids, and aircraft and ship communications. Telecommunications is important for both civilian and defence purposes. Because solar EUV emission controls the strength of the ionosphere, the EUV emission determines radio propagation conditions (the available wavelength and transmission quality) and the temperature and density of the Earth's upper atmosphere—thus the lifetime of satellites. Monitoring, studying, modeling, and predicting the changes in the solar EUV/XUV fluxes are very important tasks which are of great interest to the broader world society than just the scientific community.

Despite the aeronomical importance of EUV/XUV irradiance variations, measurements of the EUV flux were sporadic and were mostly conducted at X-ray wavelengths short-ward of 35 nm by the long-term NOAA/GOES program, the SNOE/SXP and the most recent TIMED/XPS measurements [see details by *Woods et al.,* this volume]. In addition to the full disk X-ray fluxes, images in X-ray have been provided by YOHKOH, HESSI and some of the GOES experiments. Continuous EUV monitoring in the 0.1–50 nm spectral range was started by the SEM/CELIAS experiment on SOHO in early 1996 [*Judge et al.,* 1998]. This full disk irradiance monitoring is accompanied by the SOHO/EIT experiment, taking images in four EUV wavelength ranges. It has become apparent that higher spectral and temporal EUV irradiance measurements are necessary for thermospheric modeling and to better understand the underlying mechanisms of EUV irradiance variations. The EUV irradiance monitoring experiment on the "Solar Dynamics

Observatory" (SDO) will be the first attempt to measure solar EUV irradiance with the required resolution in the 1 nm to 200 nm spectral range for aeronomical and solar physics studies.

2.1.4. Energetic particles. Energetic particle precipitation into the Earth's atmosphere is primarily affected by solar processes in the following four ways: 1) The galactic cosmic ray flux is modulated by the solar wind; 2) Solar particle events result from solar flares and coronal mass ejections; 3) Relativistic electrons often accompany solar particle events; and 4) Auroral electrons result from solar storm disturbances of the Earth's magnetosphere [see review by *Reames,* 1999]. The energy from these precipitating particles may be considered the second (but very significant) energy input to the atmosphere, primarily influencing the thermosphere and mesosphere. The incoming particles, mostly around the poles, and the bremsstrahlung radiation produced by the electron component produce significant numbers of ions in the Earth's atmosphere.

The higher energy solar protons influence the stratosphere as well as the mesosphere and create HO_x and NO_y constituents, which can then influence ozone on time scales of months and even years [see *Jackman et al.,* 2000; *Jackman and McPeters,* this volume]. The ionizations caused by galactic cosmic rays and changes in the global electric circuit caused by solar activity variations may be important in the processes controlling cloud formation and, subsequently, climate [*Tinsley and Yu,* this volume].

Particle data are available from several platforms, e.g. GOES-8, GOES-10, SOHO, ACE, CLUSTER, IMP-8, WIND, and UARS/PEM. An ongoing research initiative is to develop a so-called "particle climatology" [see *Sharber et al.,* 2002], which is based on statistical modeling to describe average spectral characteristics, precipitating particle fluxes, and ionization profiles as a function of latitude, local time, and solar activity level. It is anticipated that the GOES series of satellites will continue to provide particle data for the next several years. The STEREO set of satellites, due to be launched in 2006, will also include particle measurements.

2.2. Solar Modeling and Theory

A credible physical model is the ultimate aim of models for the solar component of Sun-climate research. The measure of our understanding of the Sun's role in climate variability and change is contained in the mathematical expression of the physical models. Beginning with the Sun, researchers must (1) establish the extent to which magnetic variability and global effects contribute to irradiance changes, and (2) identify the underlying mechanisms of rapid solar events, like coronal mass ejections and flares, that increase particle fluxes.

2.2.1. Surface magnetic activity. Since the solar irradiance monitoring experiments observe the Sun as a star, it is necessary to analyze high-resolution solar images to account for magnetic structures (like sunspots, faculae, network) contributing to irradiance variations. Solar images have been taken at various space-based and ground-based solar observatories as described by *Fox* [this volume]. One of the largest problems when analyzing solar images is the large number of images and a lack of standardized automated image analyzing systems. In other words, different research groups use different image analysis techniques, identifying features differently. Consequently, solar feature parameters will differ from author to author, with inconsistent results when empirical models are applied that depend on the applied image analysis and object classification systems used [see also *Jones et al.,* 2000; *Turmon et al.,* 2002].

Surface irradiance variations are associated with three major activity components: (1) sunspots, (2) faculae and (3) the so-called active or enhanced network. It should be noted that long-term homogeneous sunspot area and position data are lacking, due largely to the discontinuity of the Greenwich Catalogue after 1976. Since then, sunspot data for modeling the effect of sunspots on solar irradiance are taken from the NOAA/WDC Solar Geophysical Data Catalogue. The uncertainty of sunspot area in the SGD Catalogue may be as large as 50%–70%. Information on the direct area, position and contrast measurements of faculae is missing simply because their contrast in the photospheric levels is low. Because of this, chromospheric features like Ca K plages are used as faculae proxies, or simple full disk measurements (the Mg II index or the equivalent width of the He-line at 1083 nm) are considered as faculae proxies. Current irradiance models are even more handicapped by missing information on the contribution of the network component to irradiance variations. The term "active network;" is too broad and not well-defined. Its intensity, contrast and area are not determined experimentally and varying with wavelength. An additional problem in modeling irradiance is that the quiet-Sun intensity values are assumed to be constant on solar cycle or longer time scales, when it is well known that the quiet-Sun and chromospheric network components have not yet been separated. Thus, it is possible that slow secular changes occur both in the network and globally in the Sun. Both remain unmeasured at this time—introducing further uncertainties into model simulations [see *Pap and White*, 1994; *Pap*, 2003; *Fox*, this volume].

High resolution imaging of various layers from the photosphere to the corona are needed to estimate the contribution of magnetic features to irradiance variations at various wavelengths. The main research goal is to understand the appearance, evolution, and decay of photospheric magnetic fields and their relation to structures above their photospheric counterparts in the chromosphere and corona. High-resolution imaging will make it possible to better understand the spectral distribution of irradiance variations. Furthermore, connecting observed variations in the photospheric magnetic field to chromospheric and coronal changes would help to understand and predict rapid changes that cause eruptions that produce Solar Particle Events [see *Lario and Simnett*, this volume], which may also have an indirect effect on climate [see *Jackman and McPeters*; and *Tinsley and Yu*, this volume].

To develop and improve irradiance models, especially covering long time scales, analyses of modern, high-resolution images are needed, similar to images from the San Fernando Observatory, PSPT, YOHKOH, SOHO/MDI and TRACE as well as the forthcoming SOLIS, ASTM and SDO observations. Analysis of long-term historical image data sets, like the white-light sunspot observations at the Heliophysical Observatory of the Hungarian Academy of Sciences at Debrecen, the 100-year long Mt. Wilson white light and Ca K images [*Ulrich et al.*, 2002], and the Kodaikonal images [*Singh and Prabhu*, 1985] is equally important. While analysis of the white-light images will improve sunspot indices for irradiance studies, analysis of the Mt. Wilson Ca K data, in parallel with the Mt. Wilson and Kitt Peak magnetograms, is very promising in identifying the role of weak magnetic fields in longer-term irradiance changes. Introducing these parameters to the new generation of irradiance models, the so-called "synthetic models", will lead to better models rather than those based on simple empirical regression. These synthetic models derive specific intensity values at various wavelengths, they are based on radiative transfer calculations of the physics of the magnetic structure of the solar atmosphere, and incorporate, to various degrees, all known continuum, atomic and molecular opacity sources, populations and ionization stages [see *Fox*, this volume; and *Avrett*, 1998; *Fontenla et al.*, 1999].

2.2.2. Solar global effects. As mentioned before, both theoretical considerations and new measurements show that global events, such as solar dynamo fields at the bottom of the convective zone or below, temperature changes, large scale motions and flows, and radius changes may all contribute to variations in irradiance, especially on climatologically important time scales. These results underscore that new strategies are needed to study solar variability and its effect on climate. If credible projections of global warming or precipitation patterns are important, a century—which is necessary to gather long enough irradiance data sets—is too long to wait to identify solar influences. Appropriate research strategies, yet to be articulated, will expedite progress while addressing difficult issues. For example, models of the solar interior indicate that relationships between variations of the global parameters depend on the details of the Sun's internal mechanisms, such as depth and magnitude of the field [see *Sofia and Li,* this volume]. An effective research strategy should include concurrent measurements of solar irradiance, radius and shape, and solar oscillations in order to determine their relationship experimentally [see *Sofia and Li;* and *Kuhn and Armstrong,* this volume].

Various efforts have been in progress to combine radius, irradiance and helioseismic measurements from SOHO/MDI and SOHO/VIRGO [e.g. *Kuhn et al.*; 1998]. The French PICARD experiment, which is presently in its design phase, will be the first dedicated space experiment to measure simultaneously solar total irradiance (and spectral irradiance in a few spectral domains) and the solar diameter during the rising portion of solar cycle 24. For diameter measurements, the PICARD science team selected three spectral domains which are free of Fraunhofer lines. In addition, measurements will be conducted also in the Ca II (393 nm) line to monitor changes of the Sun's magnetic activity [*Thuillier*, 2002, private communication]. Since we anticipate measuring very small radius changes, redundant space radius measurements would be especially important considering the altering effects of the Earth's atmosphere [*Pap et al.*, 2001]. Development of such measurement and research strategies should be part of the plans of future international space program developments.

3. CONCLUSIONS

3.1. Necessity of Continuous and Overlapping Irradiance Measurements

To reveal the effect of irradiance variations on the Earth's atmosphere and climate system, high precision, homogeneous and long-term irradiance measurements from EUV to IR must be maintained. It is especially important to ensure irradiance measurements between SORCE and NPOESS, in the time frame of 2007–2012, to avoid data gaps in total and

UV irradiances and to make sure that for the first time spectral irradiance data in the near-UV, visible and IR will cover an entire solar cycle.

3.2. Necessity of Improved Irradiance Models

To clarify the role of solar surface activity events and global effects in irradiance variations, analyses of high resolution images at various wavelengths are necessary. A combination of image decomposition techniques and semi-empirical models, like synthesis models, is necessary to replace current empirical models based on proxy solar activity indices and regression analysis. More importantly, we need to understand the physical mechanisms that cause irradiance variability. Detailed observations which explore the energy budget around magnetic fields at the photosphere are beginning to address this requirement.

3.3. Necessity of New Measurement Strategy

Develop experiments to measure surface brightness changes with an accuracy of at least 10^{-3}. The limb shape and solar radius need to be measured in parallel with solar total irradiance. Since we are measuring small changes, redundant measurements are important.

3.4. Necessity of Indirect Studies of Solar Variability Through the Study of Sun-Like Stars

Identify bona fide solar analogs through detailed, high resolution spectroscopic comparisons between candidate stars and the solar spectrum, and monitor the activity and irradiance variability of these "solar twins". Develop techniques of measuring the p-mode spectrum of Sun-like stars. Observations of solar-type stars will expand our knowledge of how the Sun works simply by enlarging the sample to a larger set of conditions. Comparison of solar and stellar irradiance variations, together with theoretical studies, will lead to a better understanding of solar irradiance variations on long time scales.

3.5. Necessity to Improve How Climate Models Account for Solar Effects

To achieve an improved understanding of the effects of solar irradiance and particle fluxes on the upper atmosphere, and to understand the propagation of effects to the lower atmosphere and Earth's climate, physical, chemical, and dynamical global climate models must be employed in a collaboration among solar, atmosphere, and climate sciences. Further investigation and development of terrestrial proxies of solar activity in the geological and historical record should also be improved to study the long-term solar forcing on climate.

Acknowledgments. The authors express their gratitude to Drs. J. Beer, M. Giampapa, H. S. Hudson, C. Jackman, S. Jordan, L. Floyd, P. Fox, J. Kuhn, J. McCormack, D. Reames, G. Thuillier, and S. T. Wu for their help and useful comments to complete this summary.

REFERENCES

Avrett, E. H., Modeling solar variability—synthetic models, in *Solar Electromagnetic Radiation Study for Solar Cycle 22*, Proceedings of the SOLERS22 Workshop held at the National Solar Observatory, Sacramento Peak, Sunspot, New Mexico, June 17–21, 1996, edited by J. M. Pap, C. Fröhlich, and R. K. Ulrich, Kluwer Academic Publishers, 449, 1998.

Baliunas, S. and R. Jastrow, 1990, Evidence for long-term brightness changes of solar-type stars, *Nature, 348,* 520, 1990.

Bradley, R. S. and P. D. Jones, "Little Ice Age" summer temperature variations: their nature and relevance to recent global warming trends, *The Holocene, 3,* 387, 1993.

Brasseur, G., The response of the middle atmosphere to long-term and short-term solar variability, *J. Geophys. Res., 98,* 23,079, 1993.

Brueckner, G. E., K. L. Edlow, L. E. Floyd, J. Lean, and M. E. Van Hoosier, The solar ultraviolet spectral irradiance monitor (SUSIM) experiment on board the Upper Atmosphere Research Satellite (UARS), *J. Geophys. Res., 98,* 10,695–10,711, 1993.

Cebula, R. P. and M. T. DeLand, Comparisons of the NOAA-11 SBUV/2, UARS SOLSTICE, and UARS SUSIM MG II solar activity proxy indexes, *Solar Physics, 177,* 117, 1998.

Chapman, G. A., A. M. Cookson, J. J. Dobias, and S. R. Walton, An improved determination of the area ratio of faculae to sunspots, *Ap. J., 555,* 462, 2001.

Damon, P. and A. Perystikh, Solar and climatic implications of the centennial and millennial periodicities in atmospheric C14 variations, *this volume.*

DeLand, M. and R. P. Cebula, Solar Backscatter Ultraviolet, model 2 (SBUV/2) instrument solar spectral irradiance measurements in 1989–1994, 2, Results, validation, and comparison, *J. Geophys. Res., 103,* 16,251, 1998.

Fontenla, J. M., O. R. White, P. A. Fox, E. H. Avrett, and R. L. Kurucz, Calculation of solar irradiances I. Synthesis of the solar spectrum, *Astrophys. J., 518,* 480, 1999.

Foukal, P. and J. Lean, Magnetic modulation of solar luminosity by photospheric activity, *Astrophys. J., 328,* 347, 1988.

Fox, P., Solar activity and irradiance variations, *this volume.*

Fröhlich, C., B. Andersen, T. Appourchaux, et al., First results from VIRGO, the experiment for helioseismology and solar irradiance modeling on SOHO, *Solar Phys., 170,* 1, 1997.

Fröhlich, C. and J. Lean, Total solar irradiance variations, in *IAU Symposium 185: New Eyes to See Inside the Sun and Stars*, edited by F. L. Deubner, Kluwer Academic Publishers, 98, 1998.

Frölich, C., Observations of irradiance variability, *Space Sci. Rev.*, *94*, 15, 2000.

Fröhlich, C., Solar irradiance variability, *this volume*.

Fuller-Rowell, T., S. Solomon, R. Viereck, and R. Roble Impact of the EUV and X-ray variations on the Earth's atmosphere, *this volume*.

Gray, D. F. and W. C. Livingston, Monitoring the solar temperature: Spectroscopic temperature variations of the Sun, *Ap. J.*, *474*, 802, 1997.

Haigh, J., Fundamentals of the Earth's atmosphere and climate, *this volume*.

Hansen, J. E. and A. A. Lacis, Sun and dust versus greenhouse gases: as assessment of the relative roles in global climate change, *Nature*, *346*, 713, 1990.

Harder, J., G. M. Lawrence, G. Rottman, and T. Woods, Solar Spectral Irradiance Monitor (SIM), *Metrologia*, *37*, 415, 2000.

Hood, L., Effects of Solar UV variability on the stratosphere, *this volume*.

Jackman, C., E. L. Fleming, and F. M. Witt, Influence of extremely large solar proton events in a changing stratosphere, *J. Geophys. Res.*, *105*, 11659, 2000.

Jackman, C. H. and R. D. McPeters, The effect solar particles on ozone and other constituents, *this volume*.

Jones, H. P., D. D. Branston, P. B. Jones, P. B., and M. J. Wills-Davey, Analysis of NASA/NSO Spectromagnetograph observations for comparison with solar irradiance variations, *Astrophys. J.*, *529*, 1070, 2000.

Judge, D. *et al.*, First solar EUV irradiance obtained from SOHO by CELIAS/SEM, *Solar Phys.*, *177*, 441, 1998.

Kuhn, J., K. G. Libbrecht, K. G., and R. H. Dicke, The surface temperature of the Sun and changes in the solar constant, *Science*, *242*, 908, 1988.

Kuhn, J. and K. G. Libbrecht, Non-facular solar luminosity variations, *Ap. J.*, *381*, L35., 1991.

Kuhn, J., R. Bush, X. Scheick, and P. Scherrer, *Nature*, *392*, 155, 1998.

Kuhn, J. R. and J. D. Armstrong, Mechanisms of solar irradiance variations, *this volume*.

Lario, D. and G. Simnett, Solar energetic particle variations, *this volume*.

Lawrence, G. M., J. Harder, G. Rottman, T. Woods, J. Richardson, and G. Mount, Stability Considerations for a Solar Spectral Intensity Monitor (SIM), in *Proc. SPIE*, *3427*, 477, 1998.

Lawrence, G. M., G. Rottman, J. Harder, and T. Woods, Total solar irradiance monitor (TIM), *Metrologia*, *37*, 407, 2000.

Lean, J., J. Beer, and R. Bradley, Reconstruction of solar irradiance since 1610: Implications for climate change, *Geophys. Res. Lett.*, *19*, 1595, 1995.

Lean, J., The Sun's variable radiation and its relevance for Earth, *Ann. Rev. Astron. Astrophys.*, *35*, 33, 1997.

Lean, J., Solar irradiance and climate forcing in the near future, *Geophys. Res. Lett.*, *28*, 4119, 2001.

Lee III, R. B., M. A. Gibson, Wilson, R. S., an S. Thomas, Long-term solar irradiance variability during sunspot cycle 22, *J. Geophys. Res.*, *100*, 1667, 1995.

Muesheler, R., J. Beer, and P. W. Kubik, Long-term solar variability and climate change based on radionuclide data from ice cores, *this volume*.

Nesme-Ribes, E., D. Sokoloff, and R. Sadourney, Solar rotation, irradiance changes and climate, in *The Sun as a Variable Star: Solar and Stellar Irradiance Variations*, editeb by J. M. Pap, C. Fröhlich, H. S. Hudson, and S. Solanki, Cambridge Univ. Press, 244, 1994.

North, G., R., Q. Wu, and M. J. Stevens, Detecting the 11 yr solar cycle in the surface temperature field, *this volume*.

Pap, J. and White, O. R., Panel discussions on total solar irradiance variations and the Maunder Minimum, in *"The Solar Engine and its Influence on Terrestrial Atmosphere and Climate"*, edited by E. Ribes, *NATO ASI Series*, Vol. 125., 235, 1994.

Pap, J. M. and C. Fröhlich, Total solar irradiance variations, *J. Atmos. Solar Terr. Phys.*, *61*, 15, 1999.

Pap, J., M. Anklin, C. Fröhlich, Ch. Wehrli, F. Varadi, and L. Floyd, Variations in total solar and spectral irradiance as measured by the VIRGO experiment on SOHO, *Adv. Space Res.*, *24*, 215, 1999.

Pap, J., J. P. Rozelot, S. Godier, and F. Varadi, On the relation between total irradiance and radius, *Astron. Astrophys.*, *372*, 1005, 2001.

Pap, J. M., M. Turmon, L. Floyd, C. Fröhlich, and C. Wehrli, Total solar and spectral irradiance variations from solar cycles 21 to 23, *Adv. Space Res.*, *29*, 1923, 2002.

Pap, J., Total and spectral irradiance variations from near-UV to infrared, in *The Sun's Surface and Subsurface, investigating shape and irradiance,* edited by J. P. Rozelot, Springer-Verlag, p. 129, 2003.

Radick, R., Long-term solar variability: evolutionary time scales, *this volume*.

Rast, M. P, P. A. Fox, H. Lin, O. R. White, R Meisner, and B. Lites, Bright rings around sunspots, *Nature*, *401*, 678, 1999.

Reames, D. V., Particle Acceleration at the Sun and in the Heliosphere, *Space Science Revs.*, *90*, 413, 1999.

Reid, G., Solar total irradiance variations and the global sea surface temperature record, *J. Geophys. Res.*, 96 (D2), 1835, 1991.

Reid, G., Solar forcing of global climate change since the mid-17th century, *Climate Change*, *37*, 391, 1997.

Reid, G., Solar variability and its implications for the human environment, *J. Atmosp. Terr. Res.*, *61*, 3, 1999.

Ribes, E., B. Beardsley, T. N. Brown, Ph. DeLache, F. Laclare, J. Kuhn, and N. V. Leister, The variability of the solar diameter, in *The Sun in Time*, edited by C. P. Sonetti, M. S. Giampapa and M. S. Matthews, Arizona Univ., 59, 1992.

Rottman, G. J., T. N. Woods, and T. N. Sparn, Solar-Stellar Irradiance Comparison Experiment 1. I – Instrument design and operation, *J. Geophys. Res.*, *98*, 10667, 1993.

Rottman, G. J., L. Floyd, and R. Viereck, Measurement of solar ultraviolet irradiance, *this volume*.

Schlesinger, M. and N. Andronova, Has the Sun changed climate? Modeling the effect of solar variability on climate, *this volume*.

Sharber, J. R., J. D. Winningham, R. A. Frahm, G. Crowley, A. J. Ridley, and R. Link, Construction of a particle climatology for the study of the effects of solar particle fluxes on the atmosphere, *Adv. Space Res.*, Vol. 29, 1513, 2002.

Singh, J., and T. P. Prabhu, Variation in solar rotation rate derived from Calcium-K plage areas, *Solar Phys.*, *97*, 203, 1985.

Sofia, S. and L. H. Li, Solar variability caused by structural changes of the convection zone, *this volume*.

Solanki, S. K. and M. Fligge, Solar irradiance since 1874 revisited, *Geophys. Res. Lett.*, *25*, 341, 1998.

Solanki, S. K. and M. Fligge, A reconstruction of total solar irradiance since 1700, *Geophys. Res. Lett.*, *26*, 2465, 1999.

Tinsley, B. and F. Yu, Atmospheric ionization and clouds as links between solar activity and climate, *this volume*.

Turmon, M., J. M. Pap, and S. Mukhtar, Automatically finding solar active regions using SOHO/MDI photograms and magnetograms, *Astrophys. J.*, *568*, 396–407, 2002.

Ulrich, R. K., S. Evans, and J. E. Boyden, Mount Wilson Synoptic Magnetic Fields: Improved Instrumentation, Calibration, and Analysis Applied to the 2000 July 14 Flare and to the Evolution of the Dipole Field 2002, *Astrophys. J.*, *139*, 259, 2002.

Wehrli, Ch., T. Appourchaux, D. Crommelynck, W. Finsterle, and J. Pap, Solar irradiance variations and active regions observed by VIRGO experiment on SOHO, in *Sounding Solar and Stellar Interiors*, edited by J. Provost, F-X. Schmeider, Kluwer Academic Publishers, 209, 1998.

White, O. R., G. Rottman, T. N. Woods, S. I. Keil, W. C. Livingston, K. F. Tapping, R. F. Donnelly, and L. Puga, Change in the UV output of the Sun in 1992 and its effect in the thermosphere, *J. Geophys. Res.*, *99*, 369, 1994.

White, W. B., J. Lean, D. R. Cayan, and M. D. Dettinger, Response of global upper ocean temperature to changing solar irradiance, *J. Geophys. Res.*, *102*, 3255, 1997.

Wigley, T. M. L., The Science of Climate Change: Global and U. S. Perspectives. Pew Center on Global Climate Change, Arlington, VA. 48 1999

Willson, R. C., Total solar irradiance trend during solar cycles 21 and 22, *Science*, *277*, 1963, 1997.

Willson, R. C., The ACRIMSAT/ACRIM3 experiment—Extending the precision, long-term total solar irradiance climate database, *The Earth Observer*, May/June 2001, V. 13, No. 3, pp. 14, 2001.

Woods, T., G. Rottman, J. Harder, G. Lawrence, W. McClintock, G. Kopp, and C. Pankratz, Overview of the EOS SORCE mission, *SPIE Proceedings*, *4135*, 192, 2000.

Woods, T. *et al.*, Solar extreme ultraviolet and X-ray irradiance variations, *this volume*.

Zhang, Q., W. H. Soon, S. Baliunas, G. W. Lockwood, B. A. Skiff, and R. R. Radick, A method of determining possible brightness variations of the SUn in past centuries from observations of solar-type of stars, *Astrophys. J.*, *427*, L111, 1994.

W. Sprigg, Institute for the Study of Planet Earth, University of Arizona, 715 N. Park Ave., Tucson, AZ 85721, USA. (wsprigg@u.arizona.edu)

J. M. Pap, Goddard Earth Sciences and Technology Center, UMBC, c/o NASA Goddard Space Flight Center, Code 680.0, Greenbelt, MD 20771, USA. (papj@marta.gsfc.nasa.gov)